Günter Müller
Kai Rannenberg
Manfred Reitenspieß
Helmut Stiegler (Hrsg.)

Verläßliche
IT-Systeme

DUD-Fachbeiträge

herausgegeben von Andreas Pfitzmann, Helmut Reimer, Karl Rihaczek
und Alexander Roßnagel

Die Buchreihe DuD-Fachbeiträge ergänzt die Zeitschrift DuD – Datenschutz
und Datensicherheit in einem aktuellen und zukunftsträchtigen Gebiet, das für
Wirtschaft, öffentliche Verwaltung und Hochschulen gleichermaßen wichtig
ist. Die Thematik verbindet Informatik, Rechts-, Kommunikations- und
Wirtschaftswissenschaften.

Den Lesern werden nicht nur fachlich ausgewiesene Beiträge der eigenen
Disziplin geboten, sondern auch immer wieder Gelegenheit, Blicke über den
fachlichen Zaun zu werfen. So steht die Buchreihe im Dienst eines interdiszi-
plinären Dialogs, der die Kompetenz hinsichtlich eines sicheren und verant-
wortungsvollen Umgangs mit der Informationstechnik fördern möge.

Unter anderem sind erschienen:

Karl Rihaczek
Datenverschlüsselung in
Kommunikationssystemen

*Ulrich Pordesch, Volker Hammer,
Alexander Roßnagel*
Prüfung des rechtsgemäßen
Betriebs von ISDN-Anlagen

Hans-Jürgen Seelos
Informationssysteme und
Datenschutz im Krankenhaus

Heinzpeter Höller
Kommunikationssysteme –
Normung und soziale Akzeptanz

Wilfried Dankmeier
Codierung

Heinrich Rust
Zuverlässigkeit und
Verantwortung

Bernd Blobel (Hrsg.)
Datenschutz in medizinischen
Informationssystemen

Patrick Horster (Hrsg.)
Trust Center

*Albrecht Glade, Helmut Reimer und
Bruno Struif (Hrsg.)*
Digitale Signatur &
Sicherheitssensitive Anwendungen

Joachim Rieß
Regulierung und Datenschutz im
europäischen
Telekommunikationsrecht

Ulrich Seidel
Das Recht des elektronischen
Geschäftsverkehrs

Rolf Oppliger
IT-Sicherheit

*Günter Müller, Kai Rannenberg,
Manfred Reitenspieß, Helmut Stiegler
(Hrsg.)*
Verläßliche IT-Systeme

Günter Müller
Kai Rannenberg
Manfred Reitenspieß
Helmut Stiegler (Hrsg.)

Verläßliche IT-Systeme

Zwischen Key Escrow und elektronischem Geld

Die Tagung wurde freundlich unterstützt durch:
Badischer Verlag, Freiburg; Deutsche Bank Freiburg; Dresdner Bank Freiburg, Gerling-Konzern AG; Gödecke AG, Gottlieb Daimler- und Karl Benz-Stiftung, Ladenburg; Industrie- und Handelskammer Südlicher Oberrhein; Poppen und Ortmann Druckerei und Verlag, Freiburg; ROBERT BOSCH GMBH; SECUNET; SIEMENS AG; Stadt Freiburg im Breisgau; TeleTrusT Deutschland e.V.

ISBN-13: 978-3-528-05594-3   e-ISBN-13: 978-3-322-86842-8
DOI: 10.1007/978-3-322-86842-8

Vorwort

Zwischen Key Escrow und elektronischem Geld ist nicht nur der Titel der Tagung, sondern beschreibt auch die Situation der IT-Sicherheit augenblicklich weltweit und im deutschsprachigen Raum. Die von mehreren Staaten geäußerte Absicht, Key Escrow staatlicherseits vorzuschreiben und zu kontrollieren, scheint zusammen mit den Export-restriktionen der USA für Verschlüsselungstechnik das große Hemmnis für einen breiten Einsatz von Verschlüsselungs- und Sicherheitstechniken, während die Verheißungen des Elektronic Commerce mit elektronischem Geld einen großen Ansporn geben, neue technische Ansätze, die alle IT-Sicherheitsmechanismen beinhalten, zu untersuchen und einzusetzen.

Key Escrow wirft technische Probleme und neue Fragen auf. Wie sinnvoll ist eine solche Maßnahme angesichts der heutigen Erkenntnisse der Steganographie? Elektronisches Geld und Electronic Commerce allgemein beruhen auf den klassischen Erkenntnissen der Kryptologie mit Signaturen, Zertifikaten und den Protokollen mit diesen umzugehen. Früher schon andiskutierte Themen, wie z.B. der Kopierschutz, rücken verstärkt ins Blickfeld. Die klassischen Themen, wie sichere Netze, vertrauenswürdige Software mittels Evaluierung auf Basis von Kriterienkatalogen oder neuerdings auch auf Basis von Prozeß-Assessments, aber auch die Aspekte der Mobilkommunikation als Spezialfall des Ubiquitous Computing ordnen sich in das Spannungsfeld zwischen behindernden und antreibenden Kräften ein.

Die Beiträge dieses Bandes bewegen sich im vollen Spektrum der erwähnten Schlagworte, ohne es natürlich ausfüllen zu können. Zu Key Escrow und zum deutschen Signaturgesetz wurden deshalb Podiumsdiskussionen angesetzt, die die mehr tagesaktuellen oder weniger technischen, z.B. eher politischen Aspekte beleuchten sollen und naturgemäß im Tagungsband nicht enthalten sein können. Drei zusätzlich zur Abrundung der Thematik eingeladene Vorträge über PGP (Phil Zimmermann), IT-Sicherheit in Japan (Ryoichi Sasaki) und Teledienste (Alfred Büllesbach) sind ebenfalls nicht im Tagungsband enthalten, da sie zu gegebener Zeit an anderer Stelle nachgelesen werden können. Trotzdem enthält dieser Band mehr Beiträge als die vorher-gegangenen Bände von VIS-Tagungen. Dies hat zwei Gründe.

Zum einen ist aus der VIS erstmalig eine volle drei Tage Veranstaltung geworden (natürlich mit einem zusätzlichen, vorausgehenden Tag mit Tutorien für Studenten wie für kommerzielle Interessenten). Zum anderen haben wir diesmal neben den üblichen Langbeiträgen (13 - 20 Seiten) Kurzbeiträge (6 -12 Seiten) in das Programm mit aufgenommen, um zu einzelnen Themen durch Gegenpositionen mehr Raum für Diskussionen zu eröffnen. Zu dem Bericht über den Stand der schon klassischen Common Criteria wurde z.B. die Vorstellung des ECMA-Entwurfs von Kriterien für Electronic Commerce gesetzt. Beiträge wurden auch deswegen als Kurzbeiträge plaziert, um (wie es der wissenschaftliche Anspruch der Tagung erfordert) Überlappungen mit an anderer Stelle schon veröffentlichten Beiträgen zu vermeiden.

Insgesamt umfaßt der Band eine Fülle interessanter Beiträge, die aus unserer Sicht auch von ihrer Zusammenstellung profitieren. Dies gibt uns die Hoffnung, daß er viele interessierte Leser findet und diesen von Nutzen ist.

Freiburg und München im Juli 1997

Günter Müller Kai Rannenberg Manfred Reitenspieß Helmut Stiegler

Inhaltsverzeichnis

Proceedings der GI-Fachtagung VIS '97

30. 9. - 2.10. 1997
in
Freiburg/Brsg

Zwischen Key Escrow und elektronischem Geld

Programmkomitee
J. Biskup, Universität Dortmund
J. Brüggemann, Universität Hildesheim
R. Dierstein, c/o DLR Oberpfaffenhofen
M. Domke, GMD Sankt Augustin
W. Gerhardt, TU Delft
H.J. Appelrath, Universität Oldenburg
R. Grimm, GMD Darmstadt
D. Fox, Universität Siegen
H. Kurth, IABG München
M. Hegenbarth, Detecon Darmstadt
P. Horster, TU Chemnitz
A. Lubinski, Universität Rostock
G. Müller, Universität Freiburg
J. Nedon, Universität Hamburg
A. Pfitzmann, TU Dresden
H. Pohl, ISIS Essen
K. Rannenberg, Universität Freiburg
M. Reitenspieß, Siemens Nixdorf München (Vorsitz)
I. Schaumüller, Universität Linz
H. Stiegler, STI-Consulting München (Vorsitz)
K. Vogel, BSI Bonn
G. Weck, Infodas Köln

Leitung und Organisation der Tagung
Prof. Dr. G. Müller und K. Rannenberg

A Copyright Protection Environment for Digital Images

Alexander Herrigel
r^3 security engineering ag
Email: herrigel@r3.ch
Zurichstrasse 151
CH-8607 Aathal,
Switzerland.

Adrian Perrig
École Polytechnique
Fédérale de Lausanne
EPFL-LSE
CH-1015 Lausanne,
Switzerland.

Joseph J. K. Ó Ruanaidh
University of Geneva
Centre Universitaire
d'Informatique
CH-1211 Geneva,
Switzerland.

Abstract

This paper* presents a new approach for the copyright protection of digital images transmitted over the Internet. Current watermark techniques emphasise the robustness of digital watermarks only. In addition to being robust, our approach uses cryptographic protocols and public key techniques to ensure the legal binding of spread spectrum based watermark methods. Our approach allows legal action even if a watermark is not found because ownership is legally registered. We show that the copyright problem can be reduced in its complexity if the copyright verification process is associated with the consumer side of the commercial digital image distribution process.

1. Introduction

Confronted with the need for the rapid development and deployment of flexible information services for the information highway, the Internet community has recently developed new technologies, protocols, and interfaces which enable the fast provision of interactive multimedia and hypermedia services. Originally the applied communication TCP/IP protocol suite with its network services was developed with the emphasis on availability only. In the last three years additional security requirements have been identified to execute commercial business processes on the Internet. Specific security architectures such as the Secure Socket Layer Protocol [SSL] or the Simple Key Management For Internet Protocols [SKIP] have been specified and implemented to address identified threads. These security architectures enable the generation of an authenticated and trusted virtual communication channel between the client and the server system. They offer, however, no means for the copyright protection of digital images which are distributed over the Internet in a commercial environment.

The copyright protection of digital images is defined as the process of proofing the intellectual property right to a court of law against an unauthorized reproduction, processing, transformation, or broadcasting of a digital image. This process is based on a prior registration of the copyright with a trusted third party, called the copyright office. After the successful registration, the copyright ownership is legally bound with a copyright notice which was generated by the copyright office on request of the copyright holder. This copyright notice is required to notify and prove copyright ownership.

---

*This work has been funded by the Swiss National Science Foundation under the SPP program (Grant. 5003-45334).

2.   State-of-the-Art Approaches

In contrast to traditional techniques digital information can be copied in miliseconds without any loss of quality. This weakens substantially the interests of a copyright holder if he would like to distribute his images in the Internet environment. Many research teams [Craver, Kutter, NEC, Ó Ruanaidh, Pitas, Zhao] are, therefore, currently working on the watermark problem. Their stated research goals emphasize finding methods of embedding invisible digital watermarks in digital images that are very resistant to image processing and transformation techniques. The most robust approaches are based on the theory of spread spectrum communication and are perceptually adaptive. However, these publications concentrate only on the robustness of a watermark in a digital image but not on a comprehensive security architecture which enables the copyright holder to prove his intellectual property rights to a third party, such as account of law, in case of misuse. In addition, these techniques do not address different copyright issues from a legal perspective, since the embedded watermarks are not referenced by a trusted third party such as the copyright office.

3.   Notation

The expressions used in this report follow the steganographic terminology reported in [Pfitzmann]. We use the expression cover-image and stego-image. A key is applied by Alice to embed a watermark into the cover-image I, resulting in a stego-image I*. Then Bob can extract the watermark from the stego-image if has the correct key. This process is illustrated in Figure 1.

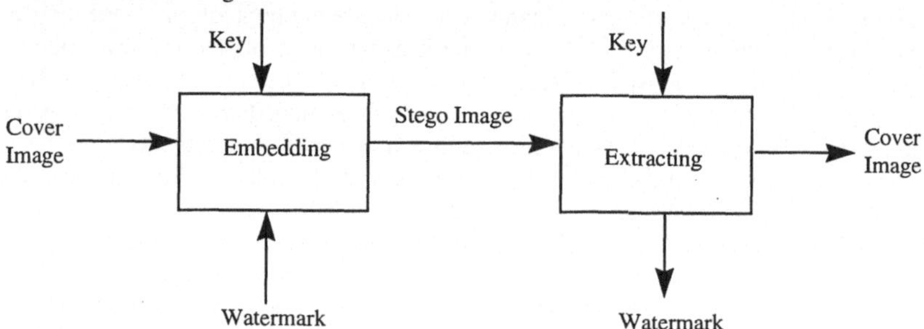

Figure 1: The Watermark Process.

A public watermark is defined as message which was embedded with a public known key, i.e. everybody has access to this key to read the public watermark. A private watermark is defined as message embedded with a cryptographic secret key, which was generated for a symmetric or assymmetric cryptographic mechanism such as DES, IDEA, and DSS or RSA. We assume that the function aplied for the embedding of the private watermark is a one-way function*, collision resistent and robust, i.e. it is for an unauthorized third party not possible to overwrite or delete this watermark without the cryptographic keying information.

---

*A function f from a set X to a set Y is called a one-way function if the computational complexity CC to find for a given y from Y a x in X such that f(x) = y is infeasible.

4.   Legal Aspects of Copyright

The copyright laws [Cinque] of different countries vary. Most countries base their copyright laws on the Berne Convention and Universal Copyright Convention which address copyright issues on the International market for the protection of Literary and Artististic Works. The Berne Convention provides a minimum of 25 years protection for photographic works. Member states may provide additional protection. The simple act of creating a work does not legally entitle an author to the copyright. For obvious reasons, before owning the copyright, it is first necessary to register that copyright with a trusted third party which can act as witness in the case of a copyright dispute. The legal entity which exists for the purpose of registering copyright is the Copyright Office (CO). The following items of information are essential to a copyright application: Full name and post address, complete information how ownership was acquired, title of the work, description of the type of work (artistic, literary, musical, dramatic) and date and place of first publication. The new copyright owner receives a certificate after having paid the fee. A Universal Copyright Convention Notice (UCCN) is placed on the work which comprises the following familiar notice: Copyright symbol ©, the term "Copyright", the year of the copyright, the name of the copyright holder, and the phrase "All Rights Reserved". Only when all the above requirements are satisfied a legally binding claim of copyright ownership can be made. The UCCN is the standard format required to notify copyright ownership.

5.   Risk Analysis and Risk Assessment

For the reader's convenience we describe the different scenarios of interest with typical people, such as Alice, Bob and Mallet. This method of explanation is consistent with previous cryptographic reports [Schneier].

Alice is in our case the copyright holder. She has taken a very valuable photograph, scanned it and would like to sell it to Bob in digital form. Mallet would also like to sell or distribute the image himself. He has a lot of computing power and he can listen and intercept/change any information that Alice and Bob transmitted on the Internet. Figure 2 shows a basic commercial scenario. Alice holds a copyright for image I. Bob wants to buy the image and Mallet tries to attack this in any possible way. We list the encountered risks for each party in Table 1. The following dialog describes the scenario between our three parties:

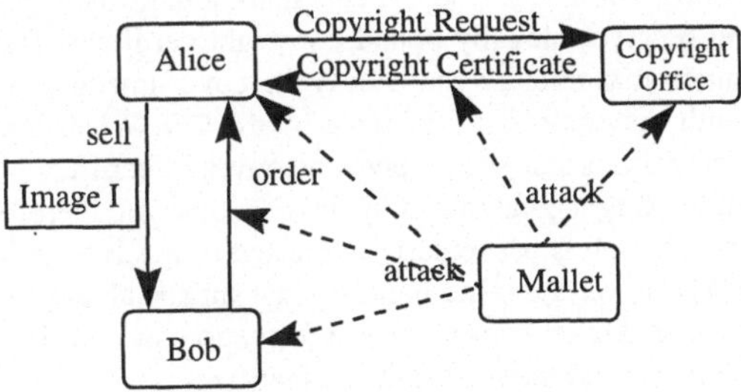

Figure 2: The Commercial Scenario.

Bob is creating a graphical brochure for a customer and needs an image of a lawn mower. After a query of the thumbnails of Alice's image database he discovered the picture he needed. Bob contacts Alice's image server and a mutual authentication is made. Bob then orders the desired image and pays for it electronically. Alice sends the image to Bob and the two parties close their connection. This is the ideal scenario. But Mallet is extremely intelligent and creative. We can therefore imagine a number of sophisticated attacks he will try.

5.1. Identified environment and parties

For the risk analysis and the rsik assessment we envision an open environment with different computers which are all interconnected by the Internet on the basis of the TCP/IP protocol suite. Users can be located anywhere and can sell or buy images. In order to receive legally binding watermarks, the Copyright Holder (CH) sends copyright information and the image to the Copyright Office (CO). After having received a copyright certificate from the CO, the copyright owner can sell his digital image. Another legal party, called the Public Key Infrastructure (PKI), is also introduced. The PKI supports specific security services between the involved parties such as mutual authentication. The communication channels between the identified parties are shown in Figure 3.

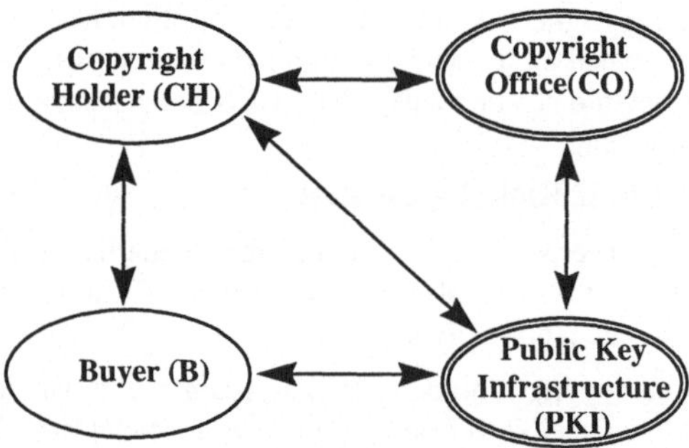

Figure 3: The Communication Channels of the Identified Parties.

5.2. Risk Analysis and Assessment

It would be beyond the scope of the paper to discuss all possible thrats in detail. We have listed them in Table 1 along with the exploitation assessment. We can now analyse some of the risks for Alice and Bob. Alice's main interest is to protect her cover-image, register it and receive a legally bound copyright certificate from the CO. She wants to have guarantee that nobody can access the cover-image and that it is saved permanently along with the copyright certificate by the CO. When registering the image, Alice wants to be certain that she receives a copyright certificate for a novel image and that the legal binding is guaranteed by the CO. She can, therefore, successfully sue fraudulent behaviour. Bob is particularly interested to receive the images he wants and he had paid for. Mallet on the other hand is very antisocial and tries to hurt Alice or Bob wherever possible. Mallet tries to steal information and sell it himself. We discuss now some of the important threats we have identified:

Threat 1: Mallet buys the image but distributes it further.

Description: Mallet has bought image I from Alice. Without Alice's approval, he sells it himself.

Threat 2:   Mallet embeds his own watermark and registers the image himself.

Description: Mallet buys image I from Alice and embeds himself a watermark . He then fills in his own copyright information and registers the image at his CO.

Threat 3:   Mallet removes the watermark.

Description: Mallet removes Alice's watermark, embeds his own and sells the image on his own.

Threat 4:   Counterfeit original attack.

Description: In [Craver] it was shown that if an invertible watermark embedding function is used, an attack called the counterfeit original, is possible.

Threat 5:   Statistical removal attack.

Description: This attack is based on the fact that we can embed multiple invisible watermarks, without degrading the image. Mallet first embeds a large number of watermarks in the image, using the same method as Alice. If he needs high confidence of successful removal, he embeds a larger number. He now starts to apply image transformations until all of his embedded watermarks disappear. With a calculable confidence, the watermark embedded by Alice also vanished.

Threat 6:   Reduced colour image attack.

Description: Alice tries to watermark a logo that consists only of very few colours. The watermarking algorithm adds more colours and they become visible in the image. Mallet only needs to apply a "de speckle" algorithm to successfully remove the watermark.

Threat 7:   Mallet performs the known plaintext attack.

Description: With any watermarking scheme, an attacker could apply the known plaintext attack to find out our key. To make this attack possible, Alice always uses the same key to embed the watermarks. But for one image, Alice was not cautious enough in hiding the cover-image and the watermark message so that Mallet could steal both. With the embedded watermark, the cover and the stego-image, Mallet can derive Alice's key, which is known as the known plaintext attack. Mallet, knowing the key, can find out which images contain Alice's watermark. He could also apply small changes to the image until the watermark is removed.

Threat 8:   Mallet destroys Alice's image or copyright certificate.

Description: Mallet wants to sell Alice's image himself. He therefore attacks Alice's system and destroys the cover-image, the secret key and the copyright certificate.

Threat 9:   Mallet steals the watermarked image (from Alice, or Bob during the transmission.

Description: Mallet tries to obtain the watermarked image without paying. He tries to attack through a security hole in Alice's or Bob's computer or read the image while it is transmitted.

Threat 10:   Watermark collision.

Description: Mallet found out that when he extracts a watermark with his key out of Alice's image, he gets a meaningful watermark message. He could then claim that Alice stole his image and embedded her watermark on top of his.

Threat Nr.	Threat Description	Exploitation Assessment
	Table 1:Identified threats and exploitation assessment	
1	Mallet buys the image but distributes it further	High
2	Mallet embeds his own watermark and registers the mage himself	High
3	Mallet removes the watermark	High
4	Counterfeit original attack	High
5	Statistical removal attack	High
6	Reduced colour image attack	High
7	Mallet performs the known plaintext attack	High
8	Mallet destroys Alice's image or copyright certificate	High
9	Mallet steals the watermarked image (from Alice or Bob) during transmission	High
10	Watermark collision	High
11	Mallet steals Alice's cover-image and/or secret key	Medium
12	Mallet intercepts and changes the image sent to Bob	High
13	The watermark was removed and Alice cannot discover fraud	High
14	Denial-of-service attacks such as Bob paid for the image but did not receive it	Low
15	The CO loses Alice's information	Low
16	The issued copyright certificate is not legally bound	Low
17	Mallet destroys or alters the copyright certificate	High
18	Mallet impersonates Alice and registers images at the CO	High
19	Alice loses her secret key	Low
20	Mallet sets up an image server in a country that does not support copyright laws and distributes "stolen" images	Low
21	Bob paid, but did not receive the correct image	Low
22	Mallet can find out which images Bob is buying	Low
23	Mallet impersonates the CO to get all images from Alice	High
24	Mallet impersonates Bob	Medium
25	Alice denies later to have send a copyright request to the CO	Low

6. Identified Security Requirements

Before establishing the security architecture, we first present the desired properties and objectives of a secure watermarking environment which enables the copyright protected trading of digital images over the Internet.

- **Scalability/non-ambiguity of watermark extraction**

We are striving for a scalable system supporting millions of different users holding billions of copyrights and images.

- **Robustness of the watermarking method**

For any watermark algorithm, we would like it to be robust against any image manipulation technique. Examples include mirroring, turning the image by a fixed angle, up-sizing, down-sizing, resampling, printing and rescanning, screen capture, non-linear image transformations, shredding, zoom in, extract parts of the image, clipping, changing colours, grey-scale representation, change the order of the colour table entries, requantisation, change the texture of some areas, compression (getting rid of redundancy) such as JPEG or FIF (Fractal Compression), and photoshop filters.

Because the number of non-linear transformations for a digital image are not countable*, any watermark function can only be robust against a specific number of possible attacks. Given a specific watermark function, it is possible to specify an additional non-linear image transformation the watermark function is not robust against. We consider, therefore, any watermarking algorithm to be not robust. It is, therefore, not possible to cover with a specific watermarking algorithm all elements of the copyright protection problem space.

Current watermarking techniques are designed to be specifically robust against particular attacks such as JPEG or RST (Rotation, Scaling, Translation) [Ó Ruanaidh].

- **Public algorithm**

The security of the watermarking environment should not rely on the secrecy of the algorithm itself (i.e. Kerkhoff's principle).

- **Speed**

Embedding and extraction should be feasible on any PC.

- **Information hiding "capability"**

The amount of information that can be hidden, without disturbing the image, is limited. Watermarking small images with low information content with long watermarks may disturb the image and when we enhance the contrast or view the image on a "false color" display, the watermark will become visible.

- **Invisibility**

The watermarking algorithm should not degrade the image quality, it should remain invisible to the human viewer, even when embedding a multitude of watermarks.

- **Watermarking function**

The following requirements have been identified:

1. One-way watermark function

---

*Since the functions are defined over a discrete space set X, the number of images generated by these non-linear transformations is countable.

2. The applied watermark technique is at least robust against compression, rotation, translation, and scaling.

3. To disable the counterfeit original attack, the watermark extraction algorithm should not be based on the cover-image.

4. The watermark function should support any image type such as digital photographs, rendered images, graphic logos, cartoons.

- **Legal concerns**

We have seen previously that a CO provides a good solution for many problems of invisible image watermarking. But in order to achieve legal binding between the CH and the cover-image, we need to establish a sound cryptographically secure basis for all communication protocols. The image registration and the issue of a copyright certificate both happen electronically. The legal binding requires strong authentication between the communicating parties and secure transmission channels. An eavesdropper must have no chance of recovering the cover-image, a password or the copyright certificate.

- **Dispute Resolution**

Alice has registered image I at the CO and sells it successfully on the Internet. One day she notes that Mallet has also registered that image and now sells it by himself. She is furious and wants to take legal action against him.

We can distinguish two distinct scenarios in this case of copyright infringement.

1. The image sold by Mallet contains Alice's watermark. Because Alice has registered the image before Mallet, the timestamp in her copyright certificate is older and therefore she is the righteous owner. She extracts the watermark and transfer to the judge who can then verify the validity of the copyright certificate and the watermark. As the watermark is legally binding, the conflict is resolved.

2. If Mallet's image only contains Mallet's watermark (he successfully removed Alice's), the judge has to look at Alice's cover-image, her copyright certificate and compare it to Mallet's image. He then decides after conventional copyright laws applied to non-digital images.

- **Public/Private Watermarks**

We can have two different kinds of invisible watermarks in an image: public and private ones. Public watermarks serve to hold public information, such as image ID or content information. Everybody knows the key to extract the public information. Private watermarks, on the other hand, provide the means for copyright protection. Only the CH holder has the secret key to extract the message. In the case of a lawsuit, the CH needs to show evidence that he was the one who embedded the watermark.

7. Derived Security Architecture

To design our copyright protected watermarking environment we aimed at a system that is usable by every person that would like to protect images. If anybody anywhere wishes to copyright an image, she only needs to download an application and every operation including the embedding of the watermark should be possible electronically. This rationale served as the basis for the design.

7.1. Applied Watermarking Algorithm

It is our aim to provide a fully integrated and secure approach to image watermarking as a means of protecting copyright. A digital watermark on its own provides no legal proof of ownership. Also, a purely cryptographic approach does not appear to be the solution to the copyright problem since once an encrypted image is deciphered it is no longer protected against copyright infringement. We propose a hybrid approach to the problem of image watermarking. Perceptually adaptive spread spectrum (PASS) techniques [Ó Ruanaidh, Nec] provide a reliable means of embedding watermarks that are robust to image modifications. In addition, spread spectrum communication is a form of symmetric cryptosystem. In order to embed a watermark in, remove or intercept (extract a watermark from) an image it is necessary to know the exact values of the seeds used to produce pseudorandom sequences used to encode a watermark. The seed is considered to be a cryptographic key for watermark generation and verification. System security can therefore be based on proprietary knowledge of the keys (or seeds). Note that since the encryption system is based on a symmetric key algorithm then if a sophisticated attacker Mallet can read a watermark then, theoretically at least, he is also capable of removing it. In addition, because spread spectrum signals are statistically independent (and therefore virtually orthogonal), it is possible to have more than one watermark resident on an image at the same time [Ó Ruanaidh, Nec]. For our purposes we can take advantage of this feature by having two very different kinds of digital watermarks with different structures: a private and a public watermark. The public watermark indicates that the document is copyright material and provide information on true ownership. At the same time there is a secure private watermark whose secrecy depends on a proprietary key. This watermark is legally registered at the copyright office and is timestamped. We embed a unique image ID (one-to-one binding with the UCCN) into a block on the basis of the random seed spread spectrum random sequence described above. Note that the image ID is embedded throughout the entire image. The deletion of large regions (image cropping) does not adversely the watermark. As mentioned before, there are several legal issues that need to be appreciated. First, if a number of different watermarks are identified within an image then there has to be some means of separating the different claims of ownership. This may be readily carried out by examining the timestamps of the copyright certificate. In all cases, it is the earliest of the timestamps which is registered with a recognised copyright office identifies the legal owner. Second, the image owner and copyright holder are represented by one legal party in our model.

7.2. Legal Binding

Our system should provide all basic functions required for achieving the legal binding between the copyright certificate and the stego-image. The CH loads a cover-image, generates the image ID, and embeds this image ID as a private watermark into the image generating the stego-image I*. I* is taken as an input for embedding a public watermark. After the stego-image I** with the private and public watermark has been generated, the CH enters the copyright information needed by the CO to issue the copyright certificate and send the stego-image I** with the assocaited registration information to the CO. The CO verifies the validity and consistency of the information and issues a copyright certificate. The CH then starts selling or distributing the stego-image.

In order to achieve legal binding between the copyright certificate and the cover-image, we need authentication, non-repudiation, integrity and confidentiality. These requirements can be fulfilled by using cryptographic methods. We are using the key agreement protocol 3 of the ISO 11770-3 standard. Using this scheme we are verify the identity of the other communicating party, in other words, we have mutual authentication. After the key agreement, both parties have established a mutual secret key, used to encrypt all following messages, therefore satisfying the confidentiality requirement. The non-repudiation property is guaranteed through the signature of every communicated packet. To achieve integrity, we use signatures and consistency checks on both sides. We have therefore established all properties required for the legal binding.

7.3. Trust Model

In our model, every entity trusts the PKI. The certificates issued by the PKI contain the public key of the legal entity and a digital signature of the PKI (We assume that the public key of the Certification Authority (CA) from the PKI is accessible and verifiable by every user.). Every certificate is signed with the CA private key and the trust is built on the validity of our authentic copy of the CA's public key. The CH also trusts the CO and is sure that critical data is handled safely and carefully, so that it is not accessible by unauthorised third parties. In addition, the CH is sure that issued watermarks remain valid until the end of the copyright validity. The CO therefore needs to guarantee that certificate information is stored securely and persistently.

7.4. Secret Information Protection

In our choice of algorithms we need to watch out for security in every aspect of the environment. For example the cover-image is very precious and we have to protect it. First, every network connection is authenticated and the exchanged data is protected against modification and interception (integrity, confidentiality), based on a trust model where all parties trust the CA. Second all important files are saved in encrypted form so an attacker with access to the files has no chance of intercepting the data.

7.5. Mutual Authentication and Key Agreement

Our solution to mutual authentication is based on the RSA public key scheme [Schneier] and on certificates described in [X509]. Every user possesses the public key of the CA and can therefore check the validity of a certificate issued by the CA. Every certificate contains the public key and other information about a legal entity (which is usually a person or an organisation). The details of mutual authentication are explained through the following dialog: Alice would like to communicate over an insecure channel with Bob. Because they are distant and the data sent are very sensitive, she would like to be absolutely certain that she communicates to Bob only. Bob on the other hand needs to be absolutely certain that the data comes from Alice only. The two need to authenticate themselves to each other and therefore sign all the messages sent. Before initiating the communication, Alice requests Bob's certificate from the CA. Because she has a copy of the CA's public key, she can verify the certificate's signature. She can then extract Bob's public key and encrypt all messages for Bob. She can now be certain that Bob is the only person on the planet that can decrypt the data, therefore Bob "authenticated" himself to Alice by being able to decrypt the data and sending back a

coherent response. To authenticate Alice, Bob requests Alice's certificate from the CA*
and extracts Alice's public key. He can now verify Alice's signature and be absolutely
certain that she is the originator of the data.

This scheme works well for small data packets but is slow for longer messages. The
reason is that encryption with a public key algorithm is very slow. The solution is to
apply a mutual authentication and establish a shared secret key between the two communicating parties. We can then use a fast symmetric encryption scheme for the remaining packets of the communication.

Table 2: Symbols and Abbreviations			
Symbol	*Description*	*Symbol*	*Description*
r	Random number	S_{AB}	Symmetric key of A and B (session key)
V_A	Public Key of A	ID_A	Unique ID of A
P_A	Private key of A	CD	Control Data with additional information (timestamp)
V_B	Public Key of B	KAT	Key Agreement Token
P_B	Private key of B		

This process is called key agreement. In our system we are using the key agreement
protocol number 3, as described in [ISO11770-3]. The key agreement, described in
Table 3, is one-pass, we have therefore only 1 message sent between A and B.

Table 3: Key Agreement Protocol			
Sender		**Receiver**	
Phase	*Operation*	*Phase*	*Operation*
1	Choose random number r	1	Receive tokenT_{AB} from A
2	Get V_B from CA	2	Resolve KAT = P_B [V_B[KAT]]
3	Compute S_{AB} = Hash(r)	3	Retrieve ID_A from KAT = < r, S_{AB} [CD], ID_A >
4	Generate KAT = < r, S_{AB} [CD], ID_A>	4	Get V_A from CA
5	Encrypt V_B[KAT]	5	Verify signature VA[þ$_A$] = Hash(V_B[KAT])
6	Compute the signature þ$_A$ = P_A [Hash(V_B[KAT])]	6	Extract r and compute S_{AB} = Hash(r)
7	Generate token T_{AB} = < V_B[KAT], þ$_A$ >	7	Decipher CD = S_{AB}[S_{AB} [CD]] and verify validity.
8	Send tokenT_{AB} to B.		

The detailed protocol is presented below. The signature in A's first data packet also results in the support of the non-repudiation security service. If we also want non-

---

*This step is not necessary if Alice sends her certificate along with the data. This solution was
adopted by [SSL].

repudiation of the remaining communication, each party has to append a signature to every information packet sent. For our system, we used the fast IDEA cipher [Schneier].

7.6.   Copyright Information and Copyright Certificate

Figure 6 shows the steps performed when the CH requests a copyright certificate for an image I. When the CH wants to request a copyright certificate for his image I, he first assigns a new unique image ID. The first 3 bytes determine the CO he is client at and the following 3 bytes determine the client ID he received from the CO. Finally the CH can freely assign last 4 bytes for each one of his images. This system gives us globally unique image ID's. The CH can now enter the other copyright information that is necessary for the issue of the copyright certificate. The "Visual Arts Form" (VA) of the US Copyright Office [USCopyright] served as a basis for the copyright information used. The main fields define the title, author, nature of authorship and other information. The next step is to embed the public and private watermarks into the cover-image. As indicated in Figure 6 the hash value of the private RSA key* is used as a secret key and the image ID is the message to embed. The key to embed the public watermark is known publicly. After the embedding of the two watermarks the CH has created the stego-image I**. If all information is consistent the CH can send the stego-image I**, the image ID and the copyright information to the CO. The CO also verifies the consistency of all information supplied, for example it will test the uniqueness of the image ID supplied. After all verification steps were successful, the CO will issue the copyright certificate. It includes a timestamp, a hash of the image, the supplied copyright information and the CO signature. The image I** and the copyright certificate are uniquely bound through the image hash function. Any customer can verify the copyright certificate by computing the image hash function and verifying the CO's signature.

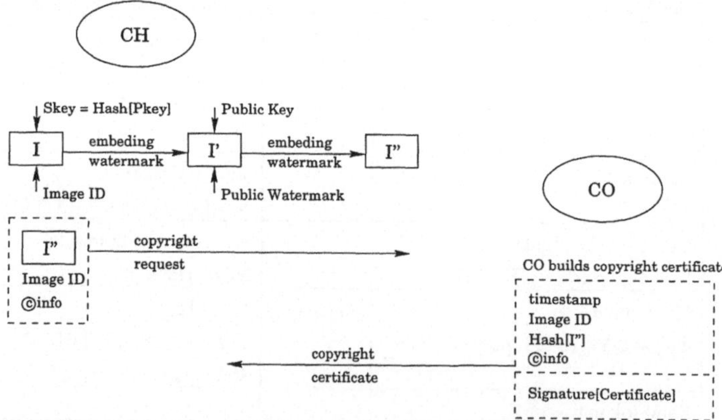

Figure 4: The Copyright Certificate Request and Granting Process

---

*Our approach is based on a public key technique. The key applied for the watermarking process is the hashed private key from the RSA key pair of the CH. It is, therefore, not necessary to disclose the private RSA key.

8.   Implementation

To prove the feasibility of our approach, we have implemented a Java based copyright protection environment for digital images. The Public Key Infrstructure (PKI), realized as a Trusted Third Party (TTP), the Copyright Holder (CH), the Copyright Office (CO), and the Buyer (B) applications all implement a Graphical User Interface (GUI) and a server, supporting both console users and other requests through a socket interface.

The CH supports the following functionality:

- Graphic display of an image
- Generate public/private key-pair, register it at CA and save it persistently
- Designate important regions in the image (rectangles)
- Entry of copyright information
- Manage copyright information
- Embed watermark it in the image
- Contact Copyright Office to get a copyright certificate
- Save the cover-image encrypted and delete the unencrypted version
- Extract watermark from an image
- Adjust the security or quality factor of the watermark
- Sell image to customer

The communication between the application and the dispatching watermarking proxy process is achieved through a socket interface, giving us a good control over communication security and application location. Flexibility is just one advantage of this technique: the database, watermarking engine and GUI can run on different and distant machines. For the first version of our implementation, we have assumed that all parts of the application are run on the same secure machine. It is, therefore, not necessary to protect the privacy of the connections used by the application internally. The proxy process resides on port 4004 and dispatches the CH requests to the watermarking engine or to the database, displayed in Figure 5.

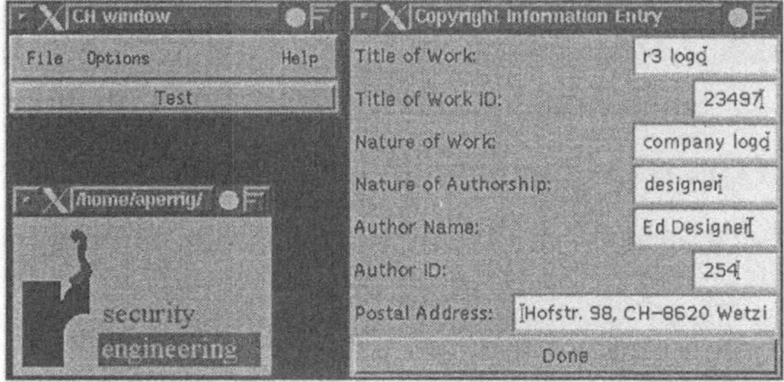

Figure 5: The Copyright Holder Application.

The copyright holder can load images, enter the copyright information, contact the CO, embed the watermark into the cover-image, save all data encrypted in a database and sell the digital image. Figure 5 illustrates an example of the loaded cover-image together with the copyright information.

We have seen previously that the CO is important to solve many of the possible attacks of fraudulent people. The CO is even indispensable for having legally bound copyright certificates. Today's copyright laws require that the CO is in possession of a copy of the cover-image. We can therefore list the following necessary functionality's of the CO application:

- Legally embedded institution backed by the government

- Provide a server to answer incoming copyright requests

- Verify consistency of all received copyright requests

- Save all information persistently and securely, since the data has to be recovered after any restart of the server.

Because all transactions are executed electronically, it is obvious that all the requests must be authenticated, secure (integrity, confidentiality), and transactional. The legal binding further requires non-repudiation. In particular, the persistence property is necessary in order to leave the copyright certificate valid, even if the legal dispute is 20 years later. The data integrity is needed so nobody could claim that the copyright certificate was forged by the CO. We can note that the trust that the CH puts into the CO is substantial. Therefore the CO should be backed by some public authority such as the chamber of commerce.

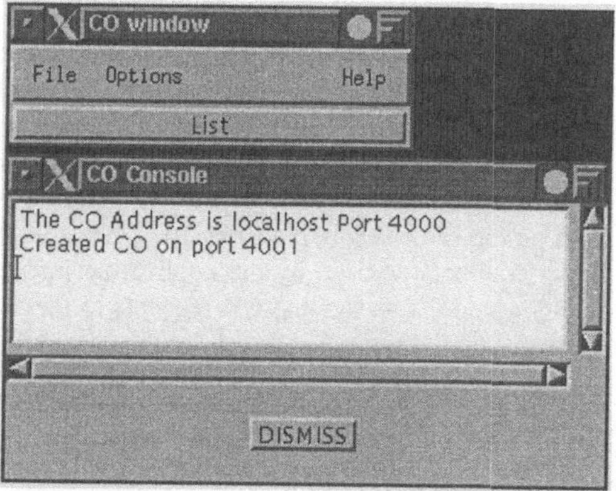

Figure 6: The Copyright Office Server Application.

Figure 6 shows a screen shot of the CO GUI. We can note on the console window that the application has created the CO server on port 4001 and is waiting for incoming copyright requests. The user can list all the copyrights issued.

The PKI, implemented as a Trusted Third Party (TTP) Server, issues all certificates for all users. On the basis of the authenticly distributed public key from the TTP, a public/private key-pair along with its associated certificate may be generated and then passed back as ciphered information to the client. These operations are based on a specific request from the CH. The request includes a password for online key generation, key distribution, and certificate handling. The request is protected on the basis of the CA's public key which was authentically distributed before.

The Buyer application can contact a CH image server residing per default at port 4003. The Buyer can enter the name of the desired image and if the image was already registered at the CO, the CH image server sends back the image.

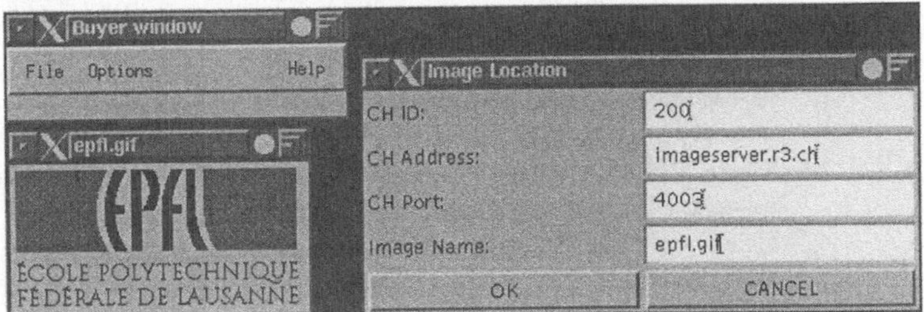

Figure 7: The Buyer Application

The application of the Buyer is shown in Figure 7. We can see the dialog between the Buyer and CH to order the image. The response of the CH is also shown. The image is saved and displayed by the Buyer.

9.   Conclusions

Based on an extensive risk analysis and risk assessment we have identified the security requirements needed to design a copyright protection environment for digital images over the Internet. We have derived an adequate security architecture and shown that the CO and PKI are indispensable in achieving the legal binding for copyright protection of digital images. In addition to being robust, our approach applies cryptographic protocols and public key techniques to ensure the legal binding of spread spectrum based watermark methods. Since the secret key for the watermarking process is based on the hash value of the private RSA key from the CH, the private RSA key has not to be revealed if the CH would like to prove his interlectual property rights to a third party, such as a court of law. An important point is that our approach allows legal action even if a watermark is not found because ownership is legally registered.

It is widely believed today that copyright protection for digital images is only possible through un-removable, un-breakable digital watermarks. We have shown that this assumption is not satisfied. In addition, the approach presented reduces the complexity of the copyright problem space, since the copyright verification process can be associated with the consumer side of the commercial digital image distribution process if every buyer verifies the ownership of a digital image with a public watermark on the basis of the specific copyright certificate.

Acknowledgments

We would like to thank Thomas Mittelhozer and Claus Rassmussen from r^3 security engineering ag for stimulating discussions. We would like to thank especially Professor Thierry Pun of the University of Geneva, CUI, Computer Vision Group, for the technical discussions, the support, and the good collaboration within the research project.

References

[Cinque]   Robert A. Cinque, "Making Cyberspace Safe for Copyright: The Protection of Electronic Works in a Protocol to the Berne Convention", 18 FORDHAM INT'L L.J., 1995, citing BerneConvention art 7(6).

[Craver]   Scott Craver, Nasir Memon, Boon-Lock Yeo and Minerva Yeung, "Can Invisible Watermarks Resolve Rightful Ownerships?", IBM Research Report, RC 20509, July 25, 1996.

[ISO 11770-3]   ISO/IEC CD International Standard 11770-3 "Information technology-Security techniques-Key management", Part 3: Mechanisms using asymmetric techniques, 1994.

[Kutter]   Martin Kutter and Frederic Jordan, "Digital Signatures of Color Images using Amplitude Modulation", SPIE-EI97 Proceedings, 1997.

[NEC]   I. Cox, J. Killian, T. Leighton and T. Shamoon, "Secure Spread Spectrum Watermarking for Images, Audio and Video", Proceedings of the IEEE Int. Conf. on Image Processing, ICIP-96, Lausanne Switzerland, 1996.

[Ó Ruanaidh]   Joseph J.K. Ó Ruanaidh and Thierry Pun, "Rotation, Scale and Translation Invariant Digital Image Watermarking", Signal Processing, January 1997.

[Pfitzmann]   Birgit Pfitzmann, Ross Anderson (Ed.), "Information Hiding", "Information Hiding Terminology", First International Workshop, Cambridge, UK, May/June, 1996, Proceedings, Springer, Lecture Notes in Computer Science, 1174.

[Pitas]   I. Pitas, "A method for Signature Casting on Digital Images", Proceedings of the IEEE Int. Conf. on Image Processing, ICIP-96, Lausanne Switzerland, 1996.

[Schneier]   Bruce Schneier, "Applied Cryptography", John Wiley and Sons Inc., 1994.

[SKIP]   A. Aziz, "SKIP extension for Perfect Forward Secrecy", Internet draft, work in progress, February 1996.

[SSL]   A. Freier and P. Karlton and P. Kocher, "SSL Version 3.0", Netscape Communications", Version 3.0, November 1996.

[USCopyright]   Copyright Act of the United States, Section 106, 1. 3. 1995, http://lcweb.loc.gov/copyright.

[X509]   R. Housley, D. Solo, and W. Ford, IETF: PKI Working Group (PKIX),: X.509 Certificate and CRL Profile, "Internet Public Key Infrastructure Part I", 03. 01. 1997.

[Zhao]   J. Zhao and E. Koch, "Embedding Robust Labels Into Images For Copyright Protection", Proc. Of the Int. Congress on Intellectual Property Rights for Specialized Information, Knowledge and New Technologies, Vienna, August 1995.

Kopierschutz durch asymmetrische Schlüsselkennzeichnung mit Signeten

Birgit Pfitzmann[*]
Universität des Saarlandes, Fachbereich Informatik
Postfach 151150, D-66041 Saarbrücken; <pfitzmann@cs.uni-sb.de>

Michael Waidner
IBM Research Division, Säumerstrasse 4
CH-8803 Rüschlikon; <wmi@zurich.ibm.com>

Zusammenfassung: *Schlüsselkennzeichnung (Traitor Tracing)* ist eine kryptographische Kopierschutztechnik für Daten, die in verschlüsselter Form an viele Empfänger verteilt werden sollen. Die 1994 von *Chor, Fiat* und *Naor* vorgestellte Grundidee besteht darin, allen Käufern zum Entschlüsseln unterschiedliche Schlüssel zu geben, so daß Raubkopierer, die ihre Schlüssel unerlaubterweise weiterverteilen, identifiziert werden können. Zwei Varianten dieses Modells sind *asymmetrische Schlüsselkennzeichnung* und *Signete.*

Im folgenden werden die Modellvarianten in einem einheitlichen Rahmen vorgestellt und eine Konstruktion für *asymmetrische Signete* beschrieben. Ein Vergleich der Modellvarianten ergibt, daß Schlüsselkennzeichnung das für den Einsatz in einem realen Rechtssystem deutlich bessere Konzept ist. Signete haben allerdings gewisse Effizienzvorteile, weshalb auch beschrieben wird, wie *asymmetrische Schlüsselkennzeichnung* aus Signeten konstruiert werden kann.

1 Einleitung

Kopierschutz für digitale Daten kann technisch unterstützt werden durch Verfahren zum *Verhindern* oder zum *Entdecken* illegalen Kopierens. Verhindern im engeren Sinne setzt spezielle manipulationsresistente (*tamper-resistant*) Hardware voraus und ist relativ teuer und umständlich im Einsatz. Es gibt daher einen wachsenden Bedarf nach Verfahren zur *Entdeckung* illegalen Kopierens.

Verfahren zur Entdeckung fallen hauptsächlich in zwei Klassen, je nachdem, wer schlußendlich abgeschreckt werden soll: Die Verwender von Raubkopien, hier *Raubnutzer* genannt, oder die *Raubkopierer*, die Kopien legal kaufen, aber dann illegal weiterverteilen. Für ersteres werden Techniken zur Registrierung legaler Benutzer und Techniken verwendet, bei denen ein für alle Kopien gleicher Copyrightvermerk in die Daten eingebettet wird (*Watermarking*). Im allgemeinen ist der Hersteller jedoch eher daran interessiert zu entdecken, von wem Raubkopien in großem Stil weiterverteilt werden. Die entsprechenden Verfahren werden hier *Datenkennzeichnung (Fingerprinting)* genannt [Wagn_83]. Die Grundidee ist, jede verkaufte Kopie individuell unauffällig so zu kennzeichnen, daß der Händler später den ursprünglichen Käufer anhand der Kopie identifizieren kann. (Siehe [Caro_95, ZhKo_95, BoRD_95, CKLS1_96] für die Einbettungsaspekte und [BlMP_86, BoSh_95] für die kryptographischen Aspekte von Datenkennzeichnung.)

---

[*]   Die hier vorgestellte Arbeit wurde vorwiegend an der Universität Hildesheim durchgeführt und von der DFG unterstützt.

Schlüsselkennzeichnung (*Traitor Tracing*) [ChFN_94] ist in gewisser Weise eine Variante von Datenkennzeichnung: Es ist speziell für solche Anwendungen gedacht, bei denen die Daten in großem Stil verkauft werden sollen, so daß aus ökonomischen Gründen alle Käufer sie in der exakt gleichen Form erhalten müssen. Beispiele sind Pay-TV via Satellit oder Kabel und die Verteilung von Software oder Nachschlagewerken auf CD-ROMs. Üblicherweise werden solche Daten verschlüsselt verteilt und dann der Entschlüsselungsschlüssel verkauft. Die Idee von Schlüsselkennzeichnung besteht darin, Datenkennzeichnung nicht auf Daten, sondern auf *Schlüssel* anzuwenden, d.h. jeder Käufer erhält einen anderen Entschlüsselungsschlüssel, und später kann der Händler anhand eines weiterverteilten Schlüssels den ursprünglichen Käufer identifizieren. Die Bedeutung von „jeder Käufer erhält eine leicht unterschiedliche Kopie" ist für Schlüssel natürlich sehr verschieden von der für andere Datentypen: die einzige Anforderung ist, daß alle Schlüssel dieselben Daten entschlüsseln. (Folglich müssen spezielle Verschlüsselungsverfahren verwendet werden; siehe die folgenden Beispiele.)

Offensichtlich ist Schlüsselkennzeichnung nur dann sinnvoll, wenn der Wert der Daten nicht besonders hoch ist im Vergleich zu den Kosten für die Verteilung: Schlüsselkennzeichnung schreckt ja nicht davon ab, die entschlüsselten Daten weiterzuverteilen — nichts in diesen Daten würde Rückschlüsse auf den Raubkopierer zulassen. Es wird daher angenommen, daß die Anwendung selbst die illegale Weiterverteilung der Daten praktisch unmöglich macht, zum Beispiel weil die Raubkopierer keinen eigenen Satellitenkanal betreiben können oder keine hinreichend billige Möglichkeit haben, CD-ROMs herzustellen. Schlüsselkennzeichnung schreckt Raubkopierer lediglich davon ab, ihre Schlüssel weiterzugeben. Diese sind üblicherweise sehr viel kürzer als die Daten an sich und daher sehr viel einfacher weiterzuverteilen. Ein Händler muß also eine Risikoanalyse erstellen: Falls er die Daten als zu wertvoll für Schlüsselkennzeichnung erachtet, so muß er Datenkennzeichnung anwenden und den Preis für die individuelle Verteilung der Kopien zahlen.

1.1 Asymmetrische Schlüsselkennzeichnung

Asymmetrische Schlüsselkennzeichnung wurde erstmals in [Pfit_96] beschrieben, basierend auf asymmetrischer Datenkennzeichnung. Effizientere Schemata wurden in [PfWa_97] vorgestellt.

Symmetrische Schlüsselkennzeichnung wie in [ChFN_94] beschrieben erzeugt wie klassische, d.h. symmetrische Datenkennzeichnung keinen Beweis, der eine dritte Partei von einer illegalen Weitergabe überzeugen könnte: Händler und Käufer kennen beide den Schlüssel des Käufers. Selbst wenn der Schlüssel irgendwo gefunden wird, wohin er offensichtlich nicht verbreitet werden durfte, so ist doch nicht klar, ob ihn der Käufer weiterverteilt hat oder jemand auf Händlerseite oder jemand, der in den Rechner des Händlers eingebrochen ist.

In diesem Sinne sind die ursprünglichen Schemata für Schlüsselkennzeichnung ebenso symmetrisch wie Nachrichtenauthentisierungscodes: Diese überzeugen den Empfänger einer Nachricht von deren Ursprung, können aber keinen Dritten davon überzeugen (d.h. sie erzielen keine *Non-Repudiation*), da Sender und Empfänger nur einen gemeinsamen geheimen Schlüssel haben. Asymmetrische Schlüsselkennzeichnung ist dann das Analogon zu digitalen Signaturen: Der Käufer hat sein eigenes

Geheimnis, das auf irgendeine Art zur Konstruktion seines Entschlüsselungsschlüssels verwendet wird, aber so, daß der Händler diesen Entschlüsselungsschlüssel nicht kennt, außer er findet ihn nach einer Weiterverteilung. Wird der Schlüssel eines Käufers irgendwo gefunden, wo er nicht sein darf, so ist dies folglich ein klarer Beweis dafür, daß der Käufer ihn weiterverteilt hat.

1.2 Signete

Signete wurden in [DwLN_96] als Alternative zur Schlüsselkennzeichnung vorgeschlagen. Allerdings sind sie eher eine Variante davon: Signete schrecken Raubkopierer ebenfalls dadurch ab, daß jeder Käufer einen anderen Entschlüsselungsschlüssel erhält. Der Unterschied besteht darin, daß bei Signeten keine Verfolgung der Raubkopierer durch die Händler stattfindet. Statt dessen zwingen Signet-Schemata den Raubkopierer B, zusammen mit dem Entschlüsselungsschlüssel ein persönliches Geheimnis sec_B an die Raubnutzer weiterzugeben, indem dieses irgendwie in den Entschlüsselungsschlüssel eingebettet wird. Die Händler müssen also darauf hoffen, daß kein Raubkopierer das Risiko eingehen möchte, sein Geheimnis weiterzuverteilen.

Signete sind zudem *symmetrisch*: Der Käufer muß sein Geheimnis dem Händler geben, damit dieser es in den Entschlüsselungsschlüssel einbetten kann.

1.3 Überblick

Abschnitt 2 beschreibt die Modelle für die drei bekannten Klassen von Schlüsselkennzeichnung etwas ausführlicher. In Abschnitt 3 stellen wir das Modell für asymmetrische Signete vor. Eine Diskussion aller Modelle folgt in Abschnitt 4, insbesondere hinsichtlich der Bedeutung der erzeugten Beweismittel und der möglichen Strafen im realen Leben. Abschnitt 5 stellt Konstruktionen vor, hauptsächlich für die neue Klasse asymmetrischer Signete und für die Herleitung asymmetrischer Schlüsselkennzeichnung aus Signeten. Beide besitzen den Hauptvorteil von Signeten gegenüber Schlüsselkennzeichnung: einen geringeren Nachrichtenaufwand. Vom Rechenaufwand her stellen sie allerdings eher konstruktive Existenzbeweise dar.

2 Modelle

Sowohl bei Datenkennzeichnung als auch bei Schlüsselkennzeichnung gibt es folgende Hauptrollen:

- *Händler*, die digitale Daten verkaufen,

- *Käufer*, die diese Daten kaufen,

- *Raubkopierer*, also unehrliche Käufer, die die Daten ohne Zustimmung des Händlers weiterverteilen,

- *Raubnutzer*, die diese Daten von Raubkopierern erhalten,

- und *Schiedsrichter*, womit jede Art von dritter Partei gemeint ist, die von einem bestimmten Sachverhalt überzeugt werden soll.

Alle Schemata für Schlüsselkennzeichnung und Signete haben folgende Unterprotokolle und Parameter gemein:

- *Schlüsselinitialisierung* ist ein probabilistischer Algorithmus, der vom Händler einmal für jeden einzeln zu verkaufenden Datensatz ausgeführt wird, z.B. für jeden einzelnen Film bei Pay-TV. Seine Eingaben sind bestimmte Sicherheitsparameter, üblicherweise

  - *coll_size*, die maximale Zahl zusammenarbeitender Raubkopierer, die das Protokoll tolerieren soll,

  - N_M, die maximale Zahl von Käufern, an die der Datensatz verkauft werden kann,

  - σ, ein Sicherheitsparameter, in dem gewisse Fehlerwahrscheinlichkeiten exponentiell klein sind,

  - k, ein Sicherheitsparameter für die kryptographischen Sicherheitsaspekte.

  Die Ausgabe dieses Unterprotokolls ist ein *Primärschlüssel*[1]. Dieser bestimmt später die persönlichen Schlüssel der einzelnen Käufer für diesen Datensatz.

- *Schlüsselkennzeichnung* ist ein Unterprotokoll[2], das der Händler mit jedem Käufer dieses speziellen Datensatzes ausführt. Der Händler hat den *Primärschlüssel* als Eingabe. Zumindest für die echte Schlüsselkennzeichnung wird ein von beiden akzeptierter Text, *text*, benötigt, der genau beschreibt, worum es bei dem Kauf geht, z.B. durch einen Verweis auf einen Lizenzvertrag.

  Als Hauptausgabe erhält der Käufer seinen *persönlichen Schlüssel*. Zumindest bei echter Schlüsselkennzeichnung erhält der Händler ebenfalls eine Ausgabe, die *Verkaufsnotiz* genannt wird.

  - Bei symmetrischer Schlüsselkennzeichnung wird Schlüsselkennzeichnung üblicherweise vom Händler alleine ausgeführt. Er sendet lediglich den persönlichen Schlüssel an den Käufer. Die Verkaufsnotiz besteht aus dem Text *text* und dem Namen und persönlichen Schlüssel des Käufers.

  - Bei asymmetrischer Schlüsselkennzeichnung ist Schlüsselkennzeichnung ein interaktives Protokoll zwischen Käufer und Händler und hat zusätzliche Ein- und Ausgaben.

  - In symmetrischen Signet-Schemata besteht Schlüsselkennzeichnung aus zwei Schritten: Der Käufer sendet sein Geheimnis sec_B an den Händler, und der Händler sendet einen Wert α_B, das Signet, zurück. Das Paar (sec_B, α_B) ist der persönliche Schlüssel. Falls ein Käufer wiederholt beim selben Händler einkauft, so kann der Händler sec_B natürlich auch einfach speichern.[3]

- *Datenverteilung.* Der Händler verteilt die echten Daten verschlüsselt mit einem *Sitzungsschlüssel,* den alle Käufer in einem der Nachricht vorausgeschickten

---

[1] In [DwLN_96] wird dies "choice of privileged information A by an authorization center" genannt.

[2] „Protokoll" bedeutet, daß alle Beteiligten lokale Programme ausführen, die aber untereinander Nachrichten austauschen können.

[3] Die Definition in [DwLN_96] erlaubt auch Schemata, in denen die nachfolgenden Vorspanne bereits zu diesem Zeitpunkt bekannt sein müssen (was die Zahl von Sitzungsschlüsseln beschränkt, die bezüglich einem persönlichen Schlüssel übertragen werden können). Die konkret beschriebenen Schemata machen von dieser Möglichkeit allerdings keinen Gebrauch.

Vorspann erhalten. Dies bedeutet, daß jeder Käufer für jeden Datensatz zwei Entschlüsselungsschritte durchführen muß: Zuerst entschlüsselt er den Vorspann mit seinem persönlichen Schlüssel und erhält den Sitzungsschlüssel; sodann verwendet er diesen, um den eigentlichen Datensatz zu entschlüsseln.[4]

Wir nennen ein Schema *gedächtnislos,* wenn die Datenverteilung unabhängig davon ist, welche Käufer die Daten bislang gekauft haben, d.h. wenn die Eingabe des Händlers lediglich aus dem Primärschlüssel und den Daten besteht. Diese Eigenschaft, die in [DwLN_96] prinzipiell vorausgesetzt zu werden scheint, ist notwendig, wenn ein solches Schema für verschlüsselte CD-ROMs verwendet werden soll.

Die Existenz von Sitzungsschlüsseln ist ein rein technisches Detail, d.h. man könnte auch eine allgemeine Definition aufstellen, in der die Sitzungsschlüssel nicht vorkommen. Allerdings würde dies die Risikoanalyse erschweren: Da die Sitzungsschlüssel für alle Käufer gleich sind, hält nichts einen Raubkopierer davon ab, diese weiterzuverteilen. Folglich muß alles, was in der Einleitung über die Weiterverteilung von Daten gesagt wurde, genaugenommen auch für die Weiterverteilung von Sitzungsschlüsseln gesagt werden (und folglich dürfen die Sitzungsschlüssel nicht sehr viel kürzer sein als die Daten). Wir werden dies in den Effizienzbetrachtungen in Abschnitt 5.3 genauer sehen.

Spezielle Klassen von Schlüsselkennzeichnung haben zusätzliche Unterprotokolle:

- *Identifizierung* wird für echte Schlüsselkennzeichnung benötigt, nicht aber für Signete. Grundannahme ist, daß der Händler einen weiterverteilten Schlüssel gefunden hat (oder auch nur ein Gerät eines Raubnutzers, das die Schlüssel in verwürfelter Form enthält und das vielleicht sogar manipulationsresistent ist) und nun versucht, wenigstens einen Raubkopierer zu identifizieren. Als weitere Eingaben hat der Händler den Primärschlüssel und alle Verkaufsnotizen für denselben Datensatz. Seine Ausgabe ist zumindest die Identität eines Käufers und der Wert *text,* der den Kauf beschreibt, zu dem der weiterverteilte Schlüssel gehört.

- *Schlüsselerzeugung für den Käufer* wird bei asymmetrischer Schlüsselkennzeichnung benötigt. Dort erzeugt ein Käufer ein Schlüsselpaar (sk_B, pk_B) und verteilt den öffentlichen Schlüssel pk_B zuverlässig. In allen bisherigen Konstruktionen ist dies das Schlüsselpaar eines Signaturschemas, das innerhalb der Konstruktion benutzt wird. Dieses Unterprotokoll muß nur einmal für jeden Käufer ausgeführt werden.

  Diese Schlüssel werden im Schlüsselkennzeichnungsprotokoll verwendet: Der Käufer gibt seinen geheimen Schlüssel und der Händler den dazugehörenden öffentlichen Schlüssel ein. Das Schlüsselkennzeichnungsprotokoll kann nun auch eine Verkaufsnotiz für den Käufer erzeugen, falls dieser eine falsche Anschuldigung später aktiv ableugnen können muß.

  Weiter erzeugt Identifizierung nun eine Ausgabe, *proof,* für den Händler, die als Beweis verwendet werden wird.

---

[4] In [DwLN_96] heißt der Sitzungsschlüssel "decryption key K". Der Vorspann wird mit z bezeichnet und "common information", "hint" und im Kontext von "incompressible functions" dann "key" genannt.

- *Verhandlung* ist ebenfalls ein spezielles Protokoll für asymmetrische Schlüsselkennzeichnung. Hier versucht ein Händler, einen Schiedsrichter davon zu überzeugen, daß ein bestimmter Käufer ein Raubkopierer ist. Seine Eingaben sind die Ausgaben von *Identifizierung*: die Identität des Käufers, dargestellt durch pk_B, der *text*, der den Kauf beschreibt, und der Beweis *proof*. Der Schiedsrichter muß ebenfalls pk_B und *text* eingeben, so daß der Kontext klar ist. In einigen Schemata muß der beschuldigte Käufer teilnehmen und versuchen, die Beschuldigung mittels seiner eigenen Verkaufsnotiz zurückzuweisen.

Die Sicherheitsanforderungen an ein solches Schema können kurz wie folgt zusammengefaßt werden: Zuallererst muß das Schema natürlich effektiv arbeiten, d.h. wenn sowohl Käufer als auch Händler ehrlich sind, muß der Käufer die Daten erfolgreich entschlüsseln können. Die Hauptforderungen sind sodann:

- *Sicherheit für den Händler.* Angenommen, es gibt höchstens *coll_size* Raubkopierer, und diese verwenden ihr gemeinsames Wissen dazu, Raubnutzern Zugriff auf die Daten zu ermöglichen. Weiter angenommen, die Raubkopierer tun dies, ohne wenigstens soviel Information zu verteilen, wie zur Verteilung der Sitzungsschlüssel nötig wäre. Dann gibt es mindestens einen Raubkopierer, der durch das Schema effektiv bedroht wird. Das heißt für symmetrische Schlüsselkennzeichnung, daß ein Raubkopierer identifiziert wird, für asymmetrische Schlüsselkennzeichnung, daß der identifizierte Raubkopierer die Verhandlung verlieren wird, und für Signete, daß das persönliche Geheimnis eines Raubkopierers an die Raubnutzer verraten werden wird.

- *Sicherheit für den Käufer.* Ehrliche Käufer dürfen keine der Nachteile haben, die Raubkopierer durch die Schemata angedroht bekommen. Das heißt, in symmetrischer Schlüsselkennzeichnung werden ehrliche Käufer nicht fälschlich als Raubkopierer identifiziert, in asymmetrischer Schlüsselkennzeichnung werden sie keine Verhandlungen mit ehrlichen Schiedsrichtern verlieren, und in Signeten wird ihr Geheimnis nicht an Raubnutzer verraten.

Weitere Varianten sind denkbar; man vergleiche z.B. die Sicherheit des Händlers gegen das Erheben falscher Anschuldigungen in [Pfit_96].

3 Asymmetrische Signete

Der Hauptnachteil der bisherigen Signete ist, daß der Käufer dem Händler ein echtes Geheimnis verraten muß; daher auch die Bezeichnung „symmetrisch".

Zunächst bedeutet dies, daß Signete nichts dagegen bewirken, daß der Käufer den Schlüssel an Freunde weitergibt: alles, was man einem Händler verraten kann, kann man sicher auch zumindest manchen Freunden anvertrauen. Allerdings ist dies kein großes Problem, da der Händler in diesen Fällen ohnehin kaum eine Raubkopie eines Schlüssels finden wird, was die Grundvoraussetzung für die Identifizierung bei Schlüsselkennzeichnung ist. All diese Schemata sind vorwiegend dazu gedacht, Massenbetrug zu verhindern.

Allerdings ist der Begriff eines persönlichen Geheimnisses, das man mit Händlern teilen kann, problematisch. Das Hauptbeispiel in [DwLN_96] sind Kreditkartennummern, die in der Tat gelegentlich im selben Sinne auch im realen Leben verwendet werden. Allerdings ist dies auch im realen Leben eine sehr zweifelhafte Praxis: Sobald Dutzende von Leuten eine bestimmte Kreditkartennummer kennen, kann man die Kenntnis dieser Nummer vernünftigerweise nicht mehr zur Authentisierung verwenden. Folglich ist auch die Motivation, diese Nummer vor weiteren Leuten geheimzuhalten, nicht sehr hoch. Zumindest mit gedächtnislosen Schemata für Signete kann man das Problem etwas vermindern, indem alle Händler ihre Primärschlüssel einigen unparteiischen Dritten geben, die dann die Schlüsselkennzeichnung ausführen. Nun hat jeder Käufer zumindest eine gewisse Auswahl, wem er bezüglich der Geheimhaltung seines Geheimnisses vertrauen möchte. Nichtsdestotrotz ist ein Geheimnis, das mit irgendeiner anderen Partei geteilt werden muß, nicht mehr wirklich geheim, insbesondere dann, wenn diese „eine Partei" in Wirklichkeit aus einem großen Rechensystem besteht und von einer Unzahl an Angestellten betrieben wird. Zur Vermeidung dieses Problems schlagen wir asymmetrische Signet-Schemata vor. Die weiteren Probleme werden in Abschnitt 4 diskutiert.

- Asymmetrische Signet-Schemata haben vier Unterprotokolle: *Schlüsselinitialisierung*, *Schlüsselkennzeichnung* und *Datenverteilung* wie alle betrachteten Schemata und zusätzlich *Schlüsselerzeugung für den Käufer,* ähnlich wie bei asymmetrischer Schlüsselkennzeichnung. Es gibt jedoch weder *Identifizierung* noch *Verhandlungen.*

  Schlüsselerzeugung für den Käufer ist etwas anders als bei asymmetrischer Schlüsselkennzeichnung, da ein externes Geheimnis sec_B verwendet werden muß. Beispielsweise kann dies ein *gegebener* geheimer Schlüssel eines Signaturschemas sein, der in einem sicheren digitalen Kreditkartenzahlungssystem zum Signieren von Zahlungsanweisungen verwendet wird (analog zu dem symmetrischen Kreditkartenbeispiel in [DwLN_96]). Allgemein nehmen wir an, daß es zu dem Geheimnis einen entsprechenden öffentlichen Wert pub_B und eine Verifizierungsprozedur gibt, die erlaubt zu entscheiden, ob Geheimnis und öffentlicher Wert zusammenpassen.[5] Sodann benötigen wir eine Registrierungsinfrastruktur, die sicherstellt, daß ein bestimmter öffentlicher Wert zu einem bestimmten Käufer gehört und in Schlüsselkennzeichnung verwendet werden kann.

- *Sicherheit für den Händler* ist genauso definiert wie in symmetrischen Signet-Schemata: Angenommen, maximal *coll_size* zusammenarbeitende Raubkopierer erlauben Raubnutzern Zugriff auf den Klartext. Dann müssen diese entweder im wesentlichen soviel Information verteilen, wie die Verteilung der Sitzungsschlüssel erfordern würde, oder die Raubnutzer erhalten das Geheimnis sec_B wenigstens eines Raubkopierers B.

- *Sicherheit für den Käufer.* Falls ein Käufer nichts weiterverteilt, dann kann selbst eine Konspiration aus Händler, Registrierungsautorität und allen anderen Käufern das Geheimnis dieses Käufers nicht erhalten. Genauer gesagt darf niemand irgendwelche Information über sec_B erhalten zusätzlich zu dem, was bereits durch pub_B und die Verwendung von sec_B außerhalb des Signet-Schemas bekannt ist.

---

[5]   Die meisten Signaturschemata haben von sich aus eine solche Verifizierungsprozedur, z.B. RSA [RSA_78] und Schemata, die auf dem diskreten Logarithmus basieren, z.B. die von ElGamal oder Schnorr [ElGa_85, Schn_91]. Man kann eine solche Prozedur leicht zu jedem beliebigen Signaturschema hinzufügen, indem man die Zufallszahlen, die zur Erzeugung des Schlüsselpaars verwendet wurden, als Teil des Geheimnisses auffaßt. Dann ist der Rest der Schlüsselerzeugung deterministisch und kann leicht verifiziert werden.

4 Modelle und Realität

Signete wurden in [DwLN_96] vorgeschlagen um, im Vergleich zu symmetrischer Schlüsselkennzeichnung, Verhandlungen und die Suche nach Raubkopien zu vermeiden. Wir halten dies jedoch zugleich für ein sehr ernsthaftes Problem, da die Vermeidung von Verhandlungen zu einer Art von Selbstjustiz führt, bei der das Strafmaß üblicherweise in keinem vernünftigen Verhältnis zur Schuld steht.

4.1 Mögliche Geheimnisse und Selbstjustiz

Ganz allgemein gibt es verschiedene Gründe, weshalb ein Käufer sein „Geheimnis" sec_B tatsächlich geheim halten möchte.

- Es kann ein wirklich persönliches Geheimnis sein. Dann kann allerdings auch niemand verifizieren, ob der Käufer ein echtes Geheimnis verwendet, und folglich kann das Signet-Schema nicht sicher für den Händler sein. Es hilft auch nichts, wenn der Käufer das Geheimnis einer Registrierungsautorität gibt und von ihr zertifiziert bekommt, da diese das Geheimnis ebensowenig verifizieren kann.

- Es kann eine vertrauliche Information sein, die bereits manchen dritten Parteien bekannt ist, wie zum Beispiel finanzielle oder medizinische Daten. In diesem Fall könnten diese dritten Parteien den Wert dieses Geheimnisses unter einer Einwegfunktion (genauer: ein *Commitment* [BrCC_88]) veröffentlichen und so das Geheimnis für den Händler verifizierbar machen. In einem asymmetrischen Signet-Schema wäre dies kein Problem für ehrliche Käufer, da die Händler keine Information über das Geheimnis erhalten würden. Nichtsdestotrotz ist es ethisch bedenklich, daß die dritten Parteien eine solche Verbindung von ihnen anvertrauten Geheimnissen mit dem Problem der Weiterverteilung von Information erlauben. Dies klingt wie eine sehr zweifelhafte Form von Bestrafung für Fehlverhalten unter Umgehung des üblichen Rechtssystems.

- Es kann Information sein, die erlaubt, irgend etwas im Namen des Käufers zu tun, z.B. ein Signaturschlüssel, der für Zahlungen verwendet werden kann. Dies scheint die noch vernünftigste Variante zu sein. Nichtsdestotrotz ist die Kopplung zwischen der Weiterverteilung eines solchen Geheimnisses mit der Weiterverteilung von Information ethisch bedenklich. Dies wäre vergleichbar mit einer Bank, die bei Kontoeröffnungen Nachschlüssel zu der Wohnung des Kunden verlangt und einen Vertrag abschließt, daß, falls dieser Kunde sein Konto überzieht, die Bank seine Wohnung ausrauben darf. Insbesondere dürfte der Gesamtwert eines solchen Geheimnisses meist sehr viel größer sein als der Wert der weiterverteilten Information.

All dies trifft auch dann zu, wenn es keinerlei Zweifel an der Schuld des Käufers gibt. Es wird jedoch noch schlimmer, sobald wir die technischen Rahmenbedingungen beim Betrieb eines solchen Systems betrachten, siehe Abschnitt 4.2.

Es ist denkbar, daß ein Schema zur asymmetrischen Schlüsselkennzeichnung zugleich ein asymmetrisches Signet-Schema ist, da die Unterprotokolle des zweiten Typs eine Untermenge des ersten darstellen. Falls die Selbstjustizaspekte explizit unerwünscht sind, kann die folgende Gegenforderung erhoben werden:

- *Verhinderung von Selbstjustiz.* Selbst dann, wenn ein Käufer Information aus dem Schema zur Schlüsselkennzeichnung weiterverteilt, so gefährdet dies keines seiner externen Geheimnisse.

Die einfachste (aber nicht die einzige) Möglichkeit, dies zu garantieren ist, in der Schlüsselerzeugung für den Käufer innerhalb des asymmetrischen Schlüsselkennzeichnungsschemas ein neues Schlüsselpaar zu generieren, selbst dann, wenn der Käufer bereits ein Schlüsselpaar des zugrundeliegenden Signaturschemas besitzt.

4.2 Sichere Geräte

Wie in der Kryptographie üblich haben wir bisher stillschweigend und sehr idealisierend vorausgesetzt, daß alles, was das Gerät des Käufers tut, auf den expliziten Wunsch des Käufers hin geschieht. Insbesondere haben wir angenommen, daß ein Käufer seinen persönlichen Schlüssel nur dann weiterverteilt, wenn er ein Raubkopierer ist. In der Realität kann dies aber auch passieren, weil

- irgend jemand den Schlüssel aus dem Gerät des Käufers stiehlt

- oder dem Käufer ein Bedienungsfehler unterläuft, was abhängig von der Benutzerschnittstelle mehr oder weniger wahrscheinlich ist,

- oder die Implementierung des Schemas (die der Käufer möglicherweise sogar vom Händler selbst bekommen hat) nicht sicher ist.

Die Konsequenzen sind im Fall von Signeten schwerwiegender als im Fall von Schlüsselkennzeichnung: Wenn ein echtes Geheimnis erst einmal bekannt geworden ist, so kann es nicht mehr zurückgenommen werden. Falls die Raubnutzer zum Beispiel den Schlüssel verwendet haben, um das Konto des Käufers zu plündern, so ist das Geld weg. Im Gegensatz dazu kann eine explizite Verhandlung nichtkryptographische Aspekte in Betracht ziehen. Insbesondere kann der Softwarehersteller für fehlerhafte Implementierungen zur Verantwortung gezogen werden. Desweiteren kann der Lizenzvertrag, der in *text* referenziert wird, genau die vom Käufer erwartete Sorgfalt spezifizieren. Zum Beispiel könnten hochwertige Daten nur an solche Käufer verkauft werden, die versichern, daß sie bereit sind, persönlich für Weiterverteilungen zur Verantwortung gezogen zu werden.[6] Andere Daten könnten auch an Käufer verkauft werden, die diese Haftung ablehnen, aber zustimmen, Schäden bis zu einer gewissen Obergrenze zu begleichen, oder der Händler könnte von solchen Käufern einen höheren Preis erheben.

In gewisser Weise kann man natürlich auch argumentieren, daß wer auch immer einen persönlichen Schlüssel aus dem Gerät des Käufers stehlen kann, sich vermutlich auch andere Geheimnisse verschaffen könnte, d.h. daß durch Signete eigentlich nichts verloren geht. Allerdings stimmt dies nur dann, wenn diese Geheimnisse im selben Gerät gespeichert und von derselben Software bearbeitet werden. Letzteres ist

---

[6] Dies sollten nur Käufer unterschreiben, die speziell gesicherte Geräte verwenden. Die Händler sollten auch fordern, daß die Käufer explizit bestätigen, daß sie solche Geräte verwenden. Andernfalls könnte ein Gericht vernünftigerweise Erklärungen zur persönlichen Haftung von anderen Käufern für ungültig nach BGB §138 (2) (Sittenwidriges Rechtsgeschäft) erklären: Man kann erwarten, daß Käufer wissen, welche Geräte sie verwenden. Aber jeder, der denkt, er könne persönlich dafür verantwortlich sein, was auf einem normalen PC mit Internet-Zugang passiert, muß als unerfahren gelten.

äußerst unwahrscheinlich. Ersteres ist ebenfalls nicht klar, da ein Signaturschlüssel z.B. in einem speziellen Sicherheitsmodul gespeichert werden könnte, wohingegen selbst der Teil eines Schemas zur Schlüsselkennzeichnung, der den Vorspann entschlüsselt, in normaler Software sein könnte.

4.3 Schlüsselverteilung

Das Schlüsselverteilungsproblem besteht darin, daß jeder den richtigen öffentlichen Schlüssel pk_B dem richtigen Käufer zuordnen können muß. Wie üblich ist dies eine Frage von Registrierungsprozeduren, Sicherheit von Zertifizierungsautoritäten gegen ihre Betreiber und deren Angestellte (die zur Zeit sicher nicht perfekt sein kann), und Haftung von Zertifizierungsautoritäten für falsche Zertifikate. Für asymmetrische Schlüsselkennzeichnung ist dies exakt dasselbe Problem wie für Signaturschemata. Bei asymmetrischen Signeten ist dies nicht so offensichtlich: Wenn der Käufer sein Geheimnis sec_B geheim hält, so bekommen es die Raubnutzer nicht, unabhängig davon, wie die Registrierung aussieht. Allerdings wird das zugrundeliegende System, aus dem die Geheimnisse kommen, meist ein ähnliches Problem haben. In gewisser Weise kann das Problem in beiden Fällen vermieden werden, falls die Bestrafung nicht auf eine externe Identität abzielt, sondern z.B. auf ein Konto, das speziell für diesen Zweck eingerichtet wurde. Allerdings sind die Händler mit dieser Möglichkeit vermutlich nicht zufriedenzustellen: falls die Kenntnis des Geheimnisses zum Plündern dieses Kontos ausreicht, hält nichts den Käufer davon ab, dieses selbst zu tun.

4.4 Finden von Raubkopien

In der Tat ist es ein Vorteil von Signeten, daß die Händler nicht nach Raubkopien suchen müssen. Allerdings können Händler bei Schlüsselkennzeichnung einen hohen Preis auf Raubkopien aussetzen, was für die Raubkopierer wieder ein ähnliches Risiko darstellt, daß die Raubnutzer sich zu ihrem eigenen Vorteil gegen sie wenden.

Ein Nachteil aller bislang vorgeschlagenen Konstruktionen von Signeten im Gegensatz zur Schlüsselkennzeichnung ist allerdings, daß keine Prozedur bekannt ist, die das Geheimnis aus manipulationsresistenten Geräten von Raubnutzern extrahiert (d.h. in einem Black-Box-Modell).

4.5 Geheimhaltung der echten Daten

Wir möchten noch einen Vorteil aller Varianten von Schlüsselkennzeichnung gegenüber Datenkennzeichnung der echten Daten erwähnen (wohingegen in anderer Hinsicht, wie in der Einleitung erläutert, Schlüsselkennzeichnung nur eine Variante für geringwertige Massendaten darstellt):

Es ist nicht notwendig, daß der Käufer, sobald er z.B. einen Videofilm angesehen hat, diesen komplett vergißt oder zumindest mit niemandem darüber redet. Aus formalen Gründen muß dies für Datenkennzeichnung gefordert werden, da die Geheimnisse dort in die echten Daten eingebettet werden und daher diese Forderung die einzige Möglichkeit zu sein scheint zu garantieren, daß der Käufer nicht jedesmal ein paar Bits seines Geheimnisses verrät, wenn er sich über diese Daten unterhält.

5 Konstruktionen

Trotz unseres Unbehagens hinsichtlich des Modells von Signeten, wie insbesondere in Abschnitt 4.1 erläutert, möchten wir doch zeigen, daß asymmetrische Signete existieren, zumindest unter der Annahme, daß das symmetrische Signet-Schema aus [DwLN_96] sicher ist. Sodann konstruieren wir, was wir für deutlich relevanter erachten, ein asymmetrisches Schema zur Schlüsselkennzeichnung aus einem Signet-Schema.

5.1 Asymmetrische Signete

Wir konstruieren ein asymmetrisches Signet-Schema aus einem nahezu beliebigen symmetrischen Signet-Schema. Wie in Abschnitt 3 beschrieben, müssen wir annehmen, daß jeder Käufer ein externes Geheimnis sec_B mit dazugehörigem öffentlichen Wert pub_B besitzt und daß ein Algorithmus bekannt ist, mit dem der Händler verifizieren könnte, daß sec_B und pub_B zusammenpassen, falls er sie jemals sehen sollte.

Wir fordern zwei Eigenschaften von dem symmetrischen Signet-Schema: Es muß gedächtnislos sein, und die Menge möglicher Geheimnisse muß groß genug sein, so daß unsere verifizierbaren externen Geheimnisse verwendet werden können. Keine der beiden Forderungen ist eine ernsthafte Einschränkung: Signet-Schemata nach der ursprünglichen Definition aus [DwLN_96] sind stets gedächtnislos, und lange Geheimnisse zuzulassen ist selbst für die symmetrischen Schemata sinnvoll. Insbesondere hat die Konstruktion aus [DwLN_96] diese Eigenschaft: der Grundbereich kann ein beliebiger Körper $\mathbb{Z}_q$ sein.

Wir verwenden ein Protokoll für sichere 2-Parteien-Berechnungen. Dies ist eine kryptographische Grundtechnik, die folgenden Zweck erfüllt: Zwei Parteien haben geheime Eingaben x_1 und x_2. Beide oder einer von beiden möchte $g(x_1, x_2)$ erfahren, wobei g eine Funktion ist, die beide Parteien kennen. Allerdings darf weder die erste Partei zusätzliche Information über x_2 bekommen, noch die zweite Partei zusätzliche Information über x_1. Ein nettes Beispiel ist das Millionärsproblem aus [Yao_82]: Die Parteien sind Millionäre, und x_1 und x_2 geben an, wie viele Millionen sie besitzen. Sie möchten herausfinden, wer von beiden reicher ist, aber ohne sich gegenseitig x_1 und x_2 zu verraten. Konkrete Konstruktionen für sichere 2-Parteien-Berechnungen werden z.B. in [Yao_86, ChDG_88, GMW_87] beschrieben. Wir erheben die sehr starke Sicherheitsforderung, daß das 2-Parteien-Protokoll durch ein vertrauenswürdiges Orakel ersetzt werden kann, das dieselbe Funktion berechnet.

Die Parameter *coll_size* und N_M sind dieselben im zu konstruierenden asymmetrischen und im zugrundeliegenden symmetrischen Signet-Schema.

- Schlüsselerzeugung für den Käufer. Da das Geheimnis sec_B bereits gegeben ist, benötigen wir lediglich eine Registrierungsautorität, die garantiert, daß ein bestimmter öffentlicher Wert pub_B (und ein Algorithmus zum Verifizieren des Geheimnisses) zu einem bestimmten Käufer gehört und zur Schlüsselkennzeichnung verwendet werden kann.

- Schlüsselinitialisierung. Der Händler wählt einen Primärschlüssel wie im zugrundeliegenden symmetrischen Signet-Schema.

- Schlüsselkennzeichnung. Das Protokoll für sichere 2-Parteien-Berechnungen wird für folgende Berechnung verwendet:

  - Der Käufer hat als geheime Eingabe sec_B.

  - Der Händler hat als geheime Eingabe seinen Primärschlüssel und (nicht notwendigerweise geheim) pub_B.

  - Innerhalb des Protokolls wird verifiziert, daß sec_B und pub_B zusammenpassen. Falls nicht, bricht das Protokoll ab.

  - Für den Primärschlüssel und sec_B wird innerhalb des Protokolls die Schlüsselkennzeichnung aus dem zugrundeliegenden symmetrischen Signet-Schema ausgeführt. Das Ergebnis, also das Signet, wird an den Käufer ausgegeben.

- Datenverteilung geht wie im zugrundeliegenden symmetrischen Signet-Schema.

Die Sicherheit dieser Konstruktion ist leicht zu sehen, wenn man das Protokoll für sichere 2-Parteien-Berechnungen durch ein vertrauenswürdiges Orakel ersetzt: Die Signete sind dieselben wie im zugrundeliegenden Schema, und niemand erhält irgendwelche zusätzliche Information.

Man beachte, daß das Protokoll für sichere 2-Parteien-Berechnungen nur zur Schlüsselkennzeichnung verwendet wird, wohingegen alle Operationen zur Datenverteilung unverändert bleiben. Allerdings sind die bekannten allgemeinen Protokolle für sichere 2-Parteien-Berechnungen nicht effizient, und wir haben noch keine spezifische effizientere Implementierung für die symmetrische Konstruktion aus [DwLN_96] gefunden. Daher betrachten wir unser Schema derzeit hauptsächlich als konstruktiven Existenzbeweis.

5.2 Asymmetrische Schlüsselkennzeichnung aus Signeten

Wir konstruieren nun ein Schema zur asymmetrischen Schlüsselkennzeichnung aus einem nahezu beliebigen gedächtnislosen Signet-Schema. Wir müssen lediglich annehmen, daß die Menge möglicher Geheimnisse groß genug ist, um als Urbildbereich für eine Einwegfunktion zu dienen. Zur Vereinfachung nehmen wir an, dieser Urbildbereich sei $\{0, 1\}^k$ für einen Sicherheitsparameter k. Als kryptographische Grundfunktionen benötigen wir diese Einwegfunktion, ein Signaturschema und ein Protokoll für sichere 2-Parteien-Berechnungen wie oben. Die Parameter $coll\_size$ und N_M sind im zu konstruierenden Schema und im zugrundeliegenden Signet-Schema identisch.

- Schlüsselerzeugung für den Käufer. Jeder Käufer erzeugt ein Schlüsselpaar (sk_B, pk_B) des gegebenen Signaturschemas und verteilt seinen öffentlichen Schlüssel pk_B zuverlässig, so daß zumindest alle Händler und potentiellen Schiedsrichter ihn bekommen können.

- Schlüsselinitialisierung. Der Händler wählt einen Primärschlüssel wie im gegebenen Signet-Schema.

- Schlüsselkennzeichnung. Der Käufer erzeugt einen Zufallswert sec_B der Länge k und berechnet $im_B := f(sec_B)$. Er signiert die Nachricht $msg_B = (text, im_B)$, wobei $text$ der Text ist, der den Kauf beschreibt. Er sendet msg_B und die Signatur sig_B zum Händler, der sig_B verifiziert. Sodann wird das Protokoll für sichere 2-Parteien-Berechnungen für die folgende Berechnung verwendet:

- Der Käufer hat als geheime Eingabe sec_B.

- Der Händler hat als geheime Eingabe seinen Primärschlüssel und (nicht notwendigerweise geheim) im_B.

- Innerhalb des Protokolls wird überprüft, ob $f(sec_B) = im_B$. Falls nicht, bricht das Protokoll ab.

- Für den Primärschlüssel und sec_B wird innerhalb des Protokolls die Schlüsselkennzeichnung des zugrundeliegenden Signet-Schemas ausgeführt. Das Ergebnis wird an den Käufer als sein persönlicher Schlüssel ausgegeben.

- Datenverteilung geht wie im zugrundeliegenden Signet-Schema.

- Für die *Identifizierung* wird angenommen, daß der Händler einen Wert sec_{B*} gefunden hat, der von einem Raubkopierer verwendet wurde. Dies ist eine stärkere Annahme als in anderen Schemata zur Schlüsselkennzeichnung, aber es ist dieselbe wie im zugrundeliegenden Signet-Schema: es wird angenommen, daß alle Raubnutzer sec_{B*} berechnen können. Folglich kann es auch der Händler, sobald er irgendeine Raubkopie eines Schlüssels erhält. Der Händler berechnet $f(sec_{B*})$ und vergleicht es mit allen gespeicherten Werten im_B, um den Käufer zu finden, zu dem es gehört. Aus der Verkaufsnotiz für diesen Käufer erhält er schließlich den Wert sig_B.

- In einer Verhandlung verwendet der Händler den Beweis $proof = (sig_B, sec_{found})$. Der Schiedsrichter berechnet $im_{found} := f(sec_{found})$ und verifiziert, daß sig_B die Signatur des beschuldigten Käufers unter $msg_{found} := (text, im_{found})$ darstellt.

Die Sicherheit dieser Konstruktion ist leicht zu sehen, wenn man das Protokoll für sichere 2-Parteien-Berechnungen durch ein vertrauenswürdiges Orakel ersetzt: Die Sicherheit des Händlers folgt direkt aus der Annahme, daß er den Wert eines Raubkopierers sec_{found} gefunden hat, und der Art, wie der Beweis konstruiert und verifiziert wird. Der Händler ist auch sicher dagegen, falsche Anschuldigungen zu erheben, selbst dann, wenn es mehr als *coll_size* Raubkopierer gibt: in diesem Fall gelingt die Identifizierung nicht, aber er schadet seinem Ansehen nicht durch das Verlieren einer Verhandlung. Ein Käufer, dessen Signaturen nicht gefälscht werden können, kann nur dann betrogen werden, wenn der Händler ein Urbild von im_B vorzeigen kann, was aufgrund der Einwegeigenschaft von f unmöglich ist, da ein ehrlicher Käufer niemals irgendwelche Information darüber preisgibt, außer der in im_B.

5.3 Arbeitsfaktoren der asymmetrischen Schlüsselkennzeichnung

Der ganze Sinn von Schlüsselkennzeichnung basiert auf der Hoffnung, daß die Verteilung für die Händler noch kostengünstig, aber die Weiterverteilung von Sitzungsschlüsseln oder echten Daten zu aufwendig für Raubkopierer ist. (Rechtzeitigkeit mag manchmal auch kritisch sein, falls die Raubkopierer nur Sitzungsschlüssel weiterverteilen, aber üblicherweise können Raubnutzer die verschlüsselten Daten puffern.) Ein wichtiger Parameter ist daher der Quotient aus der Länge eines Vorspanns und eines Sitzungsschlüssels; wir nennen dies den *Nachrichtenverteilungsfaktor*. Er beschreibt das Verhältnis zwischen den zusätzlichen Daten, die der Händler verteilen muß, und den Daten, die die Raubkopierer an die Raubnutzer verteilen müssen. Ein anderer Arbeitsfaktor, der Quotient aus dem Nachrichtenaufwand von Schlüsselkennzeichnung und Weiterverteilung von Sitzungsschlüsseln, ist datenabhängig: Je größer die Menge an Daten ist, die ein Käufer bekommen kann, wenn er einen persönlichen Schlüssel kauft, desto geringer ist der Einfluß der Schlüsselkennzeichnung.

In [DwLN_96] hat die Folge der Sitzungsschlüssel die Form $g_1{}^a$, $g_2{}^a$, ..., wobei a ein Teil des Primärschlüssels und die Werte g_i Generatoren einer Gruppe primer Ordnung sind, in der man annehmen kann, die Berechnung diskreter Logarithmen sei hart. Der Grundgedanke der Konstruktion ist, daß obwohl die Folge der Sitzungsschlüssel eine kurze Darstellung hat, nämlich a, es für die Raubkopierer unmöglich ist, diese zu berechnen. Die komplette Konstruktion, die mehreren Käufern erlaubt, solche Sitzungsschlüssel zu berechnen, sieht wie folgt aus:

- Schlüsselinitialisierung. Der Händler wählt eine geeignete Gruppe G_q, z.B. eine Untergruppe der Ordnung q von $\mathbb{Z}_p{}^*$, wobei p, q Primzahlen sind mit $q \mid (p - 1)$. Er wählt auch ein zufälliges $a \in G_q$ und ein Polynom Q vom Grade $coll\_size$ mit konstantem Koeffizienten a. Sei $Q(x) =: a + b_1x + ... + b_{coll\_size}\, x^{coll\_size}$ und $Q^* := Q - a$.

- Schlüsselkennzeichnung. Ein Käufer mit Geheimnis sec_B erhält $Q(sec_B)$. Dies würde $coll\_size + 1$ Käufern die Berechnung von a erlauben, aber weniger Käufer erhalten in diesem Schritt keine Information über a, wie in Shamirs Schema zur Geheimnisaufteilung (*secret sharing*) [Sham_79]. Der persönliche Schlüssel des Käufers ist (sec_B, $Q(sec_B)$).

- Datenverteilung. Der Vorspann enthält als erstes einen neuen, zufällig gewählten Generator g_i von G_q für diese Sitzung.[7] Als zweites enthält er $coll\_size$ Werte h_{ij} := $g_i{}^{bj}$. Wie oben erwähnt, ist $g_i{}^a$ der Sitzungsschlüssel, der dazu verwendet wird, die echten Daten zu verschlüsseln.

  Jeder Käufer kann $g_i{}^a$ wie folgt berechnen: Zuerst berechnet er $g_i{}^{Q(sec_B)}$, sodann $g_i{}^{Q^*(sec_B)}$ als das Produkt der Terme $(g_i{}^{bj})^{sec_B{}^j}$ (in einer etwas effizienteren Auswertungsreihenfolge). Als letztes berechnet er $g_i{}^a = g_i{}^{Q(sec_B)}/g_i{}^{Q^*(sec_B)}$.

Man sieht leicht, daß der Nachrichtenverteilungsfaktor in diesem Schema nur $coll\_size + 1$ beträgt: Für jeden Sitzungsschlüssel $g_i{}^a$ verteilt der Händler den Generator g_i und $coll\_size$ Werte h_{ij}. Der Nachrichtenverteilungsfaktor bleibt konstant, wenn wir dieses Signet-Schema wie in Abschnitt 5.2 beschrieben für asymmetrische Schlüsselkennzeichnung verwenden. Dies ist deutlich besser als in den zuvor bekannten Schemata, siehe [PfWa_97]. Zum Beispiel könnte es das Schema für CD-ROMs geeignet machen: Falls die Hälfte der CD für Vorspanne mit einem moderaten Wert $coll\_size$ verwendet wird, passen die dazugehörenden Sitzungsschlüssel nicht mehr auf eine Diskette.

5.4 Andere Reduktionen

Zunächst ist klar, daß ohne Effizienzverlust jedes symmetrische Signet-Schema als symmetrisches Schlüsselkennzeichnungs-Schema verwendet werden kann, zumindest falls die Geheimnisse sec_B aus einer einigermaßen großen Menge stammen, aus der

---

[7] Man kann auch mehrere Generatoren als Teil eines längeren Sitzungsschlüssels auffassen. Da aber die Länge und Anzahl von Sitzungen noch frei ist, macht dies keinen Unterschied, solange vor der eigentlichen Verschlüsselung keine zusätzliche Operation auf diesen Schlüsseln ausgeführt wird.

Elemente mehr oder weniger zufällig ausgewählt werden können: Statt daß der Käufer ein Geheimnis sec_B besitzt, wählt der Händler einen Zufallswert $rsec_B$ aus dieser Menge für jeden Käufer und speichert ihn in seiner Verkaufsnotiz. Folglich würden die Raubnutzer nun $rsec_B$ erhalten. Der Händler kann diesen Wert einem bestimmten Käufer zuordnen, aber der Wert kann zu nichts anderem verwendet werden.

Man kann auch betrachten, wie man Signete aus Schlüsselkennzeichnung konstruieren kann, gleichwohl das eher eine kryptographische Fingerübung darstellt als ein praktisch relevantes Problem: Wir kennen bislang noch keine allgemeine Reduktion. Allerdings kann man dieselben konkreten Ideen, die bisher auf Schlüssel-kennzeichnung angewandt wurden, auch auf Signete anwenden. Im Vergleich zu dem Schema aus [DwLN_96] führt dies zu beweisbarerer Sicherheit und bietet mehr Möglichkeiten, persönliche Schlüssel aus manipulationsresistenten Geräten der Raub-nutzer zu extrahieren.

Wir skizzieren dies hier nur kurz für Leser, denen die üblichen Konstruktionen von Schlüsselkennzeichnung vertraut sind: Der Händler wählt eine 2-stufige Struktur von symmetrischen Schlüsseln wie in [ChFN_94] (oder etwas anschaulicher in [PfWa_97]). Für einen Käufer B mit persönlichem Geheimnis sec_B wählt der Händler ein zufälliges Codewort $word_B$ der ersten Stufe, wählt alle Codeworte der zweiten Stufe identisch als $sec_{B_,}$ und gibt dem Käufer den zugehörigen persönlichen Schlüssel. Für dieselben Parameter wie in [PfWa_97, Abschnitt 4] gilt, daß es in jedem gültigen Schlüssel (der eine Mischung der persönlichen Schlüssel mehrerer Raub kopierer sein kann) wenigstens ein Codewort der zweiten Stufe gibt, das exakt das Geheimnis sec_B eines Raubkopierers ist. Falls die zugehörige öffentliche Information pub_B ebenfalls allen Raubnutzern bekannt ist, was ein Händler leicht sicherstellen kann, so können die Raubnutzer alle Codewörter der zweiten Stufe der erhaltenen Schlüssel durchprobieren und so herausfinden, welche davon echte Geheimnisses waren und von wem.

Literatur

BlMP_86   G. R. Blakley, C. Meadows, G. B. Purdy: Fingerprinting Long Forgiving Messages; Crypto '85, LNCS 218, Springer-Verlag, Berlin 1986, 180-189.

BoRD_95   F. M. Boland, J. J. K. Ó Ruanaidh, C. Dautzenberg: Watermarking Digital Images for Copyright Protection; 5th IEE International Conference on Image Processing and its Applications, Edinburgh, 1995, 326-330.

BoSh_95   D. Boneh, J. Shaw: Collusion-Secure Fingerprinting for Digital Data; Crypto '95, LNCS 963, Springer-Verlag, Berlin 1995, 452-465.

BrCC_88   G. Brassard, D. Chaum, C. Crépeau: Minimum Disclosure Proofs of Knowledge; Journal of Computer and System Sciences 37 (1988) 156-189.

Caro_95   G. Caronni: Assuring Ownership Rights for Digital Images; VIS '95, DuD Fachbeiträge, Vieweg, Wiesbaden 1995, 251-263.

ChDG_88   D. Chaum, I. B. Damgård, J. van de Graaf: Multiparty Computations ensuring privacy of each party's input and correctness of the result; Crypto '87, LNCS 293, Springer-Verlag, Berlin 1988, 87-119.

ChFN_94   B. Chor, A. Fiat, M. Naor: Tracing traitors; Crypto '94, LNCS 839, Springer-Verlag, Berlin 1994, 257-270.

CKLS_96   I. Cox, J. Kilian, T. Leighton, T. Shamoon: A Secure, Robust Watermark for Multimedia; Information Hiding, LNCS 1174, Springer-Verlag, Berlin 1996, 185-206.

DwLN_96　C. Dwork, J. Lotspiech, M. Naor: Digital Signets: Self-Enforcing Protection of Digital Information; 28th Symposium on Theory of Computing (STOC), ACM, New York 1996, 489-488.

ElGa_85　T. ElGamal: A Public Key Cryptosystem and a Signature Scheme Based on Discrete Logarithms; IEEE Trans. on Information Theory 31/4 (1985) 469-472.

GMW_87　O. Goldreich, S. Micali, A. Wigderson: How to play any mental game – or – a completeness theorem for protocols with honest majority; 19th Symposium on Theory of Computing (STOC) 1987, ACM, New York 1987, 218-229.

Pfit_96　B. Pfitzmann: Trials of Traced Traitors; Information Hiding, LNCS 1174, Springer-Verlag, Berlin 1996, 49-64.

PfSc_96　B. Pfitzmann, M. Schunter: Asymmetric Fingerprinting; Eurocrypt '96, LNCS 1070, Springer-Verlag, Berlin 1996, 84-95.

PfWa_97　B. Pfitzmann, M. Waidner: Asymmetric Fingerprinting for Larger Collusions; 4th ACM Conference on Computer and Communications Security, Zürich, April 1997, 151-160.

RSA_78　R. Rivest, A. Shamir, L. Adleman: A Method for Obtaining Digital Signatures and Public-Key Cryptosystems; Communications of the ACM 21/2 (1978) 120-126, reprinted: 26/1 (1983) 96-99.

Schn_91　C. Schnorr: Efficient Signature Generation by Smart Cards; Journal of Cryptology 4/3 (1991) 161-174.

Sham_79　A. Shamir: How to share a secret; Communications of the ACM 22/11 (1979) 612-613.

Wagn_83　N. Wagner: Fingerprinting; Symposium on Security and Privacy, IEEE, Oakland, California 1983, 18-22.

Yao_82　A. Yao: Protocols for Secure Computations; 23rd Symposium on Foundations of Computer Science (FOCS), IEEE Computer Society, 1982, 160-164.

Yao_86　A. Yao: How to Generate and Exchange Secrets; 27th Symposium on Foundations of Computer Science (FOCS), IEEE Computer Society, 1986, 162-167.

ZhKo_95　J. Zhao, E. Koch: Embedding Robust Labels Into Images For Copyright Protection; International Congress on Intellectual Property Rights for Specialized Information, Knowledge and New Technologies; R. Oldenbourg Verlag, Wien 1995.

Cryptographic Containers and the Digital Library

Jeffrey Lotspiech, Ulrich Kohl

IBM Research Division
Almaden Research Center
650 Harry Road
San Jose, California 95120

Marc A. Kaplan

IBM Research Division
Thomas J. Watson Research Center
30 Saw Mill River Road
Hawthorne, New York 10532

Abstract

Today, information is distributed on the Internet and other communication infrastructures mainly for free. However, once information or digital contents is assigned some value, a means is needed to protect its copyrights and control its use.

In this paper we describe an approach for rights management in the digital library. A digital library is a huge repository of digital content which is offered to the community of its users. The first step to control the use of its content is encryption. Encrypted content is not usable unless the decrypting or unlocking keys are acquired. The management of all keys involved in content transactions and their appropriate use is the task of rights management.

After a short introduction into issues of electronic publishing we present in section 2 the cryptolope™ [1] technology as a container for both encrypted content and information to purchase the content which serves as the basis for rights management. The second section will introduce the structure of cryptolopes and the procedures to use them. In the third section we will present the results of a digital library pilot project for rights management using cryptolope-like technology. The fourth section gives an outlook on advanced cryptolope transactions. We conclude the paper with a summary in section 5.

1 Introduction

Information has always been a valuable business asset. Increasingly, though, we observe the following: rather than just to *serve* for conventional business (where owners may assign some estimated value to information), information in the form of *information goods* is the *core* of pure electronic business transactions and thus has a concrete monetary value. This trend towards information goods is boosted by more and more content being digital and thus subject to electronic creation, dissemination, and trade.

Today, most of the digital content a publisher wants to distribute broadly has to be offered for free on the Internet, due to a lack of open technical measures to control the distribution and use of digital information. Many enterprises offer valuable information for free now, but it is obvious that this does not gain any return on investment. A way to make a self-sustaining business model is needed. For many content holders, this means a viable

---

1. Cryptolope and infoMarket are trademarks of the IBM Corporation. Some of the processes described herein are patented or have patents pending.

method of copyright protection. In its absence they will not trust their holdings to the digital domain. Less discussed, but equally important, is the other side of the coin: end users want to be guaranteed that their privacy will be respected, and that the information they are accessing is authentic and has not been tampered with. Searching for a term that covers both aspects, we have settled on **rights management** as being appropriately broad. It turns out that the underlying technologies that support both copy protection, privacy, and authenticity derive from a common base in cryptography.

Rights management enables a broad base of distribution technologies. Although the Internet, with its typical interactive point-to-point (i.e. client-to-server) connections and services like File Transfer Protocol (FTP) or the World Wide Web (WWW), attracts the highest attention at the moment, many other media types are also very well suited for information distribution, especially high-bandwidth and/or cheap broadcast media like satellite or cable modem, and physical media like CD-ROM or the new DVD-ROM. Especially for the broadcast media, the need for protection of the content is obvious.

Encryption offers the possibility to decouple the distribution of information and its licensing by distributing encrypted bulk data and controlling the release of content through the key management. Access to the content thus is controlled via a separate non-broadcast channel, which is basically a "key exchange" between a user's personal computer and a dedicated royalty/license clearing center. This concept, called **superdistribution** [1], allows for a very flexible way to use the most appropriate distribution infrastructure.

The basic unit of superdistribution thus consists of some part of encrypted content which is disseminated and some control information how to unlock, i.e. decrypt, it. In general, such units are called cryptographic containers. The specific form we used has been developed by Marc Kaplan, Joshua Auerbach, and Chee-Seng Chow [2] and has been given the trademark "cryptolope"[TM], a coinage based on the words *cryptographic* and *envelope* [3].

Cryptolopes are already being used in IBM infoMarket[TM] [4], a "pay per view" content delivery service. The service's underlying infrastructure, however, is single-server-based and does not take full advantage of many features cryptolopes offer. Our emphasis--the system we have built to date--has been on the particular application of cryptolopes to digital libraries, which is based on a more complex and flexible infrastructure concerning both technical and rights management issues. However, because little has been published to date on cryptolopes, we will begin by giving a brief overview of cryptolope principles.

2 Cryptolopes as the basis for trading information goods

In order to be able to disseminate a certain amount of information freely without losing the control on its usage, it is clear that the information has to be transferred in a form which is not of immediate use. This transformation has to be reversed once the owner or copyright holder of the information (or his agent) agrees to its use. The transformation and unlocking of the information is done using cryptographic algorithms.

A cryptolope is a special form of a cryptographic container which has all the information which is necessary for a patron to make a buy decision and be able to purchase the content. In addition, it contains all the information that a **clearing center** (CC) needs to perform the unlocking transformation. Of course, some of this information, like the content keys, must be themselves encrypted in a way only the clearing center can unlock.

An additional feature of the cryptolope systems that have been deployed is that the messages that flow between the purchaser and the clearing center are themselves cryptolopes. The purchaser sends a **buy** cryptolope and receives a **license** cryptolope in response. As cryptolopes are encrypted or digitally signed, the communication channel for this transaction need not be secure.

A cryptolope contains document and control files which are packed in a single file using a common file aggregation tool like zip or tar. The files of a cryptolope refer to each other in various ways. Each control file is composed of a number of specific records. Figure 1 shows the composition of the content cryptolope [5].

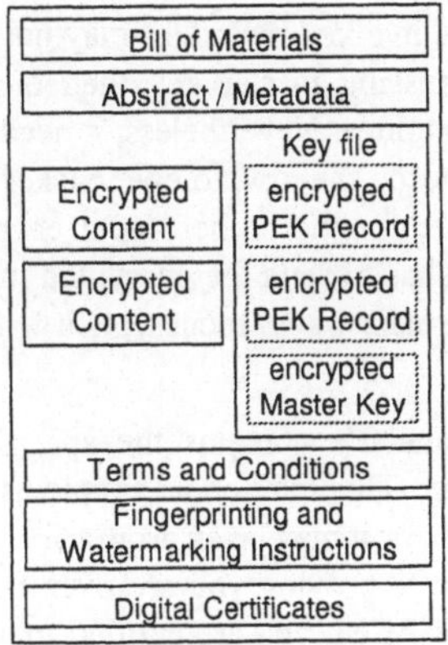

Figure 1: The Content Cryptolope Container

- The **Bill of Materials** (BOM) lists all the other components of the cryptolope, i.e. files which are included or Uniform Resource Locators (URLs) or Uniform Resource Names (URNs) to components stored outside, together with their MD5 message digests. The BOM also contains pointers to the clearing centers which are able to unlock the contents. A final record contains the digital signature of the BOM records. The signature on the BOM records guarantees that the user can check if all the intended files are contained in the cryptolope and the MD5 hashes are authentic. The MD5 hashes ensure that both the user and the clearing center can detect any manipulations of content or other files.

- The **abstract** is a clear text description of the encrypted content of the cryptolope which serves to support a user's purchase decision (and thus is sometimes called a **teaser**). In case the content consists of text, the abstract could be a summary of the text; for image, video or audio content, it could be a low quality representation.

Metadata is clear text information about the contents as a whole, e.g. author, creation date, size, or format of the content.

- The **encrypted content** files contain any form of digital information which is encrypted using a fast bulk encryption algorithm. Each file is encrypted using a different, randomly chosen part encryption key (PEK). The partitioning of the content files is left to the creator of the cryptolope, i.e. the content provider. Typical segments would be chapters of books, magazine articles and their embedded images, sound or video files. For example, a complete issue of a magazine could be packed into a single cryptolope, and the articles could be unlocked separately.

- The **key file** contains several **key records**. Each part encryption key which was used to encrypt a content file is stored in a key record. The part encryption keys are not stored in the clear, but are on their part encrypted using a master key, so it is sufficient to know the single master key in order to unlock multiple content files in a multi-stage decryption process. A master key may be encrypted using a higher-level master key and also stored in a key record, or may not be contained at all within the cryptolope. This latter, missing master key then is the highest-level key and is necessary to unlock a cryptolope. Nevertheless, it needs **not** to be transmitted during the buy transaction. Instead, the cryptolope packer uses the public key of an authorized[2] clearing center for the highest-level encryption[3]. A client wishing to purchase the content then has to send the encrypted master key record to a clearing center which decrypts the master key, reencrypts it with the public key of the client and sends it back.

- The **terms and conditions** file contains the specification describing the rights associated with the content. This may be as simple as a fixed price for the right to view or print the document, or it may be an arbitrarily complex scheme, for example: "You must have a certificate proving you are over 18. Additionally, if you have a certificate as a member of ACM, you are entitled to a 20% discount. Additionally, after January 1, 1999, this cryptolope is 50% off after all other discounts are taken. Additionally, check URL 'http://niftycontent.ibm.com' for possible special offers regarding this cryptolope. Etc.". To express such a broad range of possible specifications in a machine-readable way, so that it can be administered automatically by a CC, a **rights management language** like the Xerox DPRL [6] has to be used.

- **Fingerprinting and watermarking** are technologies for adding identifying information to documents. Whereas a watermark is readily detectable, such as a banner which is lightly printed over a document, fingerprints are subtle variations of the digital content that are not apparent to a user. Fingerprints can be added to identify the source of the information, but as each cryptolope is unlocked at the

---

2. If several clearing centers shall be authorized, a key record for each of these is created.

3. Although the "pure theory" of cryptolopes proposes that public key systems be used consistently, it is also feasible that the master key is a shared (symmetric) key between the packer and a clearing center. Especially in 1:n-relationships, where only one packer for the creation of all the cryptolopes is deployed, it makes no practical difference which encryption method is used.

request of a specific purchaser, it becomes possible that the content is fingerprinted to identify the licensee. [7]

The fingerprinting and watermarking information contained in a cryptolope comprise a post-processing algorithm which is applied to the protected content of the cryptolope as it is decrypted.

Note that fingerprinting the content to identify an end-user does not invade his privacy: it does not mean that you, as the content owner, have to keep a record that you gave it to him. The only way he can be identified is if he abuses his rights and widely redistributes the marked content.

- The **digital certificates** of the cryptolope serve the purpose that potential users can authenticate its entire contents. They contain the certificate of the cryptolope's packer and perhaps of any certification authority (CA) that issued the packer's certificate. The certificates contain the public keys with which the digital signature (on the BOM), can be checked.

- All parts of the cryptolope, except the BOM, need not be included "physically" in the cryptolope, but can be stored in separate files and be referred to logically using links such as URLs or URNs.

Three parties play an active role during cryptolope processing:

1. The **content provider** who owns information which he wants to distribute using cryptolopes.

2. The **user** who acquires the cryptolope via an arbitrary channel and wants to use the content.

3. The **clearing center** which serves as an agent between content provider and user by handling the buy and license transaction.

Figure 2 shows the major components and the interactions which are necessary to perform the purchase of a cryptolope.

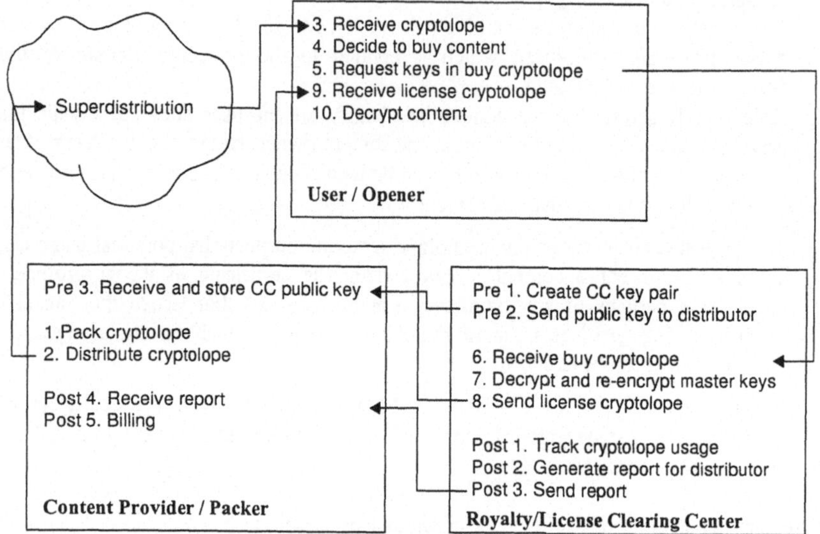

Figure 2: Cryptolope processing components and process

Before the cryptolope processing system becomes operable, each of the intended CCs has to create a public/private key pair and give the public key to the **packer** (steps Pre 1, 2 and 3)[4]. The content provider then uses the packer to create the cryptolope which contains the master keys encrypted with the public CC keys (step 1). The cryptolope is super-distributed (step 2) and eventually reaches a potential user[5] (step 3). If the user decides to buy the content (step 4), his **opener** creates a buy cryptolope and sends it to the CC (step 5). The CC receives the buy cryptolope (step 6) and checks the authorization. If the request can be fulfilled, the master key is decrypted using the CC's private key, then re-encrypted again using the user's public key (step 7) and sent back to the user in a license cryptolope (step 8). The user receives the license cryptolope (step 9) and his cryptolope opener can now decrypt the master key using the user's private key and decrypt the content files using the decrypted master key (step 10). The control of the use of the decrypted content has to be done within the opener, which is therefore subject to attacks[6]. After processing, the CC may track the cryptolope usage, generate a report and send it to the content provider (steps

---

4. In detail, each of the entities has to have not only a public key pair, but also a certificate of the public key which serves to check the authenticity and validity of the content, buy and license cryptolopes.

5. Superdistribution needs not to be asynchronous or off-line. The concept can also be used to realize secure multicast on synchronous broadcast channels by broadcasting cryptolopes which then can be stored locally until unlocking. The broadcast can be initiated on request, the broadcast being announced in advance (e.g. for books, journals or real-time media), or being repeated permanently (e.g. for stock prices).

Post 1, 2 and 3) which can use this information to bill the user (steps Post 4 and 5). Alternatively, billing can also be performed by he clearing center itself using an arbitrary electronic payment system which in turn would determine the privacy features of the overall system.

3 Digital Libraries and the ISI *Electronic Library Project*

The term *digital library* has been used to cover a lot of different concepts. Here we use it in a fairly restrictive sense: the augmenting of traditional libraries to hold or redistribute digital content. To investigate the use requirements for rights managements technologies in the digital library, we have been engaged as a technology partner in a pilot project with the Institute for Scientific Information (ISI). ISI is a widely respected publisher of navigation aids for the world's scientific periodical literature. Their products *The Science Citation Index* and *Current Contents* are standard tools in research libraries around the world.

ISI populates their data bases that support these products by actually digitally scanning the entire corpus of the world's scientific periodical literature, 25,000,000 pages (2 Terabytes) a year. Thus, in this case, one of the major traditional obstacles to digital libraries--the data capture problem--was already solved. What was not solved was the rights management aspect. None of the corpus is owned by IBM or even ISI; the rights are retained by the original journal publishers. The publishers had to be convinced that our system would protect the value of their copyrighted information.

This ISI *Electronic Library Project* has been on-line as in a beta form in a variety of different libraries around the world since September, 1995. The rights management aspects were designed by Jeffrey Lotspiech, Cynthia Dwork and Florian Pestoni. They had used all of the cryptolope components described earlier; syntactically, however, they had intertwined them with the "journal article" data model specific to this pilot. When Kaplan *et.al.* independently developed the cryptolope model, which was divorced from content type, we enthusiastically recognized it as a big step in the right direction for digital libraries. That is perhaps the main lesson we took from this pilot; however, we have learned other interesting points:

1. The predominant economic model for how content is paid for is what we technologists would call "site licensing", but it is really the very traditional model: libraries acquire content on behalf on their patrons, who themselves do not pay individually for access. All participants--the patrons, the librarians, and even the publishers themselves--are very happy with this arrangement, each for their own reasons. In the traditional library world, no one appears to want or need the "information free market" that some propose--the view of the world where all information is pay-for-use and free-market dynamics drive prices and content value.

2. Users are already overloaded with digital identities and no one is happy about

---

6. If the opener is considered to be a trusted system [6], this problem seems to be solved, but gets mapped to the (unsolvable?) problem of realizing a trusted system on the end user's PC.

learning a new userid and password just to use their library. More emphatically, librarians are trained information professionals, not userid administrators.

3.   End users naturally expect that their privacy will be respected when they visit a library; the librarians' code of ethics also demand this. No one wants to see this lost in the digital library. This we knew going in, and we took pains to protect user privacy. However, we learned that even aggregate information is sensitive: for example, in an industrial research setting like a drug company, the aggregate papers being read reveal the diseases that new drugs are being developed for, which is their single most sensitive industrial information.

The system we ended up deploying had several interesting features. First, even though the conversations with the end-users were completely encrypted, we had no userid/ password unique to the system--we interfaced to an already existing infrastructure, Lotus Notes. Secondly, we kept no logs of access that associated users with articles read. We were often told by outsiders, "you have to keep logs! What are you going to do when users complain about their bills?" The system made it easy for the user to keep a private log at his or her workstation. But in the end, if the user complained, we simply credited him. Our experience has been that this policy is not a source of fraud; users seem to expect that they will be caught if they repeatedly abuse the refund privilege. And, goodness knows, there are myriad sources of problems, outside of our control, but for which we would be willing to grant refunds: printers jam or run out of paper, fax machines break, networks go down halfway through, workstations crash.

3.1 Three-tier architecture and key management

The basic architecture for the ISI pilot is what we call **three-tiered**. The first tier is a central server located at the main ISI production facility. The second tier is the intermediate "campus" servers located at each library. The third and final tier is the end users themselves. This architecture was picked for its performance characteristics: the campus servers are connected to the end users over high bandwidth local area networks, and can cache encrypted content and perform searching locally, often without needing an interaction with the central server.

We quickly realized that this three-tier architecture had important advantages in rights management as well as in performance. The middle tier serves as a trusted platform--certainly much more trusted than the end-users' workstations. In particular, we felt the campus server could be trusted to unlock the copy-protected content. In those sites whose networks were behind firewalls, this eliminated the need for user authentication: if the end-user could connect to the clearing center, he was by definition entitled to participate in the site subscriptions. This gave us the operational characteristic of "no new userids and passwords" that librarians were telling us were so important to them.

Operationally, though, it seemed difficult to make the campus servers super-secure. Many of our pilot libraries were at universities, and we expected, at the very least, a lively attitude of experimentation on the part of the students towards a system that was enforcing copyright protection. We ended up designing a system with a hierarchy of keys. For no good reason, early in the design we assigned colors to these keys, and now we cannot help but refer to them that way. The **pink keys** are the per-content keys, the PEKs in cryptolope terms. Each document (in fact, each page in each document in the ISI pilot) is encrypted with a different pink key. The same document will be encrypted with the same key, regardless of which library it is sent to.

In true cryptolope style, the pink keys are encrypted themselves and shipped along with the content. We call the keys that encrypt the pink keys (the "keys to the keys") the **green keys.** The green keys are symmetric keys (which is possible because of the 1:n structure) and thus much more important secrets. In the ISI pilot we partition these secrets amongst the different libraries: each library has a unique green key. Thus if a green key is compromised, the entire corpus is not at risk. The only thing that is lost is the content that is currently cached at the library, plus whatever the adversary can induce us to serve before the theft is discovered. This requires us, however, to customize the cryptolopes for each library, at least that small part of the cryptolope that is the encrypted pink keys. Figure 3 shows the components of the three-tiered configuration.

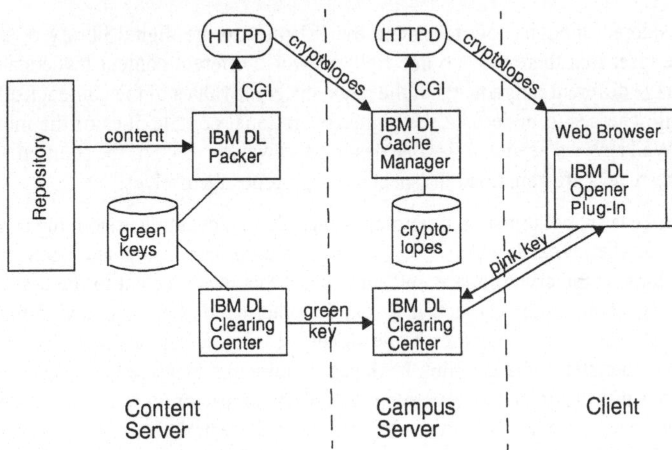

Figure 3: ISI three-tier configuration

We put an obstacle in the way of an attacker trying to learn a green key by never storing it in non-volatile store at the campus server. Thus, an adversary walking away with (copies of) the hard drives of the server has stolen nothing more valuable than the media itself. An additional complication was that the server machines, in general, would not be attended 24 hours a day, and had to automatically reboot in the case of a power loss. We settled on a fairly standard technique: every time it powers up, the server machine gets its green key dynamically from the central server. The central server will only serve the green key to the expected address. (In case of the ISI pilot, this was the other end of a leased line. If we were connecting over the open Internet, we would have put a cheap modem in the campus server and dialed it back from the central server when the campus server notified us that it had just powered on.)

After we implemented this design, we realized we had just picked one point in a continuum of possible security/operational trade-offs. For example, if the physical security of the campus server was higher, or the value of the content were lower, it might be perfectly satisfactory to store the green key on the server's disk and avoid the operational complexity of fetching it from the central server on each power-up. To avoid the complexity of customizing the encrypted pink keys for each campus, it could be possible for campuses to share keys. On the other extreme, if no campus was to be trusted, the green key secret could be kept in the central server only and all unlocking operations routed back to it.

Our design of partitioning the keys by campus server was certainly not the only approach. Another approach would be to partition the green keys by content: for example, each journal could have a different green key. This matches well with the way the content is primarily sold, by journal subscription. However, in the particular case of the ISI pilot, it is unlikely the publishers would have agreed to such a scheme. The loss of one green key, although it would only affect one journal, would be a world-wide loss to that journal's publisher.

As we considered, though, how to extend the ISI pilot to the digital library of the future, it became clear that there was no one right answer. Different content owners and custodians will have different judgements of the risks, different values of the content itself, and different willingnesses to undertake the higher operational complexities of the more secure systems. We knew we needed a flexible design, and we are proud of the solution we settled on: the green keys are contained in (special) cryptolopes themselves.

When green keys are contained in cryptolopes there is an elegant recursion: the campus clearing center, when it needs a green key, becomes a cryptolope client and opens the green key cryptolope. If the cryptolope is self-unlocking, this is equivalent to the case of the green key being stored on the server's disk. On the other hand, the green key cryptolope can require a connection to another higher-level clearing center, analogous to the power-on green key transfer in the ISI pilot. In all cases, (although this is not essential), we find it serendipitous that the green key cryptolope can carry tamper-proof terms and conditions that spell out the relationships between the campus and central clearing centers, the rights and obligations of each, and even the expected physical address of the campus center.

3.2 Authentication and authorization

As already mentioned above, the existence of a public/private key pair and a public key certificate is a prerequisite for all open cryptolope transactions in a n:m environment, to which the 1:n system described above should be extended. Although the needed technology, algorithms and services are quite well understood, many practical problems arise when it comes to build a **running** infrastructure.

It is widely believed that a perfect solution maximizes the use of standards or common software and interfaces, so the term "open" can be applied. For the definition of what an open public key infrastructure (PKI) should look alike, several standards have been or are being developed. The components of a PKI consist of certification authorities, directory servers, and clients who perform certain communication protocols and cryptographic functions for creation, exchange and management of key pairs and certificates.

The basis for every public key system is the public key algorithm and the algorithms to generate the key pairs. The most important public key algorithm is the RSA algorithm [8] which has the advantage that it is widely known and implemented, but the disadvantages that it is patented within the U.S. and, even worse, subject to export restrictions which sparked severe worldwide discussions on legal regulations of cryptographic mechanisms. (In the pilot, since the single source of encrypted data was always the central server in the U.S., we were able to obtain an export license for world-wide distribution.)

For authentication and digital certificates, the X.509 authentication framework of the former CCITT, now ITU, can be considered the standard to follow. It has been adapted from the ISO as standard 9594-8 [9] and defines protocols for simple and strong authentication, and above all, the abstract definition of certificates and a certificate infrastructure using the X.500 directory. The functions of the infrastructure are manifold and comprise not only the creation and distribution of certificates, but also the management of their revocation, for example.

Many other specifications and proposals from organizations or enterprises are based on RSA and X.509, like the Internet Public Key Infrastructure [10] or RSA, Inc.'s Public Key Cryptography Standards [11]. At the same time, first products and services which are complying with the standards are being offered. For example, Netscape Navigator V3 is able to generate a key pair which is initiated by the HTML tag <keygen>, Verisign offers a public certification service [12], and certification authority software and utilities for individual use are offered by a number of enterprises.

One is tempted to believe that all these components just can be put together to form a running PKI, where the ideal of a client owning exactly one key pair and perhaps holding multiple certificates on the public part of that pair (or even one chained certificate) was possible.

Today's situation is different, however: Public CAs are rare so there is not yet a spanning network of trust. Cheap but simple certificates just certify the existence that the certificate holder can be reached on the certified email address, for example--which is not a trustworthy basis for business transactions. Furthermore, the client software is mostly proprietary. Often the local storage formats of keys and certificates are kept secret, so each application of a user has to acquire its own specific certificates or even deploy its own PKI. Keys and certificates then may not be portable across different computers, even if the same application software is used. Beyond these application-induced problems, the concepts of certification systems also shows inherent weaknesses which are discussed in [13].

For the digital library, not all of the functions required in the PKI proposals are really needed. The way a library card is issued to a library patron today serves as a model for the handling of the library certificate. First of all, it is clear that the certification authority should be within the library, as the decision to grant a certificate has to be taken there. This is at the same place where the clearing centers are located, so the clearing center and the certification authority run on the same machine. The process of issuing a certificate can be automated so that a librarian would not have to care about issuing certificates if the decision whether a certificate request should be fulfilled can be made by a machine. As mentioned above, a simple strategy is to grant a certificate to everybody who is within the limits of a domain firewall and thus is simply able to connect to the certificate authority. More complex decisions can be based on given information like student or employee numbers, which can be cross-checked against a institutional database. For example, to get a certificate, the student or employee might need write access to a particular place on the local file server, which proves that he is an authorized user and has entered a valid password. The important point is that the security domain need **not** be administered by the librarian!

The certification network for a three-tier digital library scenario--i.e. the lattice of certification authorities on the one hand and instances who have to recognize certificates on the other hand--is a tree consisting only of the root (i.e. the top level CA), represented by the content provider (which also operates the single packer), and the leaves (low level CAs) in the libraries which operate the clearing centers. The root issues certificates to the libraries, and each library issues chained certificates that not only contain the library's certificate of the patron, but also the root's certificate of the library. The packer knows the public keys of the clearing centers and uses them for encryption of cryptolopes. The clearing centers only have to know the root's public key, but don't have to know and trust each other to check the validity of an arbitrary patron's certificate. As all clearing centers have a relationship to the content provider, they know his public key and can check in a two-step-process a patron's certificate. To unlock cryptolopes, they simply use their private keys. So a complex certificate management infrastructure can be omitted. Figure 4 depicts the public key infrastructure and the elements which use the keys and certificates.

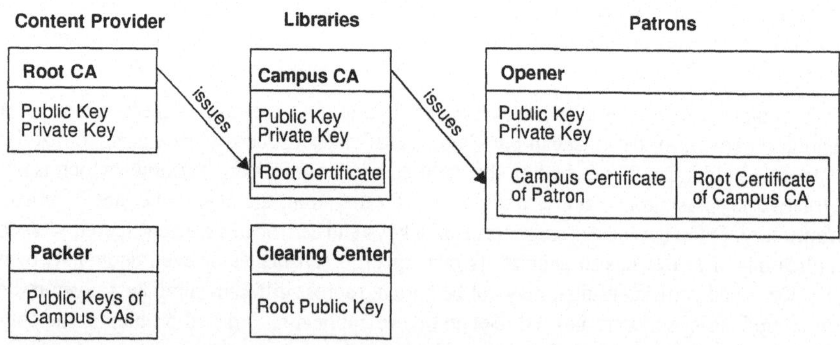

Figure 4: Public Key Infrastructure and key/certificate usage in the digital library pilot

Active certificate revocation is not needed if the certificates have certain validity restrictions based on location and time which are sufficient to achieve a similar security to a library card. "Not Before" and "Not After" dates are specified in each certificate. Also, location information can be included in the certificate so the decision on an unlocking request could be made after inspecting the address the request comes from. In general, a variety of requirements that serve for authorization checking can be specified using the attributes of a certificate [14].

An obvious problem of digital library applications is that they are especially vulnerable to insider attacks. For example, students do not have to worry about damage to themselves when they pass their secret key to a non-authorized person who can then use it to unlock cryptolopes. Because of the model of contracts and payment, the damage would stay completely with the content provider or the library. In a world where a single key pair is used for several applications, a secret key holder is deterred from passing the secret key because the receiver could possibly cause damage to the holder by using the key for other applications as well. On the other hand, a legitimate member of an organization should be able to unlock cryptolopes whenever and wherever he wants to.

A solution to this problem uses a combination of location and time attributes in the certificate. Long term certificates only are issued for use in a specific domain, e.g. within a campus, or a specific internet address, e.g. that of the patron's home computer. A long-term certificate has the same validity time like a library card. Short term certificates are issued for worldwide use, but their validity is restricted to a short time which is long enough for a business trip, for example. Insider attacks by giving away private keys of long term certificates thus can only be performed in the same domain, which is not considered a problem because it is high likely that a person who is able to use a machine in a domain also would have the library rights of this domain. Insider attacks by giving away short term keys cannot be detected accurately, but suspect patrons who request short term certificates too often can be detected by keeping a log file. Furthermore, the effort of requesting a short term certificate is, for the insider, rather the same as using his rights to unlock some content and give the decrypted content to an outsider, so the deployed mechanism can be considered to be secure enough against such "pass secret key" attacks.

4 Examples for advanced Cryptolope transactions

4.1 Repackaging

A reason the Internet became so successful is the liberal, open and decentralized manner in which it has been developed and in which most of its servers are organized and managed. The cryptolope approach ideally fits in this open environment as it allows the content and infrastructure providers to package content and superdistribute it on the Internet in a very self-determined way, but offers also a means not only to guarantee the payment of royalties for the content providers and copyright holders, but also create revenue for all providers of the superdistribution infrastructure, e.g. providers of anonymous FTP-servers, manufacturers of CD-ROMs, or providers of satellite channels. During the content packaging and superdistribution phases, content may be grouped and marked in order to achieve a more complex and flexible pricing, e.g. to state "wholesale" prices for redistributors who then take a "mark-up" to be re-imbursed for their costs.

Repackaging is a means for infrastructure providers to modify a cryptolope so they also will profit when some patron purchases a cryptolope they helped to superdistribute.

This serves as an incentive for independent service providers, e.g. FTP, CD-ROM, cable, etc., to take part in the superdistribution process on a voluntary basis. We call all such intermediaries **infrastructure providers.** Depending on the amount of trust among the providers, both a simple and a multilateral secure technical solution can be designed. Two pricing models seem to be reasonable: First, that an intermediate server adds a surcharge on the price which is claimed by the content provider (as is also usual in shareware distribution shops). Second, that the price for the patron remains constant independent from which way the cryptolope was shipped, and the content provider agrees to share the revenue with the infrastructure provider (which is the common model for business using wholesale trade).

Absence of trust

In case there is no business relationship between the content provider and the provider of a FTP server, for example, a mechanism has to be used which guarantees both parties will get their part of the revenue. A straightforward solution is that the infrastructure provider also uses a cryptolope packer, but does not pack original content. Instead, he creates a new cryptolope by copying all parts of the cryptolope of the content provider and encrypting the encrypted content again under the public keys of its own CCs and adding the necessary key records. A patron wishing to buy a document then first has to unlock the outer cryptolope using a CC authorized by the infrastructure provider before he is able to unlock the content of the inner cryptolope using a CC authorized by the content provider. This solution can be deployed without any relationship between content and infrastructure provider, but is not very efficient: the patron has to unlock two cryptolopes using two different CCs. Additionally, in a pure networked environment, this solution seems not to be feasible as each patron who finds an interesting, repackaged cryptolope will search for the original one in order to avoid the surcharges. However, if the effort to search and upload the original cryptolope is higher than the surcharge, e.g. if the patron already has a big repackaged cryptolope on CD-ROM and only has a slow and/or expensive Internet connection, this approach could be reasonable.

The realization of the second pricing model requires that the content provider confirms that he is willing to lower the price when an infrastructure provider served as an intermediate. Such a statement can be given in the terms&conditions file. The buy cryptolope which the patron sends to the CC of the content provider then has to include also the license cryptolope given from the infrastructure provider as a receipt that the patron already paid something to get the original cryptolope. The CC would check the license and lower its own price according to the given original terms&conditions.

Presence of trust

If at least one of the content and infrastructure providers trusts the other, a simpler solution using a single CC for a purchase transaction can be realized. The ISI scenario is of this nature, as the content provider trusts the libraries and their middle-tier CC. In general, however, it is more likely that the content provider does not trust an arbitrary infrastructure provider which is a potential pirate because it can sell cryptolopes without giving royalties to the content provider. Additionally, the ISI method loses efficiency with the number of CCs deployed, because the content provider has to know the public keys of the infrastructure providers.

In the case an infrastructure provider trusts the content provider, it is sufficient that it just adds an additional entry telling it took part in the distribution process to the BOM file and signs it. When a patron requests to purchase such a cryptolope at the content provider's CC, the CC recognizes the additional entry and pays the infrastructure provider. The amount of the payment either can be stated in the terms&conditions or negotiated off-line in a contract between infrastructure and content provider.

4.2 Transaction safety and privacy

Safety and privacy often are competing goals. While for transaction safety, transaction logs should be kept also in the CC, the mere existence of these logs jeopardizes the privacy of the patrons--for example, they are subject to subpoena or search warrant by law enforcement agencies. However, using public key encryption, the server logs can be designed so that their entries can only be decrypted on request of a buyer by the buyer himself. To achieve this, the phases of the buy transaction have to be secured as follows:

When the patron tells his cryptolope viewer to purchase the contents of a certain cryptolope, a nonce is generated which is stored locally and sent with the buy cryptolope to the CC. The CC unpacks the cryptolope as usual and sends the license cryptolope back, but additionally signs the transaction data, encrypts this information using the patron's public key (which also was given in the buy cryptolope) and logs this receipt record indexed by the given nonce. In case the license cryptolope did not arrive at the patron's machine, an error occurred during the opening of the cryptolope locally or even the unlocked contents are lost later, a patron can prove the purchase of a cryptolope only by providing the nonce to the CC which did the unlocking. The patron has to give the nonce to the CC and request the encrypted record. The CC finds the encrypted receipt using the nonce as an index and sends it to the patron to decrypt it using the private key. The fact that a patron can successfully decrypt the valid receipt serves as the proof that he is a license holder. This proof is accepted by a CC when the patron is trying to re-open something he has already bought. This request is then handled as usual, but without charging.

5 Summary

We introduced the cryptolope technology as a basis for rights management within digital library applications. Cryptolopes are a special form of digital envelopes which contain both encrypted content and the information how and under which conditions the content can be decrypted. We presented the structure of a cryptolope and the parts it has to contain. The components of the cryptolope processing infrastructure and the protocols needed to perform a cryptolope transaction were described. As an example of a cryptolope application we explained the architecture of the ISI Electronic Library Project and told the experiences we gained in this project. We exemplified the high flexibility of cryptolope utilization by describing the advanced mechanisms of repackaging and concluded with the description of a scheme which allows for both transaction safety and user privacy.

6 Acknowledgments

Jeff Crigler, executive for the IBM infoMarket service, has been an enthusiastic evangelist for superdistribution; Willy Chiu, his counterpart for IBM Digital Library, has likewise kept a primary focus on rights management. We would also like to thank our colleague Henry Gladney for coining the term "campus server", which we find especially apt.

References

[1]   Ryoichi Mori, Masaji Kawahara: Superdistribution: The Concept and the Architecture. Transactions of the IEICE, Vol. E 73, No. 7, July 1990. http://www.virtualschool.edu/mon/ElectronicProperty/MoriSuperdist.html

[2]   Marc A. Kaplan: IBM Cryptolopes, SuperDistribution and Digital Rights Management. Working Paper, V1.3.0, December 1996. http://www.research.ibm.com/people/k/kaplan/cryptolope-docs/crypap.html

[3]   Cryptolope Showcase Homepage. http://www.cryptolope.ibm.com

[4]   IBM infoMarket Homepage. http://www.infomarket.ibm.com

[5]   Cryptolope Container Technology. Whitepaper. http://www.cryptolope.ibm.com/white.htm, March 1997.

[6]   Mark Stefik: The Digital Property Rights Language. Manual and Tutorial. Xerox Corporation, February 1997.

[7]   Neal R. Wagner: Fingerprinting. Proc. IEEE Symposium on Security and Privacy, 1983.

[8]   Ronald Rivest, Adi Shamir, Leonard Adleman: A Method for Obtaining Digital Signatures and Public Key Cryptosystems. Communications of the ACM, Vol. 21, pp. 120-126, 1978. http://theory.lcs.mit.edu/~rivest/rsapaper.ps

[9]   ISO 1995: International Standardization Organization: Information Technology - Open Systems Interconnection - The Directory: Authentication Framwork. International Standard 9594-8, ISO, Geneva 1995. http://www.itu.ch/itudoc/itu-t/rec/x/x500up/x509_27505.html

[10]   Internet Engineering Task Force: Internet Public Key Infrastructure. http://www.ietf.org/ids.by.wg/X.509.html

[11]   Public Key Cryptosystems (PKCS) Standards Homepage. RSA Laboratories 1997. http://www.rsa.com/rsalabs/pubs/PKCS

[12]   Verisign Digital ID Center Homepage. http://digitalid.verisign.com

[13]   Edgardo Gerck: Overview of Certification Systems: X.509, CA, PGP and SKIP. http://novaware.cps.softex.br/mcg/cert.htm

[14]   Ulrich Kohl: Benutzerbezogene Datensicherheit in Kommunikationssystemen (User Oriented Security in Communication Systems). VDI Fortschrittberichte, Reihe 10, Nr. 446, VDI Verlag 1996.

Maßgeschneiderte Trust Center und elektronisches Bezahlen im Internet

Dr. Bertolt Krüger, Dr. Frank Damm
debis Information Security Services GmbH
Oxfordstraße 12-16
D-53111 Bonn
email: b-krueger@itsec-debis.de

Zusammenfassung

Die praktischen Erfahrungen bei der Konzeption, dem Aufbau und dem Betrieb verschiedener Trust Center werden in diesem Beitrag verallgemeinert und systematisiert. Dazu wird ein weit gefaßter Begriff eines Trust Centers dargelegt und einige mögliche Dienstleistungen genannt. Es wird hergeleitet, warum die Forderung nach Konfigurierbarkeit und Skalierbarkeit eines Trust Centers so wichtig ist. Es wird ein Baukastensystem dargestellt, in dessen Zusammenstellung Erfahrungen bei der Einrichtung verschiedener Trust Center eingegangen sind. Das System beruht auf der Verwendung eines Komponentenbaukastens und der Auswertung von Anforderungskatalogen, die durch Beantwortung eines Fragebogens ermittelt werden. Dieser Fragebogen wird für das Internet-Zahlungsprotokoll SET exemplarisch beantwortet.

1 Einleitung

Die praktischen Erfahrungen bei der Konzeption, dem Aufbau und dem Betrieb verschiedener Trust Center sind mittlerweile so vielfältig, daß es sich lohnt, sie zu verallgemeinern und zu systematisieren. Dies ist für die Industrie auch im Interesse einer ökonomischen Vorgehensweise notwendig. Trust Center (im folgenden auch kurz TC) müssen für verschiedenste Einsatzbereiche einerseits nach deren spezifischen Anforderungen konzipiert werden, andererseits ist zu vermeiden, daß jedesmal eine völlige Neuentwicklung erfolgt. Natürlich wird jeder Entwickler frühere Erfahrungen in die Entwicklung neuer Systeme einbeziehen. In diesem Beitrag wird jedoch eine **systematische** Vorgehensweise zur Erreichung der genannten Ziele dargestellt. Gleichzeitig mit diesem Ansatz ergibt sich eine neue Vorgehensweise in der Akquisition und Planung von Trust-Center-Anwendungen.

Überblick: In Kapitel 2 wird der zugrunde liegende Begriff eines Trust Centers dargelegt und mögliche Dienstleistungen genannt. In Kapitel 3 wird hergeleitet, warum die Forderung nach Konfigurierbarkeit und Skalierbarkeit eines TC so wichtig ist, und die Vorgehensweise skizziert. In Kapitel 4 werden die Komponentenbaukästen skizziert, in Kapitel 5 ist der für das Beispiel SET beantwortete Fragebogen abgedruckt.

Aus Gründen der Know-How-Sicherung und des Kundenschutzes konnten sowohl die Vorgehensweise als auch die Beispiele nur abstrakt oder in Form von Forderungen dargestellt werden, dennoch sollte die Relevanz der Vorgehensweise deutlich werden.

2 Trust Center

2.1 Begriffsbestimmung

Ein Trust Center ist eine Einrichtung, die Dienstleistungen im Bereich eines Systems der Informationstechnologie erbringt, welche eine sichere Handhabung, Erzeugung oder Speicherung von Daten erfordert, die ein einzelner Teilnehmer nicht leisten kann und die er daher einer solchen Instanz anvertraut. Ein typisches Beispiel ist die Berechnung von Zertifikaten für öffentliche Schlüssel, die deshalb eine vertrauenswürdige Instanz durchführen muß, weil die Teilnehmer eines Systems untereinander ihre Authentizität nicht über gesicherte Kanäle oder durch Zusammentreffen prüfen können. Ein anderes Beispiel ist die verläßliche Archivierung wichtiger Daten.

Im allgemeinen stellen solche Dienstleistungen besondere Anforderungen an die räumliche und Hardware-Ausstattung der Instanz, die einzelne Teilnehmer ebenfalls nicht realisieren können.

Die Definition ist deshalb möglichst allgemein gefaßt, weil ein Anbieter von vertrauenswürdigen Dienstleistungen, die die o. g. besondere Infrastruktur erfordern, alle Dienstleistungen anbieten wird, zu denen ihn diese einmal aufgebaute Infrastruktur befähigt. Er wird sich daher keineswegs auf die Rolle einer CA (Certification Authority) im Sinne einer Public-Key-Infrastruktur beschränken. Daher wurde bewußt der unschärfere Begriff Trust Center statt Trusted Third Party oder CA gewählt.

2.2 Mögliche Dienstleistungen eines Trust Centers

In diesem Abschnitt werden mögliche Dienstleistungen eines Trust Centers beispielhaft aufgeführt, um die Fülle darzustellen, aus der eine Auswahl zu treffen ist.

Dienstleistungen im Zusammenhang mit Zertifikaten bzw. asymmetrischen Verfahren

Die klassischen Dienstleistungen eines Trust Centers in einer Public-Key-Infrastruktur werden hier nicht genannt, soweit sie bereits im Fragebogen (siehe Kapitel 5) auftauchen. Daher folgen hier nur mögliche Zusatzdienste, die für die beispielhafte Beantwortung des Fragebogens zunächst nicht relevant waren.

- Zeitstempeldienst, Zweitunterschriften,

- Verifikation von Zertifikaten im Auftrag von Teilnehmern,

- gegenseitige Zertifizierung von Trust Centern,

- Rezertifizierung von Schlüsseln aus verschiedenen Gründen:
    * Verlängerung der Gültigkeitsdauer,
    * Berücksichtigung geänderter Teilnehmerdaten (z. B. Anschrift),
    * Änderungen im Zertifikatsformat des Systems,

* Zur Verwendung des Schlüssels in mehreren Systemen mit verschiedenen Zertifikatsformaten,

• Erstellung von Attributzertifikaten (aus ähnlichen Gründen).

Dienstleistungen, die über die klassische Public-Key-Infrastruktur hinausgehen

Diese Dienstleistungen zu erwähnen, liegt den Verfassern besonders am Herzen, auch wenn sie im Beispiel SET keine Rolle spielen und aus Platzgründen nicht weiter ausgeführt werden können. Sie zeigen besonders deutlich, daß es aus praktischer Sicht eine völlig willkürliche Abgrenzung wäre, ein Trust Center auf die klassischen Aufgaben im Rahmen einer Public-Key-Infrastruktur festzulegen.

• Erzeugung, Speicherung und Verteilung von Schlüsseln aller Art,

• Archivierungsdienst,

• Non-Repudiation Dienste (hier nicht im Detail aufgeführt),

• Authentikationsdienste (hier nicht im Detail aufgeführt),

• Gewährleistung von Anonymität,

• Registrierung von Software-Benutzern, Freischalten von Software,

• Registrierung von Daten (z. B. Quellcode) zwecks Urhebernachweis.

• Gesicherte Online-Personalisierung von Geräten in Sicherheitssystemen (z. B. KFZ-Diebstahlschutz).

Dienstleistungen, die mit der Funktion des Trust Centers selbst oder mit seiner Beziehung zum Auftraggeber oder Teilnehmer zusammenhängen

• Zur-Verfügung-Stellen von Testdaten,

• Online Recherche in Log-Daten oder zur Identifikation von Komponenten (in technischen Systemen).

3 Begründung und Skizzierung der Vorgehensweise

3.1 Anforderungen an spezielle Trust-Center-Anwendungen

Natürlich hängt die Gestaltung einer TC-Anwendung wesentlich davon ab, welche Dienstleistungen gewünscht werden, und welche Volumina dabei zu erwarten sind. Daß diese Parameter eine Skalierbarkeit der TC-Anwendung erfordern, ist klar.

Auch die folgenden Überlegungen zeigen die Notwendigkeit der Skalierbarkeit und der Konfigurierbarkeit: Für die meisten Anwendungen besteht die Erfordernis der Kompatibilität zu diversen Normen und Modellen. Ebenso gibt es die Erfordernis der

Aufwärts-Kompatibilität z. B. zu höheren Sicherheitsanforderungen. Stichworte sind: Signaturgesetz, Internet-Standards, Standards internationaler Normungsgremien.

Der konzeptionell nächstliegende Weg, die größtmögliche Normgerechtheit und -offenheit und das höchste denkbare Sicherheitsniveau **immer** zu implementieren, ist nicht gangbar. Manche dieser Anforderungen bewirken nämlich hohe Aufwände. Hohe Aufwendungen entstehen beispielsweise durch alle manuellen Tätigkeiten, die etwa bei der Teilnehmeridentifikation, der Verteilung von PIN-Briefen und PSE's anfallen, wie sie in den Katalogen zum Signaturgesetz zur Zeit diskutiert werden. Solche Abläufe kann man in geschlossenen Systemen viel leichter handhaben.

Fazit: Eine generische Vorgehensweise für Trust Center Anwendungen muß Skalierbarkeit und Konfigurierbarkeit beinhalten.

3.2 Grundsätzliche Vorgehensweise

Die Skalierbarkeit und Konfigurierbarkeit werden durch einen modularen Aufbau erreicht, der es erlaubt, aus einem **Baukasten** von möglichen Komponenten diejenigen zusammenzufügen, die benötigt werden. Dies bezieht sich nicht nur auf die Hard- und Softwarekomponenten des Systems, für die Modularität schon bisher eine selbstverständliche Forderung ist. Die Entwicklung eines skalierbaren Trust Centers erfordert damit zwei wesentliche Schritte:

1.   Die Erstellung eines Konzeptes für ein generisches Trust Center, dieses stellt den Baukasten für die Zusammenstellung spezieller TC-Anwendungen dar.

2.   Die Erstellung eines Bewertungskataloges - für potentielle Kunden in Form eines Fragebogens -, der die Skalierung des generischen Konzeptes zur konkreten TC-Anwendung ermöglicht, also die Auswahl der Bausteine.

Selbstverständlich ist eine Vorgehensweise nach ISO9000/9001. Darüber hinaus benötigt man eine Absicherung von Konfigurationsänderungen und Einbringen neuer Komponenten mit kryptographischen Mitteln, damit Anwendungen nicht unbefugt umkonfiguriert werden können.

Die wesentlichen Schritte bei der Einrichtung eines TC (Planung, Entwicklung, physische Errichtung, Testbetrieb, Wirkbetrieb) werden also zweimal durchgeführt: einmal generisch, dann für jeden Kunden maßgeschneidert unter Rückgriff auf den generischen Teil. Dabei werden nicht alle Schritte in beiden Fällen vollständig vorkommen: Bei der generischen Entwicklung wird es keinen Wirkbetrieb geben, auf der anderen Seite wird das einmal gewählte Rechenzentrum im allgemeinen für alle speziellen TC-Anwendungen verwendet.

Besonderheiten bei der Entwicklung der Bausteine im Rahmen des generischen Entwicklungsvorganges

Zu beachten ist, daß schon bei der Erstellung des generischen Konzeptes und der Entwicklung von Bausteinen die möglichen Parameter eingehen können: Um dies zu berücksichtigen, gibt es für jeden Baustein zwei Wege: Der erste Weg besteht darin, den entsprechenden Baustein schon in der generischen Phase nach der absehbaren Maximalanforderung zu entwickeln. Der zweite Weg besteht darin, mit der Konzeption zunächst an einer Stelle stehenzubleiben, bis zu der sie mit vertretbarem Aufwand durchführbar ist.

Welchen dieser Wege man im Einzelfall wählt, ist eine schwierige Abwägung, für die man kein einfaches Kriterium angeben kann. Ein Beispiel für eine solche Abwägung ist: Soll man heute schon sein Trust Center so entwickeln, als sei das Signaturgesetz bereits in Kraft, für die eigenen Anwendungen relevant und in Form der BSI-Anforderungskataloge mit detaillierten Anforderungen unterlegt? Oder soll man zwar versuchen, diesen Anforderungen gegenüber offen zu bleiben, aber ihre vollständige Umsetzung zunächst zu unterlassen?

3.3 Beispiele für Maßnahmen zur Unterstützung der Skalierbarkeit

Die heute allgemein anerkannten Vorgehensweisen zur objektorientierten Softwareentwicklung sind genau die Voraussetzungen für das angestrebte Baukastenprinzip, da sie zum Beispiel Wiederverwendbarkeit, Modularität und Weiterentwickelbarkeit ermöglichen. Analog kann man eine modulare und aufrüstbare Hardwareausstattung (verschiedene Hardware für Schlüsselerzeugung, PSE-Personalisierung, Directory Services, Briefausdruck, Kommunikation...) realisieren und verschiedene Kommunikationsformate unterstützen. Das System sollte möglichst robust und einfach zu konfigurieren sein, d. h. Systemanpassungen nicht durch Neukompilierung von Software, sondern durch Änderung von Konfigurationseinstellungen im System ermöglichen (natürlich gegen unbefugte Manipulation kryptographisch abgesichert).

Das Sicherheitskonzept sollte genauso wie das Gesamtsystem zunächst generisch erstellt werden, um konkrete Sicherheitskonzepte für einzelne Anwendungen daraus abzuleiten.

Die Organisation der Abläufe erfolgt unter Effektivitätsgesichtspunkten. Beispiele: Schlüsselerzeugung auf Vorrat, Vorabregistrierung von Benutzern auf Vorrat und ggf. auch Erzeugung von z. B. PINs auf Vorrat (in geschlossenen Benutzergruppen), die Aufteilung der PSE-Personalisierung in mehrere Schritte, Vorpersonalisierung (z. B. mit Urschlüsseln), (kryptographische) Prüfung der Eingabedaten einer RA (Registration Authority, die Stelle, die die Identifizierung und Registrierung des einzelnen Teilnehmers durchführt und evtl. auch die Übermittlung von PSE und/oder PIN-Brief übernimmt). Es erfolgt eine organisatorische Planung der Abläufe dergestalt, daß eine Skalierbarkeit des Personaleinsatzes möglich ist und ein Ausnutzen von

Synergie- und Rationalisierungseffekten durch Betrieb verschiedener Trust Center in einer Hand in einem oder mehreren großen sicheren RZ.

Bemerkung: Eine Rezertifizierung beispielsweise kann in bestimmten Fällen automatisiert ablaufen, nämlich dort, wo die Prüfung des vorhandenen Zertifikats als Legitimation für die Erstellung des neuen Zertifikats ausreicht. Also muß man nicht jedesmal eine Neuerfassung des Teilnehmers bzw. Neuerzeugung von Schlüsseln vorsehen. Auch der Übergang zu längeren Schlüsseln oder Schlüsseln für neue Algorithmen kann automatisiert ablaufen (mit Legitimation durch Prüfung des alten Zertifikats) sofern Schlüssel der alten Länge noch nicht als kompromittiert gelten müssen.

Unbedingte Notwendigkeit ist die Einstellbarkeit der Kompatibilität zu Formaten relevanter Normen und bekannter Anwendungen (bei Zertifikaten, PSEs, Nachrichten, Directories, Namen,...):X.509, PKCS, PEM, MailTrusT, ISO-Normen zur Schlüsselgenerierung, EDI. Wünschenswert ist auch, daß Schnittstellen, die eigentlich innerhalb des Systems liegen, normgemäß bzw. offen definiert sind, um Fremdprodukte einbinden zu können.

„Entwicklungsbegleitende Evaluierung" ist ein bekanntes Stichwort, ein neuer Vorschlag wäre (passend zum Baukastensystem) eine Art generischer Evaluierung, in deren Rahmen Listen von Sicherheitszielen und -niveaus gleichzeitig untersucht werden (im Sinne der für die Common Criteria vorgesehenen Profile könnte man also „parametrisierbare Profile" vorschlagen). Diese Überlegungen sind aber zur Zeit noch nicht in die Praxis umgesetzt, so daß die Evaluierung der einzige in diesem Bericht behandelte Bereich ist, bei dem man mit einer Multiplizierung des Aufwandes mit der Zahl der zu entwickelnden Anwendungen rechnen muß (außer wenn identische Anwendungen mehrfach erstellt werden). Dies gilt natürlich nur für Systemevaluationen, während die Ergebnisse von Produktevaluationen durchaus wiederverwertbar sind.

Weitere Maßnahmen sind automatische Abrechnung in Verbindung mit der Protokollierung, Abrechnung nach verschiedenen Konzepten (Volumen, Zeit, Zugriffszahl, Pauschalen...). Die Vertragsgestaltung sollte ebenfalls anpaßbar sein, die Kalkulation sowohl der Kosten der Planung, des Aufbaus und des Betriebs eines konkreten Trust Centers als auch die Preise einzelner Dienstleistungen (z. B. Schlüsselpreise) sollten im Idealfall nach Eingabe der Parameter ermittelbar sein.

4 Baukästen

Im folgenden sind die Baukästen skizziert, aus denen Bausteine für die maßgeschneiderte Realisierung eines Trust Centers gewählt werden. Diese Baukästen sind aus Platzgründen, aber auch aus Gründen des Know How Schutzes, bewußt stichwortartig gehalten. Insbesondere werden bei Hard- und Software keine Produktnamen angegeben, obwohl sie in den praktisch genutzten Listen natürlich enthalten sind. Dies hat zusätzlich den Grund, daß die Firma der Verfasser einerseits gute Zusammenarbeit mit

Hard- und Softwareherstellern pflegt, bei der beschriebenen Vorgehensweise andererseits aber auch Herstellerunabhängigkeit bewahrt werden soll.

4.1 Baukasten für Konzepte und Dokumente

Selbst erstellte Dokumente lassen sich in wesentlichen Teilen für viele Trust-Center-Anwendungen nutzen. Dazu gehören das (funktionale) Systemkonzept, das Sicherheitskonzept, Ausführungsvorschriften, Bedienhandbücher, Vertrags- und Angebotsmuster, Software- und Hardwaredokumentation (bei Eigenentwicklung), Softwarespezifikation, Hardwarearchitektur, Arbeitspläne und Stufenpläne.

Zu den erforderlichen Dokumenten aus externen Quellen gehört Soft- und Hardwaredokumentation von Zulieferern, Normen und andere Literatur, externe Spezifikationen.

4.2 Baukasten für Kryptoalgorithmen und Kompatibilität

Auf den ersten Blick sind die Gebiete der Kryptoalgorithmen einerseits und der Kompatibilität zu Normen oder anderen Anwendungen andererseits ganz verschiedene Dinge. Die Erfahrung zeigt aber, daß bei der praktischen Einrichtung eines Systems die Auswahl und Parametrisierung von Kryptoalgorithmen einerseits und die Erreichung von Schnittstellen- und Formatkompatibilität untrennbar verbunden sind (vergleiche auch die Ausführung wenige Zeilen weiter).

- Einstellbare Formate von Zertifikaten, PSEs, Nachrichten, Directories und Kommunikationsschnittstellen. Zu beachten ist hier, daß manche Formate auch im laufenden Betrieb noch leicht angepaßt werden können, etwa Kommunikationsformate. Andere dagegen sind eher schwer zu ändern, insbesondere Zertifikatsformate, da dort die Änderung ggf. umfangreiche Rezertifizierungen zur Folge hat. Hinzu kommt, daß die Software der beteiligten Sicherheitsmodule im allgemeinen von einer solchen Änderung betroffen ist. Die schwer anpaßbaren Formate müssen von vornherein so flexibel wie möglich gehalten werden.

- Algorithmen: Alle anerkannten Algorithmen vorhalten, ebenso die Anpaßbarkeit von Parametern, insbesondere Schlüssellängen.

4.3 Einstellbare Formate von Zertifikaten, PSEs, Nachrichten, Directories und Kommunikationsschnittstellen.

Bei der baulichen Infrastruktur wird es naheliegenderweise die geringste Skalierbarkeit geben. Denn ein TC-Dienstleister wird sich Räumlichkeiten verschaffen, die höchste Anforderungen erfüllen, und auch TC-Anwendungen dort betreiben, die diese gar nicht benötigen. So verfügt die Firma der Verfasser über Rechenzentren im Münchner Raum, die alle Anforderungen an Abstrahlsicherheit, Verfügbarkeit (z. B. ununterbrechbare Stromversorgung, Brandschutz), Backupmöglichkeit und Zutrittskontrolle erfüllen.

Was die Ausstattung der Gebäude betrifft, so gibt es dagegen Möglichkeiten zur Skalierbarkeit, etwa die Möglichkeit zur räumlichen Abtrennung voneinander isolierter Sicherheitsbereiche, wie sie etwa für den gleichzeitigen Betrieb verschiedener TC-Anwendungen, aber auch innerhalb einzelner Anwendungen benötigt werden können.

4.4 Baukasten für Personal und Organisation

Da es sich bei den verwendeten Rechenzentren unserer Firma um Backup-Rechenzentren handelt, waren die folgenden Bausteine bereits vor Aufnahme des TC-Betriebs organisiert:

- Sicherheitsüberprüftes Personal,

- Alarmbereitschaft,

- hohe Verfügbarkeit.

Es ist also nur noch die Personalstärke nach Volumina und zeitlicher Verfügbarkeit der gewünschten Dienste (z. B. Zahl gewünschter Zertifikate, aber auch 24-Stunden-Verfügbarkeit) bei Einhaltung des Vieraugenprinzips zu skalieren.

4.5 Hardwarebaukasten, Kommunikationsbaukasten

Chipkarten, weitere Kryptohardware für kryptographische Berechnungen und Schlüsselspeicherung, Zufallszahlengeneratoren, normale Rechnerhardware, Kommunikationshardware, Spezialmaschinen (Briefverschickung, PSE-Personalisierung, Archivierung,...).

4.6 Softwarebaukasten

- Fremdsoftware: z. B. Krypto-API, Directory-Dienste, Software in komplett gekaufter Kryptohardware
- Eigensoftware: z. B. Ablaufsteuerung, Bedienoberflächen, Anpassung erworbener Software, Logging-Mechanismen, Änderungen an erworbener Software
- Vorhalten von (kryptographisch abgesicherten) Online-Updatemöglichkeiten für die Software wesentlicher Systemkomponenten

4.7 Baukasten von Maßnahmen zur Gewährleistung der Vertrauenswürdigkeit

Die folgenden Punkte geben Maßnahmen an, die bei konkreten Projekten das Vertrauen des Auftraggebers bzw. der Teilnehmer in das Trust Center sichern können. Dabei wird im Geiste des Baukastenprinzips nicht jede Maßnahme für jedes Projekt ergriffen, beispielsweise aufwendige Zulassungen.

- Vorlegen bereits ausgearbeiteter Sicherheitskonzepte,

- Angabe von Referenzprojekten und -kunden, die die Kompetenz im Bereich der IT-Sicherheit und von Rechenzentrumsdienstleistungen nachweisen; die folgende Liste gibt Beispiele von Referenzprojekten:
  * Betrieb eines Trust Centers zur Erzeugung und Zertifizierung asymmetrischer Schlüssel für vertraulichen Dokumentenaustausch, künftig auch für Signaturen, innerhalb des Daimler-Benz Konzerns,
  * Konzeption, Realisierung und Betrieb eines Trust Centers zur Erzeugung und Verteilung symmetrischer Schlüssel für integre Kommunikation zwischen Signalkomponenten im technischen System eines Transportunternehmens, das höchste Anforderungen an die Verfügbarkeit und korrekte Funktion der Komponenten stellt,
  * Konzeption und Realisierung eines Trust Centers zur Schlüsselerzeugung, -verteilung und -speicherung im Rahmen der Personalisierung von ec-Chipkarten,
  * Konzeption und Realisierung eines Trust Centers für Erzeugung und Verteilung asymmetrischer Schlüssel für technische Komponenten in einem Bankensystem,
  * Konzeption von Trust Centern zur Schlüsselerzeugung und -zertifizierung für die Anwendung asymmetrischer und Hybrid-Verfahren im Gesundheitswesen,
  * Konzeption und Unterstützung bei der Realisierung eines Trust Centers zur Erzeugung, Verwaltung und Verteilung geheimer und anderer Parameter für ein System des Fahrzeug-Diebstahlschutzes, einschließlich Online-Personalisierung von Ersatzkomponenten, Online-Recherchen durch Sicherheitsbeauftragte etc.
- ITSEC-Evaluation von Systemen und Komponenten,
- Zulassung des Systems nach Signaturgesetz,
- Hinweis auf Zugehörigkeit zu einer namhaften Firma mit Erfahrung bei professionellem Rechenzentrumsbetrieb, der unter anderem
  * eine hohe Verfügbarkeit,
  * die Einhaltung von Datenschutzgesetzen und
  * die Einhaltung langfristiger Verträge

  erfordert.

Alle bisher genannten Bausteine sind im wesentlichen für Geschäftskunden (Industrie, Banken, Organisationen etc.) wichtig.

Der folgende Punkt ist primär für die einzelnen Teilnehmer eines Systems von Belang, insbesondere also für private Kunden in offenen Systemen. (Nur am Rande sei bemerkt, daß er aber auch für Geschäftskunden interessant sein kann, da z. B. von den amerikanischen Regelungen zum „key escrowing" die Rede geht, sie sollten auch zur Wirtschaftsspionage dienen.)

- Kein Zugriff staatlicher Stellen (z. B. Geheimdienste) auf Trustcenterdaten.

Dies ist natürlich kein Baustein, den der TC-Betreiber selbst liefern kann, er setzt vielmehr die politische Verhinderung einer analogen Gesetzgebung für Trust Center voraus, wie es sie für Telekommunikationsdienste bereits gibt. An der Beliebtheit des Systems PGP, das die Einrichtung von CAs nicht vorsieht, sieht man unter anderem die Sorge vor dem Einfluß zentraler Instanzen. Daher ist der letztgenannte Baustein vermutlich das wesentliche Vertrauensargument für Einzelpersonen, die man in offenen Systemen als Kunden gewinnen will. Eine Stelle, der man nachsagen kann, daß sie Daten an Geheimdienste weitergibt, wird mit Sicherheit nicht als **Trust** Center angesehen. In dem Fall, daß sich eine Kryptogesetzgebung, die die Forderung nach generellem „key escrowing" enthält, nicht verhindern läßt, könnte die Rolle eines Trust Centers wenigstens darin bestehen, einen willkürlichen und unbefugten Zugriff auf Teilnehmerdaten kryptographisch abgesichert zu verhindern. Dies setzt allerdings wiederum voraus, daß nur eine Anfrage vom Format einer richterlichen Anordnung die Herausgabe von Daten legitimiert. Die Erklärung der Notwendigkeit durch Stellen der staatlichen Exekutive darf nicht ausreichen. Von dieser Diskussion nicht betroffen ist natürlich der Fall, daß Kunden ein key escrowing zum Zweck der Vermeidung von Datenverlusten ausdrücklich wünschen.

5 Der Fragebogen für Trust Center am Beispiel SET

5.1 Feststellung der Anforderungen mittels eines Fragebogens

Um für ein geplantes Trust Center die Analyse der benötigten Konfiguration und Skalierung zu erleichtern, wird ein Fragebogen ausgefüllt, der die erfahrungsgemäß wichtigen Fragen zu den Anforderungen der betrachteten Anwendung zusammenfaßt. Einen solchen Fragebogen mit detaillierten Fragen zur Gestaltung des Trust Centers wird man natürlich in dieser Form nur Geschäftskunden aus Industrie, Bankenwelt oder anderen Organisationen vorlegen, nicht jedoch Privatpersonen, die als Einzelteilnehmer an einem offenen Netz gewonnen werden sollen. Bei der Konzeption eines Trust Centers für ein offenes Netz, das den Normalverbraucher als Einzelkunden gewinnen will, muß sich der Dienstleister diese Fragen durch Marktanalysen selbst beantworten.

Wichtig für das Verständnis der Fragestellung ist, daß die Fragenliste von Dienstleister und Auftraggeber im allgemeinen gemeinsam durchgegangen werden sollte. Aus diesem Grunde ist es zulässig, daß die Fragen zum Teil Voraussetzungen an das Vorwissen und die Präzision der Vorstellungen machen, die bei einem nicht mit der Materie vertrauten Auftraggeber nicht unbedingt vorhanden sind.

Der folgende Fragebogen ist ein Beispiel, das bereits unter Kenntnis bestimmter Voraussetzungen erzeugt wurde. Insbesondere war bekannt, daß die angesprochenen Kunden Zertifikatsdienstleistungen für eine Sicherheitsinfrastruktur mit asymmetrischen Verfahren wünschen. Wie anfangs erläutert, ist dies keineswegs unsere generelle Voraussetzung. Durch diese Kenntnis wurden insbesondere die Fragen zu den gewünschten Dienstleistungen modifiziert, zu einer voraussetzungsfreien Liste der möglichen Dienstleistungen siehe 2.2.

Zur Verdeutlichung der Fragen sind die Antworten für eine konkrete Anwendung des Fragebogens mit angegeben. Dazu wurde das Protokoll SET ausgewählt. Da die SET-Spezifikation einige, für das Trust Center wesentliche Festlegungen (noch) nicht macht, wurden bestimmte nach dem derzeitigen Entwicklungsstand plausible Annahmen gemacht. Diese Annahmen liegen vornehmlich im organisatorischen Bereich:

- Die Realisierung der Client-Software auf Kundenseite erfolgt in Software in einem PC des Kunden, insbesondere erfolgen kryptographische Berechnungen in Software, sind die Schlüssel auf der Festplatte oder einer Diskette gespeichert, wobei sie mit einem Kennwort verschlüsselt sind; auch die Schlüsselerzeugung erfolgt in Software.
  Diese Annahme ist zwar nicht erfreulich, da Krypto-Hardware eine erheblich höhere Sicherheit bietet, nach unserer Erfahrung ist sie aber in einer ersten Entwicklungsphase realistisch.

- Die Identifizierung und Registrierung von Teilnehmern erfolgt durch den jeweiligen Betreiber des Systems, der hier als Kreditkartenherausgeber oder Großbank angenommen wird. Dieser Systembetreiber ist Auftraggeber im Sinne des Fragebogens.

- Bezüglich der Zeitabläufe der Entwicklung wurden realistische Schätzungen aus der bisherigen Erfahrung gemacht.

Frage	Antwort am Beispiel SET
Gewünschte Dienstleistungen des TC	
Zertifizierung von Schlüsseln,	ja
Personalisierung des Trägermediums (PSE - personal secure environment), Aushändigung PSE und der sie schützenden Geheimzahl (PIN) an Teilnehmer,	Wegen Softwarelösung erfolgt dies in Form einer Datenübertragung zum Teilnehmer, organisatorischer Ablauf der PIN-Absicherung noch nicht klar.
Verzeichnisdienst (Verzeichnis der Zertifikate, Sperrlisten, etc.),	ja
Billing, also Abrechnung von an Teilnehmer erbrachten Dienstleistungen gegenüber diesen oder dem Auftraggeber,	Über den Ablauf dieser Dienstleistungen werden keine Annahmen gemacht.
Protokollierung sicherheitsrelevanter Vorgänge (Logging, Auswertung der Protokolldaten),	nicht ausdrücklich festgelegt, aber selbstverständlich
Erzeugung von Schlüsseln für die Teilnehmer (alternativ ist dezentrale Schlüsselerzeugung durch Hardware bei den Teilnehmern),	Schlüsselerzeugung durch die Teilnehmersoftware
Identifizierung/Registrierung von Antragstellern,	Nach den Annahmen Übernahme der Teilnehmerdaten vom Systembetreiber

Frage	Antwort am Beispiel SET
Betreuung der Teilnehmer bei der Anwendung der Schlüssel,	nicht festgelegt
Entwicklung, Vertrieb und Support für Anwendungssoftware bei den Teilnehmern, bzw. Beratung der Teilnehmer bei der Auswahl von Software,	nicht festgelegt
Sollen die Verzeichnisse der öffentlichen Schlüssel und die Sperrlisten Online zur Verfügung stehen oder reicht eine Offline-Übertragung (z. B. über Disketten) aus?	online (Internet)
Ist ein 24-Stunden-Telefonservice (etwa für Sperrungen verlorener Karten) gewünscht?	nicht festgelegt, aber vorgesehen
Für welche Zwecke sollen die Teilnehmerschlüssel verwendet werden: Direkte Verschlüsselung, Schlüsselaustausch bei Hybridverfahren, Signaturen, Authentikation?	alle
Sollen die Schlüssel im TC zu Backupzwecken gespeichert werden (kommt für Signatur- oder Authentikationsschlüssel i. a. nicht in Frage)?	nein
Gibt es bereits Vorgaben des Auftraggebers über den Ablauf einzelner dieser Dienstleistungen?	teils in der Spezifikation bereits festgelegt, ansonsten nach den Wünschen des Auftraggebers
Fragen zum Grad des Vertrauens und der generellen Zielrichtung	
Soll die digitale Signatur in der Anwendung einen Sicherheitsgrad vergleichbar zur Unterschrift haben oder soll sie üblichen Geschäftsabläufen dienen, wie Handschlag oder der Verkauf über die Ladentheke?	letzteres
Sind die Teilnehmer Personen oder Geräte (Clients, Prozesse)? Liegen offene oder geschlossene Benutzergruppen vor?	Personen, Kunden eines Betreibers, aber tendenziell offen für alle
Ist der Auftraggeber eine Organisation, die für eine eigene geschlossene Benutzergruppe TC-Dienste benötigt oder sind die Auftraggeber einzelne Teilnehmer, die an einem offenen System teilnehmen?	Organisation (Betreiber)
Ist der Auftraggeber eine Organisation: Welche Aufgaben soll der Dienstleister für das Trust Center übernehmen: Planung und Konzeption („plan"), den Aufbau („build") bzw. Betrieb („run")?	alle möglich
Wenn der Dienstleister den Betrieb übernimmt: Soll er ein im Besitz des Auftraggebers befindliches Trust Center betreiben oder die Dienstleistung mittels eines eigenen Trust Centers erbringen?	eigenes Trust Center

Frage	Antwort am Beispiel SET
Volumen (die folgende Frage steht stellvertretend für die Abschätzung der Volumina, die für jede einzelne Dienstleistung vorgenommen werden muß)	
Wie groß ist die Anzahl der Schlüssel bzw. Zertifikate in der Startphase?	1000 Käufer, 50 Händler bei einem Betreiber
bzw. im Wirkbetrieb (z. B. pro Jahr)?	zur Zeit nicht zu schätzen, zwischen einigen tausend und mehreren Millionen Käufern, wobei es mehrere Betreiber geben wird
Zeitplanung im Projekt	
Welcher Zeitraum ist für die Konzeption geplant?	ein Jahr
Für welchen Zeitraum ist ein Testbetrieb bzw. eingeschränkter Wirkbetrieb geplant?	ein Jahr
Wann ist der volle Wirkbetrieb geplant? Soll eine Stufenplanung vereinbart werden?	nach zwei Jahren, ja
Verfügbarkeit	
Soll es außer dem eventuellen 24-Stunden-Sperrdienst (siehe auch Dienstleistungen) weitere Dienstleistungen geben, die außerhalb der normalen Arbeitszeiten zur Verfügung stehen sollen?	keine bekannt
Wie lange nach einer Bestellung (etwa eines Zertifikats) soll es geliefert werden?	vom Trust Center online, vorab jedoch Registrierung beim Betreiber erforderlich
Diverse Randbedingungen	
Wie lange sollen die Zertifikate gültig sein?	vom Betreiber nach dem Testbetrieb festzulegen
Sollen die Schlüssel für Verschlüsselung einerseits und digitale Signaturen andererseits getrennt werden?	Trennung
Ist die Einbettung des TC in eine Zertifizierungshierarchie, also die Zertifikatserteilung durch oder an andere TC geplant bzw. soll diese Möglichkeit vorgesehen werden?	ja
Zum organisatorischen und Personalaufwand	
Ist eine Durchsetzung des Vier-Augen-Prinzips unter Mitwirkung durch MA des Kunden vorzusehen oder soll es allein durch MA des TC realisiert werden?	eher letzteres
Gibt es aus Sicht des Auftraggebers Vorschriften für die Protokollierung der Vorgänge im TC und Aufbewahrungsfristen für die Vorgänge und Protokolle?	nicht bekannt, die allgemeinen rechtlichen Vorgaben für das Bankwesen sollten ausreichen

Frage	Antwort am Beispiel SET
Übernimmt der Auftraggeber oder eine sonstige Stelle die Rolle der RA?	nach unseren Annahmen ja
Räumliches	
Ist eine Betriebsumgebung wie im Infrastrukturbaukasten beschrieben als Standort angemessen? (Bemerkung: diese wird einem Kunden natürlich auf Wunsch vorgeführt).	ja
Kompatibilität	
Sind bereits Normen im Gespräch, an die man sich halten möchte, wie z.B.: X500/X509, PEM, PKCS, MailTrusT, sonstige?	PKCS#7, X509v3
Sind alle Schnittstellen des Systems offen zu gestalten und so zu dokumentieren, daß beliebige Hersteller z. B. Client-Software für Teilnehmer herstellen können?	ja
Gibt es bereits Teilsysteme (etwa Client-Anwendungen oder Chipkarten, die als PSEs vorgesehen sind) oder Spezifikationen, die berücksichtigt werden sollen?	bisher nur Testbetrieb
Gibt es bestehende Systeme, zu denen Abwärtskompatibilität erforderlich ist oder geplante Systeme, zu denen Aufwärtskompatibilität erforderlich sein wird?	nicht relevant
Zulassung und Sicherheitsstandards	
Ist eine Zulassung des Systems oder von Komponenten des Systems nach bestimmten Sicherheitskriterien geplant oder soll das System zumindest in diese Richtung ausbaufähig sein? [Solche Kriterien sind die ITSEC (Information Technology Security Evaluation Criteria), die künftigen CC (Common Criteria), Kriterien des ZKA (Zentraler Kreditausschuß) oder die Anforderungen des geplanten Signaturgesetzes.]	nicht entschieden, vom Betreiber noch festzulegen
Gibt es spezifische Sicherheitsziele des Auftraggebers bezüglich Vertraulichkeit, Nicht-Manipulierbarkeit oder Verfügbarkeit, die sich aus dem Wert der abzusichernden Daten oder Nachrichten oder durch besondere Risiken ergeben?	die Maßstäbe der Kreditwirtschaft
Kryptographische Algorithmen	
Für Signaturen bzw. Zertifikate und die Verschlüsselung werden Verschlüsselungsverfahren und Hashfunktionen benötigt. Eine „klassische" Kombination wäre RSA (als asymmetrisches Verschlüsselungsverfahren), Triple-DES (als symmetrisches Verschlüsselungsverfahren) und eine Hashfunktion (RIPE-MD 160 oder SHA-1). Ist der Einsatz dieser Kombination gewünscht?	bisher ist einfacher DES spezifiziert, ansonsten ja

Frage	Antwort am Beispiel SET
Ist es geplant, das System für weitere Verfahren offenzuhalten bzw. diese fest vorzusehen?	vermutlich ja
Kommunikation	
Soll die erforderliche elektronische Kommunikation (insbesondere zwischen TC und Teilnehmern) von Beginn an über Online-Verbindungen erfolgen oder ist ein Start mit Datenträgern vorzusehen?	von vornherein über Internet
Gibt es bereits eine Kommunikationsinfrastruktur für das geplante System (z. B. Firmennetze, aber auch das Internet), oder ist diese mit zu entwickeln?	Internet
Wie sollen PIN-Briefe, PSE's und Anträge übermittelt werden?	Nach unserer Annahme werden Anträge und PIN-Briefe vom Betreiber gehandhabt (z. B. am Bankschalter), für PSE wegen Softwarelösung nicht relevant

6 Fazit

Der im vorigen Kapitel dargestellte und exemplarisch beantwortete Fragebogen zeigt die Vielfalt der Parameter, die bei Planung, Einrichtung und Betrieb eines Trust Centers berücksichtigt werden müssen.

Es wurde eine Vorgehensweise aufgezeigt, die ausgehend von der Ermittlung dieser Parameter und unter Verwendung von vorhandenen Bausteinen in den verschiedenen konstituierenden Teilen eines Trust Centers (Kryptographische Algorithmen und Kompatibilität, Personal und Organisation, Infrastruktur, Hardware, Software, Konzepte und Dokumente...) Trust Center für konkrete Anwendungen „maßschneidert".

7 Literatur

[1]   LINN, J. ET AL., „Privacy Enhancement for Internet Electronic Mail, Part I-IV",
       RFC1421-1424, February 1993.

[2]   X.509v1-v3: „The Directory - Authentication Framework", CCITT bzw. ITU-T Re-
       commendation X.509 (1988, 1993, 1996).

[3]   Signaturgesetz, in: „Entwurf eines Gesetzes zur Regelung der Rahmenbedingungen für
       Informations-und Kommunikationsdienste (IuKDG)", Bundesratsdrucksache 966/96,
       20.12.96

[4]   Signaturverordnung: „Verordnung zur digitalen Signatur (SigV)", Entwurf vom
       20.12.1996, http://www.iid.de/rahmen

[5]   RSA Laboratories. PKCS#1: RSA Encryption Standard. Version 1.5, November 1993.

[6]   RSA Laboratories. PKCS#7: Cryptographic Message Syntax Standard, Version 1.5,
       November 1993.

[7]   F. BAUSPIEß, „MailTrusT-Spezifikation", Version 1.1, *TeleTrusT Deutschland e. V.,
       (Geschäftsstelle: Eichendorffstraße 16, 99096 Erfurt)*, 18.12.96

[8]   TH. HUESKE , „MailTrusT-Infrastrukturbeschreibung", Version 1.0, *TeleTrusT
       Deutschland e. V., (Geschäftsstelle: Eichendorffstraße 16, 99096 Erfurt)*, 15.09.1995.

[9]   ISO/IEC 9735, -1997, Electronic data interchange for administration, commerce and
       transport (EDIFACT) - Application level syntax rules [10] Secure   Electronic   Tran-
       saction   (SET)   Specification   -   Book   1-3,   Draft   Version,   24.6.1996;
       http://www.visa.com/cgi-bin/vee/sf/set.

Freier und sicherer elektronischer Handel mit originalen, anonymen Umweltzertifikaten

Markus Gerhard [1], Alexander W. Röhm [2]

[1] Universität Gießen
Volkswirtschaftslehre
Markus.Gerhard-1@wirtschaft.uni-giessen.de

[2] Universität GH Essen
Wirtschaftsinformatik
roehm@wi-inf.uni-essen.de

Zusammenfassung

Elektronische Märkte in offenen Netzen stellen neue Möglichkeiten der Leistungskoordination mit geringeren Transaktionskosten zur Verfügung [MaYB_87, Wiga_95] und ermöglichen so neue Lösungen in vielen Bereichen. In dieser Arbeit wird die Abwicklungsphase eines offenen elektronischen Marktes spezifiziert, der dazu verwendet werden kann, das Modell der Umweltzertifikate effizient in einer Umweltregion zu installieren. Es wird ein Konzept mit originalen und anonymen Lizenzen vorgestellt, auf dem eine technische Realisierung basieren könnte. Dieses Konzept ist eine grundsätzliche Überlegung und ist auf alle digitalen Waren in einem offenen Netz übertragbar, die nur dann wertvoll sind, wenn sie als Original vorliegen und deren Käufer zudem vollständig anonym bleiben soll.

1 Einleitung

Die menschlichen Eingriffe in die Natur haben inzwischen ein Ausmaß angenommen, daß die Belastungsgrenzen vieler Ökosysteme der Erde erreicht, in einigen Bereichen sogar bereits überschritten sind. Die bekanntesten Beispiele solcher antropogener Überbeanspruchungen der Natur sind derzeit das sogenannte Ozonloch, der Treibhauseffekt, das Waldsterben sowie die Verschmutzung von Luft, Gewässern und Böden mit den verschiedensten Schadstoffen. Forderungen nach einer neuen Erdpolitik, die am Leitbild einer dauerhaftumweltgerechten Entwicklung (Sustainable Development) orientiert ist, sind daher in aller Munde. Aus anthropozentrischer Sicht geht es dabei um das Ziel, die Umwelt, die als eine unersetzliche Grundlage jeder sozio-ökonomischen Entwicklung angesehen wird, dem Menschen zu erhalten, damit auch zukünftige Generationen ihre Bedürfnisse befriedigen können. In der umweltpolitischen Praxis der Bundesrepublik Deutschland und anderer westlicher Industrieländer bedient man sich zur Bewirtschaftung von Umweltgütern traditionell ordnungsrechtlicher Lenkungsstrategien [GaHa_95, Breu_87]. Dieser Instrumentierungsansatz wird von Ökonomen seit langem sehr kritisch eingeschätzt: Das ordnungsrechtliche Instrumentarium sei - so wird postuliert - durch einen Mangel an ökologischer Effektivität und ökonomischer Effizienz charakterisiert. Vor diesem Hintergrund wurde von wirtschaftswissenschaftlicher Seite immer wieder eine instrumentelle Neuorientierung der Umweltpolitik angemahnt, bei der sogenannte marktorientierte Strategien im Mittelpunkt des Interesses stehen. Als wohl prominentestes marktsteuerndes Instrument hat die Umweltökonomik die sog. handelbaren Emissionsrechte (Zertifikate und Kompensationen) hervorgebracht, die in der U.S.-amerikanischen Luftreinhaltepolitik erstmals in der umweltpolitischen Praxis implementiert wurden. Auch in der deutschen Luftreinhaltepolitik findet man Elemente einer Kompensationsregelung. In der modelltheoretischen Analyse zeitigen diese mengensteuernden Allokationsverfahren sehr gute Ergebnisse: Sie versprechen Umweltschutzziele sicher und „wirtschaftlich", d.h. zu minimalen volkswirtschaftlichen Kosten zu erreichen und darüber hinaus dynamische Anreize zur Entwicklung und Einführung umwelttechnischen Fortschritts zu induzieren.

Als ökonomische Funktionsbedingung für Zertifikatlösungen wird insbesondere die Verfügbarkeit eines möglichst kostenlosen Koordinationsmechanismus zur Anbahnung und zum Abschluß von Zertifikattransaktionen für erforderlich gehalten, mit anderen Worten die Abwesenheit von Transaktionskosten [EwGa_94]. Sind nämlich die Transaktionskosten gering, steigt auf einem Zertifikatmarkt - unterstellt man divergierende Kostenstrukturen der Emittenten - das Transaktionsvolumen und damit die ökonomische Effizienz.

Elektronische Märkte in offenen Netzen eröffnen nun gerade Möglichkeiten der Leistungskoordination mit geringen Transaktionskosten [MaYB_87, Wiga_95]. Unter einem *elektronischen Markt* versteht man einen mit Hilfe der Informationstechnologie realisierten Markt zum Handel mit Produkten, in dem mindestens eine der Phasen des Kaufes – Informations-, Vereinbarungs- und Abwicklungsphase – unterstützt werden [Schm_93].

In dieser Arbeit sollen die technischen Voraussetzungen für die Abwicklungsphase eines elektronischen Zertifikatmarkts herausgearbeitet werden. Dazu wird ein System spezifiziert, in dem ein elektronischer Markt als Koordinationsmechanismus bei der Umsetzung des Umweltzertifikatemodells eingesetzt wird.

Die Arbeit ist wie folgt aufgebaut: Im 2. Kapitel wird das Grundmodell der Zertifikatelösung kurz charakterisiert und nach ökonomischen Maßstäben beurteilt. Kapitel 3 thematisiert die Anforderungen, die an die technische Realisierung eines elektronischen Markts gestellt werden müssen. Das Konzept der originalen und anonymen Lizenzen wird im 4. Kapitel spezifiziert. Nachdem dann in Kapitel 5 das organisatorische Umfeld diskutiert wurde, folgt die Bewertung und ein Ausblick im 6. Kapitel.

2 Das Grundmodell der Emissionszertifikate

Der auf DALES und CROCKER [Dale _68, Croc_1966] zurückgehende Ansatz der Umweltzertifikate bzw. -lizenzen besteht darin, Märkte für die Inanspruchnahme von Umweltgütern zu schaffen. Da funktionsfähige Märkte die Existenz von privaten Eigentumsrechten voraussetzen, Eigentumsrechte für Umweltgüter aufgrund ihrer Kollektivguteigenschaften jedoch nicht institutionalisiert werden können, beziehen sich die Rechte auf spezifische Nutzungsformen, z.B. eine genau festgelegte Inanspruchnahme der Umweltgüter als Schadstoffaufnahmemedium.

Im Modell der Emissionszertifikate definiert der Staat zunächst ein Umweltqualitätsziel (Immissionsziel) für eine bestimmte Umweltregion, etwa für einen See. Die Fixierung des Immissionsstandards kann entweder an ökonomischen Nutzen-Kosten-Kriterien („optimale Umweltqualität") oder Nachhaltigkeitskriterien (Immissions-quantum darf Assimilationskapazität der Umweltmedien nicht übersteigen) orientiert werden. Aus dem Immissionsziel wird die maximale Emissionsmenge abgeleitet und in regional differenzierte Emissionsgrenzen für einzelne Schadstoffe in einem bestimmten Raum umgesetzt. Diese Emissionskontingente für die einzelnen Schadstoffe werden dann in kleine Teilmengen gestückelt und in Zertifikaten bzw. Lizenzen verbrieft. Umweltschädigende Emissionen sind nur dann erlaubt, wenn ein Emittent entsprechende Zertifikate besitzt. Ein solches Verschmutzungszertifikat gestattet die Emission einer festgelegten Menge eines bestimmten Schadstoffs innerhalb eines abgegrenzten Gebiets für einen genau definierten Zeitraum.

Die Anfangsausstattung der Wirtschaftssubjekte mit Emissionsrechten kann durch Versteigerung, Verkauf zu einem staatlich fixierten Festpreis oder durch freie Vergabe (sog. Grandfathering-Variante) erfolgen. In der ökonomischen Fachliteratur wird aus politischen und rechtlichen Praktikabilitätserwägungen das Primärallokationsverfahren des Grandfathering präferiert. Hier werden die Emissionsrechte zunächst an die Altemittenten frei vergeben. Durch anschließende Abwertung der Zertifikate wird dann das politisch fixierte Umweltziel schrittweise erreicht. Die weitere Allokation der Emissionsrechte wird durch Angebot und Nachfrage an der Umweltbörse gelenkt, d.h. die Zertifikate sind innerhalb der jeweiligen Regionen frei handelbar. Es entstehen also Märkte, auf denen sich Zertifikatpreise bilden.

Orientieren sich die Emittenten an ihren individuellen Kosten zur Vermeidung von Emissionen, werden sie Zertifikate kaufen, wenn ihre Grenzvermeidungskosten über dem Zertifikatpreis liegen. Sind die Zertifikate teurer als eigene schadstoffreduzierende Maßnahmen, werden sie sich zu Vermeidungsmaßnahmen entschließen. Für den einzelnen Emittenten lohnt es sich also solange Verschmutzungsrechte zu erwerben, bis die individuellen Grenzvermeidungskosten dem Marktpreis für Zertifikate entsprechen. Auf diese Weise wird sichergestellt, daß der politisch fixierte Umweltstandard volkswirtschaftlich kosteneffizient realisiert wird: Letzlich befinden sich die Zertifikate im Besitz der Unternehmen mit den höchsten Vermeidungskosten, während Schadensvermeidungsaktivitäten dort durchgeführt werden, wo die geringsten Kosten anfallen. Durch Abwertungen der Zertifikate resp. durch eine Offenmarktpolitik der Umweltbehörde ist es möglich Emissionszielwerte zu verschärfen, etwa wenn neue ökologische Erkenntnisse vorliegen.

Die theoretische Beurteilung von Lizenzlösungen im Umweltschutz erfolgt in der Regel nach den folgenden drei zentralen Evaluierungskriterien:

- Kriterium der *ökologischen Effektivität*: Hier wird gefragt, inwieweit das umweltpolitische Instrument in der Lage ist, ein vorgegebenes Umweltqualitätsziel sicher, schnell und dauerhaft zu realisieren.

- Kriterium der *ökonomischen Effizienz*: Hier wird geprüft, ob ein Instrument den Umweltqualitätsstandard zu volkswirtschaftlich minimalen Kosten erreicht.

- Kriterium der *Innovationseffizienz*: Hier wird die Potenz von Instrumenten geprüft, umwelttechnischen Fortschritt zu induzieren. Durch diesen wird es möglich, gleiche Umweltstandards zu geringeren volkswirtschaftlichen Kosten bzw. mit gleichen Kosten höhere Umweltstandards zu realisieren.

Die anhand dieser Evaluationskriterien vorgenommene theoretische Beurteilung von Emissionszertifikaten fällt äußerst positiv aus. Die Anzahl der ausgegebenen Zertifikate ist auf den Emissionszielwert begrenzt, der bei rechtmäßigem Verhalten der betroffenen Wirtschaftssubjekte nicht überschritten werden kann. Ökologische Effektivität ist - bei hinreichender Kontroll- und Sanktionsflanke - in hohem Maße zu erwarten. Auch hinsichtlich des Kriteriums der ökonomischen Effizienz läßt eine Zertifikatlösung gute Ergebnisse erwarten. Ein rationaler Emittent vermeidet solange Emissionen, wie die Emissionsvermeidungskosten für eine zusätzlich vermiedene Schadstoffeinheit unter dem Zertifikatkurs liegt. Bei diesem Verhalten ergibt sich bei funktionierenden Märkten und fehlenden Transaktionskosten eine volkswirtschaftlich kostenminimale Partitionierung der Emissionsvermeidungsaktivitäten. Für die Emittenten besteht darüber hinaus ein permanenter Anreiz, kostensparende und emissionsmindernde technische Neuerungen zu entwickeln und einzuführen, um damit Ausgaben für Zertifikate zu sparen. Kosteneinsparungen und Erlöse aus dem Verkauf von freigewordenen Zertifikaten bilden den ökonomischen Anreiz zur Innovation. Innovationen führen zu einer Verschiebung der gesamtwirtschaftlichen Grenzvermeidungskostenfunktion nach unten, d.h. gleiche Umweltstandards sind nun zu geringeren volkswirtschaftlichen Kosten bzw. höhere Umweltstandards sind zu gleichen Kosten zu erreichen. Eine Verbesserung der Vermeidungstechniken führt im Zeitablauf dazu, daß die Nachfrage nach Zertifikaten und somit deren Kurs sinkt. Damit reduziert sich die Anreizwirkung dieses Instruments. Diesem Effekt kann der Staat durch Aufkauf, Abwertung oder verminderte Erneuerung zeitlich befristeter Zertifikate begegnen.

Angesichts dieser theoretischen Leistungsfähigkeit dieses Instrumententypus ist zu fragen, warum die umweltpolitische Praxis bislang nur sehr zögerlich versucht, Zertifikatelemente in die bestehende Politik zu integrieren. Aus der Sicht der umweltpolitischen Praxis wird dies zumeist mit der „Praxisschwäche" ökonomischer Politikvorschläge begründet: Die obigen Effizienzaussagen basieren - so wird häufig postuliert - auf stark idealisierten Modellen, die insbesondere die institutionellen Rahmenbedingungen der umweltpolitischen Realität zu wenig berücksichtigen. Die ökonomische Theorie bemüht sich daher seit geraumer Zeit um die Berücksichtigung spezifischer Anwendungsvoraussetzungen und -restriktionen von Zertifikatlösungen. Es werden hier zumeist ökologische, ökonomische, technische und rechtliche Anwendungsrestriktionen untersucht [Huck_93].

Im Mittelpunkt der ökonomischen Restriktionsanalyse, steht das Problem der Transaktionskosten. Transaktionskosten sind sämtliche Kosten, die bei Anbahnung, Aushandlung, Vollzug und Kontrolle von Verträgen anfallen. Empirische Studien zeigen, daß auf Zertifikatmärkten aufgrund technischer und marktlicher Intransparenz hohe Informations- und Transaktionskosten entstehen, welche die Ausschöpfung vorhandener Austauschpotentiale begrenzen [MaGa_92]. Elektronische Märkte mindern tendenziell diese Transaktionskosten und verbessern damit die Funktionsfähigkeit eines Zertifikatmaktes.

3 Technische Anforderungen

Schon 1987 diskutierten Malone, Yates, und Benjamin die These, daß durch die Verschmelzung von Telekommunikation und Informatik zur Telematik in verschiedenen Bereichen eine Verschiebung von hierarchischen zu marktmäßigen Koordinationsmechanismen stattfinden wird, da die Transaktionskosten durch die neuen Technologien in elektronischen Märkten stärker sinkt als in den entsprechenden elektronischen Hirarchien [MaYB_87]. Die in ihrer Arbeit vorgebrachten Argumente werden heute allgemein anerkannt.

In manchen Bereichen ist die Entwicklung von elektronischen Märkten bereits weit vorangeschritten. Erfolgreiche elektronischen Märkte zeichnen sich oft dadurch aus, daß eine Produkt- bzw. Leistungsbeschreibung unkompliziert und der Vergleich von Angeboten anhand des Preises und weniger Eigenschaften des Produktes möglich ist. Beispiele solcher elektronischer Märkte im Internet sind Buchmärkte (www.amazon.com, www.books.com, www.bookpool.com u.s.w.) und die Tourismusbranche (www.ltu.de, www.british-airways.com u.s.w.). Handelbare Rechte, wie die in dieser Arbeit behandelten Lizenzen im Umweltzertifikathandel, haben die gleichen Vorzüge. Außerdem sind sie – vorausgesetzt geeignete rechtliche Rahmenbedingungen existieren – vollständig digital repräsentierbar. Dies impliziert, daß die Transaktionskosten besonders in der Abwicklungsphase stark reduziert werden können.

Warum ist dieser Bereich im allgemeinen noch unterrepräsentiert im Internet? Worauf ist die noch mangelnde Akzeptanz zurückzuführen? Grundsätzlich ist durchaus eine Tendenz vorhanden elektronische Märkte für handelbare Rechte im Internet zu akzeptieren. Abschreckend wirkt auf die Beteiligten der Mangel an Datensicherheit, der fehlende Datenschutz und ungenügende Rechtssicherheit im elektronischen Handel [Weil_96]. Unser Ziel ist es, einen elektronischen Markt für Lizenzen im Umweltzertifikatemodell zu spezifizieren, der die oben genannten Kriterien ökologische Effektivität, ökonomische Effizienz und Inovationseffizienz erfüllt. Dieses Ziel kann nur zusammen mit den oben genannten Anforderungen Datensicherheit, Datenschutz und Rechtssicherheit erreicht werden, da nur so Akzeptanz bei den umweltpolitschen Akteuren (Politiker, Verwaltung, Emittenten etc.) herbeigeführt werden kann.

Markteffizienz kann theoretisch erreicht werden, wenn die Annahmen für einen idealen Konkurrenzmarkt erfüllt werden [Endr_94]. Dazu gehört, daß eine ausreichend große Teilnehmerzahl an dem System partizipiert. Für Zertifikatmärkte leitet sich daraus die Forderung ab, daß neben den Emittenten auch andere Akteure – etwa Umweltschutzverbände – am Handel mit Emissionsrechten teilnehmen können. Damit wird auch erreicht, daß die Präferenzen der Bürger stärker berücksichtigt werden. Es wird deshalb als Vorteil angesehen, wenn der elektronische Markt frei zugänglich ist. Das Internet ist durch seine offene Struktur, einer sehr großen potentiellen Teilnehmerzahl offen zugänglich und ist deshalb geeignet einen solchen elektronischen Markt zu beherbergen [Hans_96].

Durch die Offenheit des Netzes, die aus den oben genannten Gründen vorteilhaft ist, entstehen jedoch spezifische Bedrohungen der Sicherheit, denen mit entsprechenden Sicherheitsmechanismen entgegen gewirkt werden muß. Neben den elementaren Bedrohungen: Verletzung der Vertraulichkeit, der Integrität und der Authentizität übertragener Nachrichten, ist in dem hier betrachteten Kontext die Bedrohung vorhanden, daß Lizenzen kopiert und doppelt verwendet werden. Diesen Bedrohungen zu begegnen und den Teilnehmern ein angemessenes Maß an Sicherheit zu gewährleisten steht im Zentrum dieser Arbeit.

Neben dem Aspekt Datensicherheit besteht ein weiterer Weg zur Förderung der Akzeptanz des elektronischen Marktes in der Verbesserung des Datenschutzes, indem das System die *Anonymität* der Teilnehmer beim Kauf einer Lizenz bewahrt. Man kann verschiedene Grade der Anonymität unterscheiden [PfWP_90]. Hier ist gemeint, daß die Daten der Lizenz weder Rückschluß auf frühere noch den momentanen Besitzer zulassen und daß keine dritte Partei ermitteln kann, wer eine Lizenz erworben hat und welche Menge des Schadstoffes er tatsächlich legal eingeleitet hat. Anonymität mag im Interesse von schadstoffemittierenden Unternehmen liegen, deren Image beim Verbraucher heute wesentlich von Faktoren des Umweltschutzes abhängt. Andererseits ist es in einem freien Umweltzertifikatemarkt für Umweltschutzinitiativen möglich, Lizenzen ohne die Absicht Schadstoffe zu emittieren, zu kaufen. Beim Kauf zum Zwecke der Verhinderung von Schadstoffemissionen mögen auch sie das Interesse haben anonym zu bleiben, da sie so Einfluß auf das erreichte Maß an Umweltqualität nehmen können ohne sich politischem Druck anderer gesellschaftlicher Gruppen auszusetzen. Ein Nachteil der Anonymität ist, daß marktmächtige und finanzstarke Unternehmen Zertifikate in großem Ausmaß kaufen und horten könnten, um andere aus dem Markt zu drängen. Man kann in einem System ohne Anonymität dieses Blockadeverhalten auch nicht verhindern, weil dieses Verhalten mittels Strohmännern immer möglich ist.

Ein sehr wichtiger Teil und zugleich eines der größten Probleme der Umweltpolitik ist die Realisierung einer angemessenen Kontrolle. Angemessen bedeutet, daß ein Kompromiß zwischen den Kosten und einer möglichst lückenlosen Kontrolle gefunden werden muß. Eine lückenlose Kontrolle im herkömmlichen Sinne ist sehr kostenintensiv, während mangelnde Kontrolle dazu führt, daß durch unkontrollierte Emissionen das Umweltziel verfehlt wird. Jede Realisierung benötigt daher eine geeignete Kontrolltechnologie. Die im folgenden präsentierte Spezifikation skizziert dafür einen möglichen Weg.

Während die *ökonomische Effizienz* stark von den Kosten und der Akzeptanz abhängt, ist für die Anforderung *ökologische Effektivität* wesentlich, daß keine gefälschten oder illegal kopierten Lizenzen in Umlauf kommen. Ein wichtiges Mittel zur Erreichung der *Innovationseffizienz* ist die befristete Gültigkeitsdauer der Lizenzen.

Während der Gültigkeitsdauer sollte die Lizenz eine rechtsgültige und damit einklagbare Erlaubnis für Emissionen sein. Dies kann nur durch zusätzliche organisatorische Maßnahmen und neue Gesetze erreicht werden. So ist z.B. die Einbindung in Zertifikatinfrastrukturen nach dem *Informations- und Kommunikationsdienste Gesetz* wichtig, da dort die Rechtsverbindlichkeit der *digitalen Signatur* geregelt werden wird [Bund_97]. In Abbildung 1 wird zusammenfassend dargestellt, wie aus den allgemeinen Zielen der Umweltpolitik Anforderungen an die Technik abgeleitet werden können. Eine genaue Untersuchung über die Kosten eines solchen Systems kann in dieser Arbeit nicht durchgeführt werden.

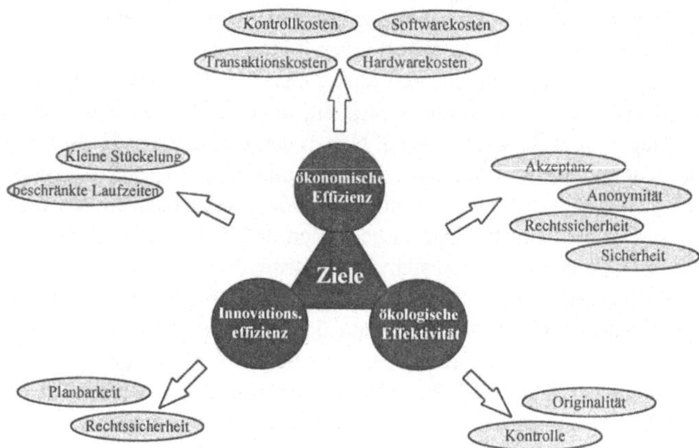

Abbildung 1: Ziele der Realisierung

4 Anonyme originale Lizenzen

In diesem Kapitel wird eine mögliche technische Realisierung der Abwicklunsphase eines elektronischen Marktes für originale, anonyme Lizenzen (Umweltzertifikate) beschrieben. Es wurden bisher verschiedene Vorschläge für originale und anonyme Kommunikation in offenen Netzwerken gemacht [Chau_85]. In dieser Arbeit orientieren wir uns für die Realisierung an der Notwendigkeit der Emissionskontrolle.

Im Mittelpunkt der Realisierung des elektronischen Marktes zum Handel mit Umweltlizenzen steht das Konzept der anonymen, originalen Lizenzen. Eine *anonyme originale Lizenz* - im folgenden kurz *Lizenz* genannt - ist ein digitales Dokument, das nur in einer ausgezeichneten Instanz (Original) gültig ist und dessen Inhalt keinen Rückschluß auf die Identitäten seiner Eigentümer zuläßt. Alle – sowohl der jetzige als auch alle vorangegangenen – Eigentümer der Lizenz bleiben in diesem Sinne anonym. Zudem wird beim Kauf einer Lizenz die Adresse des Käufers keinem Dritten bekannt.

4.1 Kryptographische Verfahren

Zur Realisierung der anonymen, originalen Lizenzen werden kryptographische Verfahren verwendet, welche die Sicherheitsdienste Vertraulichkeit und Integrität erbringen [ISO1_89]. Durch deren Einbindung in eine Sicherheitsinfrastruktur kann Verbindlichkeit gewährleistet werden [ISO_95]. Die Eigenschaften Anonymität und Originalität werden durch Zusammenwirken dieser Sicherheitsdienste erreicht.

Sowohl asymmetrische als auch symmetrische Kryptographie wird bei den vorgeschlagenen Lizenzen eingesetzt. Während sich für die symmetrische Verschlüsselung zwei Kommunikationspartner einen geheimen Schlüssel (*secret-key*) teilen und deshalb einander vertrauen müssen, hat jeder Teilnehmer des asymmetrischen Verfahrens ein eigenes Schlüsselpaar, bestehend aus einem öffentlich bekannten Schlüssel (*public-key*) und einem Geheimen, die zueinander komplementär sind, so daß man mit dem Einen entschlüsseln kann, was mit dem Anderen verschlüsselt wurde. In den folgenden Abschnitten werden die in Abbildung 2 erläuterten Schreibweisen für die kryptographischen Verfahren verwendet:

Abk.	Bedeutung
$e_A(M)$	Public-Key-Verschlüsselung der Nachricht M mit öffentlichem Schlüssel der Partei A
$d_A(C)$	Public-Key-Entschlüsselung des Chiffrats C mit dem geheimen Schlüssel der Partei A
ρ	Zufallszahlengenerator
$\sigma_A(M)$	Digitale Signatur der Partei A zur Nachricht M
$\upsilon_A(M)$	Prüfen der Digitalen Signatur der Partei A an der Nachricht M
$s_K(M)$	Symmetrische Verschlüsselung der Nachricht M mit dem Schlüssel K
$s_K(C)$	Symmetrische Entschlüsselung des Chiffrats C mit dem Schlüssel K
$+$	Operator zur Konkatination von Daten

Abbildung 2: Schreibweise für kryptographische Verfahren

Symmetrische kryptographische Verfahren wie IDEA [LaMa_91] oder DES [NBS_77] haben in ihrem bisherigen Einsatz noch keine extremen Schwächen gezeigt, obwohl der DES seit 20 Jahren bekannt ist und angewendet wird. Für die praktische Anwendung sollten zur Zeit mindestens Schlüssel der Länge 128 Bit verwendet werden. Obwohl auch der Triple-DES eine Schlüssellänge in dieser Größenordnung aufweist, ist es wegen der höheren Effizienz der Software sinnvoll IDEA zu verwenden.

Der am weitesten verbreitete Mechanismus für asymmetrischen Verschlüsselung ist RSA [RSA_78]. Nach heutigem Kenntnisstand und mit der zur Zeit vorhandenen Rechenleistung von Computern ist der RSA bei 1024 Bit Schlüssellänge sicher. Soll er zusätzlich für die digitale Signatur angewendet werden, wird ein kryptographisches Hash-Verfahren wie z.B. MD5 [Rive2_92] benötigt, das die Nachricht auf einen „Fingerabdruck" reduziert, welcher mit dem geheimen Schlüssel verschlüsselt die digitale Signatur ergibt.

Ein weiteres wichtiges kryptographisches Element des vorgeschlagenen Systems ist ein kryptographischer Zufallsgenerator, der einen Session-Key erzeugt [BlBS_86]. Dieser Session-Key wird jeweils nur während einer Übermittlung verwendet. Die Originalität der Lizenzen beruht unter anderem auf der praktischen Nicht-Vorhersagbarkeit, des durch diesen kryptographischen Zufallszahlengenerator erzeugten Session-Keys.

4.2 Parteien des Systems

Die Konzeption des elektronischen Marktes für originale, anonyme Lizenzen basiert auf einer zentralen Partei *S*, deren Aufgabe sowohl die Erstellung gültiger Lizenzen, als auch die Gewährleistung der Originalität der Lizenzen ist. Ihr muß deshalb ein besonders hohes Maß an Vertrauen entgegengebracht werden können [FoHK_95]. Der Staat wird diese Aufgabe in der Regel durch eine Umweltbehörde erfüllen.

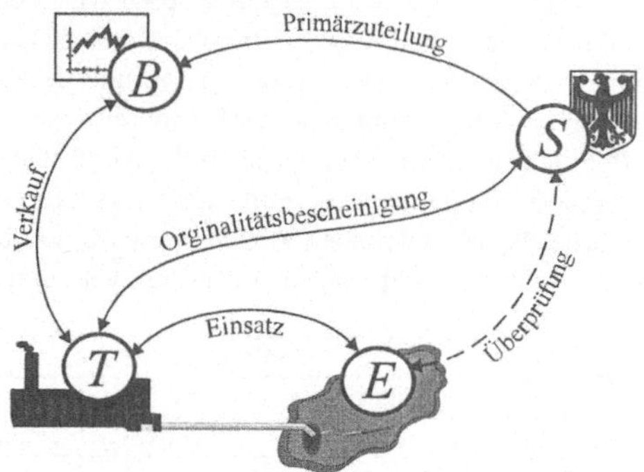

Abbildung 3: Kommunikationsmodell

Neben der zentralen *trustet third party (TTP) S*, ist bei jedem Emittenten *T* eine Komponente *E* am Emissionsort installiert, deren Hard- und Software vertraut werden muß [PPSW_95]. *E* wird nachfolgend Emissionsstelle genannt. Sie mißt die tatsächliche Emission und muß daher geeicht und verplombt sein. Verantwortlich für die Integrität dieses Gerätes ist die ausführende Umweltbehörde. Allen anderen Teilnehmern *T* und Komponenten muß nicht vertraut werden, damit von einem fairen und sicheren Handel ausgegangen werden kann.

In Abbildung 3 sind die Kommunikationsverbindungen zwischen den Parteien, die an dem System teilnehmen, eingezeichnet, wobei die gestrichelt gezeichnete Verbindung zwischen der Emissionsstelle *E* und der staatlichen Stelle *S* bedeutet, daß die Mitarbeiter der staatlichen Stelle die Emissionsstelle regelmäßig überprüfen müssen, aber keine Online-Verbindung vorhanden ist. Als Handelsplatz steht eine Börse *B* zur Verfügung, die Lizenzen auf die gleiche technische Weise an- und verkauft wie die Teilnehmer *T*.

4.3 Die Lizenz

Eine *gültige Lizenz* besteht in dieser Realisierung aus zwei Teilen. Der *Lizenz L*, die den Lizenzgegenstand wiedergibt und einem Originalitätstoken O_L. Die *Originalität* der Lizenz wird durch den Besitz des zugehörigen Originalitätstoken gewährleistet, der immer nur einem Teilnehmer gleichzeitig bekannt ist. Der Lizenzinhaber kann jederzeit den Besitz des Originalitätstoken nachweisen. Technisch ausgedrückt bedeutet *Anonymität* der Lizenz, daß die Daten der Lizenz und des Originalitätstokens keine Informationen über die Identität des momentanen oder der früheren Besitzer enthalten. Beim Kauf einer Lizenz darf die Identität des Käufers keiner dritten Partei bekannt werden.

Jede Lizenz *L* hat eine Seriennummer, die fortlaufend vergeben wird und die sie eindeutig identifizierbar macht. Zudem enthält sie die Beschreibung des Umweltziels *U* in Mengeneinheiten pro Zeiteinheit. Die absolute Emissionsmenge pro Zeiteinheit, deren Emission durch die Lizenz erlaubt wird, ist der Prozentsatz u der Gesamtmenge U. Ein Käufer kann im Streitfall seine Rechte mit der Digitalen Signatur von S einklagen. Die Lizenz ist während eines bestimmten Zeitraums (v bis b) gültig. Ist dieser Zeitraum abgelaufen, verliert die Lizenz ihre Gültigkeit.

Lizenz L
Seriennummer n
Anteilswert am Umweltziel u in %
Umweltziels U in ME/ZE
Zeitraum von v und bis b
Digitale Signatur $\sigma_S(n+u+U+v+b)$

Originalitätstoken O_L
Seriennummer n
Versionsnummer t
Digitale Signatur $\sigma_S(n+t)$

Abbildung 4: Inhalt einer Lizenz

Zu jeder Lizenz L gehört ein Originalitätstoken O_L, der die gleiche Seriennummer n wie L trägt und in dem eine Versionsnummer t vermerkt ist. Für jede ausgegebene Lizenz ist jeweils nur ein Originalitätstoken gleichzeitig gültig. Der Besitz dieses Originalitätstokens zeichnet den momentan rechtmäßigen Besitzer der Lizenz aus. Prüfbar ist die Gültigkeit des Originalitätstokens durch die Versionsnummer, die sich bei jedem Wechsel des Besitzers ändert und nur dem neuen Besitzer mitgeteilt wird. Wenn also zwei Originalitätstoken die gleiche Versionsnummer t tragen würden, wird einer als Kopie behandelt. Damit entfällt das Interesse des rechtmäßigen Besitzers den Originalitätstoken unrechtmäßig weiterzugeben, da er dabei selbst den Verlust zu tragen hätte. Die Daten n und t des gültigen Originalitätstokens werden bei der staatlichen Stelle S in einer Datenbank gespeichert. So kann sie bei jeder Transaktion die zu transferierende Lizenz auf ihre Originalität prüfen.

4.4 Primärzuteilung

Zunächst werden die Lizenzen an die Börse B gegeben, wo die Teilnehmer T die Lizenzen erwerben können. Dieses Vorgehen wurde gewählt, damit auch der erste Käufer einer Lizenz gegenüber dem Staat anonym bleibt und seine Internet-Adresse nicht preisgeben muß. Ausgestellt wird die Lizenz von der staatlichen Stelle. Sie generiert die Lizenz L zusammen mit ihrem zugehörigen Originalitätstoken O_L und sendet beides, mit dem öffentlichen Schlüssel von B verschlüsselt $e_B(L,O_L)$, an die Börse. Durch die Verschlüsselung ist sichergestellt, daß nur B die Lizenz durch Entschlüsseln erhalten kann. Bei der staatlichen Stelle werden alle Originalitätstoken gespeichert. Mittels des gespeicherten O_L kann bei der nächsten Transaktion die Originalität der Lizenz L geprüft werden. Dieses Verfahren der direkten Übertragung kann auch bei dem Primärallokationsverfahren des Grandfathering angewendet werden, indem die gültigen Lizenzen anstatt an die Börse B verschlüsselt an die Teilnehmer T versendet werden $e_T(L,O_L)$.

An der Börse werden die Lizenzen zum Kauf angeboten und jeder im Netzwerk befindliche Benutzer kann Lizenzen anonym kaufen. Die Börse bildet somit einen Marktplatz an dem sich Anbieter und Nachfrager anonym treffen können. Die Preisbildung kann technisch mittels einer Versteigerung realisiert werden. Für elektronische Märkte wurden verschiedene Modelle der Preisbildung diskutiert, die jeweils damit beginnen, daß die einzelnen Nachfrager je ein erstes Gebot abgeben [KlLa _94].

4.5 Verkauf

Während des Verkaufs einer Lizenz sind drei Parteien involviert: Die staatliche Stelle S, der Käufer und der Verkäufer. Da der Vorgang immer gleich abläuft wird er hier beispielhaft beschrieben, indem die Börse B die Rolle des Verkäufers und der Teilnehmer T die des Käufers übernimmt.

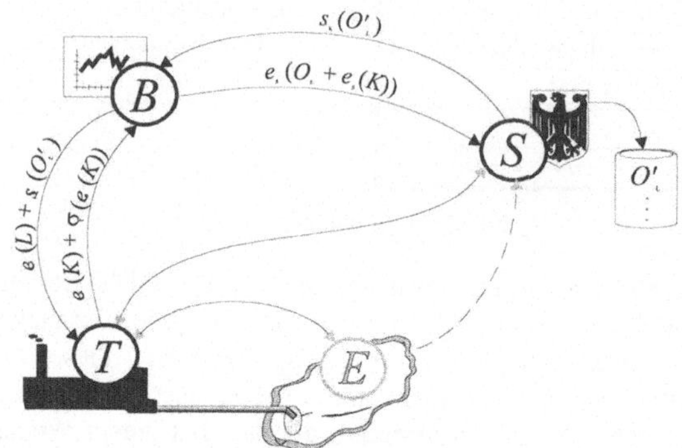

Abbildung 5: Verkauf der Lizenz

Die Verhandlung zwischen Käufer und Verkäufer werden hier nicht näher betrachtet. Es wird vorausgesetzt, daß ein Vertragsabschluß beziehungsweise eine rechtsgültige Übereinkunft des Käufers und des Verkäufers existiert und ein fairer Austausch von Lizenz und Zahlungsmittel möglich ist (z.B. Sendenachweis für die Lizenz durch die TTP). Die Überführung der Lizenz in den Besitz des Käufers T geschieht auf, die in Abbildung 5 dargestellte, Weise.

Zunächst sendet T an den Verkäufer B einen, von ihm mittels eines kryptographischen Zufallszahlengenerators erzeugten, symmetrischen Schlüssel K. Der Schlüssel ist mit einem asymmetrischen Verfahren unter Zuhilfenahme des öffentlichen Schlüssels der staatlichen Stelle S verschlüsselt, so daß nur S den Schlüssel K entschlüsseln kann. Die Nachricht muß zudem von T unterzeichnet sein, da sonst kein Schutz gegen einen *man-in-the-middle* Angriff vorhanden wäre [RiSh_84]. Um die Gefahr eines *Replay* Angriffes zu vermeiden, verschlüsselt B das verschlüsselte K zusammen mit dem Originalitätstoken nochmals und sendet $e_S(O_L + e_S(K))$ an die staatliche Stelle [NeSc_78].

Die staatliche Stelle kann jetzt die Gültigkeit des Originalitätstokens prüfen, indem sie zunächst ihre eigene digitale Signatur verifiziert und damit feststellt, ob sie selbst O_L ausgestellt hat. Danach sucht sie in der Datenbank die Seriennummer n, die in O_L angegeben ist und erhält dadurch die zugehörige Versionsnummer t. Nur falls t mit der Versionsnummer im erhaltenen Originalitätstoken übereinstimmt, ist der Originalitätstoken gültig. Wenn dies zutrifft, generiert die staatliche Stelle einen neuen Originalitätstoken O'_L, der die gleiche Seriennummer n aber eine neue mit dem Zufallszahlengenerator erzeugte Versionsnummer $t=\rho$ enthält und verschlüsselt ihn mit dem symmetrischen, geheimen und nur dem Käufer bekannten Session-Key K. Sie sendet das Ergebnis an den Verkäufer und speichert den neuen Originalitätstoken O'_L mit der neuen Versionsnummer t in ihrer Datenbank. Ausschließlich S und T kennen den Session-Key K, da er verschlüsselt zu S gesendet wurde. Aus diesem Grund wird hier keine weitere Authentikation benötigt.

Der Verkäufer sendet den von S erhaltenen verschlüsselten Originalitätstoken O'_L zusammen mit dem Lizenzinhalt L an den Käufer. Dieser kann beides entschlüsseln und jeweils die digitale Signatur der staatlichen Stelle zu den Dokumenten L und O'_L prüfen.

Nr	Partei	Inhalt	Beschreibung
1	$T \leftrightarrow B$	Preis, Lizenzdaten	Vertragsabschluß
2	$T \rightarrow B$	$\sigma_T(e_S(K)) + e_S(K)$	Verschlüsselter und digital signierter Session-Key
3	$B \rightarrow S$	Wenn $\sigma_T(e_S(K))$ gültig: $e_S(O_L + e_S(K))$	Prüfen der Signatur. Falls dies erfolgreich war: zusammen mit O_L verschlüsseln
4	S	$d_S(e_S(O_L + e_S(K))$ Prüfe O_L Wenn O_L gültig: O'_L $(n=O_L.n, v=\rho)$	Entschlüsseln. Originalitätstoken prüfen. Falls erfolgreich: Neuen Originalitätstoken mit gleicher Seriennummer und zufälliger Versionsnummer erstellen und speichern
5	$S \rightarrow B$	$s_K(O'_L)$	Symmetrisch verschlüsselter Originalitätstoken
6	$B \rightarrow T$	$e_T(L) + s_K(O'_L)$	Asymmetrisch verschlüsselter Lizenzinhalt Originalitätstoken mit Session-Key K verschlüsselt.
7	T	$L=d_T(e_T(L))$ und $v_S(L)$ $O'_L = s_K(s_K(O'_L))$ und $v_S(O'_L)$	Entschlüsseln des Lizenzinhalt und prüfen der digitalen Signatur der staatlichen Stelle. Entschlüsseln des Originalitätstokens mit dem symmetrischen Session-Key K und Prüfen der Signatur.

Abbildung 6: Ablauf des Verkaufs einer Lizenz von B an T

4.6 Verwendung einer Lizenz

Eine Emission auf Grund der Lizenz L ist dem Eigentümer T nur dann gestattet, wenn die Verwendung der Lizenz bei seiner Emissionsstelle E angemeldet wurde. Dies kann nur mittels des in Abbildung 7 beschriebenen Protokolls geschehen.

Damit der Emittent die Lizenz, während er sie für Emissionen verwendet, nicht weiter verkaufen kann, muß er den Originalitätstoken an die Emissionsstelle abgeben. Dieser Wechsel des Standortes des Originalitätstokens verläuft genauso wie beim Verkauf einer Lizenz. Die Emissionsstelle ist direkt bei dem Teilnehmer installiert und es reicht eine *point-to-point* Verbindung zu dem Rechner von T aus. Bei der Übermittlung $T \rightarrow S$ wird analog zu der Vorgehensweise beim Verkauf keine digitale Signatur verwendet. Eine Maskerade eines Angreifers ist hier deshalb nicht zu befürchten, da kein Anderer außer T nach Voraussetzung den Originalitätstoken, der von der staatlichen Stelle digital signiert ist, besitzen kann.

Nach der Anmeldung der Verwendung besitzt nur die Emissionsstelle den aktuell gültigen Originalitätstoken. Die Emissionsstelle mißt zu jedem Zeitpunkt die tatsächliche Emissionsmenge des Emittenten und speichert nur im Falle eines Verstoßes – soll heißen einer Überschreitung der maximalen Schadstoffmenge, die durch die zum Einsatz angemeldeten Lizenzen bestimmt wird – die Daten, die dann die Beweisbasis einer Beschuldigung des Emittenten bilden. Das Gerät darf deshalb nur nach einer Authentifikation von außen gelesen werden können. Sein geheimer Schlüssel darf niemandem bekannt werden und die gespeicherten Daten über die Emissionen sollten nicht an Dritte geraten bzw. nicht vom Emittenten manipuliert werden können. Es sollte eine Alarmfunktion oder ein Auditing-Mechanismus besitzen, der Manipulationsversuche dokumentiert.

Nr	Partei	Inhalt	Beschreibung
1	$E{\rightarrow}T$	$\sigma_E(e_S(K)) + e_S(K)$	Session-Key für staatliche Stelle verschlüsselt und digital signiert.
2	$T{\rightarrow}S$	Wenn $\sigma_E(e_S(K))$ gültig: $e_S(O_L + e_S(K))$	Prüfen der Signatur. Falls dies erfolgreich war: zusammen mit dem Originalitätstoken verschlüsseln u. an die staatl. Stelle senden
3	S	$d_S(e_S(O_L + e_S(K))$ Prüfe O_L Wenn O_L gültig: O'_L $(n{=}O_L.n, v{=}\rho)$	Entschlüsseln. Originalitätstoken prüfen. Falls erfolgreich: Neuen Originalitätstoken mit gleicher Seriennummer und zufälliger Versionsnummer erstellen und speichern
4	$S{\rightarrow}T$	$s_K(O'_L)$	Symmetrisch verschlüsselter Originalitätstoken
5	$T{\rightarrow}E$	$e_E(L) + s_K(O'_L)$	Asymmetrisch verschlüsselter Lizenzinhalt Originalitätstoken mit dem Session-Key K verschlüsselt.
6	E	$L{=}d_E(e_E(L))$ und $v_S(L)$ $O'_L{=}s_K(s_K(O'_L))$ und $v_S(O'_L)$	Entschlüsseln des Lizenzinhalt und prüfen der digitalen Signatur der staatlichen Stelle. Entschlüsseln des O'_L mit dem symmetrischen Session-Key K und Prüfen der Signatur.
7	E		Messung der emittierten Menge des Schadstoffes.

Abbildung 7: Ablauf der Anmeldung des Einsatzes einer Lizenz

Damit die Lizenz wieder gehandelt werden kann, muß ihre Verwendung an der Emissionsstelle explizit beendet werden. Dazu wird keine Einbindung der TTP benötigt, da der richtigen Funktionsweise von E ohnehin vertraut werden muß. Also wird auch hier angenommen, daß sie die Lizenz in ihrem Speicher löscht bzw. sie nicht weitergibt. Die verschlüsselte Übermittlung des Originalitätstoken reicht darum aus.

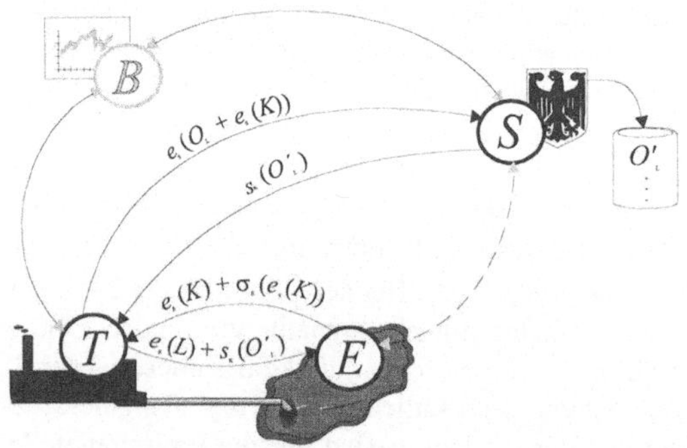

Abbildung 8: Einsatz einer Lizenz

5 Organisatorische Maßnahmen

Die zur verläßlichen Funktionsfähigkeit des Systems nötigen organisatorischen Maßnahmen lassen sich durch den Zeitpunkt ihres Auftretens in drei Phasen unterteilen. Die drei verschiedenen Phasen Herstellung, Installation und Betrieb sind in der Abbildung 9 zusammen mit den wichtigsten Maßnahmen pro Phase dargestellt.

Die Anwender müssen der Hard- und Software vertrauen können, daß sie tatsächlich nur die spezifizierte Funktionen ausführt. Während der Herstellung sollten deshalb Sicherheitsrichtlinien beachtet werden, damit die Produkte ein Sicherheitszertifikat auf Basis eines Kriterienkatalogs (z.B. Common Criteria [CCEB_96]) erhalten können. Davon wäre die Software aller Parteien beziehungsweise die Hardware der TTP und der Emissionsstelle betroffen. Da die Vertrauenswürdigkeit der TTP und der Emissionsstelle für die Sicherheit des elektronischen Marktes eine wichtige Rolle spielt, sollten sie ein hohes Niveau der Zertifizierung erreichen. Zur Zeit ist die Zertifizierung von Software jedoch noch sehr aufwendig und dadurch relativ teuer. Hersteller versuchen diese Kosten zu umgehen, indem sie Softwareprodukte digital signieren und mit ihrem „guten Namen" für die Verläßlichkeit des Systems gerade stehen. Die digitale Signatur eines Softwarepakets hat unabhängig davon, daß sie kein vollständiger Ersatz für Zertifizierung sein kann, den Nutzen, daß die Produkte vor nachträglichem Einspielen von Viren und Trojanischen Pferden geschützt sind.

- Zertifizierung aller Komponenten nach Common Criteria:
    - Software für TTP, Emissionsstelle, Börse, Zertifizierungsstelle und Teilnehmer
    - Hardware für TTP und Emissionsstelle

- Hard- und Software der TTP installieren und Schlüssel personalisieren
- Emissionsstelle installieren, eichen und Personalisierung der Schlüssel
- Abnahme und Überprüfung

- Zugangs- und Zugriffskontrolle zur TTP
- Zugriffskontrolle und mechanischer Einbruchschutz an der Emissionsstelle
- Autorisierter Zugriff durch Personal der Behörde z.B. Lesen der Protokolldaten

Abbildung 9: Phasen organisatorischer Maßnahmen

Während der Installationsphase wird der Computer, der die Rolle der TTP spielt, samt der Software in Betrieb genommen. Hierbei sind zwei mögliche Ansatzpunkte für einen Angriff vorhanden: Manipulation der Software und Manipulation bei der Personalisierung der Schlüssel. Diesen Angriffen kann dadurch begegnet werden, indem die Vertrauenswürdigkeit des Personals geprüft wird. Das Gerät an der Emissionsstelle muß installiert und anschließend geeicht werden. Auch das Gerät an der Emissionsstelle benötigt in dieser Realisierung ein public-key Schlüsselpaar. Es muß sichergestellt werden, daß keiner außer dem Gerät selbst den secret-key erhält aber jeder den zugehörigen public-key authentisch erhalten kann. Eine mögliche Lösung wäre, das Schlüsselpaar bei der TTP zu generieren und dort ein Zertifikat für den public-key auszustellen. Die Übergabe des geheimen Schlüssels an das Gerät, die sogenannte Personalisierung könnte mittels des Vier-Augen-Prinzips geschehen. Zwei Personen müssen jeder ein Paßwort eingeben, bevor der Schlüssel zum Beispiel von einer Chipkarte auf das Gerät übertragen wird. Die TTP und das Gerät an der Emissionsstelle dürfen erst nach einer Überprüfung in Betrieb genommen werden. Die Verläßlichkeit dieser Systeme bestimmt den Beweiswert im Falle eines Gerichtsverfahrens. Ein Beispiel für diesen Zusammenhang sind die Urteile zu ec-Automaten [OLGH_96].
Während des Betriebs müssen die TTP und das Gerät an der Emissionsstelle gegen nicht autorisierten Zugriff geschützt werden. Bei beiden Geräten hat das zunächst eine physische Dimension. Der zentrale vertrauenswürdige Rechner TTP der staatlichen Stelle muß zum

Beispiel durch bauliche Maßnahmen vor physischen Manipulationen an dem Computer selbst geschützt werden. Der physische Zugang zu diesem Rechner sollte auf einen kleinen Personenkreis reduziert und der Zugriff über das Netzwerk nur nach den spezifizerten Protokollregeln möglich sein.

Die Geräte an den Emissionsstellen müssen gegen mechanische Beschädigungen geschützt werden und von der staatlichen Stelle verplombt sein, so daß eine Manipulation im nachhinein festgestellt und geahndet werden kann. Die Protokolldaten des Gerätes müssen regelmäßig daraufhin überprüft werden, ob illegale Emissionen gemessen wurden. Autorisierter Zugriff durch Personal der Behörde zum Lesen der Protokolldaten muß deshalb ermöglicht werden. Durch den Einsatz von Chipkarten zur Authentisierung gegenüber dem Computer und eines Schlüssels für das Gehäuse kann dies sicher geschehen. Falls das Gerät an der Emissionsstelle einen Verstoß anzeigt, wird ein geeignetes Verfahren zur Beweissicherung auf einem mobilen Computer benötigt.

Die Rahmenbedingung für die Rechtsgültigkeit der digitalen Signatur ist das Informations- und Kommunikationsdienste Gesetz, das vermutlich in Kürze in Deutschland verabschiedet werden wird. Dort ist die Funktionsweise und Struktur der Zertifizierungsstellen beschrieben [Bund_97]. Durch die Zertifikate dieser Zertifizierungsstellen wird die Zugehörigkeit eines öffentlichen Schlüssels zu dem entsprechenden Teilnehmer rechtsverbindlich bestätigt. Für die breite Akzeptanz des Systems ist es von Vorteil, öffentliche Zertifizierungsstellen zu verwenden. Eine spezielle Lösung mit eigenen Zertifizierungsstellen würde die Einstiegsschwelle in das System erhöhen. Jeder Teilnehmer hat deshalb selbst für die Zertifizierung seines öffentlichen Schlüssels zu sorgen oder kann seinen bisherigen Schlüssel weiter verwenden. Die Generierung der Schlüssel der Emissionsstellen und ihre Zertifizierung wird von der staatlichen Stelle S durchgeführt.

6 Bewertung und Ausblick

Es wurde in dieser Arbeit die Abwicklungsphase eines sicheren elektronischen Marktes für digitale, originale und anonyme Lizenzen im Umweltzertifikatmodell spezifiziert. Das Konzept der originalen und anonymen Lizenzen ist prinzipiell auf jedes digitale Produkt anwendbar, für das die Sicherheitsdienste Originalität und Anonymität erbracht werden sollen. In der Spezifikation wird den Bedrohungen der Sicherheit insbesondere Bedrohungen der Integrität und der Vertraulichkeit im offenen Netz (z.B. Internet) begegnet, indem kryptographische Mechanismen angewendet werden. Beim Verkauf der Lizenz wird die Internet-Adresse des Verkäufers – im Gegensatz zu der des Käufers – bei der staatlichen Stelle bekannt. Eine stärkere Form der Anonymität, bei welcher auch die Absenderadresse des Verkäfers nicht bekannt wird, kann theoretisch durch die Verwendung von MixedNets erreicht werden [PfWa_86]. Offen ist zur Zeit noch, ob bzw. wie eine Evaluation der Sicherheit von Hard- und Software, die auch in dem vorgestellten Beispiel eine wichtige Rolle spielt, mit wirtschaftlich vertretbaren Kosten realisierbar ist.

Die im Umweltzertifikathandel entstehenden Transaktionskosten sind durch die Anwendung des Koordinationsmechanismus elektronischer Markt wahrscheinlich niedriger als bei einer herkömmlichen Implementierung [Wiga_95]. Diese Senkung der Transaktionskosten könnte einen entscheidenden Beitrag für die Kosteneffizienz des Modells leisten. Dazu trägt auch die in der vorgeschlagenen Implementierung mögliche freie Stückelung der Lizenzen bei.

Schwer vorhersehbar sind jedoch die Kosten des Gerätes an der Emissionsstelle. Für manche Schadstoffe ist preisgünstige Meßtechnologie verfügbar für andere nicht. Falls preisgünstige Meßtechnik verfügbar ist, können die Kosten des Staates bei der Kontrolle mit dem vorgestellten System gering gehalten werden, da die Geräte an den Emissionsstellen die Daten über Verstöße protokollieren. Dadurch kann mit geringen Personalkosten eine lükkenlose Kontrolle durchgesetzt werden.

Aus der Sichtweise der Phasen des Kaufes wurde hier ein Protokoll beschrieben, das die Abwicklungsphase unterstützt. Ein weiterer Aspekt der Abwicklungsphase ist der Zahlungsverkehr. Auch heute sind die Kosten, die durch Finanztransaktionen entstehen, noch relativ hoch. Gerade bei kleinen Stückelungen der Lizenzen und kleinen Preisen sollte die Bezahlung mittels digitalen Geldes direkt im Internet abgewickelt werden können [PeRö_97].

Zur Zeit findet eine Diskussion über die zukünftige Rolle der Intermediäre, die im elektronischen Markt Cybermediäre genannt werden, statt [SaBS_95]. Ein solcher Intermediär ist in dem spezifizierten elektronischen Markt für Umweltzertifikate die Börse. Noch zu klären ist in diesem Zusammenhang, ob eine Börse der effizienteste Weg bezüglich der Koordination in der Informationsphase ist, oder ob ein System aus verschiedenen elektronischen Agenten hierfür besser geeignet ist.

7 Literatur

BaVB_95　Backhaus, K.; Voeth, M.; Bendix, K. B.:*Die Akzeptanz von Multimedia Diensten.* Arbeitspapier Nr.19/1995; Herausgeber: Backhaus, K.; Universität Münster; 1995.

BlBS_86　Blum, L.; Blum, M.; Schub, M.: *A Simple Unpredictable Pseudo-Random Number Generator.* SIAM J. Computing, 15/2, 1986, S. 364-383.

Bund_97　Beschluß des Bundeskabinetts: *IuKDG Informations- und Kommunikationsdienste Gesetz.* DuD Datenschutz und Datensicherheit 21; Verlag Vieweg; Wiesbaden; 1/1997; http://www.iid.de/rahmen/iukdg_3.html (Zugriff: Jan. 1997)

CCEB_96　*Common Criteria for Information Technology Security Evaluation.* Version 1.0; 1996. http://www.tno.nl/instit/fel/refs/cc.html#download (Zugriff: Nov. 1996)

Chau_85　Chaum, D.: *Security without Identification: Card Computers to make big Broher Obsolete.* Communications of the ACM, vol. 28, no. 10; Oktober 1985. http://www.digicash.com/publish/bigbro.html (Zugriff: Jan. 1997)

Croc_66　Crocker, T. D.: *The Structuring of Atmospheric Pollution Controll Systems.* in: Wolozin, H. (Hrsg.): The Economics of Air Pollution, New York 1966, S. 61-86.

Dale_68　Dales, J. H.: *Land, Water and Ownership.* In: Canadian Journal of Economics 1, Vol. 1.

Endr_94　Endres, A.: *Umweltökonomie - Eine Einführung.* Darmstadt; 1994; S.6 ff.

EwGa_94　Ewringmann, D.; Gawel, E.: *Kompensationen im Imissionschutzrecht: Erfahrungen im Kannenbäcker Land.* Baden-Baden; 1994; S. 38.

FoHK_95　Fox, Dirk; Hoster, Patrick; Kraaibeek, Peter: *Grundüberlegungen zu Trust Centern.* In: Horster, P. (Hrsg.): Trust Center. Proceedings der Arbeitskonferenz Trust Center 95, Verlag Vieweg, Braunschweig 1995, S. 1-10.

GaHa_95　Gawel, E.; Hansmeyer, K.-H.: *Umweltauflagen.* In: Junkernheinrich, M.; Klemmer, P.; Wagner, G. R. (Hrsg.) Handbuch zur Umweltökonomie; Berlin; 1995.

Hans_96　Hansen, H. R.: *Klare Sicht am Info-Highway - Geschäfte via Internet & Co.* Ovac-Verlag; 1996.

Huck_93　Huckestein, B.: *Umweltlizenzen - Anwendungsbedingungen einer ökonomisch effizienten Umweltpolitik durch Mengensteuerung.* In: ZfU 1/1993, S. 1-29.

ISO_95　International Organisation for Standardization (ISO): *Information processing systems - Guidelines for the Use and Management of Trusted Third Parties - Part 2: Technical Aspects.* International Standard ISO/IEC Draft 14516-2; Genf; 1995.

ISO1_89   International Organisation for Standardization (ISO): *Open Systems Interconnection - Basic Reference Model - Part 2: Security Architecture.* International Standard ISO 7498-2 (E); Genf; 1989.

KlLa _94   Klein, S.; Langenohl, T.: Coordination Mechanisms and Systems Architectures in Electronic Market Systems. In: Schertler, W; Schmid, B.; Tjoa, A M.; Werthner, H. (eds.) Information and Communications Technologies in Tourism, Wien, New York: Springer-Verlag, 1994, S. 262-270.

LaMa_91   Lai, X.; Massey, J.: *A Proposal for a New Block Encryption Standard.* Advances in Cryptology - Eurocrypt '90; Springer Verlag; Berlin; 1991.

MaYB_87   Malone, Thomas W.; Yates, Joanne; Benjamin, Robert I.: *Electronic Markets and electronic Hierarchies.* CACM vol. 30 No. 6; 1987; S.484-497.

MaGa_92   van Mark, M.; Gawel, E.; Ewringmann, D.: *Kompensationslösungen im Gewässerschutz, Umwelt und Ökonomie.* Bd. 6; Heidelberg; 1992.

NBS_77   National Bureau of Standards (NBS): *Data Encryption Standard (DES).* Federal Information Processing Standards Publication (FIPS-PUB) 46-1, US Department of Commerce, 1/1977.

NeSc_78   Needham, Roger M.; Schroeder, Michael D.: *Using Encryption for Authentication in Large Networks of Computers.* Communications of the ACM; Vol. 21; No. 12; 1978; S. 993-999.

OLGH_96   Oberlandesgericht Hamm: *AZ 31 U 72/96.* Quelle: ARD-Ratgeber Technik; 1996

PeRö_97   Pernul, Günther; Röhm, Alexander W.: *Neuer Markt - Neues Geld?* Wirtschaftsinformatik 4/97; Verlag Vieweg; Wiesbaden; 1997.

PfWa_86   Pfitzmann, A.; Waidner, M.: *Networks without user observability -- design options.* Eurocrypt '85 LNCS 219; Springer-Verlag; Berlin; 1986; S. 245-253.

PfWP_90   Pfitzmann, B.; Waidner, M.; Pfitzmann, A.: *Rechtssicherheit trotz Anonymität in offenen digitalen Systemen (Teil 1).* Datenschutz und Datensicherheit 5/90; Verlag Vieweg; Wiesbaden; S. 243-253.

PPSW_95   Pfitzmann, A.; Pfitzmann, B.; Schunter, M.; Waidner, M.: *Vertrauenswürdiger Entwurf portabler Benutzerendgeräte und Sicherheitsmodule.* In: Brüggemann, H.-H.; Gerhardt, W. (Hrsg.): Proceedings der Fachtagung Verläßliche IT-Systeme VIS '95. DuD-Fachberichte, Verlag Vieweg, Braunschweig 1995, S. 329-350.

Rive2_92   Rivest, Ronald L.: The MD5 Message-Digest Algorithm. Request for Comments (RFC) 1321, Network Working Group, 4/1992, S. 1-21.

RSA_78   Rivest, Ronald L.; Shamir, Adi; Adleman, Leonard: A Method for obtaining Digital Signatures and Public Key Cryptosystems. Communications of the ACM, Bd. 21, Nr. 2, 1978, S. 120-126.

RiSh_84   Rivest, Ronald L.: *How to expose an eavesdropper.* Communications of the ACM; vol. 27 no. 4; 1984; S. 393-395.

SaBS_95   Sarkar, M.; Butler, B.; Steinfeld, C.: *Intermediaries an Cbermediaries: A Continuing Role for Mediating Players in the Electronic Marketplace.* JCMC; Vol. 1; No. 3; 1995; http://jcmc.huji.ac.il/vol1/issue3 (Zugriff: Nov. 1996)

Schm_93   Schmid, Beat: Elektronische Märkte. Wirtschaftsinformatik 35 6/93; Vieweg-Verlag, Wiesbaden; 1996.

Weil_95   Weiler, R.M.: *Money, transactions, and trade on the Internet.* MBA thesis, Imperial College London; 1995; http://graph.ms.ic.ac.uk/results

Wiga_95   Wigand, R. T.: *Electronic Commerce and Reduced Transaction Costs.* In: Alt, R.; Zbornik, S. (Hrsg.) EM - Electronic Markets; No. 16/17; Vol. 5; Competence Centre Electronic Markets St. Gallen; 1995.

Firewallsysteme und neue Sicherheitstechnologien im Internet

Kai Martius
Institut für Medizinische Informatik und Biometrie an der TU Dresden

Mit der zunehmenden Verbreitung des Internet, insbesondere im kommerziellen Bereich, spielen Sicherheitsfragen bei der Anbindung an dieses weltweite Netz eine herausragende Rolle. Dieser Entwicklung Rechnung tragend, werden derzeit eine Vielzahl neuer Sicherheitstechnologien für dieses Netz diskutiert und entwickelt. Wie diese dazu beitragen können, die Sicherheit bisheriger Firewallsysteme entscheidend zu verbessern, soll in diesem Beitrag gezeigt werden.

1 Einleitung

Die Internet-Protokolle wurden zunächst nicht unter Sicherheitsaspekten bezüglich Vertraulichkeit, Authentizität und Integrität der zu übertragenden Daten entwickelt, sondern das Augenmerk galt lediglich besonders hoher Ausfallsicherheit.

Um den Anschluß der lokalen Netze an das Internet abzusichern, sind Technologien entwickelt worden, die es ermöglichen, bestimmte Einschränkungen bezogen auf den Ursprung, das Ziel, die Art und die Richtung des Datenverkehrs durch Festlegung von Regeln vorzunehmen. Diese Regeln werden von **Firewall-Systemen** umgesetzt.

Diese Regeln können auf der Netzwerkschicht (Paketfilter), der Verbindungsschicht (Circuit Level Gateways) oder auf Anwendungsebene (Application Level Gateways, ALG) wirken. Im Allgemeinen kommen Kombinationsformen dieser Filtertechniken in einem Firewallsystem zum Einsatz. Eine umfangreiche Beschreibung ist in [ChesBello 94] zu finden.

2 Bisher genutzte Identifikations- und Authentikationsformen

2.1 Einfache Mechanismen

Um innerhalb eines Paketfilters oder Circuit / Application Level Gateways entscheiden zu können, welche Regel angewandt werden soll, ist eine möglichst sichere Identifizierung des Zugreifenden notwendig. Am einfachsten (und auch bei weitem am häufigsten) wird entsprechend der Source- und Destination-IP-Adresse und der zugehörigen Ports entschieden. Diese Lösung hat folgende Nachteile:

- IP-Adressen können gefälscht werden. An Systemen ohne abgestufte Zugriffsrechte (DOS, Windows) kann jeder die eigene IP-Adresse durch eine andere (bzgl. der Rechte auf dem Firewall höher priorisierte) IP-Adresse des eigenen Subnetzes ersetzen. Dies funktioniert problemlos, wenn das System, dem die IP-Adresse ‚gestohlen‘ wurde, nicht in Betrieb ist oder netzwerkseitig überlastet wird.

Schwieriger, jedoch keinesfalls unmöglich wird dieser Angriff, wenn von außen ein Zugriff mit einer gefälschten IP-Adresse erfolgt, da dann die zwischenliegenden Router manipuliert werden müssen, sofern diese Router Pakete mit Source Routing (Vorgabe der Route im IP-Paket) nicht weiterleiten.

- s.g. *wellknown Ports* sind Ports, auf denen **normalerweise** fest definierte Dienste bereitgestellt werden, wie z.B. Port 23 für Telnet. Diese Zuordnung ist jedoch keinesfalls zwingend. Noch schwieriger gestaltet sich die Filterung bzgl. des Client-Ports. Diese sind bei vielen Diensten nicht vorher bestimmbar, da ein Client den eigenen Port meist zufällig auswählt. In den Filterregeln müssen dann ganze Bereiche freigeschalten werden.

- Bei der Verwendung dynamisch vergebener IP-Adressen versagt dieser Mechanismus fast vollständig, da eine bestimme Maschine irgendeine IP-Adresse aus einem vorgegebenen Bereich erhalten kann. Zum einen müßte dann eine enge Kopplung zwischen dem Host, der die IP-Adressen verteilt und den Filterregeln des Firewall-Systems bestehen. Zum anderen müßten Router und ALG dynamische Filterregeln unterstützen, was heute bei fast keinem kommerziellen Produkt verfügbar ist.

Statt IP-Adressen können auch symbolische Namen eingesetzt werden, wobei dann durch einen DNS-Server die Zuordnung zu IP-Adressen erfolgt.

- Die Verwendung von DNS-Namen ist noch unsicherer, da der Angreifer nur den DNS-Server manipulieren muß, um ein falsches Mapping zwischen Namen und IP-Adresse zu erreichen. Er kann dann den Zugriff auf einen bestimmten Namen gezielt auf seine Maschine lenken, ohne irgendwelche Routen manipulieren zu müssen.

Die genannten Filterregeln können im übrigen nur Hosts und Ports berücksichtigen, nicht jedoch einzelne Nutzer der entsprechenden Maschinen. Das wirkt sich besonders in einem Multiuser-System wie UNIX oder Windows NT aus, da nicht festgestellt werden kann, wer gerade den Dienst nutzen möchte (abgesehen von den priorisierten Ports unter 1024, die - eigentlich - nur der Administrator öffnen kann; aber wer weiß, wer der Administrator eines Systems ist?). Noch weniger kann jedoch auf die Authentizität des Nutzers an einer DOS / Windows (3.x / 95) oder Apple-Maschine vertraut werden, da diese nicht über eine Nutzerauthentifizierung verfügen.

Diesem Problem steht gegenüber, daß durch die Sicherheits-Policy und die daraus abgeleiteten Filterregeln in erster Linie einzelne Nutzer reglementiert werden sollen und nicht irgendwelche Computer.

2.2   Zusätzliche Authentisierungsmechanismen

Aus diesen Gründen stellen auch klassische Firewall-Systeme teilweise Mechanismen bereit, die eine ‚richtige' Nutzerauthentisierung gegenüber dem Firewall-System ermöglichen. Diese erfordern jedoch auch einen höheren administrativen Aufwand. Mehrere verschiedene Systeme sind üblich:

2.2.1 Kerberos

Kerberos ist ein auf dem Client-Server-Prinzip beruhendes Protokoll zur Authentisierung von Nutzern. Den Kern bildet ein vertrauenswürdiger Kerberos-Server (Trusted Third Party), der eine Datenbank mit den Clients (Nutzern oder Dienste) und deren **geheimen** Schlüsseln unterhält (für einen ‚menschlichen‘ Nutzer wäre der geheime Schlüssel ein - verschlüsseltes - Paßwort). Auf den Protokollablauf soll an dieser Stelle nicht im Detail eingegangen werden – dazu sei auf [Schneier 1996] verwiesen. Die grundlegenden Sicherheitsprobleme von Kerberos sollen jedoch zusammengefaßt dargestellt werden:

- Während der Gültigkeitsdauer eines s.g. Tickets (Zulassung eines Clients für einen bestimmten Service), die i.A. 8 Stunden beträgt (!), können diese durch einen Angreifer wieder eingespielt werden (Reply-Attack), soweit durch den Server keine Vorkehrungen durch Zwischenspeicherung bereits ‚gebrauchter‘ Tickets erfolgt.

- Die Sicherheit von Kerberos hängt durch die Verwendung von Zeitstempeln stark von einer synchronisierten, sicheren Zeit zwischen Clients, Servern und dem Kerberos-Server ab, da ansonsten auch abgelaufene Tickets wiedereingespielt werden könnten. Durch das nicht besonders sichere Network Time Protocol, über das Systemzeiten synchronisiert werden, ist es einem Angreifer relativ einfach möglich, die Zeit zu fälschen.

- Einige Tickets werden ohne Zufallsanteil erstellt. Das macht Password-Guessing-Attacks möglich, wenn genügend solcher Tickets durch einen Angreifer gesammelt wurden.

- Der Kerberos-Server und seine Software sind ein Single Point of Failure, ein herausragender Angriffspunkt für Angreifer. Insbesondere die Abspeicherung geheimer Schlüssel macht sie verwundbar.

2.2.2 Token-Systeme

Token-Systeme (z.B. SecureID) werden oft im Zusammenhang mit Challenge-Response-Protokollen eingesetzt, die zum Vergleich eines gemeinsamen Geheimnisses dienen, ohne dieses selbst übertragen zu müssen. Dazu generiert der ‚Herausforderer‘ einen Zufallswert, den er mit einem bestimmten Algorithmus verarbeitet, in den auch das Geheimnis eingeht. Der Zufallswert wird dem Kommunikationspartner geschickt, der selbst das Geheimnis und den Algorithmus kennt. Er kann nun den Funktionswert des Zufallswertes berechnen und zurückschicken. Der ‚Herausforderer‘ vergleicht diesen mit dem selbst berechneten, wobei der Vergleich mit (extrem hoher Wahrscheinlichkeit) nur positiv ausfällt, wenn der andere wirklich das gleiche Geheimnis kennt.

Der Algorithmus muß im Übrigen so gestaltet sein, daß seine Inverse praktisch unmöglich zu berechnen ist, da sonst das Geheimnis von einem Angreifer wieder regeneriert werden kann. Hash-Algorithmen bieten sich hier geradezu an.

Als Token wird nun ein Gerät bezeichnet, das einen geheimen Schlüssel (also das Geheimnis) und einen Algorithmus beinhaltet. Evtl. kann das Token selbst noch durch eine PIN gesichert sein. Dieses Token entlastet nun den Anwender zum einen von der Aufgabe, sich selbst das Geheimnis (bzw. bei PIN-Sicherung sich nur ein vergleichsweise kurzes Geheimnis) zu merken. Zum anderen braucht er nicht mehr selbst einen Algorithmus durchzuführen, der das Geheimnis transformiert.

Das Problem: Auch bei diesem Verfahren muß ein Authentisierungsserver geheime Daten speichern, was ihn gegen Angriffe besonders verwundbar macht. Desweiteren sind diese Token-Systeme recht unhandlich, wenn keine direkte Kommunikation zwischen Token und Computer möglich ist. Dann nämlich muß der Anwender die erhaltene Zufallszahl (Challenge) selbst in das Token und den berechneten Wert wieder am Computer eingeben. Ist das Token nicht durch eine PIN geschützt, kann es bei Verlust oder Diebstahl einfach mißbraucht werden.

2.2.3 Allgemeine Ansätze

Die IETF unterhält in ihrer Security Area eine Arbeitsgruppe ‚Authenticated Firewall Traversal‘. Dort wurde bereits vor einiger Zeit ein Protokoll SOCKS als RFC standardisiert, das einen generischen Authentisierungsdienst bereitstellen soll [SOCKS]. Der eigentliche Mechanismus wird dabei nicht spezifiziert, es wird nur ein Rahmen bereitgestellt. Alle SOCKS nutzenden Dienste werden über diesen SOCKS-Host vermittelt. Er stellt damit einen (je nach implementierten Protokoll) sicheren Proxy dar, der meist neben anderen "normalen" Proxies laufen wird. Zudem findet die Authentisierung nur auf Applikationsebene statt.

2.2.4 Vorgestellte Mechanismen in einem an OSI angelehnten Schichtenmodell

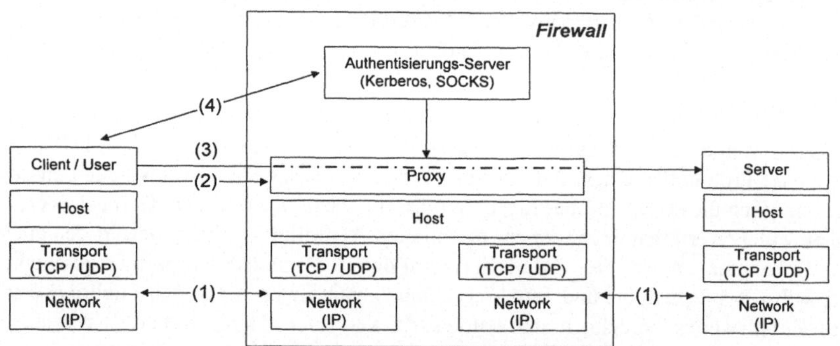

Abbildung 1 - herkömmliches Firewallsystem

(1) Netzwerkseitige ‚Zugangskontrolle‘ an Hand von IP-Adresse / Port
keine kryptographische Sicherheit, allg. Probleme s.o.
(2) Userauthentikation gegenüber Proxy-Service
selten verwendet, meist nur Username / Password-Authentisierung, einseitige Authentisierung des Clients / Users gegenüber dem Proxy

(3) Für Dienste mit eigenen Schutzmechanismen anschließende
    Authentisierungsprozedur (Telnet: Username / Password, meist im Klartext)
(4) Falls vorhanden, vorherige Authentisierung gegenüber einem
    Authentisierungsserver (bei bisher angebotenen Verfahren: Probleme s.o.)

Anmerkungen:

- Die IP-Adreß- / Portüberprüfung ist genaugenommen nicht ausschließlich im
  Network-Layer anzusiedeln, da Ports erst im Transport-Layer bereitgestellt werden.
  Eine echte Verbindungskontrolle findet aber ebenfalls nicht statt, da jedes Paket
  separat geprüft wird.

- Eine Host-Authentisierung ist nicht unbedingt mit der Überprüfung der IP-Adresse
  (selbst wenn diese kryptographisch sicher gelingen würde) gleichzusetzen, da:
  – ein Host mehrere IP-Adressen besitzen kann
  – ein Host zu verschiedenen Zeiten verschiedene IP-Adressen besitzen
    kann (dynamische Adreßvergabe)
  – die Identität eines Hosts im Allgemeinen über einen Namen, nicht über
    eine Adresse festgelegt sein sollte
  ⇒ **Secure DNS**

2.2.5 Fazit

Die bisher in Firewallsystemen eingesetzten Authentisierungs-Schemata basieren
überwiegend auf symmetrischen Kryptosystemen, mit deren Nachteilen:

- Geheimnisse müssen an einer zentralen Stelle gespeichert werden.

- diese Systeme sind relativ unflexibel bzgl. Erweiterung um zusätzliche Nutzer

- eine verteilte Lösung ist kaum realisierbar, da geheime Daten zwischen mehreren
  Authentisierungsservern ausgetauscht werden müßten.

- angebotene Token-Systeme sind meist proprietär und / oder unpraktisch.

Es existiert keine Lösung, die Sicherheit auf allen beteiligten Ebenen bietet.

Wünschenswert wäre dagegen ein System, das

- sichere und flexible Authentisierung auf allen beteiligten Ebenen

- ein möglichst einheitliches, einfaches und schichtenübergreifendes Key-
  Management

- verschiedene Sicherheitsstufen bezüglich der Anforderungen an Vertraulichkeit,
  Authentizität und Integrität

bieten würde.

3   Neue Entwicklungen im Internetumfeld

In Anbetracht der zunehmenden Forderungen nach Sicherheit, Vertraulichkeit,
Authentizität und Integrität von Kommunikationsbeziehungen im Internet sind in den

letzten Jahren sozusagen auf allen Ebenen (bzgl. der Kommunikation) Vorschläge ausgearbeitet und teilweise als Standards (RFC's) verabschiedet worden.

3.1   IPv6 / IPSec

Für das ständig wachsende Internet mußte die bisherige IP-Version 4 erweitert werden. Im Hinblick auf die neuen Anforderungen bzgl. Sicherheit wurde nunmehr bereits im Designprozeß den Sicherheitsaspekten ein breiter Raum eingeräumt, die in den RFC's 1825-27 [IPSec] [AH] [ESP] ihren Niederschlag finden. IPSec bietet dann:

- Authentication Header (AH)
  Nicht veränderliche IP-Header-Daten (z.B. IP-Source- und Destination-Adresse) sowie der Payload (User-Daten) werden durch eine Signatur geschützt, so daß deren Integrität und Authentizität geprüft werden kann, je nach verwendeten Verfahren nur vom rechtmäßigen Kommunikationspartner oder auch von Dritten. Optional kann ein AH-Tunnelmode eingesetzt werden, der jedoch einen zusätzlichen IP-Header verlangt. Dieser Modus ist für s.g. Security-Gateways eingeführt worden.

- Encapsulated Security Payload (ESP)
  Daten übergeordneter Protokolle (TCP, UDP) können verschlüsselt im IP-Datagramm abgelegt werden. Optional kann auch das gesamte IP-Paket incl. Header verschlüsselt und als Payload in ein neues Paket verpackt werden (ESP-Tunnelmode). Neuerdings bietet auch ESP optional die Möglichkeit, Authentisierungsinformationen bereitzustellen.

Die Verwaltung von Security-relevanten Informationen erfolgt in s.g *Security Associations* (SA), ein Pointer auf eine solche SA stellt der *Security Parameter Index* (SPI) dar, der in AH und ESP mitgeführt wird. SA's werden über ein auf Anwendungsschicht bereitzustellendes Key-Management-Protokoll etabliert.

Folgendes Problem ist momentan zu erkennen:

Aus Performancegründen wird eine Implementierung der Authentisierungsmechanismen als Public-Key-Verfahren auf absehbare Zeit nicht erfolgen. Geforderter Mechanismus ist ein ‚Keyed Hash'-Verfahren, HMAC-MD5. Damit ist eine Verifizierung des AH nur auf dem Zielsystem möglich (Bei auftretender Fragmentierung eines Paketes wäre eine Verifizierung auf zwischenliegenden Systemen nicht einmal mit Public-Key-Verfahren realisierbar, außer es wird Path-MTU-Discovery angewandt).

3.2   ISAKMP

Das Key-Management kann z.B. von dem Protokoll-Vorschlag ISAKMP (Internet Security Association and Key Management Protocol) bereitgestellt werden [ISAKMP]. Es stellt Protokoll-Mechanismen auf Anwendungsebene zum Etablieren von Sicherheitsdiensten als eine Art "Quality of Service" zwischen verschiedenen Instanzen im Netz bereit. ISAKMP ist ein sehr allgemeiner und flexibler Protokollansatz, über den verschiedenste Dienste in unterschiedlichen Kommunikationsebenen Sicherheitsparameter ($\Rightarrow$ Security Associations) etablieren können. SA's beinhalten z.B. Authentisierungs- und Key-Exchange-Algorithmen, Host- und

User-Zertifikate. Sie stellen eine Art Vertrag über die Sicherheitseigenschaften einer Verbindung dar. SA's sind unidirektional auf den Empfänger bezogen.

Ein erster konkreter Entwurf für ein Protokoll unter Verwendung der ISAKMP-Spezifikation ist ISAKMP/Oakley [Oakley]. Mit ISAKMP/Oakley werden folgenden Eigenschaften unterstützt:

- SA-Etablierung (Authentisierung, Schlüsselaustausch) für VPN- / Remote-User-Szenarien
- Proxy-Negotiation
- Entsprechend ISAKMP 2 Phasen (hier Modes genannt)
- Perfect Forward Secrecy

3.3 SSL (Secure Socket Layer)

SSL bietet einen recht allgemeinen Ansatz für die Absicherung von Kommunikationsbeziehungen im Internet auf **Transportebene**. Die Entwicklung von SSL wird insbesondere von der Firma NETSCAPE COMMUNICATIONS vorangetrieben. Es liegt mittlerweile als Internet-Draft in der Version 3 vor [SSL]. Da SSL den Kern der Standardisierungsbemühungen einer eigenen Arbeitsgruppe ‚Transport Layer Security' der IETF-Security-Area bildet, ist damit zu rechnen, daß es schnell zu einem anerkannten Standard für gesicherte Verbindungen im Internet avancieren wird.

SSL kann im OSI-Schichtenmodell über der Transportschicht eingeordnet werden. Bezogen auf TCP/IP liegt es dann zwischen TCP und der Anwendungsschicht.

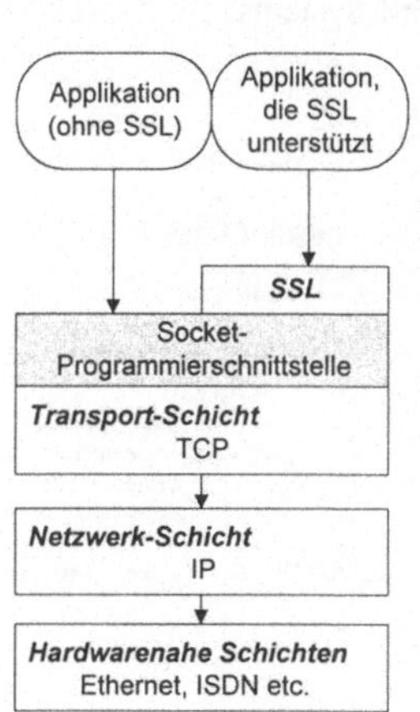

Die meisten gängigen Anwendungen (sowohl client- als auch serverseitig) nutzen die *Socket-Schnittstelle* als (programmiertechnischen) Zugang zu einem TCP/IP-Netzwerk. Auch SSL bedient sich dieser Schnittstelle u.a. zur Übertragung der verschlüsselten Daten. Aus der Sicht dieser Socket-Schnittstelle zeigt sich damit SSL wie jede andere Anwendung, z.B. Netscape Navigator.

Sowohl der Key-Exchange- als auch der Authentisierungsmechanismus von SSL basiert auf Public-Key-Verfahren (RSA oder DH). Die Bereitstellung der öffentlichen Schlüssel erfolgt durch X.509-Zertifikate, setzt also geeignete Infrastrukturen zur Ausgabe und Prüfung von Zertifikaten voraus.

3.4 Secure DNS

Das Domain Name System bietet die Möglichkeit der Verwendung symbolischer Namen für Hosts, die technisch gesehen nur über eine IP-Adresse erreichbar sind. Ein

Domain Name Server kann über eine Datenbank einen solchen Namen in eine IP-Adresse umsetzen. Da weder die Datensätze selbst noch die Übertragung gesichert ist, kann man bisher diesen Antworten nur bedingt trauen. Einige Angriffe der jüngsten Zeit basierten auf diesem Umstand (DNS-Spoofing).

Secure DNS ist ein Ansatz, in dem die Datensätze durch Signaturen gesichert werden können. Damit ist an beliebigen Stellen im Netz die Integrität und Authentizität nachprüfbar. Da es sich um öffentliche Daten handelt, sind keine Optionen für Zugangsschutz und Vertraulichkeit vorgesehen. Secure DNS bietet damit die Möglichkeit, Zertifikate für beliebige Entitäten online bereitzustellen.

4   Nutzung der neuen Sicherheitstechnologien in Firewall-Systemen

Die beschriebenen Verfahren sind nur eine Auswahl aus der Vielzahl der in Entwicklung befindlichen Sicherheitsprotokolle. Keines der Protokolle ist jedoch speziell für den Einsatz in Firewall-Systemen entwickelt worden. Vielmehr sind sie für beliebige Client-Server-Kommunikationsbeziehungen vorgesehen. Es stellt sich nun die Frage, wie diese neuen Technologien zur Verbesserung der Funktionalität und Sicherheit von Firewall-Systemen eingesetzt werden können. Denn auch in Zukunft werden diese "Tore zum Internet" Bestand haben, um durch geeignete Sicherheits-Policies das eigene Netz zu schützen.

4.1   Überblick

Abbildung 2 zeigt eine mögliche Einordnung der wichtigsten Sicherheitsprotokolle in ein Kommunikationssystem unter Einbeziehung eines Firewall-Systems.

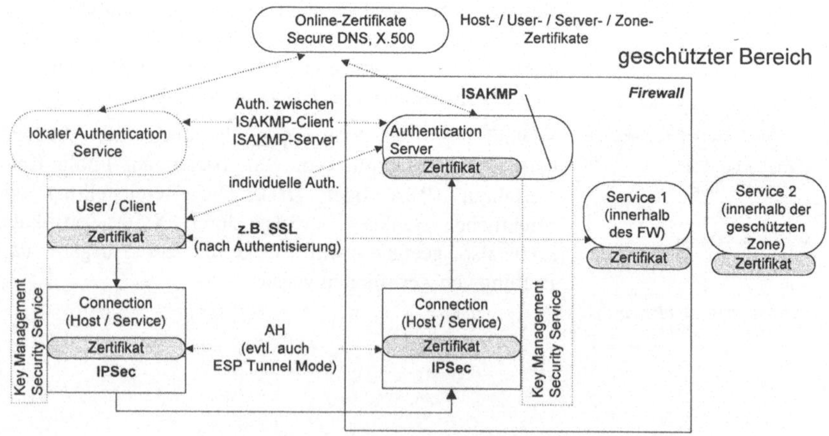

Abbildung 2 - neue Sicherheitsprotokolle in einem Kommunikationssystem

Wie sich zeigt, bieten diese neuen Technologien die Möglichkeit, ein Firewall-System flexibler und sicherer zu gestalten. Mit den in Abschnitt Fazit genannten Forderungen bildet ein Authentisierungs- und Key-Management-Service den zentralen Baustein eines solchen Systems. Dieser Service muß Zugriffe kryptographisch sicher authentisieren und je nach geforderter Sicherheitsstufe in den verschiedenen Kommunikationsschichten eine paketweise und/oder dienste- / nutzerorientierte Authentisierung (evtl. auch Verschlüsselung) etablieren.

4.2 Kommunikationsablauf

Eine Kommunikation über ein solches Firewallsystem könnte dann folgendermaßen ablaufen:

1. Schritt (Verbindungsaufbauphase)

- Client baut zu einem Authentisierungsservice im Firewall eine zunächst unge- schützte Verbindung auf
- Gegenseitige Authentisierung zwischen Client und Authentisierungsserver
- Austausch von Parametern zur Sicherung der Verbindungsaufbauprozedur (Sicherung von Integrität und Authentizität, optional Vertraulichkeit)
- Eigentlicher Service Request des Clients (welcher Service auf welchem Zielsystem soll genutzt werden)
- Prüfung der Zugriffsrechte durch den Authentisierungsserver

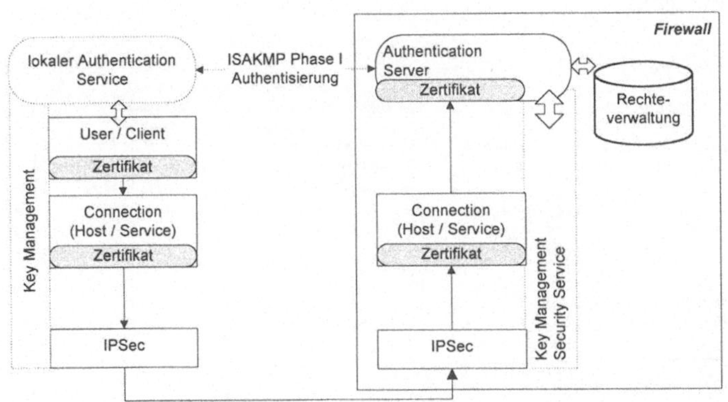

Abbildung 3 - Verbindungsaufbauphase

Anmerkungen:

Zunächst ist nur der Authentisierungsservice des Firewall-Systems uneingeschränkt erreichbar. Das macht ihn natürlich auch für Angriffe, insbesondere Denial-of-Service-

Angriffe ‚interessant'. Grundsätzlich muß dieser Dienst aber uneingeschränkt erreichbar sein, da zum Zeitpunkt des Verbindungsaufbaues noch keinerlei Informationen über die Gegenstelle vorliegen. Auch in der ISAKMP-Spezifikation wird dieses Problem behandelt. Lösbar scheint dieses Problem nur, indem möglichst sehr wenig Systemressourcen für diese Phase verbraucht werden. Dies kann über s.g. Anti-Clogging-Token (ACT) geschehen [Karn].

Ein ISAKMP-Daemon erfüllt bereits einen Großteil dieser Anforderungen, indem er in einer ersten Phase (Phase I Exchange) eine Authentisierung durchführt und optional einen gesicherten Kanal bereitstellt.

Grundsätzlich erscheint eine zertifikatbasierte Authentikation mittels asymmetrischer Kryptoverfahren die flexibelste und sicherste Methode darzustellen, weshalb in den Abbildungen die entsprechenden Entitäten bereits mit einem Zertifikat versehen wurden. Zertifikate können Online über Directory Services oder Secure DNS bereitgestellt werden.

2. *Schritt (Service Request,* Abbildung 4, Abbildung 5*)*
- Verbindungswunsch des Clients zu Service 1, optional geschützt durch ISAKMP-SA (SA zwischen ISAKMP-Client und –Server)
- Mitteilung der Sicherheitsanforderungen an den Client, allgemein: geforderte SA-Parameter des gewünschten Service
- Etablierung der entsprechenden SA's zwischen Firewall und Client (z.B. AH, ESP-Tunneling, SSL)
- Mitteilung der Security Parameter an die Firewall-Schichten, sozusagen "Freischalten" einer Verbindung mit den geforderten Parametern zum entsprechenden Service
- Client etabliert geforderte Sicherheitsparameter zum Endsystem (geschützt durch SA mit Firewall)

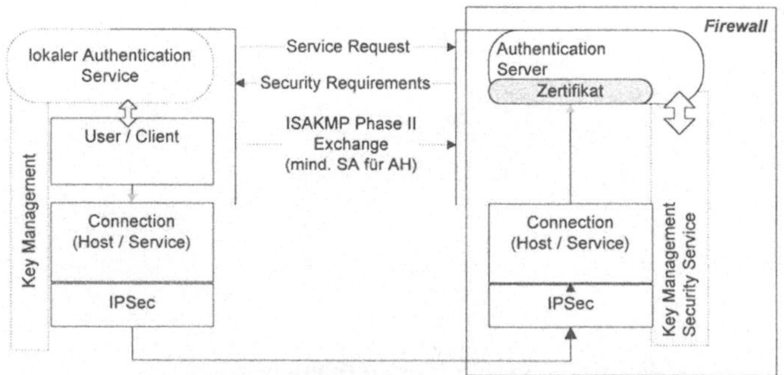

Abbildung 4 - Authentisierung gegenüber Firewall-System

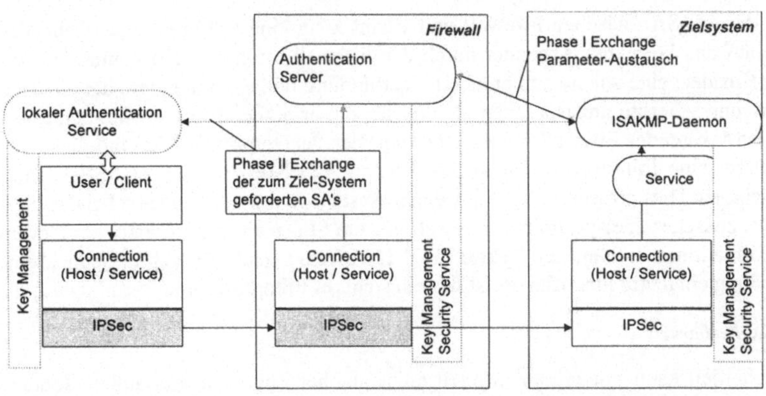

Abbildung 5 - (geschützte) Authentisierung zum Zielsystem

Anmerkungen:

Die Authentisierung gegenüber dem Firewall-System und die Etablierung einer SA zu diesem dient dazu, daß der Firewall nachfolgenden Verkehr zum Zielsystem überprüfen und filtern kann. Je nach Sicherheitspolicy ist ohne eine solche Authentisierung keine Kommunikation über das Firewall-System möglich.

3. Schritt (endgültige Kommunikationsbeziehung, Abbildung 6*):*

- Aufbau der Verbindung zum eigentlichen Service mit den etablierten (geforderten) Sicherheitsparametern, z.B. IP-Security ("Host-Keying" und Verbindungsverschlüsselung mit SSL)

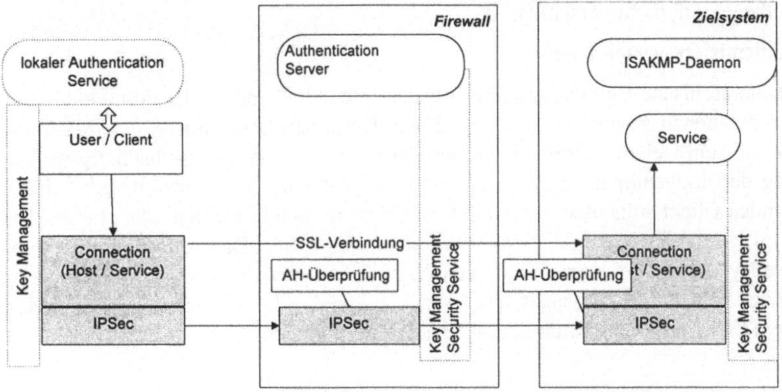

Abbildung 6 - geschützte Kommunikation zum Zielsystem

Anmerkungen:

Durch die AH-SA zwischen Firewall und Client kann eine Prüfung der eingehenden Pakete auf den korrekten Absender durch den Firewall erfolgen. Nun könnte ein Angreifer (Insider) eine solche authentisierte Verbindung nutzen, um andere als die Zielsysteme und –dienste anzusprechen als die, für die er authentisiert wurde, denn über die AH-SA kann der Firewall nur die *Herkunft* der Pakete prüfen. Eine mögliche Lösung wäre, eine Filterung an Hand des Vektors {Ziel-IP-Adresse, Ziel-SPI-Value} vorzunehmen. Dies erfordert allerdings einen Austausch des / der entsprechenden SPI-Werte(s) zwischen Zielsystem und Firewall. Da ein SPI-Wert keine sicherheitsrelevanten Informationen enthält, der Vektor {Ziel-IP-Adresse, Ziel-SPI-Value} jedoch eine Verbindung eindeutig identifiziert, ist das ein sicheres Filterkriterium.

4.4 Skalierbarkeit

Die Sicherheit kann mit diesem Modell nicht nur bei Zugriffen von außen, sondern auch bei Nutzung des Firewalls von internen Anwendern bereitgestellt werden. Je nach Aufbau des Systems ist dann für interne Nutzer ein eigener Authentisierungsserver vorzusehen. Die Nutzung anonymer externer Dienste wird dabei i.A. über Proxies realisiert werden, da dabei der Datenfluß besser kontrolliert werden kann (Virenscanner). Bei Zugriff auf geschützte Dienste (z.B. Server des eigenen Unternehmens an einem anderen Standort, Stichwort VPN) kann über die Authentisierungsserver der dann beidseitig vorhandenen Firewallsysteme ein geschützter Kanal aufgebaut werden, die IPSec-Tunnelmodi bieten sich dafür an.

Oftmals ist die Bereitstellung anonymer Dienste gewünscht, die keine (eingehende) Authentisierung erfordern. Eine serverseitige Authentisierung ist meist jedoch trotzdem notwendig. Das vorgeschlagene Firewall-System unterstützt verschiedene Sicherheitsstufen. Vereinfacht dargestellt, können diese folgendermaßen unterteilt werden:

- nicht authentisiert, nicht verschlüsselt

- authentisiert, nicht verschlüsselt

- authentisiert, verschlüsselt

Auch hierarchische Firewallsysteme werden unterstützt, indem der Authentisierungsserver des ersten Firewalls ‚weiß‘, daß zur Kommunikation mit einem bestimmten Zielservice ein weiterer Firewall zu überqueren ist. Er kann dann selbständig die Etablierung der notwendigen SA's durchführen. Will der nächste Firewall selbst den zugreifenden Client authentisieren, muß dem Client mitgeteilt werden, daß er eine eigene SA mit dem zweiten Firewall zu etablieren hat. Die Information, welche Firewallsysteme "auf dem Weg" zwischen Client und Zielservice liegen, könnte entweder über ICMP-Messages als "Rückkopplung" oder entsprechende Records im Secure DNS vor der Kommunikation bereitgestellt werden.

4.5 Übergangslösungen und Prototypimplementierungen

4.5.1 Projekthintergrund

Ausgangspunkt für die Auseinandersetzung mit der Problematik ist das im Rahmen des FuE-Programmes der Berkom geförderte Telemedizinprojekt REGKOM. In diesem Projekt sollen medizinische Daten über ein WWW-Interface erfaßt und übertragen werden. Dabei sind die besonderen Sicherheitsbelange sowohl bei der Übertragung als auch beim Anschuß eines Netzes mit derartigen Daten zu berücksichtigen. Es erfolgte zunächst eine Evaluierung der bisherigen und aktuellen Entwicklungen im Internet, um eine möglichst allgemeine Aussage über die Möglichkeiten und Probleme verfügbarer Systeme zu erhalten. Unter Einbeziehung der sich abzeichnenden Standards soll ein System entworfen werden, das die Probleme und Schwächen herkömmlicher Ansätze umgeht.

4.5.2 Probleme

Momentan ist zwar eine Vielzahl von Protokollentwürfen in Bearbeitung, es ist jedoch nicht davon auszugehen, daß diese kurzfristig als Implementierung auf verschiedenen Plattformen bereitstehen. Zudem müßte dann die "Verzahnung" zwischen den verschiedenen Schichten noch realisiert werden. Es kann jedoch bereits mit vorhandenen Protokollimplementierungen und Plattformen ein vergleichbares Maß an Flexibilität und Sicherheit geschaffen werden.

Ein weiteres Problem sind die momentan noch restriktiven Exportbestimmungen der USA bzgl. der Verwendung kryptographischer Verfahren. Eine Vielzahl von Protokollimplementierungen oder darauf aufsetzender Applikationen bieten nur unzureichende Schlüssellängen (z.B. SSL-Implementierung in WWW-Browsern). Trotzdem könnten Standard-Client- und -Server-Applikationen einsetzbar sein, wenn generische Dienste auf den entsprechenden Ebenen zwischengeschaltet werden, die einen sicheren Kanal zwischen unsicheren Client und Server schalten.

4.5.3 Lösungsansätze

SSL zur Realisierung von Authentication Service und Connection Security

Eine in Quellcode vorliegende, außerhalb Nordamerikas implementierte Version von SSL ist SSLeay des Australiers Eric Young. Diese bildet die Grundlage zum Aufbau des Authentisierungsdienstes sowie der Sicherung der Verbindungsschicht.

SSL verwendet X.509-Zertifikate, deren Aufbau standardisiert ist. Damit können Zertifikate beliebiger Zertifizierungsstellen unterstützt werden, die sich an diesem Standard orientieren. Je nach zertifizierter Entität kann SSL zur Host-, Server- oder User-Authentisierung verwendet werden. Eine Online-Bereitstellung bzw. –Abfrage der Zertifikate wird derzeit jedoch noch nicht unterstützt.

Da eine ISAKMP-Implementierung momentan nicht bereitsteht, muß der Austausch der Informationen zwischen Client und Authentisierungsserver zunächst proprietär erfolgen.

Um vorhandene Client- und Serversoftware, wie Netscape Navigator und Commerce Server, zu nutzen, übernimmt zunächst ein generischer SSL-Proxy-Dienst die Sicherung dieser Dienste. Dieser kann zudem die Funktion eines lokalen Authentisierungsservers übernehmen.

Die Bereitstellung der Zertifikate sowie die Operationen mit dem geheimen Schlüssel erfolgen über eine Chipkarte bzw. ein Sicherheitsmodul der TELESEC. Problematisch ist die derzeit durch die verfügbaren Kryptoprozessoren begrenzte Schlüssellänge von 512 Bit für RSA.

IPv4 vs. IPv6 / IPSec

Auch die breite Einführung von IPv6 und den entsprechenden Security-Diensten auf dieser Ebene wird noch einige Zeit dauern. Damit ist zunächt eine paketweise Prüfung der Daten auf dem Firewall-System nicht möglich. Übergangsweise können jedoch durch den Einsatz dynamischer IP-Filtermechanismen die Risiken auf dieser Ebene minimiert werden. Anstatt jedes einzelne Paket zu prüfen (wie bei IPSec), wird dann ein IP-Filter für genau die geforderte Verbindung freigeschalten. Source- und Destination-Adresse / Port müssen dann explizit in der Authentisierungsphase auf Anwendungsebene gesichert übermittelt werden, da die IP-Pakete selbst (d.h. die Headerdaten) nicht gesichert sind.

Für Linux ist bereits eine IPSec-Implementierung für IPv4 verfügbar. Diese soll ebenfalls evaluiert werden. Linux bietet zudem eine einfache Möglichkeit des Einsatzes dynamischer Filtermechanismen durch eine Programmierschnittstelle im Kernel, wodurch eine einfache Prototypimplementierung möglich ist. Aus diesem Grunde wird Linux als Hostsystem für den Prototyp-Firewall eingesetzt.

Zunächst ohne Authentisierungsserver wurde bereits erfolgreich eine SSL-gesicherte Verbindung zwischen einem Standard-Web-Browser und –Server hergestellt (Clientsystem: Windows, Serversystem: Linux). Die Proxy-Applikationen werden dabei bewußt klein und flexibel gehalten, um für die verschiedensten Dienste einfach angepaßt werden zu können.

Abbildung 7 - Protoimplementierung

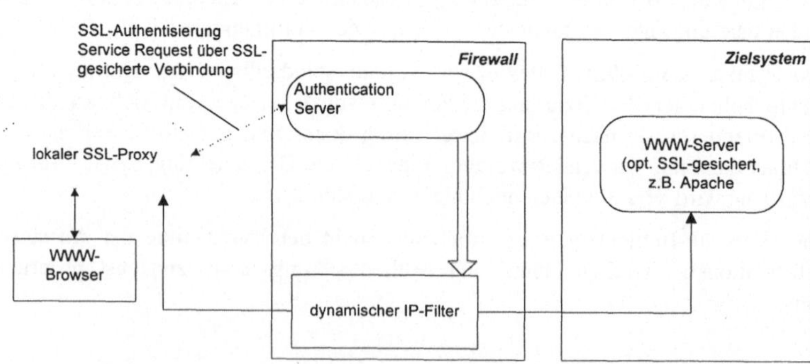

5 Literatur

[AH]   Atkinson, R., "IP Authentication Header", RFC 1826, NRL, August
                        1995.

[ChesBello 94]   Cheswick, W., Bellowin, S., Firewalls and Internet Security, Addison
                        Wesley, Reading, Massachusetts, 1994

[DNSSEC]   Eastlake III, D., Kaufmann, C.W., "Domain Name System Security
                        Extensions", Work in Progess, draft-ietf-dnssec-secext-10.txt, August
                        1996

[Ellermann 96]   Uwe Ellermann, "'IPv6 and Firewalls'", Proceedings of ECURICOM ---
                        14th International Congress on Computer   and Communications
                        Security Protection, Paris, Juni 1996.
                        (http://www.fwl.dfn.de/eng/~ue/fw/ipv6fw/)

[ESP]   Atkinson, R., "IP Encapsulating Security Payload", RFC 1827, NRL,
                        August 1995.

[IPSEC]   Atkinson, R., "Security Architecture for IP", RFC 1825, NRL, August
                        1995.

[ISAKMP]   Maughan, Schertler, Schneider, Turner, "Internet Security Association
                        and Key Management Protokoll", Work in Progress, draft-ietf-ipsec-
                        isakmp-07.txt, Februar 1997

[Karn]   Karn, P. and B. Simpson, "The Photuris Session Key Management
                        Protocol", Internet-Draft: draft-simpson-photuris-11.txt, Work in
                        Progress, Juni, 1996.

[Oakley]   Harkins, D., Carrel, D., "The resolution of ISAKMP with Oakley", Work
                        in Progress, draft-ietf-ipsec-isakmp-oakley-03.txt, Februar 1997

[Schneier 96]   Bruce Schneier, Applied Cryptography, Second Edition,
                        John Wiley & Sons, New York, NY, 1996

[SOCKS]   Leech M. et al., "SOCKS Protocol Version 5", RFC 1928, März 1996

[SSL]   Freier et al., "The SSL-Protokoll Version 3", Work in Progress, draft-
                        ietf-tls-ssl-version3-00.txt, November 1996

Formale Spezifikation von Sicherheitspolitiken für Paketfilter

Rainer Falk
Lehrstuhl für Datenverarbeitung
TU München
falk@e-technik.tu-muenchen.de

Zusammenfassung

Wird ein Rechnernetz mit dem Internet verbunden, kann jeder Rechner des lokalen Netwerks vom Internet aus erreicht werden. Dadurch kann auch jeder Rechner vom Internet aus angegriffen werden und muß deshalb gegen diese Angriffe geschützt werden. In der Praxis ist dies meist nicht möglich, da der Administrationsaufwand dafür sehr groß ist. Hinzu kommt, daß einige, im lokalen Netz nützliche Dienste wie z. B. NFS über unzureichende Sicherheitsmechanismen verfügen. Deshalb werden für die Kopplung zwischen dem Intranet und dem Internet Firewalls eingesetzt, die nur bestimmte Dienste – evtl. angereichert um zusätzliche Sicherheitsmechanismen wie z. B. leistungsfähige Authentifizierungsverfahren – zulassen.

Eine wichtige Rolle beim Aufbau von Firewalls spielen Paketfilter. Deren Konfiguration erweist sich als komplex und fehleranfällig. In dieser Arbeit stellen wir ein formales Modell vor, um die von einem Paketfilter durchzusetzende Sicherheitspolitik auf einer abstrakten, benutzerfreundlichen Ebene zu spezifizieren, und zeigen, wie diese automatisch in herkömmliche Filterregeln übersetzt werden kann.

1   Einleitung

Paketfilter spielen eine wichtige Rolle beim Aufbau von Firewalls. Ein Paketfilter kann als Firewall eingesetzt werden oder er wird als Komponente zum Aufbau einer Proxy-basierten Firewall verwendet. Da Paketfilter für die Anwendungen transparent sind und einen hohen Durchsatz erzielen, bieten sie sich ebenfalls an, um firmeninterne Firewalls aufzubauen, die zwischen einzelnen Unternehmensbereichen nur bestimmte Dienste zulassen.

Paketfilter lassen nur Pakete mit bestimmten Eigenschaften passieren. Die Filterentscheidung basiert auf folgenden Informationen[1]:

- Source- und Destination-IP-Adresse

- Protokoll (TCP, UDP, ...)

- bei TCP und UDP Source- und Destination-Ports

- bei TCP Unterscheidung zwischen Verbindungsaufbaupaketen (Ack=0) und Paketen, die zu einer bestehenden Verbindung gehören (Ack=1)

---

[1]Es gibt auch Paketfilter, bei denen nicht alle hier angegebe Kriterien überprüft werden können; diese sind jedoch für den Einsatz als Firewall schlecht geeignet [Fuh96] und werden hier nicht näher betrachtet.

Filterregeln bestehen aus einem Bedingungsteil und einem Aktionsteil. Die Liste der Regeln wird der Reihe nach abgearbeitet, bis eine Regel gefunden wird, deren Bedingungsteil zutrifft, und die entsprechende Aktion wird ausgeführt, d. h. das Paket wird weitergeleitet (accept) oder verworfen (deny).

Filterregeln sind jedoch schwierig zu konfigurieren, modifizieren, warten und zu testen [Cha92]. Dies ist zum einen durch die einfache Syntax bedingt, die auf leichte Parsebarkeit durch den Router ausgelegt ist, und zum anderen durch Abhängigkeiten innerhalb eines Filtersatzes. D. h. daß einzelne Blöcke nicht unabhängig voneinander zusammengesetzt werden können, sondern daß Abhängigkeiten mit anderen Regeln zu beachten sind. Bei umfangreicheren Regelsätzen (einige 10 bis 100 Regeln) sind diese Abhängigkeiten nur schwer überschaubar, wodurch sich leicht Fehler einschleichen. Besonders schwierig erweist sich die Änderung einer vorhandenen Konfiguration, da Abhängigkeiten, die bei der Entwicklung des Filtersatzes berücksichtigt wurden, dabei leicht übersehen werden.

Systemverwalter denken nicht in einzelnen Paketen, sondern in Diensten und Verbindungen [Cha92]. Die Umsetzung von Sicherheitsanforderungen in Filterregeln erfordert ein tiefes Verständnis der zugrundeliegenden Protokolle. Was eigentlich spezifiziert wird, ist, welche Dienste von welchen Rechnern aus zu welchen Zielen genutzt werden dürfen.

In Abschnitt 2 stellen wir ein (formales) Modell vor, mit dem Sicherheitspolitiken für Paketfilter auf Dienstebene und für jeden Dienst unabhängig von anderen Diensten spezifiziert werden können, und zeigen in Abschnitt 3, wie diese formale Spezifikation automatisiert in herkömmliche Filterregeln übersetzt werden kann. Abschnitt 4 geht auf mögliche Erweiterungen herkömmlicher Paketfilter ein, um auch schwer zu filternde Dienste einer Paketfilterung zugänglich zu machen. In Abschnitt 5 werden weitere Möglichkeiten vorgestellt, durch Werkzeuge den Entwurf von Filterregeln zu unterstützen. Abschnitt 6 nennt verwandte Arbeiten, und Abschnitt 7 endet mit einer Zusammenfassung.

2   Formales Modell für Sicherheitspolitiken

Eine Sicherheitspolitik, wie sie hier betrachtet wird, legt fest, welche Dienste von wo aus zu welchem Ziel genutzt werden dürfen. Weitere Aspekte wie z. B. ein Notfallplan oder die Überwachung des laufenden Betriebs (Logging) werden hier nicht betrachtet. Benutzerspezifische Sicherheitspolitiken lassen sich mit Paketfiltern nicht umsetzen; deshalb werden im folgenden auch keine Benutzer betrachtet.

Ein Beispiel für eine derartige Sicherheitspolitik ist in Abbildung 1 dargestellt, die die erlaubten Kommunikationsbeziehungen zeigt:

- Vom Intranet aus darf WWW genutzt werden (HTTP).

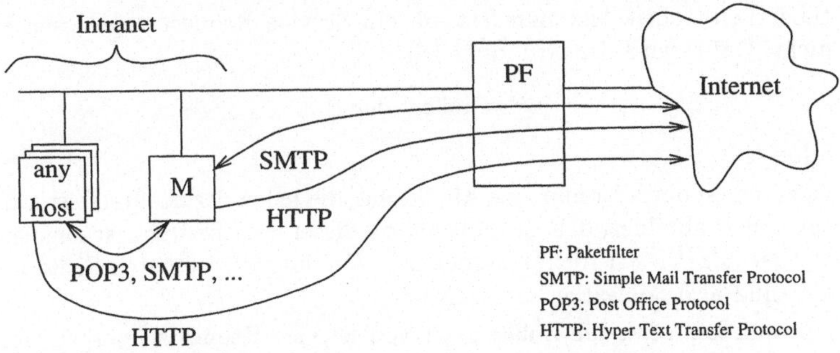

Abbildung 1: Beispiel für eine Sicherheitspolitik

- Es gibt einen zentralen Mail-Server M, über den alle Mails zwischen Intranet und Internet ausgetauscht werden (SMTP). Dieser Mail-Server befindet sich im Intranet.

2.1 Formales Modell

Um eine Sicherheitspolitik auf dieser Abstraktionsebene formal zu beschreiben, werden folgende Grundbereiche eingeführt:

S Die Menge aller Dienste (service)
C_i Die Menge aller Komponenten der Innenseite (Intranet)
C_o Die Menge aller Komponenten der Außenseite (Internet)

Eine Komponente kann sich nur im Internet oder im Intranet befinden, aber nicht in beiden, d. h.

$$C_i \cap C_o = \emptyset$$

Um die Semantik einer Sicherheitspolitik zu definieren, wird der Begriff *Service Request* eingeführt. Mit ihm wird die Anforderung eines Dienstes modelliert, der je nach Sicherheitspolitik entsprochen wird oder nicht. Damit wird sowohl von einzelnen Paketen als auch von Verbindungsphasen (z. B. Verbindungsaufbau) abstrahiert; die Nutzung eines Dienstes wird als eine Einheit betrachtet.

Ein solcher Service Request besteht aus folgenden Komponenten:

- Der Dienst, der angefordert wird,

- die Komponente, von der aus dieser Dienst angefordert wird,

- die Zielkomponente, an die die Anforderung gerichtet ist.

Eine Sicherheitspolitik legt nun fest, ob ein Service Request zugelassen wird oder nicht. Dafür wird der Grundbereich

$$P = \{\text{allow}, \text{deny}\}$$

eingeführt.

Eine Sicherheitspolitik ist nun eine Abbildung, die jeden Service Request auf ein Element von P abbildet, d. h. sie gibt an, ob dieser Service Request zugelassen (allow) oder abgewiesen (deny) werden soll. Die hier betrachteten Sicherheitspolitiken sind also boolesche Funktionen[2].

Eine Firewall kann nur die Zulässigkeit von Service Requests überprüfen, die die Grenze zwischen Intranet und Internet überschreiten. Es werden die Mengen SR_i der eingehenden Service Requests und SR_o der ausgehenden Service Requests definiert:

$$SR_i \;=\; S \times C_o \times C_i$$
$$SR_o \;=\; S \times C_i \times C_o$$

Es werden zwei Politiken definiert, eine für eingehende Service Requests und eine für ausgehende Service Requests[3]:

$$P_i \;:\; SR_i \to P$$
$$P_o \;:\; SR_o \to P$$

Ein Paketfilter, so wie er hier betrachtet wird, ist damit vollständig definiert durch ein 5-Tupel (S, C_i, C_o, P_i, P_o).

2.2 Beispiel

Diese Modellierung soll anhand des am Anfang von Abschnitt 2 angegebenen Beispiels verdeutlicht werden.

Die Grundbereiche sind:

$$S \;=\; \{\text{HTTP}, \text{SMTP}, \perp_S\}$$
$$C_i \;=\; \{\text{M}, \perp_i\}$$
$$C_o \;=\; \{\perp_o\}$$

Die Dienste HTTP (WWW) und SMTP (E-Mail) werden explizit erwähnt, als Repräsentant für alle übrigen Dienste steht $\perp_S$. Komponenten des Intranets sind der Mail-Server M und alle übrigen Hosts des Intranets, die durch das Element $\perp_i$ repräsentiert werden. Ebenso steht $\perp_o$ für alle Hosts im Internet.

Die Sicherheitspolitik sieht dann aus wie in Tabelle 1 angegeben.

---

[2]Hier werden Logging oder Alarmmeldungen nicht betrachtet.

[3]Diese Politiken sind nicht mit Inbound- und Outbound-Filtern der Paketfilter zu verwechseln. Wird ein Dienst in *eine* Richtung zugelassen, erfordert das i. a. einen Paketfluß in *beiden* Richtungen.

$$
\begin{array}{llll}
P_i(\mathsf{HTTP}, \perp_o, \mathsf{M}) &=& \mathsf{deny} & \qquad P_o(\mathsf{HTTP}, \mathsf{M}, \perp_o) = \mathsf{allow} \\
P_i(\mathsf{HTTP}, \perp_o, \perp_i) &=& \mathsf{deny} & \qquad P_o(\mathsf{HTTP}, \perp_i, \perp_o) = \mathsf{allow} \\
P_i(\mathsf{SMTP}, \perp_o, \mathsf{M}) &=& \mathsf{allow} & \qquad P_o(\mathsf{SMTP}, \mathsf{M}, \perp_o) = \mathsf{allow} \\
P_i(\mathsf{SMTP}, \perp_o, \perp_i) &=& \mathsf{deny} & \qquad P_o(\mathsf{SMTP}, \perp_i, \perp_o) = \mathsf{deny} \\
P_i(\perp_S, \perp_o, \mathsf{M}) &=& \mathsf{deny} & \qquad P_o(\perp_S, \mathsf{M}, \perp_o) = \mathsf{deny} \\
P_i(\perp_S, \perp_o, \perp_i) &=& \mathsf{deny} & \qquad P_o(\perp_S, \perp_i, \perp_o) = \mathsf{deny}
\end{array}
$$

Tabelle 1: Beispiel einer Sicherheitspolitik

Beispielsweise ist eine HTTP-Verbindung von einem beliebigen Host des Intranets zu einem beliebigen Host des Internets erlaubt, da

$$
\forall c_i \in C_i, c_o \in C_o : P_o(\mathsf{HTTP}, c_i, c_o) = \mathsf{allow}
$$

gilt.

2.3 Benutzerschnittstelle

In Abschnitt 2.1 wurde der zugrundeliegende Formalismus vorgestellt. Um effizient einsetzbar zu sein, muß aber eine benutzerfreundliche Schnittstelle angeboten werden, damit ein Systemverwalter sich nicht mit dem zugrundeliegenden Formalismus beschäftigen muß.

Eine graphische Darstellung wie in Abbildung 1 ist übersichtlich und leicht verständlich. Eine anwenderfreundliche, textuelle Spezifikation ist aber ebenfalls möglich. Das Beispiel aus Abschnitt 2 ist in Abbildung 2 in einer leicht faßbaren Notation dargestellt. Weiterhin sind Gruppierungskonstrukte für Dienste

```
DOMAIN Intranet = M, other_i ENDDOMAIN
DOMAIN Internet = other_o ENDDOMAIN
SERVICES WWW, SMTP, other_S ENDSERVICES

DEFAULT DENY
ALLOW
   WWW FROM Intranet TO Internet
   SMTP FROM M TO Internet
   SMTP FROM Internet TO M
ENDALLOW
```

Abbildung 2: Benutzerfreundliche Notation (Beispiel)

und Komponenten vorhanden. Ein Domänenbezeichner (hier `Intranet` und `Internet`) kann wie in diesem Beispiel als Bezeichner für die Gruppe aller Komponenten dieser Domäne verwendet werden. Für diese Filtersprache muß jedoch die Semantik formal definiert werden, d. h. es muß jede gültige Filterspezifikation *formal* in ein äquivalentes 5-Tupel (S, C_i, C_o, P_i, P_o) überführt werden können. Dadurch erhält man eine Methode, die sowohl leicht anwendbar ist, die aber trotzdem die Exaktheit und Unmißverständlichkeit von formalen Methoden besitzt.

2.4 Erweiterungen

Bei dem vorgestellten Formalismus wurden bisher neben dem Zulassen oder Abweisen von Service Requests keine weiteren Aktionen betrachtet. Es kann jedoch eine Menge A aller möglichen zusätzlichen Aktionen (z. B. Logging, SNMP-Trap, E-Mail, ...) eingeführt werden. Die Sicherheitspolitik ist dann eine Abbildung, die jeden Service Request auf ein Element von P und zusätzlich auf eine Menge von Aktionen abbildet, die bei dem Auftreten dieses Service Requests ausgeführt werden sollen. Die Semantik dieser Aktionen ist im formalen Modell allerdings undefiniert.

Der Formalismus kann offensichtlich auf mehrere Domänen erweitert werden, d. h. es gibt nicht nur das Intranet und das Internet, sondern eine beliebige Anzahl von Domänen, zwischen denen eine Sicherheitspolitik definiert werden kann. Dadurch lassen sich dann Multi-Port-Router, Firewall-Architekturen mit mehreren Paketfiltern sowie ganze Firmennetze, bei denen Paketfilter zur Trennung einzelner Unternehmensbereiche verwendet werden, in *einem* Modell beschreiben, aus dem die Filterlisten für mehrere Paketfilter generiert werden können.

3 Übersetzung in Filterregeln

Die Verwendung des in Abschnitt 2 vorgestellten Formalismus erlaubt es, eine Sicherheitspolitik formal auf Dienstebene zu spezifizieren. Das alleine ist bereits nützlich, da so präzise festgelegt ist, was erlaubt sein soll und was nicht.

Der praktische Nutzen der abstrakten Spezifikation einer Sicherheitspolitik steigt jedoch, wenn sie automatisch in herkömmliche Filterregeln übersetzt werden kann. Die Umsetzung hängt von den Möglichkeiten des verwendeten Paketfilters ab. Hier werden Paketfilter betrachtet, die – wie in Abschnitt 1 beschrieben – nach Source- und Destination-IP-Adresse, Protokoll, bei TCP und UDP nach Source- und Destination-Port und bei TCP nach dem Ack-Flag filtern können.

Abbildung 3 zeigt den Ablauf der Übersetzung. Zuerst wird – unter Verwendung von Dienstmodellen – eine Spezifikation auf Paketebene ermittelt[4]. Anschließend werden Filterregeln synthetisiert, die diese Politik umsetzen.

---

[4]Falls nicht durch das verwendete Format die Konsistenz der Spezifikation bereits sichergestellt ist, so ist zusätzlich eine Konsistenzprüfung durchzuführen

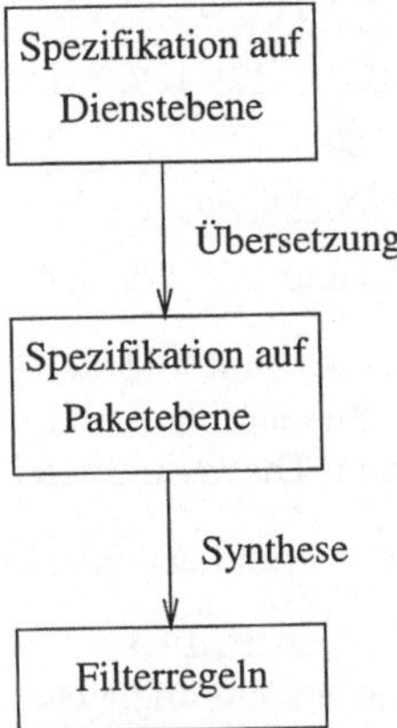

Abbildung 3: Übersetzung der formalen Spezifikation in Filterlisten

3.1 Dienstmodelle

Für die Umsetzung sind Dienstmodelle erforderlich, die für jeden Dienst angeben, welche Pakettypen dafür zugelassen werden müssen.

Das Dienstmodell für den WWW-Dienst von A nach B gibt an, daß folgende Pakete durchgelassen werden:

$$servicemodel(\text{HTTP})_{A \to B} = (TCP, A, B, > 1023, 80, \{0, 1\})$$
$$servicemodel(\text{HTTP})_{B \to A} = (TCP, B, A, 80, > 1023, 1)$$

D. h., daß von A mit Quellport grösser als 1023 zu B mit Zielport 80 alle TCP-Pakete durchgelassen werden, von B mit Quellport 80 zu A mit Zielport grösser als 1023 werden TCP-Pakete, die das Ack-Bit gesetzt haben, also keine Verbindungsaufbaupakete, zugelassen. Für SMTP ergebibt sich ein ähnliches Dienstmodell, nur die Portnummer des Servers ist 25 statt 80.

3.2 Übersetzung

Für die Übersetzung muß jeder Komponente eine IP-Adresse bzw. eine Menge von IP-Adressen zugeordnet werden. Dies geschieht durch die Abbildung

$$ip : C_i \cup C_o \to \wp IP$$

Dabei ist IP die Menge aller möglichen IP-Adressen, $\wp$ ist das Symbol für die Potenzmenge.

Für das vorgestellte Beispiel werden folgende IP-Adressen verwendet:

$$
\begin{aligned}
ip(\bot_i) &= \{129.187.240.0, \dots, 129.187.240.255\} \setminus \{129.187.240.120\} \\
ip(M) &= \{129.187.240.120\} \\
ip(\bot_o) &= \{0.0.0.0, \dots, 255.255.255.255\} \setminus \\
& \quad \{129.187.240.0, \dots, 129.187.240.255\}
\end{aligned}
$$

Daraus wird nun eine Spezifikation auf Paketebene ermittelt. Die Semantik eines Paketfilters wird auf Paketebene als Abbildung jedes möglichen Pakets auf $P = \{\text{allow}, \text{deny}\}$ beschrieben. Die Grundbereiche sind

- IP-Adresse: $IP = \{0.0.0.0, \dots, 255.255.255.255\}$

- Protokoll: $Protocol = \{\text{TCP}, \text{UDP}, \bot_P\}$
  Dabei steht $\bot_P$ für alle Protokolle ungleich TCP und UDP.

- Portnummer: $Port = \{0, 1, \dots, 2^{16} - 1\}$

- Ack-Flag: $Ack = \{0, 1\}$

Es gibt drei Arten von Paketen, TCP-Pakete, UDP-Pakete und die zu anderen Protokollen gehörenden Pakete, die zusammen die Menge *Packets* aller möglichen Pakete ergeben[5]:

$$
\begin{aligned}
TCP\_Packets &= \{\text{TCP}\} \times IP \times IP \times Port \times Port \times Ack \\
UDP\_Packets &= \{\text{UDP}\} \times IP \times IP \times Port \times Port \\
Other\_Packets &= \{\bot_P\} \times IP \times IP \\
Packets &= TCP\_Packets \cup UDP\_Packets \cup Other\_Packets
\end{aligned}
$$

Die Politik auf Paketebene wird für eingehende Pakete (also vom Internet zum Intranet) durch pf_i und für ausgehende Pakete durch pf_o beschrieben:

$$
\begin{aligned}
pf_i &: Packets \to P \\
pf_o &: Packets \to P
\end{aligned}
$$

Die Umsetzung erfolgt nun wie in Abbildung 4 beschrieben. Dabei bezeichnet *In* die Menge der zugelassenen eingehenden Pakete und *Out* die Menge der zugelassenen ausgehenden Pakete. Ein Paket $p \in Packets$ ist genau dann ein Element der Menge *In*, falls $pf_i(p) = \text{allow}$ gilt. Entsprechendes gilt für *Out* und $pf_o(p)$:

$$
\begin{aligned}
\forall p \in Packets &: \quad p \in In \Leftrightarrow pf_i(p) = \text{allow} \\
\forall p \in Packets &: \quad p \in Out \Leftrightarrow pf_o(p) = \text{allow}
\end{aligned}
$$

---

[5]Hier werden nur das TCP- und das UDP-Protokoll betrachtet. Die anderen Protokolle (z. B. Routing-Protokolle, ICMP) werden hier nicht betrachtet, sie lassen sich aber in den Formalismus ebenso einbetten. Dazu muß die Menge *Protocol* entsprechend erweitert werden, an der Funktionsweise der Übersetzung ändert sich nichts.

`translateToPacketLevel` erzeugt aus den Sicherheitspolitiken P_i und P_o auf Dienstebene die korrespondierenden Beziehungen auf Paketebene, d. h. die Mengen In und Out werden bestimmt und damit die Funktionen pf_i und pf_o.

```
 1   algorithm translateToPacketLevel
 2    begin
 3       In := ∅
 4       Out := ∅
 5       forall  (service, src, dst) mit Pi(service, src, dst) = allow
 6           In := In ∪ servicemodelA→B(service)[statt A: ip(src),
                                                   statt B: ip(dst)]
 7           Out := Out ∪ servicemodelB→A(service)[statt A: ip(src),
                                                    statt B: ip(dst)]
 8       forall  (service, src, dst) mit Po(service, src, dst) = allow
 9           Out := Out ∪ servicemodelA→B(service)[statt A: ip(src),
                                                    statt B: ip(dst)]
10           In := In ∪ servicemodelB→A(service)[statt A: ip(src), statt B: ip(dst)]
11   end
```

Abbildung 4: Übersetzung von Dienstebene in Paketebene

Am Anfang werden die Mengen In und Out auf die leere Menge gesetzt, d. h. es werden keine Pakete zugelassen. Dann werden alle erlaubten Dienste durchgegangen. Jeder erlaubte Dienst erfordert, daß – entsprechend den Dienstmodellen – bestimmte Pakettypen zugelassen werden. Die Mengen In und Out werden um die entsprechenden Pakettypen erweitert. $servicemodel_{A\to B}(service)$ liefert die Pakete des Dienstes $service$, die in $A{\to}B$-Richtung zugelassen werden müssen, $servicemodel_{B\to A}(service)$ die Pakete in Gegenrichtung. Die allgemeinen Adressen, die in den Dienstmodellen noch vorhanden sind, werden durch die konkreten Adressen substituiert. Es ist keine Zustandsexplosion zu befürchten, da es für die interne Darstellung der Mengen nicht notwendig ist, jedes Paket einzeln darzustellen, sondern es kann eine Datenstruktur verwendet werden, die jeweils eine Menge von Paketen überdeckt, d. h. es werden nicht einzelne IP-Adressen und Portnummern verwendet, sondern Mengen von IP-Adreßbereichen resp. Portnummernbereichen. Für das Beispiel ergibt sich auf Paketebene die in Tabelle 2 angegebene Sicherheitspolitik, dabei wird – um die Übersichtlichkeit zu erhöhen – die Schreibweise $ip(component)$ verwendet, um die Menge der IP-Adressen der Komponente $component$ zu bezeichnen.

Diese Übersetzung liefert eine Spezifikation eines Paketfilters, der nur die Pakettypen zuläßt, die für die Nutzung der erlaubten Dienste notwendig sind. Durch die begrenzten Möglichkeiten des betrachteten Paketfilters lassen sich manche Sicherheitspolitiken nur unvollkommen umsetzen. Ein Problemfall sind RPC-basierte Dienste, denen keine feste Portnummer zugeordnet ist. Wird ein RPC-Dienst zugelassen, müssen Pakete zu allen in Frage kommenden Ports zugelassen werden und damit zu *allen* RPC-basierten Diensten. Evtl. werden

	Prot	Src-IP	Dst-IP	Src-Port	Dst-Port	Ack
In:	TCP	$ip(\perp_o)$	$ip(M)$	25	>1023	1
	TCP	$ip(\perp_o)$	$ip(M)$	>1023	25	{0,1}
	TCP	$ip(\perp_o)$	$ip(\perp_i)$	80	>1023	1
	TCP	$ip(\perp_o)$	$ip(M)$	80	>1023	1

	Prot	Src-IP	Dst-IP	Src-Port	Dst-Port	Ack
Out:	TCP	$ip(M)$	$ip(\perp_o)$	>1023	25	{0,1}
	TCP	$ip(M)$	$ip(\perp_o)$	25	>1023	1
	TCP	$ip(\perp_i)$	$ip(\perp_o)$	>1023	80	{0,1}
	TCP	$ip(M)$	$ip(\perp_o)$	>1023	80	{0,1}

Tabelle 2: Beispiel: Sicherheitspolitik auf Paketebene

sogar noch weitere Dienste dadurch ermöglicht. In Abschnitt 4 werden mögliche Erweiterungen konventioneller Paketfilter dargestellt, die es erlauben, auch die schwierigen Dienste effektiv zu filtern.

Die Übersetzung der Spezifikation auf Dienstebene in die auf Paketebene sollte auch eine Rückwärts-Analyse durchführen, d. h. sie sollte untersuchen, welche Pakettypen zugelassen wurden und welche Dienste *zusätzlich* zu den ursprünglich spezifizierten dadurch möglich sind, vgl. auch Abschnitt 5.

Die generierten Filter lassen evtl. mehr zu als spezifiziert ist. Dies ist kein Mangel der Übersetzung, sondern durch die beschränkten Möglichkeiten des betrachteten Paketfilters bedingt. Es kann nun sinnvoll sein, bestimmte Dienste explizit zu verbieten, obwohl eigentlich nur die erlaubten Dienste betrachtet werden. Ein Beispiel wäre, Verbindungen zu X-Window (Ports 6000–6000+n, n hängt von der max. Anzahl laufender X-Server ab) zu verbieten, obwohl z. B. für FTP Verbindungen zu Ports >1023 zugelassen werden. Dann funktioniert zwar FTP nicht, falls zufällig doch einmal eine der gesperrten Portnummern gewählt wird, aber es ist auf keinen Fall eine Verbindung zu einem X-Server möglich. Dies erfordert aber, die Politiken P_i und P_o als Abbildungen auf ein erweitertes P, auf $P' = \{$allow, deny, explicit_deny$\}$, zu definieren. Es kann dann unterschieden werden zwischen Diensten, die nicht unterstützt werden sollen, und Diensten, die explizit verboten werden. Die Semantik dieser Unterscheidung ist auf Dienstebene allerdings unklar. Eine Bedeutung ergibt sich erst bei der Übersetzung in eine Spezifikation auf Paketebene. Es werden nur Pakete zugelassen, die zum einen zu Diensten gehören, die erlaubt sind (allow), die zum anderen aber nicht zu Diensten gehören, die explizit verboten sind (explicit_deny).

3.3 Filtersynthese

Hat man nun die Spezifikation auf Paketebene vorliegen, so sind Filterregeln zu erstellen, die diese Politik realisieren. Dabei wird die in Abbildung 5 gezeigte Struktur erzeugt. Durch diese spezielle Struktur ist sichergestellt, daß

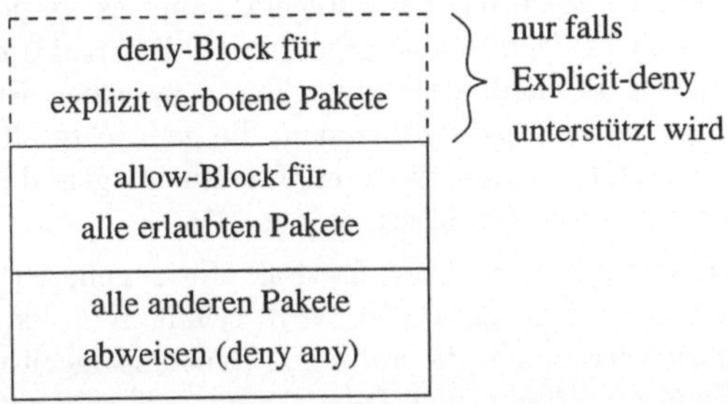

Abbildung 5: Struktur der generierten Filterregeln

alle explizit verbotenen Pakete abgewiesen werden, und daß ansonsten nur die erlaubten Pakete zugelassen werden. Da jeweils zwei aufeinanderfolgende Regeln mit der gleichen Aktion (also falls beide Regeln allow-Regeln oder beide deny-Regeln sind) vertauscht werden können, ist es ohne Auswirkung auf den generierten Filter, in welcher Reihenfolge die Regeln innerhalb jedes Blocks erzeugt werden.

Der Algorithmus der Synthese der Filterregeln ist in Abbildung 6 dargestellt.

```
1   algorithm filtersynthesis
2   begin
3       forall  explizit verbotene Paketbereiche
4           erzeuge deny-Regel(n) für Paketbereich
5       forall  erlaubte Paketbereiche
6           erzeuge allow-Regel(n) für Paketbereich
7       erzeuge Regel, die alles verbietet (deny all)
8   end
```

Abbildung 6: Synthese von Filterregeln

In Filterregeln kann bei IP-Adressen die Anzahl der zu betrachtenden Bits angegeben werden, wodurch ganze Adreßbereiche auf einmal angesprochen werden können. Für Ports sind die Vergleiche =, < und > möglich. Um nun die Regeln für einen Paketbereich zu erstellen, ist dieser Paketbereich zu zerlegen in Bereiche, die jeweils durch eine Regel dargestellt werden. Die Portnummern sind bereits in den Dienstmodellen so angegeben, daß sie direkt umgesetzt werden können. Die Source- und Destination-IP-Adressen des Paketbereichs müssen zusammengesetzt werden aus Bereichen, die jeweils in den oberen Bits übereinstimmen.

Die mit diesem Algorithmus generierten Filterregeln sind zwar korrekt[6], die Filterregeln sind aber nicht optimal, d. h. es gibt kürzere Filtersätze, die die gleiche Funktion erfüllen. Relativ viele Regeln ergeben sich dadurch, daß jeder IP-Adressenbereich durch Bereiche zusammengesetzt wird, die jeweils in den oberen Bits übereinstimmen. Es werden nur allow-Regeln generiert (bzw. für die explizit verbotenen Pakete nur deny-Regeln), aber es werden keine allow- und deny-Regeln gemischt. Dadurch ist schon durch die Struktur der erzeugten Regeln verhindert, daß Reihenfolgeabhängigkeiten zwischen Regeln beachtet werden müssen. Bessere Synthesealgorithmen, die gemischte allow- und deny-Regeln generieren, sind Gegenstand weiterer Untersuchungen, da sich mit ihnen kürzere Filterlisten synthetisieren lassen sollten.

Wenn nicht Filterlisten für konventionelle Paketfilter generiert werden, sondern falls ein neuer Paketfilter entworfen wird, erhält man den Freiheitsgrad, eine Filtersprache zu verwenden, die auf einen Synthesealgorithmus ausgelegt ist. Die Filtersprache muß dann auch kein Kompromiß sein zwischen leichter Parsebarkeit durch den Paketfilter und einer noch für Menschen verständlichen Darstellung, sondern es kann die Filtermaschine auf maximale Effizienz hin ausgelegt werden, da der Code für sie automatisch aus einer verständlichen Spezifikation generiert wird. Eine solche Maschine – für einen verwandten Anwendungsfall – ist in [MJ93] beschrieben.

4 Erweiterungen herkömmlicher Paketfilter

Mit herkömmlichen Paketfiltern ist es nicht möglich, jede Sicherheitspolitik umzusetzen. Im folgenden wird aufgezeigt, wo die Probleme liegen und um welche Fähigkeiten Paketfilter erweitert werden müssen, damit auch diese „schwierigen" Dienste einer Paketfilterung zugänglich werden.

4.1 Verbindungslose Dienste

Da UDP ein verbindungsloser Dienst ist, gibt es keine Verbindungsaufbaupakete, die die Richtung anzeigen, in der die Verbindung aufgebaut wird. Deshalb können UDP-basierte Dienste nicht nur in einer Richtung zugelassen werden. Abhilfe schaffen dynamische Regeln. Soll eine UDP-Kommunikation von A nach B nur von A initiiert werden dürfen, so gibt es eine statische Regel, die UDP-Pakete von A nach B zulässt. Falls nun ein Paket von A nach B übertragen wird, so wird eine dynamische (zeitlich befristet gültige) Regel erzeugt, die die Antwortpakete von B nach A zulässt. Damit wird die Paketfilterung zustandbehaftet, d. h. die Entscheidung, ob ein Paket weitergeleitet wird, hängt nicht nur von diesem Paket ab, sondern auch von früheren Paketen.

---

[6]Die Korrektheit der Umsetzung wurde noch nicht bewiesen, das ist aber Gegenstand weiterer Untersuchungen.

Das gleiche Prinzip lässt sich auch bei TCP-Verbindungen anwenden. Bei statischen Regeln werden in Gegenrichtung alle Pakete zugelassen, die zu einer bestehenden Verbindung gehören (d. h. das Ack-Flag ist gesetzt), unabhängig davon, ob diese Verbindung tatsächlich aufgebaut wurde. Man verlässt sich darauf, daß ein Host ein solches Paket verwerfen wird, falls keine Verbindung aufgebaut wurde. Mit Hilfe dynamischer Regeln wäre es möglich, die Pakete in Gegenrichtung erst zuzulassen, nachdem die Verbindung von der Innenseite aus aufgebaut wurde.

4.2 RPC-Dienste

RPC-Diensten sind keine festen Portnummern zugeordnet, sondern sie bekommen einen freien Port vom Portmapper zugewiesen. Soll eine Anfrage an einen RPC-Dienst gerichtet werden, so muß zuerst die Portnummer vom Portmapper erfragt werden. Bei herkömmlichen Paketfiltern können nur feste Portnummern angegeben werden, d. h. wenn ein RPC-Dienst genutzt werden soll, so muß ein ganzer Portnummernbereich freigegeben werden. Dadurch sind dann nicht nur die gewünschten Dienste erlaubt, sondern alle RPC-Dienste und evtl. noch weitere.

Eine Lösungsmöglichkeit ist die Verwendung von variablen Portnummern in Filterregeln (also z. B. UDP from host A port >1023 to host B port rpc 100012v1, um Verbindungen zum spray-Dienst, einem Netzwerk-Testprogramm, zuzulassen). Jedem RPC-Dienst ist eine eindeutige RPC-Nummer (hier 100012 für spray) zugeordnet, zusätzlich können verschiedene Versionen eines Programmes unterschieden werden (hier Version 1).

Wertet der Paketfilter eine solche Regel aus, muß er beim Portmapper des betreffenden Rechners die Portnummer erfragen, die zu diesem RPC-Dienst gehört.

4.3 Aufbau eines zweiten Kanals in Gegenrichtung

Bei normalem FTP [PR85] und bei rsh (sowie bei weiteren BSD-r-Diensten) baut nicht nur der Client eine Verbindung zum Server auf, sondern auch umgekehrt der Server eine Verbindung zum Client. Bei FTP wird die zweite Verbindung zur Datenübertragung genutzt, bei rsh für den stderr-Kanal. Dazu muß ein Aufbau einer Verbindung in Gegenrichtung zugelassen werden.

In der Anwendungsschicht überträgt der Client zum Server die Port-Nummer, auf der er die Verbindung erwartet. Wenn die Firewall die in der Anwendungsschicht übertragenen Daten überwacht, kann dynamisch eine Regel erzeugt werden, die die Verbindung in Gegenrichtung nur zu diesem Port zulässt. Ohne dynamische Regeln muß ein Verbindungsaufbau von außen zu allen in Frage kommenden Zielports zugelassen werden.

Sollen Daten der Anwendungsschicht nicht analysiert werden, kann bei FTP eine spezielle Variante, PASV-FTP [Bel94], eingesetzt werden, bei der alle Verbindungen vom Client aus aufgebaut werden. Die BSD-r-Dienste sind so unsicher, daß sie nicht einmal in einem lokalen Netz eingesetzt werden sollten [CZ95]. Für sie existieren Alternativen wie z. B. die Secure Shell [Ylö].

5   Weitere Werkzeugunterstützung

Eine auf Dienstebene erstellte Sicherheitspolitik kann automatisch (oder auch von Hand) umgesetzt werden in eine Sicherheitspolitik auf Paketebene. Es stellt sich nun die Frage, was „sichere" und was „unsichere" Sicherheitspolitiken sind. Es gibt jedoch keine einfache Unterscheidung zwischen sicher und unsicher, sondern es findet eine Gratwanderung statt. Ein Beratungssystem kann die Entwicklung von Sicherheitspolitiken unterstützen, indem es die spezifizierte Sicherheitspolitik und deren Umsetzung in Filterregeln untersucht.

Mögliche Analysepunkte sind:

- Die Sicherheit jedes Dienstes an sich, d. h. verfügt er über leistungsfähige oder eher schwache/keine Sicherheitsmechanismen, wo liegen seine Schwächen (z. B. Authentifizierung auf Basis von IP-Adressen, Übertragung von Paßwörtern im Klartext).

- Eignung der Dienste für Paketfilterung, d. h. ohne die Filterlisten generieren zu müssen, ist klar, daß z. B. RPC-basierte Dienste oder FTP Schwierigkeiten bereiten, d. h. daß viele Pakettypen zusätzlich zu denen, die eigentlich vorkommen, zugelassen werden müssen.

- Eine Analyse des generierten Filtersatzes liefert alle (dem System bekannten) Dienste, die möglich sind, obwohl sie nicht erlaubt wurden.

Was auch schon weiterhelfen würde, wäre die Entwicklung von Filterregeln von Hand durch Werkzeuge zu unterstützen. Derzeit werden die Filterregeln von Hand entworfen und dann in einen Paketfilter eingespielt. Ein Werkzeug, das den Paketfilter nur simuliert, erlaubt es, den Filtersatz zu testen, d. h. die Reaktion auf Testpakete zu untersuchen ohne diese Pakete real erzeugen zu müssen, was i. a. ein eigenes Testnetz und damit einen gewissen Aufwand erfordert. Mit Hilfe der in dieser Arbeit vorgestellten Dienstmodelle lassen sich automatisch Testpakete generieren für bestimmte Dienste zwischen bestimmten Komponenten.

6   Verwandte Arbeiten

Manche Paketfilter [SSH93, MS96] verfügen über eine benutzerfreundlichere Konfigurationssprache als es Filterregeln sind. Der in [SSH93] vorgestellte Paketfilter *Drawbridge* verfügt über einen Compiler, der die Spezifikation von einer leicht verständlichen Sprache in Filtertabellen übersetzt. Dabei können Dienste angegeben werden. Dies wird aber dadurch erkauft, daß *alle* TCP-Pakete, die nicht Verbindungsaufbaupakete sind, grundsätzlich zugelassen werden, ebenso wie *alle* UDP-Pakete in ausgehender Richtung. Da die meisten Dienste (Gegenbeispiele sind FTP und die BSD-r-Dienste) nur eine einzige Verbindung verwenden, fehlt für diese Dienste nur noch genau ein Pakettyp, der zugelassen werden muß. Dadurch sieht die Konfiguration so aus, als ob sie einzelne Dienste zulässt, tatsächlich wird aber nur der noch fehlende Pakettyp zugelassen.

Die *sf-Firewall* [MS96] implementiert einige Neuerungen wie z. B. dynamische Regeln oder das Suchen nach PORT-Befehlen in FTP-Verbindungen. Die Konfiguration erfolgt aber über konventionelle, paketbezogene Regeln.

In [Cal96] wird ein Werkzeug vorgestellt, das die Konfiguration von Cisco-Routern unterstützt, die Konfiguration der Filterregeln erfolgt aber wie bisher auf Paketebene.

Dem Autor sind keine Arbeiten bekannt, die zum Ziel haben, aus einer benutzerfreundlichen Spezifikation herkömmliche Filterregeln zu generieren. Auch scheint der Einsatz von formalen Methoden für die Spezifikation von Sicherheitspolitiken für Paketfilter neu zu sein.

7 Zusammenfassung

In dieser Arbeit wurde ein Ansatz vorgestellt, bei dem die von einem Paketfilter durchzusetzende Sicherheitspolitik auf einer abstrakten, benutzerfreundlichen Ebene spezifiziert wird. Bei der Umsetzung von dieser Ebene in Filterregeln für herkömmliche Paketfilter zeigt sich, daß manche Dienste nur schlecht zu filtern sind, d. h. es muß mehr zugelassen werden als eigentlich gewünscht ist. Dies ist Anlaß, Erweiterungen von Paketfiltern zu entwickeln, durch die auch die problematischen Dienste wirkungsvoll gefiltert werden können.

Die in diesem Beitrag dargestellten Ergebnisse stammen aus einem laufenden Forschungsvorhaben. Weitergehende Bemühungen gelten der Verbesserung der vorgestellten Methoden und der Erstellung von Werkzeugen, die die in diesem Beitrag vorgestellten Methoden einsetzen, um den Entwurf von Filterlisten zu unterstützen. Weiterhin sollen Architekturen für Paketfilter entwickelt werden, die eine möglichst vollständige Umsetzung von Spezifikationen auf Dienstebene ermöglichen. Durch die Generierung von Filtercode, der nicht für Menschen verständlich sein muß, aus einer Hochsprachenbeschreibung erhält man Freiheitsgrade, die versprechen, den Durchsatz von Paketfiltern zu erhöhen.

Die hier vorgestellten Verfahren sind insbesondere dann wichtig, falls Paketfilter zur Trennung von Bereichen innerhalb eines LANs eingesetzt werden. Für deren Management sind leistungsfähige Methoden notwendig, da mehrere Paketfilter effizient verwaltet werden müssen in einer Umgebung mit sich regelmässig ändernden Anforderungen durch neu installierte oder entfernte Komponenten und Dienste.

Literatur

[Bel94]   S. Bellovin. Firewall-Friendly FTP. RFC 1579, Februar 1994.

[Cal96]   Christopher J. Calabrese. A Tool for Building Firewall-Router Configurations. *Computing Systems*, 9(3):239–253, Summer 1996.

[CB94]   William R. Cheswick und Steven M. Bellovin. *Firewalls and Internet Security: Repelling the Wily Hacker*. Addison-Wesley, Reading, MA, 1994.

[Cha92]   D. Brent Chapman. Network (In)Security Through IP Packet Filtering. In USENIX Association, Hrsg., *UNIX Security III Symposium*, Seiten 63–76, Berkeley, CA, September 1992. USENIX.

[CZ95]   D. Brent Chapman und Elizabeth D. Zwicky. *Building Internet Firewalls*. O'Reilly & Associates, Inc., Newton, MA, 1995.

[Fuh96]   Kai Fuhrberg. Sicherheit im Internet. In *Der Weg zur sicheren Informationstechnik – ausgewählte Beispiele; Sicherheitsseminar des Bundesamtes für Sicherheit in der Informationstechnik für die Landes- und Kommunalverwaltungen von Baden-Württemberg und Bayern, Universität Ulm*, Bonn, November 1996. Bundesamt für Sicherheit in der Informationstechnik.

[MJ93]   Steven McCanne und Van Jacobson. The BSD Packet Filter: A New Architecture for User-level Packet Capture. In *Proceedings of the Winter USENIX Conference*, San Diego, CA, Januar 1993.

[MS96]   Robert Muchsel und Roland E. Schmid. Der 80:20-Firewall; Konzeption und Entwicklung eines "Packet Filters". In Prof. Dr. Kurt Bauknecht, Prof. Dr. Dimitris Karagiannis und Dr. Stephanie Teufel, Hrsg., *Sicherheit in Informationssystemen, Proceedings der Fachtagung SIS '96*, Wien, März 1996. vdf Hochschulverlag.

[PR85]   John Postel und Joyce Reynolds. File Transfer Protocol. RFC 959, Oktober 1985.

[Ran93]   Marcus J. Ranum. Thinking about Firewalls. Technical report, Trusted Information Systems, Inc., Glenwood, Maryland, 1993.

[SSH93]   David R. Safford, Douglas Lee Schal und David K. Hess. The TAMU Security Package: An Ongoing Response to Internet Intruders in an Academic Environment. In *Proceedings of the Fourth Usenix UNIX Security Symposium*, Seiten 91–118, Santa Clara, CA, Oktober 1993.

[WC94]   John P. Wack und Lisa J. Carnahan. Keeping your site comfortably secure; an introduction to Internet firewalls. NIST special publication Computer security 800-10, U.S. Dept. of Commerce, Technology Administration, National Institute of Standards and Technology, Gaithersburg, MD, 1994.

[Ylö]   Tatu Ylönen. SSH (Secure Shell) Remote Login Program. http://www.cs.hut.fi/ssh/.

On the development of a security toolkit for open networks
- New security features in SECUDE

U. Faltin, P. Glöckner, U. Viebeg, A. Berger,
H. Giehl, D. Hühnlein, S. Kolletzki, T. Surkau
GMD - TKT.SIT Security Technology
Dolivostr. 15, D-64293 Darmstadt

ABSTRACT: In this article we will discuss the requirements of security toolkits for open networks, explain some important technical details and give a perspective on modern security technology. To illustrate these issues we will focus on the current and future development of SECUDE. We will give a brief overview of the SECUDE [16] structure, emphasize the latest developments and new security APIs, such as improvements in the CRYPT-API, the integration of new smartcards, the Directory access via **LDAP**, the support of **X.509v3** certificates and new security features like **GSSv2, PKCS#7,10, S/MIME, BAKO** and **SURE**.

1 Introduction

Since global and local network-services are increasingly being used by the general public there is a strong demand for authentic, confidential and non-repudiable communication. These key issues of open telecooperation can be achieved through cryptographic primitives, like *encryption* and *digital signatures*. The trustworthiness of these mechanisms rest in the public control of the algorithms, rather than in hiding the design principles. Therefore SECUDE provides a variety of well known and intensively studied crypto-algorithms as security basis for higher level applications. Furthermore there is a need for the secure storage of private keys. SECUDE provides two possible solutions of a **P**ersonal **S**ecurity **E**nvironment (PSE) for this purpose; an encrypted directory (SW-PSE) and interfaces to a variety of smart-cards (SC-PSE). It is preferable to have a technology-independent interface used by higher level applications, because the security of the underlying algorithms and the chipcard-technology is subject to change due to further research. SECUDE contains the SECURE API, which provides access to the secure processing and secure storage module. Another important API is needed to achieve authentic communication. The functions for the management of public-keys are accessible through the Authentification Framework API. On top of these basic modules there are higher level APIs such as e.g. PEM, PKCS, GSS and S/MIME. Finally these functions are utilized in security-plugins for existing software products, like SAP/R3 or MS-Exchange, just to name the prominent ones.

These aforementioned issues lead canonically to the structure of SECUDE as shown in the following figure.

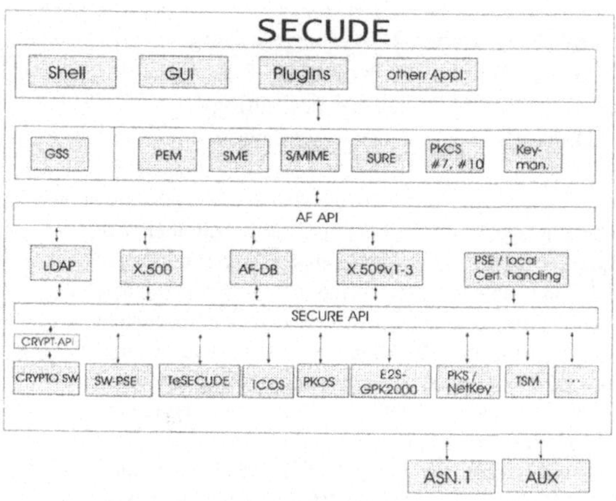

2 New Security-Features in SECUDE

The functionality of these APIs is addressed in the following, where we focus on the latest developments, new features and give perspectives for further research.

2.1 CRYPT API

This API provides multi-precision integer arithmetic, modulo arithmetic, random number generation and implementations of symmetric and asymmetric cryptoalgorithms.

The available symmetric algorithms[1] are DES, Triple DES and IDEA. The public key algorithms contained in the CRYPT API are the Diffie Hellman Key Agreement, the NIST DSS and the RSA algorithm. Furthermore there is a need for cryptographically strong hash functions for signature generation. SECUDE provides the hash-functions MD2, MD4, MD5, SHA-0, SHA-1 and RIPE-MD160 [14]. MD2 and MD4 are totally unsuitable for signature generation, MD5 is suspected to be broken soon (c.f. [2], [3]) and SHA-0 bears some weaknesses in the expand-function. Therefore we recommend SHA-1 and RIPE-MD160 as the most secure hash functions for signature generation. The other hash-functions are left in SECUDE, to keep compatibility with standards and yesterdays signatures.

2.2 SECURE - API

Like mentioned in the introduction, SECUDE provides a technology independent SECURE - API, which connects the CRYPT-API and the PSE-handling to the higher-

---

[1] A description and further references for all crypto-algorithms (except RIPE MD160) may be found in [15].

level API's. SECUDE supports two PSE-types. That is an encrypted directory (SW-PSE) and interfaces to different smartcards (SC-PSE's). Both PSE-types are PIN-protected. The used PSE- and smartcard-type(s) may be selected during the configuration process. Only the desired SC-interface is linked dynamically.

At the time of writing SECUDE supports the STARCOS and the TCOS smartcard systems. Interfaces to the GEMPLUS smartcard GPK2000 and the G&D PKOS smartcard are in the development phase.

2.3 Authentication Framework API

The AF module adds X.509 certification functionality to SECUDE. Former SECUDE versions supported the X.509 version 1 certificates, while the next SECUDE release will contain the X.509 version 3 certificates.

Both local (i.e. PSE-located) certificates and Directory-located certificates can be addressed. Obtaining public security information, like public keys, certificates, crosscertificates and certificate revocation lists used to be done using the X.500 Directory Access Protocol (DAP). SECUDE 5 now uses the Lightweight Directory Access Protocol (LDAP) instead of DAP to retrieve security information from Directory servers.

2.3.1 X.509 version 3 certificates

The experience gained in attempts to deploy X.509 v1 certificates made it clear that the v1 and v2[2] certificate formats are deficient and too restrictive. With X.509 v3, which is standardized in [20], most of the requirements of RFC 1422 [10] can be addressed using certificate extensions, without a need to restrict the CA structures used. In particular, the certificate extensions relating to certificate policies obviate the need for **P**olicy **C**ertification **A**uthorities and the constraint extensions obviate the need for the name subordination rule, because the certificate contains information (*basic constraints* - field) to distinguish between user- and CA-certificates. The certificate may contain *alternative names*, like mail-adresses or URLs for the issuer (CA) and the subject (user). The distribution and retrieval of Certificate **R**evocation **L**ists is made easier by storing *CRL distribution points* in the certificate and the *Key usage* may be restricted. Besides this standard-extension, which are discussed in [20] more detailed, it is possible to use *private extensions* for application specific needs.

2.3.2 The Lightweight Directory Access Protocol

The Lightweight Directory Access Protocol (LDAP [12],[13]) was designed to overcome the problems resulting from the requirements of the X.500 DAP. While queries and answers are still encoded using ASN.1, LDAP makes restrictions on the type and format. Another simplification is the use of string notation in most of the attributes. On the transport side TCP connections are usually used to communicate

---

[2] The difference between v1 and v2 - certificates is just the presence of two more fields for Directory access control.

with an LDAP server, eliminating the need for an OSI protocol stack. This all leads to smaller code and more acceptance on the implementors' side.

In SECUDE there are two possibilities to access X.500 Directories. Either using the X.500 ISODE ICR2.1 library or the access via LDAP. The latter is possible, if the server uses an LDAP-to-X.500 adapter. This allowes the client software to remain "light", moving the overhead upstream to the server. LDAPv2 is supported in SECUDE's various Unix ports using the LDAP reference implementation library of the University of Michigan. In the SECUDE for Windows NT/95 version the Dynamic Link Library (DLL) of the same package is used [9].

2.4 Generic Security Services API

The Network Working Group of the Internet Engineering Task Force defined in 1993 a general interface to security systems. The *Generic Security Services* (GSS) API is a set of functions and data structures to incorporate security into a program independent of the underlying security and communication protocols [4], [5].

There are (to our knowledge) currently three underlying security mechanisms available, which support the GSS-API. First the well known *Kerberos V5* using a trusted authentication-server and DES-encryption, *Simple Public Key Mechanism* [6] using X.509 certificates and a variety of algorithms for authentication and confidentiality. Finally there is the SECUDE-mechanism, which also uses X.509 certificates, but is restricted to RSA, DES and IDEA. The a priori restriction to certain algorithms removes the inherent negotiation-overhead in SPKM. The GSS-API and this three underlying mechanisms are discussed in [8] more detailed. Currently the SECUDE-mechanism is available[3] and SPKM is under development. Kerberos won't be supported, because public-key-mechanisms in this context are generally preferable.

2.5 PKCS API

While the family of PKCS-standards [11] comprises standards for RSA encryption, DH Key Agreement and other cryptographic primitives that are already available in earlier versions of SECUDE, we will focus on PKCS#7 and PKCS#10.

2.5.1 Signing and encrypting data with PKCS #7

The PKCS #7 standard describes a general syntax for data that may have cryptography applied to it, such as digital signatures and digital envelopes. The syntax admits recursion, so that, for example, one envelope can be nested inside another, or one party can sign some previously enveloped digital data. It also allows arbitrary attributes, such as signing time, to be authenticated along with the content of a message, and provides for other attributes such as countersignatures to be associated with a signature. A degenerate case of the syntax provides means for disseminating certificates and certificate-revocation lists. The PKCS #7 API of SECUDE consists of functions for creating, developing, and verifying such enveloped data. The types

---

[3] E.g. it is used to secure the SAP / R3 application program.

Signed and *Signed-and-enveloped* are interoperable with Privacy Enhanced Mail [10]. SECUDE includes conversion functions for these types.

2.5.2 Certification Requesting with PKCS #10

The PKCS #10 standard describes a syntax for certification requests. A certification request consists of a distinguished name, a public key, and optionally a set of attributes, collectively signed by the entity requesting certification. Certification requests are sent to a certification authority, who transforms the request to an X.509 public-key certificate. SECUDE includes several programs and functions dealing with PKCS #7 ContentInfo and PKCS #10 CertificationRequests format.

2.5.3 Signing and encrypting multi media messages with S/MIME

The S/MIME Message Specification [17] combines the security enhancement of PEM and the multi-purpose content-types of MIME. It defines the MIME content types application/x-pkcs7-mime and application/x-pkcs7-signature for cryptographically enhanced MIME bodies according to PKCS #7 and application/x-pkcs10 for submitting a certification request. SECUDE provides API functions for producing and parsing such S/MIME messages. The main advantage compared to PGP is the scaleable certification infrastructure. I.e. it is possible to use either a hierarchical certification infrastructure combined with cross-certificates or a network of trust like in PGP.

3 Applications

3.1 The European ICE-TEL Project - TrustFactory Digital ID Center

Within the European research project 'Interworking Public Key Certification Infrastructure for Europe' (ICE-TEL) more than 17 countries are working together to establish a common certification infrastructure not only for the R&D community, but also for interested partners in the governments, administrations and industry.

In this context GMD is running a Certification Authority; the 'TrustFactory Digital ID Center' as a European counterpart to the US american 'VeriSign Digital ID Center'.

3.1.1 Privacy Enhanced Mail Plugin for Microsoft Exchange

Another add-on application is the PEM plugin for MS Exchange where outgoing e-mail may be signed and encrypted and incoming messages may be decrypted and validated. The plugin provides full attachment support and the german MailTrusT PEM Specification [19]. This includes correct processing of raw binary data.

3.1.2 Privacy Enhanced Mail Shell Extension for Microsoft Windows Explorer

The same PEM functionality as described in (3.1.2) is avaiable as a shell extension for the new file manager of MS Windows, the Windows Explorer. Local files may be signed and encrypted or decrypted and validated.

3.2 Security Add-on for SAP R/3

In cooperation with SAP an interface for the R/3 Client/Server system was developed, which makes the use of security technology via GSS-API possible.

3.3 BAKO with SURE extension

3.3.1 Basic cooperation protocol

Another application of the SECUDE toolkit is BAKO [1] - a basic cooperation protocol for secure business transactions over open networks. In contrast to host-to-host session- or packet-based security mechanisms, BAKO can be applied where complete transactions need to be authentic and non-repudiable, and where *documents* need to be produced that are integer and confidential, like bank-orders for example. There will be BAKO - plugin for the WWW available to achieve secure web-transactions.

3.3.2 Signed Unique References - a BAKO extension

The original BAKO has two unsolved issues: minimizing the danger of replay attacks, and minimizing the network load. Even if the information is security-enhanced, a single or multiple resending of protocol units may cause trouble on both sides if the session becomes insecure. The second issue is caused by BAKO's nesting of complete transaction steps. If large objects, e.g. images, are to be transported, a high information overhead occurs: objects are sent twice or even three times. A proposed BAKO extension specifies the replacement of objects with signed and unique references ('SURE') [18] to the object as soon it has been transmitted or received. The reference additionally contains a timestamp consisting of a negotiated time base and the protocol data unit's time-to-live.

4 Conclusion

In this paper we briefly discussed the requirements of a security toolkit for open networks, gave an overview of the security features available in SECUDE, illustrated some current and future developments and finally discussed a few secure end user-applications. SECUDE is designed to be a portable **SECU**rity **D**evelopment Environment and therefore well suited for security-plugins to any kind of applications. We will continue to upgrade SECUDE with upcoming new standards, crypto-algorithms, smartcard systems, and application requirements. An open research area, for instance, is the design of Public Key Infrastructures (PKIs) for various purposes, including cross-certification between different PKIs, and the interface between end user applications and the PKI. The SECUDE development will reflect the current standardization process in the IETF.

References:

[1]   S. Kolletzki: „Secure Internet Banking with Privacy Enhanced Mail", Computer Networks and ISDN Systems 28 (1996) 1891-1899

[2]   H. Dobbertin: „Welche Hash-Funktionen sind für digitale Signaturen geeignet?", Tagungsband „Digitale Signaturen", Vieweg-Verlag, 1996, ISBN 3-528-05548-0, pp. 81-92

[3]   H. Dobbertin: „Digitale Fingerabdrücke - Sichere Hashfunktionen für digitale Signaturen", DuD 2/97, Vieweg, pp. 82-87, 1997

[4]   J. Linn: „GSS API" RFC's 1508 and 1509 (C-bindings), Sep. 93

[5]   J. Linn: „The GSS API Version 2" RFC 2078, Jan 97

[6]   C. Adams: „The Simple Public-Key GSS-API Mechanism (SPKM)", RFC 2025, Jan 96

[7]   J. Kohl, C. Neumann: "The Kerberos Network Authentication Service (V5)", RFC 1510, Sep. 1993
        J. Linn: " The Kerberos Version 5 GSS-API Mechanism ", RFC 1964, Juni 1996

[8]   D. Hühnlein: "Generische Sicherheit - Die GSS-API und drei ihrer Mechanismen", to appear in FIFF-communication, 3/97

[9]   University of Michigan Information Technology Division: „LDAP servers, client library and sample text based UNIX clients"
        ftp://terminator.rs.itd.umich.edu/x500/ldap/ldap-3.3.tar.Z
        „Windows Binary Distribution (contains LDAP32.DLL, LIB and header files)"
        ftp://terminator.rs.itd.umich.edu/x500/ldap/windows

[10]   J. Linn: „Message Encryption and Authent. Procedures" RFC 1421, Feb 93
        S. Kent: „Certificate Based Key Management" RFC 1422, Feb 93
        D. Balenson: „Algorithms modes and identifiers" RFC 1423, Feb 93
        B. Kaliski: „Key Certification and related Services" RFC 1424, Feb 93

[11]   RSA: „PKCS#1-#11: Public Key Cryptography Standards", http://www.rsa com, revised Nov. 1993

[12]   M. Wahl, T. Howes, S.Killie: „Lightweight Directory Access Protocol (v3)", 10/1996
        ftp://ds.internic.net/internet-drafts/draft-ietf-asid-ldapv3-protocol-03.txt

[13]   W. Yeong, T. Howes, S. Killie: „CURRENT LDAP Version2", March1995
        ftp://ds.internic.net/rfc/rfc1777.txt

[14]   H. Dobbertin, A. Bosselaers, B. Preneel: „RIPEMD-160: A strengthened version of RIPEMD", Fast Software Encryption, Cambridge Workshop, LNCS 1039, Springer, 1996, pp. 53-69 , corrected version via
        ftp://esat.kuleuven.ac.be/pub/COSIC/bosselae/ripemd/

[15]   B. Schneier: "Applied Cryptography - Protocols, Algorithms and Source Code in C", John Wiley & Sons, New York, 1994, ISBN 0-471-59756-2

[16]   GMD: „SECUDE 5.0 - Hyperlink Documentation", 1996,
        http:www.darmstadt.gmd.de/secude/doc/index.htm

[17]   RSA: „S/MIME Message Specification", Feb 96, *smime-editor@rsa.com*

[18]   P. Glöckner, S. Kolletzki, M. Wichert: „Signed Unique References", to appear in the proceedings of JENC8

[19]   F. Bauspieß (ed.): „MailTrusT Spezifikation, Version 1.1", 12/96

[20]   ISO/IEC JTC 1/SC 21/WG 4 and ITU-T Q15/7: „Final Text of Draft Amendment 1 to ISO/IEC 9594-8 on Certificate Extensions", December 1996

Ein effizientes *und* sicheres digitales Signatursystem

Dirk Fox

Universität Siegen, fox@nue.et-inf.uni-siegen.de

Zusammenfassung

Das am 13. Juni 1997 vom Bundestag beschlossene Signaturgesetz weist digitalen Signatursystemen eine besondere Rolle bei der Entwicklung technischer Systeme für elektronischen Rechtsverkehr und digitalen Handel zu. Sowohl die Sicherheit (im Sinne von Unfälschbarkeit) als auch die Effizienz von Erzeugung und Prüfung digitaler Signaturen sind daher für die Wahl geeigneter digitaler Signaturverfahren von entscheidender Bedeutung.

Die heute verbreiteten Verfahren zur Erzeugung digitaler Signaturen, bspw. der DSS und das RSA-Verfahren, verwenden asymmetrische Kryptographie. Sie zeichnen sich durch hinreichende Effizienz aus, genügen aber keiner strengen Sicherheitsdefinition. Ohne die Verwendung von Hashfunktionen oder den Einsatz von Redundanzschemata können sowohl einzelne DSS- als auch RSA-Signaturen leicht gefälscht werden.

In diesem Beitrag wird ein an das GMR-Signatursystem angelehntes digitales Signatursystem vorgestellt, dessen Sicherheit auf der Schwierigkeit beruht, diskrete Logarithmen zu bestimmen. Das Verfahren erlaubt sehr effiziente Implementierungen, u.a. auf Elliptischen Kurven über endlichen Körpern, und genügt der stärksten Sicherheitsdefinition; es wird die Äquivalenz von Signaturfälschung und der Bestimmung diskreter Logarithmen bewiesen. Abschließend werden Aufwandsvergleiche mit herkömmlichen Signatursystemen angestellt und Meßergebnisse einer prototypischen Implementierung angegeben.

1 Einleitung

Seit der wegweisenden Veröffentlichung von Diffie und Hellman, die 1976 erstmals die Idee eines digitalen Signatursystems auf der Basis eines asymmetrischen Kryptoverfahrens vorstellten [DiHe_76], wurde eine Vielzahl unterschiedlicher Verfahren zur Realisierung digitaler Signatursysteme entwickelt. Die wichtigsten darunter sind das verbreitete RSA-Signatursystem und das Verfahren von ElGamal [RiSA_78, ElGa_84]. Vor drei Jahren wurde in den USA das erste digitale Signatursystem, angelehnt an eine Variante des ElGamal-Verfahrens von Schnorr, als *Digital Signature Standard* (DSS) genormt [Schn_91, NIST_94]. DSS und RSA sind inzwischen sehr weit verbreitet; es kann daher erwartet werden, daß diese beiden Verfahren auch in den nächsten Jahren in den meisten Anwendungen als digitales Signatursystem eingesetzt werden.

Mit dem Mitte Juni vom Bundestag verabschiedeten Signaturgesetz (SigG, Art. 3 IuKD-Gesetz) [IuKD_97] ist die Erwartung verbunden, daß digitalen Signaturen in absehbarer Zeit dieselbe rechtliche Bindungs- und Beweiskraft beigemessen wird, die heute allein eigenhändige Unterschriften besitzen. Dies ist eine der zentralen Voraussetzungen für die weitere Entwicklung des elektronischen Rechtsverkehrs. Um die Beweiskraft einer eigenhändigen Unterschrift zu erlangen, müssen die technischen Verfahren, die zur Realisierung digitaler Signaturen eingesetzt werden, allerdings sehr hohen Sicherheitsanforderungen genügen.

In Kapitel 2 wird nach einer Einführung des Sicherheitsmodells gezeigt, daß die heute verbreiteten digitalen Signatursysteme nicht im strengen Sinne unfälschbar sind: Bei einem aktiven Angriff können RSA- und DSS-Signaturen existentiell bzw. sogar selektiv gefälscht werden. Kapitel 3 stellt drei digitale Signaturverfahren vor, für die die existentielle Unfälschbarkeit von Signaturen unter einer Komplexitätsannahme bewiesen ist, und die zudem effizient implementiert werden können; eingeschlossen das erste solche Verfahren von Goldwasser, Micali und Rivest (GMR) [GoMR_88, FoPf_91].

Eine neue Variante des GMR-Signatursystems, DLP-GMR, wird in Kapitel 4 vorgestellt. Dessen Konstruktion lehnt sich an das GMR-Signatursystem an; die Sicherheit dieses Verfahrens beruht jedoch nicht auf dem Faktorisierungs-, sondern dem diskreten Logarithmusproblem. Die Äquivalenz von Signaturfälschung und Bestimmung diskreter Logarithmen wird nachgewiesen (Kapitel 5); es folgen Vorschläge für eine effiziente Implementierung (Kapitel 6).

Kapitel 7 faßt die wesentlichen Eigenschaften der vorgestellten Verfahren vergleichend zusammen und schließt mit Meßergebnissen prototypischer Implementierungen.

2 Fälschungssicherheit digitaler Signatursysteme

Digitale Signatursysteme, die im Kern eine *trapdoor*-Funktion verwenden, d.h. eine Funktion, deren Umkehrung nur mit Kenntnis eines Geheimnisses effizient möglich ist, wie das RSA-Signatursystem und der DSS, sind höchstens **kryptographisch** fälschungssicher: Ein Fälscher darf praktisch, d.h. mit begrenzter Rechenleistung, nicht in hinreichend kurzer Zeit gültige (neue) Signaturen erzeugen können. Informationstheoretische Sicherheit ist prinzipiell nicht erreichbar, denn der öffentliche Prüfschlüssel ist bekannt: Ein Fälscher mit unbegenzter Zeit und Rechenleistung könnte alle Signierschlüssel durchprobieren, bis er den passenden fände.

Die Sicherheit eines digitalen Signatursystems sollte auf einer möglichst gesicherten Annahme über die Komplexität eines bekannten Problems beruhen, d.h. dem durchschnittlich erforderlichen Aufwand, um dieses Problem mit Hilfe eines Computers zu lösen. Der Zusammenhang von Komplexität des Problems und Fälschungssicherheit des Signatursystems sollte zudem eine Äquivalenzbeziehung sein: Das Fälschen einer Signatur darf nicht leichter sein als die Lösung des Problems (und umgekehrt).

Zu den heute in diesem Zusammenhang wichtigen, überwiegend zahlentheoretischen Problemen zählen vor allem die Zerlegung großer Zahlen in ihre Primfaktoren (**Faktorisierungsproblem**: Bestimme x, y zu n mit $n = x{\cdot}y$) und die Bestimmung von Logarithmen in einem Restklassenring (**Diskretes Logarithmusproblem**: Finde x mit $a^x \bmod p = y$). Beide sind seit Jahrzehnten Gegenstand intensiver mathematischer Forschung. Die Existenz effizienter Lösungsalgorithmen für hinreichend große Zahlen gilt als sehr unwahrscheinlich, auch wenn dies bis heute nicht bewiesen ist.

Eine Bestimmung der Fälschungssicherheit eines digitalen Signatursystems erfordert auch eine genaue Beschreibung des Angreifermodells. Nach der Klassifikation von Angriffstypen und Fälschungen von Goldwasser, Micali und Rivest besitzt ein Signatursystem die größte Fälschungssicherheit genau dann, wenn gilt [GoMR_84]:

> *Selbst bei einem adaptiven, aktiven Angriff, bei dem der Fälscher endlich oft zueiner Nachricht seiner Wahl die passende digitale Signatur erhält, gelingt diesem nicht einmal eine existentielle Fälschung, d.h. die Erzeugung einer einzigen gültigen neuen digitalen Signatur zu irgendeiner beliebigen, nicht notwendig sinnvollen Nachricht.*

Die heute verbreiteten digitalen Signatursysteme wie das RSA-Verfahren [RiSA_78] und der *Digital Signature Standard* [NIST_94], genügen einem strengen Sicherheitsbegriff nicht. Erstens ist eine Fälschung nicht nachgewiesen äquivalent der Lösung des zugrundeligenden zahlentheoretischen Problems: So ist zwar das Fälschen einer RSA-Signatur nicht schwieriger als die Faktorisierung des Moduls n; die umgekehrte Aussage, daß RSA-Signaturen nur durch Faktorisierung von n gefälscht werden können, ist bis heute nicht bewiesen. DSS-Signaturen kann leicht fälschen, wer diskrete Logarithmen effizient berechnen kann; die Umkehrung dieser Aussage ist jedoch nicht nachgewiesen. Zweitens sind RSA-Signaturen, wenn weder ein Redundanzschema noch eine Hashfunktion verwendet wird, durch einen passiven Angriff existentiell, mit einem aktiven sogar selektiv fälschbar (siehe Übersicht in [Fox1_97]). DSS-Signaturen lassen sich bei einem aktiven Fälschungsangriff existentiell fälschen [ElGa_84].

3 Existentiell unfälschbare digitale Signatursysteme

Goldwasser, Micali und Rivest stellten 1984, in einer Überarbeitung 1988 das erste digitale Signatursystem vor, das ihrer eigenen, strengen Sicherheitsanforderung (siehe oben) genügte [GoMR_84,GoMR_88]. Die zentrale Idee des GMR-Signatursystems ist, eine Signatur nicht nur abhängig von einem geheimen Signierschlüssel und der Nachricht, sondern zusätzlich auch von einem nur einmalig verwendbaren Zufallswert, Referenz genannt, zu berechnen. Ist diese Referenz authentisch, ist sie dem Zugriff eines Fälschers entzogen; selbst durch einen adaptiven aktiven Angriff gewinnt ein Fälscher keine verwertbaren Informationen.[1]

Die Authentisierung dieser Referenzen erfolgt beim GMR-Verfahren durch Anhängen an die Blätter eines binären Referenzenbaumes der Tiefe d; alle Referenzen inklusive der Hilfsreferenzen an den Knoten des Baumes werden (ebenso wie die Nachricht) bezüglich des jeweiligen Elternknotens mit einer digitalen Signatur authentisiert (siehe Bild 3-1). Damit ist die Zahl der mit einem Schlüsselpaar erzeugbaren digitalen Signaturen auf 2^d begrenzt.

Eine GMR-Signatur zu einer Nachricht m bezüglich einer Referenz Ref_i besteht also neben der Nachrichtensignatur aus allen Hilfsreferenzen mit den zugehörigen Signaturen auf dem Pfad des Referenzbaums von der (öffentlichen) Wurzelreferenz R_ε bis zur i-ten Referenz Ref_i.

Das Verfahren galt lange Zeit als für praktische Zwecke ungeeignet, obwohl es mit einer Reihe von Verbesserungen eine sehr effiziente Implementierung erlaubt [Gold_86, FoPf_91]. Ein Nachteil des Verfahrens ist jedoch die erhebliche Signaturlänge.

Inzwischen wurden zwei interessante Modifikationen des GMR-Signatursystems vorgeschlagen, das Dwork/Naor-Verfahren (DN) und das Verfahren von Cramer/Damgård (CD) [DwNa_94, CrDa_96], die beide die Signaturlänge durch die Verwendung von Referenzenbäumen mit jeweils l Nachfolgern je Knoten erheblich reduzieren (siehe Bild 3-2). Damit liegt die Baumtiefe, und so auch die Signaturlänge, um einen Faktor $\lfloor \log_2 l \rfloor$ unter der des GMR-Verfahrens. Beispielsweise genügt bei einem Sicherheitsparameter von $k = 1024$ bit im DN-System eine Baumtiefe $d = 2$, um über eine Million Signaturreferenzen authentisieren zu können. GMR benötigt für dieselbe Referenzenzahl eine Baumtiefe von $d = 20$. Eine vergleichende Darstellung dieser drei Verfahren findet sich in [Fox3_97].

Es gibt noch einige weitere Verfahren, die unter einer wohldefinierten Komplexitätsannahme existentiell unfälschbar sind, wie z.B. *fail-stop*-Signaturen, eingesetzt als „normale" Signaturen [Pfit_96, HePe_92], eine (dem in Kapitel 4 vorgestellten Verfahren ähnliche) GMR-Variante auf der Basis von „Signaturprotokollen" [CrDa_94], die one-time-Signaturen von Merkle mit beweisbar kollisionsresistenten Hashfunktionen nach Damgard [Merk_87, Damg_88] oder das Signatursystem von Bos und Chaum [BoCh_92]. Die meisten dieser Signatursysteme sind aber für die Praxis nicht hinreichend effizient.

---

[1]   Eine ähnliche Idee liegt dem ElGamal-Signatursystem zugrunde; siehe auch Kapitel 4 [ElGa_84].

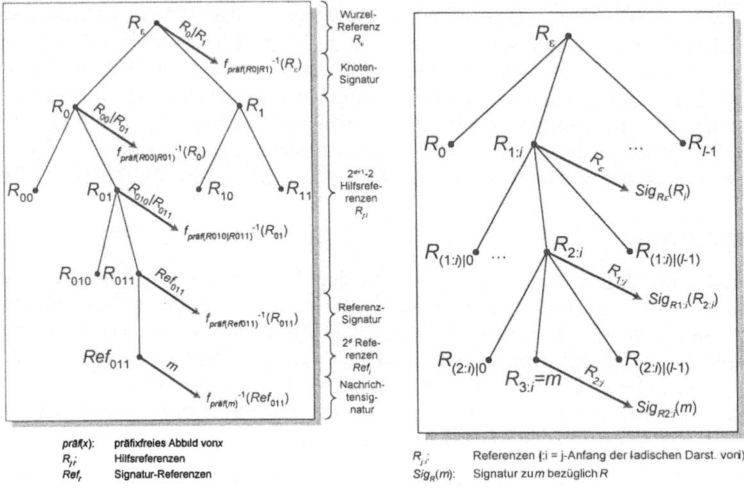

Bild 3-1: GMR-Referenzenbaum[2]   *Bild 3-2: DN-Referenzenbaum*

4 Das DLP-GMR-Signatursystem

Wie erwähnt ist der wesentliche Nachteil des GMR-Signatursystems mit $2(d+1)\,k$ bit die Länge der Signaturen. Das DN- und das CD-Verfahren verringern diese durch eine Vergrößerung der Baumbreite l. Dadurch läßt sich bei gleichbleibender Referenzenzahl mit einer um den Faktor $\lfloor \log_2 l \rfloor$ kleineren Baumtiefe d arbeiten; die Signaturänge verkürzt sich um denselben Faktor.

Eine andere Möglichkeit zur Verringerung der Signaturlänge besteht in der Verkürzung der Authentisierungen von Nachricht und Referenzen. Soll dabei die Fälschungssicherheit des Verfahrens erhalten bleiben, muß die Signierfunktion des GMR-Verfahrens gegen eine geeignete Funktion getauscht werden, die bei demselben Sicherheitsniveau kürzere Authentisierungen erzeugt. Dies gelingt mit Funktionen auf der Basis des diskreten Logarithmus-Problems (DLP), die Signaturen in Untergruppen kleinerer, primer Ordnung bilden, wie beispielsweise der DSA [NIST_94]. Sie erlauben zudem eine Implementierung auf Elliptischen Kurven über endlichen Gruppen mit einem um etwa einen Faktor 6 kürzeren Sicherheitsparameter k [FoRö_96]. Dabei steigt zugleich die Effizienz von Signier- und Prüffunktion deutlich.

Die zentrale Idee des DLP-GMR-Signatursystems beruht daher auf einer Ersetzung der Signierfunktion des originalen GMR-Signatursystem – dort aus klauenfreien Permutationenpaaren konstruiert – durch eine ElGamal-artige Funktion [ElGa_84]. Diese Auswahl der Signierfunktion fußt auf der Beobachtung, daß im originalen ElGamal-Signatursystem (ebenso in einigen Varianten wie dem DSS [NIST_94]), ein Teil der Signatur, nämlich der Wert [3]

$$r = \alpha^x \bmod p$$

---

[2]   Jede Nachricht muß vor dem Signieren präfixfrei abgebildet werden; näheres siehe [GoMR_88, FoPf_91].

[3] α ist Primitivwurzel in Z_p^*, p Primzahl und x ein nur einmalig verwendbarer, zufälliger Wert aus Z_{p-1}.

eine ähnliche Rolle spielt wie die Referenzen an den Blättern des GMR-Referenzenbaumes: Ein Angreifer kann ElGamal- bzw. DSS-Signaturen fälschen, wenn r nicht authentisch ist (oder der Signierer x wiederverwendet). Vereinfacht ausgedrückt authentisiert das im weiteren vorgestellte DLP-GMR-Verfahren die Signaturteile r ElGamal-artiger Signaturen durch die Verwendung eines GMR-artigen, binären Referenzenbaumes (d.h. $l = 2$).

Für eine etwas formalere Beschreibung des DLP-GMR-Verfahrens werden nun, in Anlehnung an die Notation in [GoMR_88], vier Phasen unterschieden:

Vorausberechnung

Der Sicherheitsparameter k wird festgelegt. Ein probabilistischer Polynomzeitalgorithmus $G(1^k)$ liefert ein Tripel (p, q, α) aus einer k bit langen Primzahl p, einer Primzahl q und der Primitivwurzel $\alpha := h^{(p-1)/q} \bmod p$ (mit $0 < h < p$, h Primitivwurzel von Z_p^*, und $\alpha > 1$), d.h. einem Generator der Untergruppe primer Ordnung q von Z_{p-1}, es gilt also: $q|(p-1)$. Die Länge von q ist unabhängig von k; q sollte, um Schutz vor „Geburtstags"-Angriffen zu bieten, wie beim DSS mindestens 160 bit lang sein [NIST_94, Dobb_97].[4] G ist dabei so definiert und k so gewählt sind, daß es praktisch unmöglich ist, diskrete Logarithmen in Z_p bezüglich α zu bestimmen.[5]

Initialisierungsphase

Die Tiefe des Referenzenbaumes d, damit auch die maximale Anzahl möglicher Signaturen je Schlüssel (2^d), wird festgelegt. Nun wählt der Signierer seinen geheimen Signierschlüssel, das Tripel $(k_S, k_{SR}, r_\varepsilon)$, zufällig aus Z_q^*. Zu diesem Signierschlüssel bestimmt er die zugehörigen Werte des öffentlichen Prüfschlüssels:[6]

$$k_V = \alpha^{k_S} \bmod p, \quad k_{VR} = \alpha^{k_{SR}} \bmod p \quad \text{und} \quad R_\varepsilon = \alpha^{r_\varepsilon} \bmod p \qquad (4.1)$$

und veröffentlicht $(p, \alpha, k_V, k_{VR}, R_\varepsilon)$ auf authentische Weise, z.B. unter Verwendung von Signaturschlüssel-Zertifikaten nach dem Signaturgesetz [Fox2_97].

Signieren

Die i-te Signatur (mit $0 \le i \le 2^d-1$) zu einer k bit langen Nachricht m berechnet der Signierer, indem er zunächst die Authentikatoren zu den Referenzen auf dem i-ten Pfad im Referenzenbaum (analog dem originalen GMR-Verfahren) bestimmt. Dazu startet er einen Suchalgorithmus $Path(i, d)$, der auf Eingabe von Signaturnummer i und Baumtiefe d den Pfad im geheimen Referenzenbaum von der Wurzel-Referenz r_ε zum i-ten Blatt $r_i := r_{d:i}$ zurückliefert (siehe Bild 4-1). Dieser „Referenzenpfad" besteht aus einer Liste von d geheimen Referenzen-Paaren $(r_{j:i}|_0, r_{j:i}|_1)$, jeweils beiden Nachfolgeknoten einer Referenz $r_{j:i}$ an einem Knoten des Pfades, und einer geheimen Nachrichtenreferenz ref_i. Alle Referenzen $r_{j:i}$ und ref_i sind dabei zufällig gewählte (und paarweise verschiedene) Elemente aus Z_q^*.[7]

---

4   Die Wahl der primen Untergruppe von Z_{p-1} hat beweistechnische Gründe; siehe Kapitel 5.

5   Eine formalere Definition dieser Annahme findet sich in Kapitel 5.

6   Der Sicherheitsbeweis erfordert die Verwendung zweier Schlüsselpaare; siehe Kapitel 5.

7   Der Index $j:i$ bezeichnet hier die höchstwertigen j Bits der Binärdarstellung von i, z.B.: $2:5 = 10$.

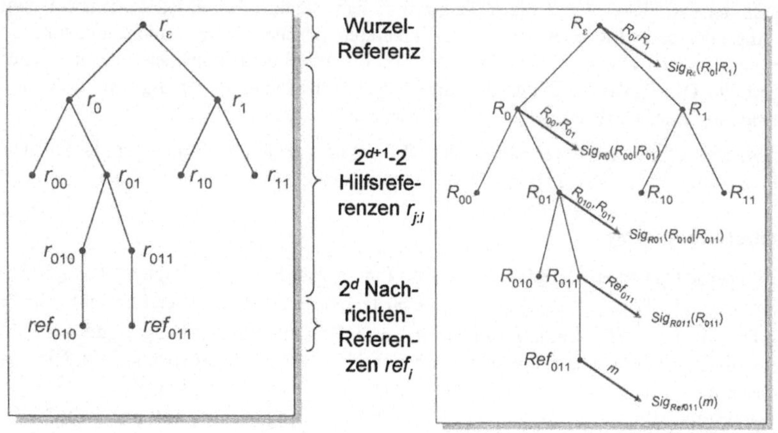

Bild 4-1: Referenzenbaum (Tiefe d=3)   *Bild 4-2: Authentisierungsbaum (Tiefe d=3)*

Für jedes Paar geheimer Referenzen $(r_{j:i|0}, r_{j:i|1})$ bestimmt der Signierer nun die zugehörigen öffentlichen Referenzen $R_{j:i|0}$ und $R_{j:i|1}$ und authentisiert beide bezüglich deren Elternknoten $R_{j:i}$, indem er für $j = 0, ..., d\text{-}1$ berechnet (siehe auch Bild 4-2):

$$R_{j:i|0} = \alpha^{r_{j:i|0}} \bmod p, \quad R_{j:i|1} = \alpha^{r_{j:i|1}} \bmod p \tag{4.2}$$

$$Sig_{R_{j:i}}(R_{j:i|0}, R_{j:i|1}) = k_{SR} \cdot hash(R_{j:i|0} | R_{j:i|1}) + r_{j:i} \bmod q \tag{4.3}$$

Auf dieselbe Weise erhält und authentisiert er die öffentliche Nachrichtenreferenz Ref_i bezüglich der Referenz R_i am i-ten Blatt des Authentisierungsbaums: [8]

$$Ref_i = \alpha^{ref_i} \bmod p \tag{4.4}$$

$$Sig_{R_i}(Ref_i) = k_{SR} \cdot hash(Ref_i) + r_{d:i} \bmod q \tag{4.5}$$

Schließlich berechnet er die Signatur zur Nachricht m bezüglich der Nachrichtenreferenz Ref_i:

$$Sig_{Ref_i}(m) = k_S \cdot hash(m) + ref_i \bmod q \tag{4.6}$$

Ist $m \geq q$, muß der Signierer die Nachricht m zuvor mit einer kollisionsresistenten Hashfunktion $hash(\,)$ [Dobb_97] auf einen Wert $hash(m) < q$ abbilden.

Für jeden Authentisierungsschritt im Referenzenbaum sind demnach zwei Exponentiationen in Z_p^* (allerdings mit lediglich 160 bit langen Exponenten) und eine Multiplikation in Z_q erforderlich (ohne Berücksichtigung des Aufwands der Hashfunktion). Die Authentisierung der Nachricht erfordert eine weitere Multiplikation in Z_q. Damit liegt der Aufwand (ohne Optimierung) bei[9]

$$2d \cdot E_k + (d+1) \cdot M_{|q|} \tag{4.7}$$

Der tatsächlich notwendige Rechenaufwand für die Bestimmung einer Signatur liegt erheblich darunter; er ist durchschnittlich sogar unabhängig von der Baumtiefe d (siehe Kapitel 6).

---

[8]   Wie im originalen GMR-Signaturverfahren sind hier aus beweistechnischen Gründen zusätzliche authentische Referenzen Ref_i für die Nachrichtensignaturen („*bridge-items*" genannt) erforderlich.

[9]   Wesentliche Operationen sind Exponentiation (E), Multiplikation (M) und Inversenbestimmung (I).

Die Länge einer Signatur wächst allerdings linear mit der Tiefe d des Authentisierungsbaums. Ohne weitere Optimierungen setzt sich eine Signatur (ohne Nachricht) zusammen aus $2d$ Referenzen $R_{j \cdot i}$ ($j > 0$), der Nachrichtenreferenz Ref_i, den zugehörigen $d+1$ Signaturen sowie der Nachrichtensignatur, zusammen also

$$(2d+1)\, k + (d+2)\, 160 \text{ bit} \tag{4.8}$$

Prüfen

Um die Gültigkeit einer Signatur zu überprüfen, beginnt der Empfänger bei der Nachrichtensignatur $Sig_{Ref_i}(m)$ und prüft, ob gilt:

$$\alpha^{Sig_{Ref_i}(m)} = k_V^{\,hash(m)} \cdot Ref_i \bmod p \tag{4.9}$$

$$\alpha^{Sig_{R_{d \cdot j}}(Ref_i)} = k_{RV}^{\,hash(Ref_i)} \cdot R_{d \cdot j} \bmod p \tag{4.10}$$

Anschließend wird nach (4.10) die Authentizität der Nachrichtenreferenz Ref_i geprüft, und schließlich testet der Empfänger, ob für alle j mit $0 \le j \le d\text{-}1$ gilt:

$$\alpha^{Sig_{R_j \cdot i}(R_{j \cdot j|0}, R_{j \cdot j|1})} = k_{RV}^{\,hash(R_{j \cdot j|0}|R_{j \cdot j|1})} \cdot R_{j \cdot j} \bmod p \tag{4.11}$$

Sind alle Gleichungen erfüllt, akzeptiert er die Signatur.

In jedem Überprüfungsschritt müssen zwei Exponentiationen (mit je 160 bit langen Exponenten) und eine Multiplikation in $Z_p^{\,*}$ durchgeführt werden (ohne Berücksichtigung der kollisionsresistenten Hashfunktion). Für den gesamten Pfad summiert sich der Berechnungsaufwand einer Signaturprüfung auf

$$2\,(d+2)\cdot E_k + (d+2)\cdot M_k \tag{4.12}$$

5 Fälschungssicherheit des DLP-GMR-Verfahrens

Die Fälschungssicherheit des in Kapitel 4 vorgestellten DLP-GMR-Signatursystems genügt der strengen Anforderung von Goldwasser, Micali und Rivest [GoMR_88], d.h. es ist existentiell unfälschbar auch unter einem adaptiven aktiven Angriff – vorausgesetzt, die folgende Komplexitätsannahme gilt:

> **Diskretes-Logarithmus-Problem (DLP):** Sei k ein Sicherheitsparameter. Gegeben eine Primzahl p der Länge k bit, α eine Primitivwurzel, die eine Untergruppe primer Ordnung von $Z_{p \cdot 1}$ erzeugt, und q die Ordnung dieser Untergruppe; (p, q, α) sind zufällig gewählt von einem Generierungsalgorithmus $G(1^k)$. Gegeben sei weiter ein zufällig gewählter Wert y aus der von α erzeugten Untergruppe von $Z_{p \cdot 1}$. Dann gilt für hinreichend große k: Es gibt keinen Polynomzeitalgorithmus, der mit nicht-vernachlässigbarer Erfolgswahrscheinlichkeit $\log_\alpha y$ bestimmt, d.h. ein $x \in Z_q$ findet, sodaß gilt: $\alpha^x \bmod p = y$.

Der Beweis für diese Behauptung wird im folgenden durch Widerspruch geführt: Es wird gezeigt, daß ein Angreifer, dem es mit nicht-vernachlässigbarer Wahrscheinlichkeit gelingt, auch nur eine Signatur zu fälschen, den diskreten Logarithmus $\log_\alpha y$ einer beliebigen, vorgegebenen Zahl y aus der von α erzeugten primen Untergruppe von $Z_{p \text{-} 1}$ bestimmen kann.

Im Kern verwendet der Beweis einen „simulierten Signierer", der von einem tatsächlichen Signierer nicht unterschieden werden kann. Wie im Beweis von Goldwasser, Micali und Rivest [GoMR_88] werden zwei verschiedenen Simulationen mit jeweils dem „halben Geheimnis" konstruiert; zu einer dieser beiden fälscht der Angreifer dann eine Signatur. Daraus läßt sich dann ein Widerspruch zum DLP herleiten. Die Simulation wird dabei abhängig vom Typ der Signaturfälschung (siehe unten) ausgewählt.

Sei k ein (geeigneter) Sicherheitsparameter. Gegeben seien eine Primzahl p der Länge k bit, die prime Ordnung q einer Untergruppe von $Z_{p\text{-}1}{}^*$ und eine Primitivwurzel α, die die multiplikative Untergruppe von $Z_p{}^*$ der Ordnung q erzeugt, sowie ein zufällig gewählter Wert y aus der von α erzeugten primen Untergruppe von $Z_{p\text{-}1}$. Dann können die beiden folgenden „Simulationen" eines Signierers konstruiert werden, die von einem realen Signierer ununterscheidbar sind:

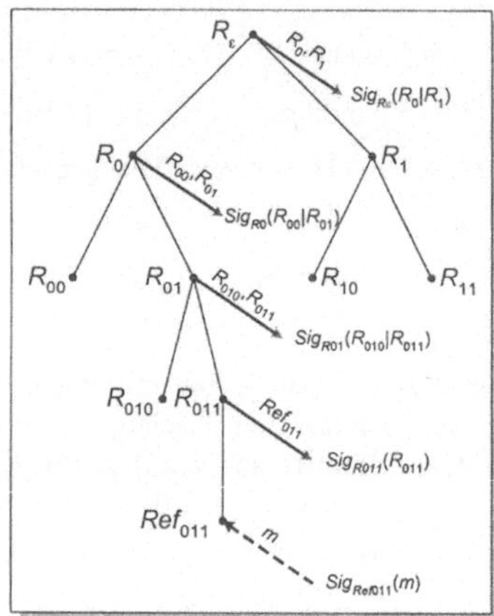
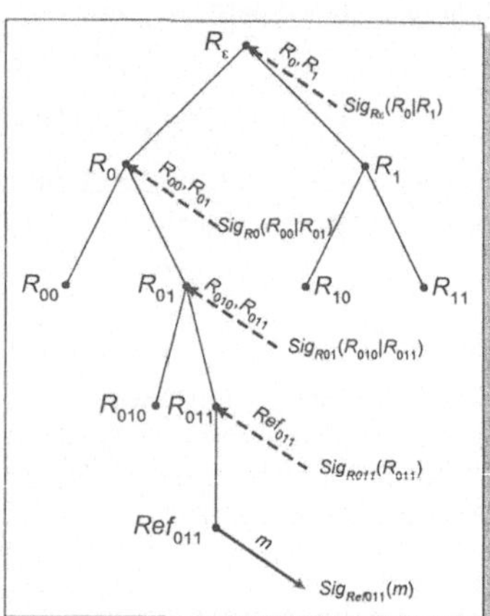

Bild 5-1: „Top down"-Simulation   *Bild 5-2: „Bottom up"-Simulation*

„Top down"-Simulation

Für die Konstruktion einer *„top down"*-Simulation wählen wir, wie in der in Kapitel 4 beschriebenen Initialisierungsphase, ein geheimes Schlüsselpaar (k_{SR}, r_ε) und bestimmen daraus die zugehörigen öffentlichen Schlüssel-Werte (k_{VR}, R_ε). Als öffentlichen Schlüssel k_V wählen wir den *zufälligen* Wert y aus der von α erzeugten primen Untergruppe von $Z_{p\text{-}1}$ und veröffentlichen $(p, \alpha, k_V, k_{VR}, R_\varepsilon)$. Das Ergebnis dieses Schlüsselgenerierungsprozesses ist nicht von dem der in Kapitel 4 beschriebenen Initialisierungsphase zu unterscheiden, da α ein Generator der multiplikativen Untergruppe von $Z_p{}^*$ der Ordnung q ist und daher Zufälligkeit und Verteilung von y und $\alpha^{k_s} \bmod p$ übereinstimmen.

Soll der so „simulierte Signierer" nun eine digitale Signatur zu einer Nachricht m erzeugen, konstruieren wir zunächst, genau wie im Falle eines realen Signieres, den i-ten Pfad im Authentisierungsbaum *„top down"* (siehe Bild 5-1) von der Wurzelreferenz R_ε bis zum i-ten Blatt $R_{d:i}$: Alle benötigten geheimen Referenzen $r_{j:i}$ werden zufällig aus $Z_q{}^*$ gewählt; die zugehörigen Referenzen $R_{j:i}$ werden wie in (4.2) bestimmt und jeweils paarweise nach (4.3) authentisiert.

Abweichend vom Signiervorgang in Kapitel 4 wird nun die Nachrichtensignatur $Sig_{Ref_i}(m)$ zufällig aus Z_q^* gewählt und die dazu passende Nachrichtenreferenz Ref_i wie folgt bestimmt:

$$Ref_i = \alpha^{Sig_{Ref_i}(m)} \cdot k_V^{-hash(m)} \bmod p \qquad (5.1)$$

Unabhängig davon, ob die Berechnung der Referenz Ref_i „aufwärts" oder „abwärts" erfolgt, bleibt die Verteilung von Ref_i erhalten. Schließlich wird Ref_i wie in (4.5) authentisiert.

Die resultierende Signatur ist ununterscheidbar von der eines „realen" Signierers. Der einzige, für den Sicherheitsbeweis wichtige Unterschied ist, daß der Signierer jede Signatur ohne Kenntnis des geheimen Schlüsselteils k_S und der Nachrichtenreferenzen ref_i erzeugt.

„Bottom up"-Simulation

Die zweite Simulation beginnt mit der Wahl von k_S als geheimem Schlüssel und der Berechnung des zugehörigen öffentlichen k_V. Anders als bei einem „realen" Signierer wird als öffentlicher Schlüssel-Parameter k_{VR} der *zufällige* Wert y aus der von α erzeugten primen Untergruppe von Z_{p-1} gewählt. Dies ist, wie bei der „*top down*"-Simulation, ununterscheidbar von der Bestimmung von k_{VR} aus einem zufällig gewählten k_{SR}. Dann wird, abweichend vom Ablauf der Initialisierungsphase eines tatsächlichen Signierers, vor der Veröffentlichung des öffentlichen Schlüssel-Tupels der gesamte Authentisierungsbaum von unten nach oben („*bottom up*") vorausberechnet. Dazu werden zunächst die geheimen Nachrichtenreferenzen ref_i zufällig aus Z_q^* gewählt und die zugehörigen öffentlichen Referenzen Ref_i nach (4.4) bestimmt.

Anschließend werden die „inneren" Referenzen $R_{j:i}$ und die zugehörigen Authentisierungen $Sig_{R_{j:i}}(R_{j:i|0}, R_{j:i|1})$ ohne Kenntnis von k_{SR} und der geheimen Referenzen $r_{j:i}$ des Referenzenbaumes wie folgt bestimmt (siehe auch Bild 5-2): Für jede Nachrichtenreferenz Ref_i wird eine Signatur $Sig_{R_{d:i}}(Ref_i)$ zufällig aus Z_q^* gewählt und R_i daraus berechnet:

$$R_{d:i} = \alpha^{Sig_{R_{d:i}}(Ref_i)} \cdot k_{VR}^{-hash(Ref_i)} \bmod p \qquad (5.2)$$

Schließlich werden für $j = d{-}1, \dots, 0$ jeweils paarweise zwei Referenzen $R_{j:i|0}$ und $R_{j:i|1}$ „aufwärts" signiert, indem die Signatur $Sig_{R_{j:i}}(R_{j:i|0}, R_{j:i|1})$ zufällig gewählt und die zugehörige Referenz $R_{j:i}$ daraus folgendermaßen berechnet wird:

$$R_{j:i} = \alpha^{Sig_{R_{j:i}}(R_{j:i|0}, R_{j:i|1})} \cdot k_{VR}^{-hash(R_{j:i|0}|R_{j:i|1})} \bmod p \qquad (5.3)$$

Der letzte Berechnungsschritt (für $j = 0$) liefert die Wurzelreferenz R_ε, die nun zusammen mit den anderen Schlüsselparametern als öffentlicher Schlüssel $(p,\ \alpha,\ k_V,\ k_{VR},\ R_\varepsilon)$ veröffentlicht werden kann.

Diese Schlüsselgenerierung ist ununterscheidbar von der in Kapitel 4 beschriebenen Initialisierungsphase, da die Verteilung der zufällig gewählten Signaturen $Sig_{R_{j:i}}$ durch (5.2) bzw. (5.3) nicht verändert wird; die Referenzen $R_{j:i}$ sind also ebenso zufällig wie bei einer Berechnung aus zufällig gewählten $r_{j:i}$.

Soll nun eine i-te Nachricht m signiert werden, sucht der „simulierte Signierer" alle auf dem i-ten Pfad von der Wurzelreferenz R_ε bis zur Nachrichtenreferenz Ref_i liegenden Referenzen und zugehörigen Signaturen heraus. Dies ist ebenfalls ununterscheidbar von einem „realen" Signierer, denn die Verteilung der Signaturen und Referenzen ändert sich nicht durch die Vorausberechnung. Abschließend wird die Nachrichtensignatur $Sig_{Ref_i}(m)$ zur Referenz Ref_i gemäß (4.6) berechnet und die gesamte Signatur ausgegeben.

Die resultierende Signatur ist nicht unterscheidbar von der eines tatsächlichen Signierers. Der für den Sicherheitsbeweis wichtige Unterschied ist, daß der Signierer weder den geheimen Schlüsselteil k_{SR} noch eine der Referenzen $r_{j:i}$ kennt.

Nun können wir den folgenden zentralen Satz beweisen (die Umkehrung gilt trivialerweise):

> **Theorem**: Das in Kapitel 4 vorgestellte digitale Signatursystem ist existentiell unfälschbar selbst unter einem adaptiven, aktiven Angriff, wenn das Diskrete-Logarithmus-Problem gilt.

Beweis: Angenommen, es gäbe einen probabilistischen Polynomzeitalgorithmus A, der mit nicht-vernachlässigbarer Erfolgswahrscheinlichkeit $P(k) \geq 1/Q(k)$ eine DLP-GMR-Signatur (für unendlich viele $k > k_0$ und ein Polynom Q) fälschen kann.

Es wird nun geziegt, daß wir dann einen probabilistischen Polynomzeitalgorithmus B konstruieren können, der (unter Verwendung von A) den diskreten Logarihmus eines beliebigen, vorausgewählten Wertes y aus der von α erzeugten primen Untergruppe von Z_{p-1} mit nicht-vernachlässigbarer Erfolgswahrscheinlichkeit bestimmt. Dies ist ein Widerspruch zu unserer Komplexitätsannahme, dem DLP.

Hat Algorithmus A eine DLP-GMR-Signatur gefälscht, muß einer der folgenden drei „Fälschungstypen" vorliegen:

- **Typ 1**: Die Signatur enthält eine Nachrichtensignatur $Sig_{Ref_i}(m)$ zu einer Nachricht m, die noch nicht signiert wurde, bezüglich einer Nachrichtenreferenz Ref_i, die bereits authentisiert wurde. Das bedeutet, daß bereits eine Nachricht bezüglich Ref_i signiert wurde.

- **Typ 2**: Die Signatur enthält **entweder** die Authentisierung $Sig_{R_{j:i}}(R_{j:i|0}, R_{j:i|1})$ zweier Referenzen $R_{j:i|0}$ und $R_{j:i|1}$ (mit $0 \leq j < d$) **oder** $Sig_{R_{d:i}}(Ref_i)$, jeweils bezüglich einer bereits authentisierten Referenz $R_{j:i}$ im Authentisierungsbaum. Dabei ist es für den weiteren Beweis unerheblich, ob $R_{j:i}$ bereits für eine Authentisierung „genutzt" wurde oder noch „frei" ist.

- **Typ 3**: Der Fälscher konnte eine Kollision für die Hashfunktion $hash(\ )$ finden und so eine Nachrichtensignatur für eine neue Nachricht m^* oder (im Authentisierungsbaum) eine Authentisierung für eine neue Referenz Ref_i^* oder $R_{j:i}^*$ verwenden.

Daher gilt für den Fälschungsalgorithmus A einer der folgenden drei Fälle:

- **Fall 1**: Für unendlich viele k kann Algorithmus A Signaturen durch einen Typ-1-Fälschungsangriff mit einer Erfolgswahrscheinlichkeit $P_1(k) > 1/3 \cdot Q(k)$ fälschen.

- **Fall 2**: Für unendlich viele k kann Algorithmus A Signaturen durch einen Typ-2-Fälschungsangriff mit einer Erfolgswahrscheinlichkeit $P_2(k) > 1/3 \cdot Q(k)$ fälschen.

- **Fall 3**: Für unendlich viele k kann Algorithmus A Signaturen durch einen Typ-3-Fälschungsangriff mit einer Erfolgswahrscheinlichkeit $P_3(k) > 1/3 \cdot Q(k)$ fälschen.

Trifft Fall 3 zu, liegt direkt ein Widerspruch zum DLP vor, sofern eine Hashfunktion eingesetzt wurde, für die die Äquivalenz von Kollisionsresistenz und DLP nachgewiesen ist, z.B. eine der von Damgård vorgeschlagenen Funktionen [Damg_87].[10]

Für die Fälle 1 und 2 können wir nun, abhängig vom jeweils vorliegenden Fall, zwei Varianten des folgenden probabilistischen Polynomzeitalgorithmus konstruieren, der mit nicht-vernachlässigbarer Wahrscheinlichkeit diskrete Logarithmen in Z_p^* bestimmt:

---

[10] Wird aus Effizienzgründen eine andere Hashfunktion (z.B. RIPEMD-160, SHS-1 [DoBP_96, NIST_95]) verwendet, dann beruht die Sicherheit des Verfahrens auch auf der Kollisionsresistenz dieser Hashfunktion.

Algorithmus B:

- **Eingabe**: Tripel (p, q, α) aus einer Primzahl p der Länge k bit, einer Primitivwurzel α (erzeugendes Element der multiplikativen Untergruppe der Ordnung q in Z_p^*) und der Ordnung q dieser Untergruppe; sowie ein (beliebiger) Wert y aus der von α erzeugten primen Untergruppe von Z_{p-1}.

Wie in der Initialisierungsphase (siehe Kapitel 4) wählen wir nun eine Baumtiefe d. Nun lassen wir, abhängig vom vorliegenden Fall, den Fälschungsalgorithmus A gegen einen der simulierten Signierer laufen:

- **Fall 1**: Der Fälschungsalgorithmus A wird gegen die „*bottom up*"-Simulation gestartet.

Nach der Initialisierungsphase der Simulation liefert der „simulierte Signierer" dem Angreifer (Algorithmus A) endlich viele Signaturen ($< 2^d$) zu von diesem gewählten Nachrichten. Liefert A nun eine Signaturfälschung von Typ 1, dann gilt für eine i-te Nachrichtensignatur $SigRef_i(m_1)$ bezüglich einer Referenz Ref_i zu einer (bisher nicht signierten) Nachricht m_1:

$$Ref_i = \alpha^{Sig_{Ref_i}(m_1)} \cdot y^{-hash(m_1)} \bmod p \tag{5.4}$$

Da die Referenz Ref_i schon authentisiert wurde, muß die Simulation zuvor schon eine Nachricht m_2 bezüglich Ref_i signiert haben; es gilt also:

$$Ref_i = \alpha^{Sig_{Ref_i}(m_2)} \cdot y^{-hash(m_2)} \bmod p \tag{5.5}$$

Damit folgt:

$$\log_\alpha y = \frac{Sig_{Ref_i}(m_1) - Sig_{Ref_i}(m_2)}{hash(m_1) - hash(m_2)} \tag{5.6}$$

- **Fall 2**: Der Fälschungsalgorithmus A wird gegen die „*top down*"-Simulation gestartet.

Nach der Initialisierungsphase liefert die Simulation dem Angreifer endlich viele Signaturen ($< 2^d$) zu von A gewählten Nachrichten. Bestimmt A eine Signaturfälschung von Typ 2, dann gilt **entweder** für eine Referenz $R_{j:i}$ auf dem Pfad einer i-ten Signatur im Authentisierungsbaum:

$$R_{j:i} = \alpha^{Sig_{R_{j:i}}(R_{j:i|0}{}^*, R_{j:i|1}{}^*)} \cdot y^{-hash(R_{j:i|0}{}^*|R_{j:i|1}{}^*)} \bmod p \tag{5.7}$$

bezüglich zweier „neuer", d.h. nicht von der Simulation bezüglich $R_{j:i}$ authentisierter Referenzen $R_{j:i|0}{}^*$ und $R_{j:i|1}{}^*$, **oder**: für eine Referenz $R_{d:i}$ am i-ten Blatt des Authentisierungsbaums:

$$R_{d:i} = \alpha^{Sig_{R_{d:i}}(Ref_i{}^*)} \cdot y^{-hash(Ref_i{}^*)} \bmod p \tag{5.8}$$

bezüglich einer „neuen", d.h. vom simulierten Signierer nicht bezüglich R_i authentisierten Nachrichtenreferenz $Ref_i{}^*$. Da die Simulation in der Initialisierungsphase bereits den gesamten Authentisierungsbaum vorausberechnet hat, existiert in beiden Fällen bereits eine Signatur bezüglich $R_{j:i}$ bzw. $R_{d:i}$. Es gilt also analog Fall 1 **entweder**:

$$\log_\alpha y = \frac{Sig_{R_{j:i}}(R_{j:i|0}{}^*, R_{j:i|1}{}^*) - Sig_{R_{j:i}}(R_{j:i|0}, R_{j:i|1})}{hash(R_{j:i|0}|R_{j:i|1}) - hash(R_{j:i|0}{}^*|R_{j:i|1}{}^*)} \tag{5.9}$$

oder:

$$\log_\alpha y = \frac{Sig_{R_{d:i}}(Ref_i{}^*) - Sig_{R_{d:i}}(Ref_i)}{hash(Ref_i{}^*) - hash(Ref_i)} \tag{5.10}$$

- **Ausgabe**: Diskreter Logarithmus $\log_\alpha y$.

Beide Varianten von Algorithmus **B** berechnen also für unendlich viele $k > k_0$ und ein Polynom Q diskrete Logarithmen zu einem gegebenen y aus der von α erzeugten primen Untergruppe von Z_{p-1} mit der Wahrscheinlichkeit:

$$P(k) > 1/3 \cdot Q(k) \qquad \text{(q.e.d.)}$$

6 Effiziente Implementierung des DLP-GMR-Systems

Die Speicheranforderungen einer DLP-GMR-Signatur und der Aufwand für das Signieren einer Nachricht lassen sich mit einigen Optimierungen gegenüber einer „naiven" Implementierung signifikant verringern.

Signieren

Durch Vorausberechnung aller Referenzen $R_{j\cdot i}$ des Authentisierungsbaums aus zufällig gewählten $r_{j\cdot i}$ kann der größte Teil des Rechenaufwands bei der Bestimmung einer digitalen Signatur, nämlich je Signatur bis zu $2d{+}1$ Exponentiationen, eingespart werden. Der Speicheraufwand für die vorausberechneten $2^{d+1}{-}1$ Referenzen $R_{j\cdot i}$ und 2^d Referenzen $r_{d\cdot i}$ liegt bei:

$$(2^{d+1}{-}1)\cdot k + 2^d\cdot|q| \text{ bit} \qquad (6.1)$$

Jede Referenz $R_{j\cdot i}$ im Authentisierungsbaum kann darüberhinaus $2^{d\cdot j}$-mal verwendet werden. Durch die Speicherung von maximal d Signaturen $Sig_{R_{j\cdot i}}(R_{j\cdot i|0}|R_{j\cdot i|1})$ der Länge k bit kann der Rechenaufwand weiter reduziert werden: Durchschnittlich sind je Signatur, wie in [FoPf_91] für das GMR-Signatursystem gezeigt wurde, nur die Berechnung von drei Authentisierungen, zusammen also drei Multiplikationen in Z_q^* erforderlich.[11]

Berechnet der Signierer auch noch die Nachrichtenreferenz Ref_i und deren Authentisierung $Sig_{R_{d\cdot i}}(Ref_i)$ voraus, z.B. in „idle"-Zuständen des Rechners, genügt die online-Bestimmung der Nachrichtensignatur $Sig_{Ref_i}(m)$ bezüglich Ref_i – eine einzige Multiplikation in Z_q^*:

$$1 \text{ M}_{|q|} \qquad (6.2)$$

Der Speicherbedarf für die vorausberechneten Werte kann erheblich reduziert werden, denn es genügt, nur jeweils die für die nächste Signatur erforderlichen Referenzen und Authentisierungen vorauszuberechnen und zu speichern. Das sind zusammen d Referenzenpaare $(R_{j\cdot i|0}, R_{j\cdot i|1})$; hinzu kommen maximal d, durchschnittlich $d/2$ Referenzen $r_{j\cdot i}$, die noch keine Nachfolgeknoten besitzen, und daher für deren Authentisierung aufbewahrt werden müssen, sowie die Nachrichtenreferenzen ref_i und Ref_i. Weiter werden insgesamt $d{+}1$ Authentisierungen vorausberechnet und gespeichert. Damit summiert sich der (durchschnittliche) Speicherbedarf beim Signierer (ohne geheime Schlüssel) auf:

$$(3d{+}2)\cdot k + (d/2{+}1)\cdot|q| \text{ bit} \qquad (6.3)$$

Verifizieren

Wie von Goldreich für das GMR-Signatursystem vorgeschlagen und auch im Verfahren von Cramer und Damgård genutzt, kann die Länge einer Signatur weiter verringert werden [Gold_86, CrDa_96]: Durch eine geringfügige Änderung der Signaturprüfung erhält der Empfänger alle Referenzen auf dem Authentisierungspfad, $R_{j\cdot i}$ (für $j = 0, ..., d$), $R_{d\cdot i}$ und Ref_i, als

---

[11] Es ist leicht zu sehen, daß für alle Signaturen mit ungeradem i, also jede zweite, die Authentisierungen aller Referenzen inklusive der Blätter des Authentisierungsbaums wiederverwendet werden können; nur Ref_i und die Nachricht m müssen online signiert werden.

Zwischenergebnisse der Prüfung (ähnlich der „*bottom up*"-Konstruktion der Simulationen in Kapitel 5):

$$Ref_i = \alpha^{Sig_{Ref_i}(m)} \cdot k_V^{-hash(m)} \bmod p \tag{6.4}$$

$$R_{d:i} = \alpha^{Sig_{R_{d:i}}(Ref_i)} \cdot k_{VR}^{-hash(Ref_i)} \bmod p \tag{6.5}$$

$$R_{j:i} = \alpha^{Sig_{R_{j:i}}(R_{j:i\|0},R_{j:i\|1})} \cdot k_{VR}^{-hash(R_{j\cdot i\|0}|R_{j.i\|1})} \bmod p \tag{6.6}$$

Zuletzt vergleicht der Empfänger $R_{0:i}$ mit dem öffentlichen R_ε. Daher genügt es, statt der Referenzenpaare in einer Signatur nur die Referenz mitzuschicken, bezüglich der keine nachfolgenden Referenzen authentisiert werden. Die Länge einer Signatur reduziert sich damit auf:

$$2 \cdot (d+1) \cdot |q| \text{ bit} \tag{6.7}$$

Die Berechnung der in (6.4), (6.5) und (6.6) je Referenz erforderlichen „Doppelexponentiation" kann ebenso wie beim DSS wesentlich effizienter als die Berechnungen (4.9), (4.10) und (4.11) implementiert werden [Fox_93].

Implementierung auf elliptischen Kurven

Alle bis heute vorgeschlagenen, auch bei adaptiven aktiven Angriffen existentiell unfälschbaren digitalen Signatursysteme, für die die Äquivalenz von Fälschbarkeit und Widerspruch zu einer wohldefinierten Komplexitätsannahme mathematisch gezeigt ist, und die zudem effizient und speichersparend implementiert werden können, setzen die Gültigkeit der RSA-Annahme oder der Faktorisierungsannahme voraus.

Angesichts der ständigen Entwicklung verbesserter Faktorisierungsalgorithmen mit subexponentiellem Aufwand sind jedoch digitale Signatursysteme wünschenswert, die Sicherheit gegen adaptive aktive Fälschungsangriffe auf der Basis des diskreten Logarithmusproblems (DLP) bieten. Denn für die Lösung des DLP auf Elliptischen Kurven sind bis heute keine subexponentiellen Algorithmen bekannt. DLP-basierte digitale Signatursysteme können daher über Elliptischen Kurven mit erheblich kürzeren Schlüssel- und Signaturlängen ($k = 160$ statt 1024 bit, d.h. Faktor 6) arbeiten und erlauben so sehr effiziente Implementierungen [FoRö_96].

Varianten des vorgeschlagenen Signatursystems

Die Signierfunktion des vorgeschlagenen Signatursystems verwendet eine Variante des ElGamal-Verfahrens. Eine Übersicht und Verallgemeinerung von verschiedenen Varianten des ElGamal-Verfahrens wurde 1994 von Horster, Petersen und Michels veröffentlicht [HoPM_94]. Durch eine „Permutation" der Exponenten kann das vorgeschlagene Verfahren auf ebensolche Weise verallgemeinert werden, ohne seine Sicherheitseigenschaft zu verlieren.

Var.	Signaturberechnung	Aufwand	Signaturprüfung	Aufwand
1	$Sig_R(m)=(m-k_S)\cdot r^{-1}$	$I_{|q|} + M_{|q|}$	$\alpha^m = k_V \cdot R^{Sig_R(m)}$	$2\,E_k + M_k$
2	$Sig_R(m)=(m-r)\cdot k_S^{-1}$	$M_{|q|}$	$\alpha^m = k_V^{Sig_R(m)} \cdot R$	$2\,E_k + M_k$
3	$Sig_R(m)= k_S + r \cdot m$	$M_{|q|}$	$\alpha^{Sig_R(m)} = k_V \cdot R^m$	$2\,E_k + M_k$
4	$Sig_R(m)=(1-k_S \cdot m)\cdot r^{-1}$	$I_{|q|} + 2\,M_{|q|}$	$\alpha = k_V^{\,m} \cdot R^{Sig_R(m)}$	$2\,E_k + M_k$
5	$Sig_R(m)=(1-r \cdot m)\cdot k_S^{-1}$	$2\,M_{|q|}$	$\alpha = k_V^{Sig_R(m)} \cdot R^m$	$2\,E_k + M_k$

Tabelle 6-1: Varianten des vorgeschlagenen digitalen Signatursystems

Tabelle 6-1 zeigt fünf Varianten der in Kapitel 4 vorgestellten Signierfunktion. Die letzte Spalte gibt jeweils die erforderlichen arithmetischen Operationen an. Der tatsächliche Aufwand der einzelnen Operationen hängt von dem jeweils verwendeten endlichen Körper ab; das Verhältnis zwischen den einzelnen Operationen unterscheidet sich beispielsweise in $GF(p)$, $GF(2^n)$, $E(GF(p))$ und $E(GF(2^n))$ erheblich; siehe z.B. [Mene_93, FoRö_96]. [12]

Die multiplikativen Inversen, die für die Signaturberechnung in den Varianten 1 und 4 benötigt werden, können vorausberechnet werden. Die Berechnung der multiplikativen Inversen des geheimen Schlüssels k_S wurde nicht berücksichtigt, da diese während der Schlüsselgenerierung einmalig vorab erfolgen kann; es genügt für diese Varianten, k_S^{-1} statt k_S zu speichern.

7 Vergleichende Zusammenfassung und Ausblick

Das in Kapitel 4 vorgestellte DLP-GMR-Signatursystem ist auch bei adaptiven aktiven Angriffen existentiell unfälschbar; die Äquivalenz von Signaturfälschung und Lösung des DLP wurde in Kapitel 5 bewiesen. Kapitel 6 zeigt, daß das Verfahren sehr effiziente Implementierungen erlaubt: Mit Vorausberechnungen läßt sich die Erzeugung einer Signatur auf eine modulare Multiplikation reduzieren. Ein für ausgewählte Anwendungen („Scheckbuch", „TAN-Liste") möglicherweise interessanter Aspekt ist die endliche Zahl von Signaturen je Schlüsselpaar.

Verfahren	DSS	RSA	GMR	DN	CD	DLP-GMR
Signaturen	beliebig	belieb.	2^d	k^d	l^d	2^d
Signieren	$2\,M_{|q|}$	E_k	$E_k + M_k$	$E_k + 80\,M_k$	$2{\cdot}E_k + M_k$	$M_{|q|}$
$|k_S|$ [bit]	k bit	k bit	$3(d{+}1){\cdot}k$ bit	$(d{+}k){\cdot}k$ bit	$(d{+}l{+}1){\cdot}k$ bit	$3(d{+}4){\cdot}k$ bit
$|Sig|$ [bit]	$2|q|$ bit	k bit	$2(d{+}1){\cdot}k$ bit	$d{\cdot}k$ bit	$(d{+}1){\cdot}k$ bit	$2(d{+}1){\cdot}|q|$ bit
Prüfen	$2\,E_k + I_{|q|} + 2\,M_{|q|}$	E_k	$2(d{+}1){\cdot}k\,M_k$	$d{\cdot}E_k + d{\cdot}80\,M_k$	$(d{+}1){\cdot}(2E_k + M_k)$	$2(d{+}1){\cdot}E_k + (d{+}1){\cdot}M_k$
$|k_V|$ [bit]	$3k{+}|q|$ bit	$2{\cdot}k$ bit	$2{\cdot}k$ bit	$2(k{+}1){\cdot}k$ bit	$3{\cdot}k$ bit	$5{\cdot}k$ bit

Tabelle 7-1: Aufwandsvergleich digitaler Signatursysteme

Tabelle 7-1 gibt den durchschnittlichen Aufwand des DLP-GMR-Signatursystems im Vergleich zu den in Kapitel 3 vorgestellten digitalen Signatursystemen, dem RSA-Verfahren und dem DSS an. Dabei sind die Kosten für die Generierung der Zufallswerte und der Hashfunktion nicht berücksichtigt – bei Verwendung kryptographisch starker Verfahren kann dies deutlichen Zusatzaufwand verursachen [BlBS_86, Damg_87]. Vorausgenerierte Zufallszahlen und die Hashfunktionen SHS-1 und RIPEMD-160 fallen hingegen nicht ins Gewicht.

Tabelle 7-2 gibt Ergebnisse von Testläufen prototypischer Implementierungen der betrachteten Verfahren unter Verwendung einer optimierten Langzahlarithmetik [Fox_91] auf einem 166 MHz-Pentium-PC wieder. Die verwendete Implementierung des CD-Verfahrens nutzt nicht alle möglichen Optimierungen; das RSA- und das DN-Signatursystem arbeiten mit nur

---

[12] GF bezeichnet ein Galois-Feld, E eine Elliptische Kurve über einem endlichen Körper.

32 bit langen öffentlichen Exponenten. Die Messungen des DLP-GMR-Verfahrens beziehen sich auf eine Implementierung über GF(p) – von einer Implementierung über E(GF(q)) dürfen weitere Effizienzsteigerungen erwartet werden.

Verfahren	DSS	RSA	GMR	DN	CD	DLP-GMR
k [bit]	768	1024	1024	1024	1024	768
Signaturen	bel.	bel.	$2^{10} = 1024$	$l^1 = 1024$	$l^1 = 1024$	$2^{10} = 1024$
Sign [s]	« 0,01	0,39	0,39	0,45	0,77	« 0,01
$\|k_S\|$ [kbyte]	0,09	0,13	4,13	128	128,25	3,94
$\|Sig\|$ [byte]	40	128	2.816	128	256	440
Verify [s]	0,52	0,06	25,8	0,17	5,7	9,1
$\|k_V\|$ [kbyte]	0,3	0,25	0,25	256,25	0,38	0,47

Tabelle 7-2: Messergebnisse einer PC-Implementierung (Pentium-PC, 166 MHz)

Dank

Birgit Pfitzmann verdanke ich die entscheidende Anregung zu dem hier vorgeschlagenen Verfahren und sehr viele wertvolle und hilfreiche Diskussionen und Kommentare zu Vorfassungen dieses Beitrags. Für viele kleine, wichtige Korrekturen gilt mein Dank Heike Neumann. Die Implementierung des DN-Verfahrens besorgte Niko Schweitzer, und das CD-Verfahren programmierte Wasili Wasilewski.

Literatur

BlBS_86   Blum, L.; Blum, M.; Schub, M.: *A Simple Unpredictable Pseudo-Random Number Generator.* SIAM J. Computing, 15/2, 1986, S. 364-383.

BoCh_92   Bos, Jurjen; Chaum, David: *Provably Unforgeable Signatures.* In: Brickell, E.F. (Hrsg.): Proceedings of Crypto '92, LNCS 740, Springer, Berlin 1993, S. 1-14.

BoDP_97   Bosselaers, Antoon; Dobbertin, Hans; Preneel, Bart: *The RIPEMD-160 Cryptographic Hash Function.* Dr. Dobb's Journal, 1/97, S. 24-28.

CrDa_94   Cramer, Ronald; Damgård, Ivan Bjerre: *Secure Signature Schemes based on Interactive Protocols.* BRICS report series, RS-94-29, September 1994, Aarhus University.

CrDa_96   Cramer, Ronald; Damgård, Ivan Bjerre: *New Generation of Secure and Practical RSA-Based Signatures.* Koblitz, N. (Hrsg.): Proceedings of Crypto '96, LNCS 1109, Springer, Berlin 1996, S. 173-185.

Damg_87   Damgård, Ivan Bjerre: *Collision free Hash Functions and Public Key Signature Systems.* In: Chaum, D.; Price, W.L. (Hrsg.): Proceedings of Eurocrypt '87, LNCS 304, Springer, Berlin 1988, S. 203-216.

DiHe_76   Diffie, Whitfield; Hellman, Martin E.: *New Directions in Cryptography.* IEEE Transactions on Information Theory, Bd. IT-22, Nr. 6, 1976, S. 644-654.

Dobb_97   Dobbertin, Hans: *Digitale Fingerabdrücke. Sichere Hashfunktionen für digitale Signatursysteme.* Datenschutz und Datensicherheit (DuD), 2/97, S. 18-23.

DoBP_96   Dobbertin, Hans, Bosselaers, Antoon; Preneel, Bart: *RIPEMD-160: A Stengthened Version of RIPEMD.* Fast Software Encryption, LNCS 1039, Springer, Berlin 1996, S. 71-82.

DwNa_94   Dwork, Cynthia; Naor, Moni: *An Efficient Existentially Unforgeable Signature Scheme and ist Applications.* In: Desmedt, Y. G. (Hrsg.): Proceedings of Crypto '94, LNCS 839, Springer, Heidelberg 1994, S. 234-246.

DwNa_95   Dwork, Cynthia; Naor, Moni: *An Efficient Existentially Unforgeable Signature Scheme and ist Applications.* Technical Report CS95-28, Weizmann Institute, 1995.

ElGa_84   ElGamal, Taher: *A Public Key Cryptosystem and a Signature Scheme Based on Discrete Logarithms.* In: Blakley, G.R.; Chaum, D. (Hrsg.): Proceedings of Crypto '84, LNCS 196, Springer, Berlin 1995, S. 10-18.

FoPf_91   Fox, Dirk; Pfitzmann, Birgit: *Effiziente Software-Implementierung des GMR-Signatursystems.* In: Pfitzmann, A.; Raubold, E. (Hrsg.): Verläßliche Informationssysteme. Proceedings der Fachtagung VIS '91, Informatik Fachberichte Nr. 271, Springer, Heidelberg 1991, S. 329-345.

FoRö_96   Fox, Dirk; Röhm, Alexander W.: *Effiziente Digitale Signatursysteme auf der Basis Elliptischer Kurven.* In: Horster, P. (Hrsg.): Digitale Signaturen. Proceedings der Arbeitskonferenz Digitale Signaturen 96, Vieweg Verlag, Braunschweig 1996, S. 201-220.

Fox_93   Fox, Dirk: *Der 'Digital Signature Standard': Aufwand, Implementierung und Sicherheit.* In: Weck, G.; Horster, P. (Hrsg.): Verläßliche Informationssysteme, Proceedings der GI-Fachtagung VIS '93, Verlag Vieweg, Braunschweig 1993, S. 333-352.

Fox1_97   Fox, Dirk: *Fälschungssicherheit digitaler Signaturen.* Datenschutz und Datensicherheit (DuD) 2/97, S. 5-10.

Fox2_97   Fox, Dirk: *Signaturschlüssel-Zertifikat.* Gateway, Datenschutz und Datensicherheit (DuD), 2/97, S. 106.

Fox3_97   Fox, Dirk: *Sichere digitale Signatursysteme.* In: Bundesamt für Sicherheit in der Informationstechnik (Hrsg.): Tagungsband 5. Deutscher IT-Sicherheitskongreß des BSI, SecuMedia Verlag, 1997, S. 61-76.

Gold_86   Goldreich, Oded: *Two Remarks concerning the Goldwasser-Micali-Rivest Signature Scheme.* In: Odlyzko, A.M. (Hrsg.): Proceedings of Crypto '86, LNCS 263, Springer, Berlin 1986, S. 104-110.

GoMR_84   Goldwasser, Shafi; Micali, Silvio; Rivest, Ronald L.: *A „Paradoxical" Solution to the Signature Problem.* In: Proceedings of 25th Symposium on Foundations of Computer Science (FOCS), 1984, S. 441-448.

GoMR_88   Goldwasser, Shafi; Micali, Silvio; Rivest, Ronald L.: *A Digital Signature Scheme Secure against Adaptive Chosen Message Attacks.* SIAM Journal on Computing, Bd. 17, Nr. 2, 1988, S. 281-308.

HePe_92   Heyst, E. van; Pedersen, T.P.: *How to make efficient fail-stop signatures.* In: Rueppel, R.A. (Hrsg.): Proceedings of Eurocrypt '92, LNCS 658, Springer, Berlin 1993, S. 366-377.

HoPM_94   Horster, Patrick; Petersen, Holger; Michels, Markus: *Meta-ElGamal signature schemes.* Proceedings of the 2nd ACM Conference on Computer and Communications Security, Fairfax, 1994, S. 96-107.

IuKD_96   *Informations- und Kommunikationsdienste-Gesetz (IuKD-G).* Beschluß des Bundestages vom 13.06.97, Bundestagsdrucksache 13/7934.

Mene_93   Menezes, Alfred J.: *Elliptic Curve Public Key Cryptosystems.* SECS 234, Kluwer Academic Publishers, Massachusetts, 1993.

Merk_87   Merkle, Ralf C.: *A Digital Signature based on a Conventional Encryption Function.* In: Pomerance, C. (Hrsg.): Proceedings of Crypto '87, LNCS 293, Springer, Berlin 1998, S. 369-378.

NIST_94   National Institute of Standards and Technology (NIST): *Digital Signature Standard (DSS).* Federal Information Processing Standards Publication 186 (FIPS-PUB), 19. Mai 1994.

NIST_95   National Institute of Standards and Technology (NIST): *Secure Hash Standard (SHS-1).* Federal Information Processing Standards Publication 180-1 (FIPS-PUB), 17. April 1995.

Pfit_96   Pfitzmann, Birgit: *Digital Signature Schemes. General Framework and Fail-Stop Signatures.* LNCS 1100, Springer 1996.

RiSA_78   Rivest, Ronald L.; Shamir, Adi; Adleman, Leonard: *A Method for obtaining Digital Signatures and Public Key Cryptosystems.* Communications of the ACM, Bd. 21, Nr. 2, 1978, S. 120-126.

Schn_91   Schnorr, Claus P.: *Efficient Signature Generation by Smart Cards.* Journal of Cryptology, Bd. 4, Nr. 3, 1991, S. 161-174.

Effizientes faires Geld mit skalierbarer Sicherheit

Holger Petersen[1]

Laboratoire d'informatique, Ecole Normale Supérieure
45, rue d'Ulm, F-75005 Paris, petersen@dmi.ens.fr

Zusammenfassung

Seit Einführung des Konzepts der blinden Signatur im Jahr 1982 durch David Chaum wurden zahlreiche Protokolle zur Realisierung von anonymem, elektronischen Geld auf dieser Basis vorgestellt. Obwohl diese Systeme einen hohen Schutz der Privatsphäre für die Benutzer gewähren, haben sie den Nachteil, daß sich die Anonymität von Kriminellen für ein sogenanntes perfektes Verbrechen mißbrauchen läßt. Um diesem entgegenzuwirken, wurden faire elekronische Zahlungssysteme entworfen, bei denen die Anonymität des Geldes im Falle eines Betruges durch einen Treuhänder aufgehoben werden kann.

Im Beitrag stellen wir ein effizientes faires Zahlungssystem vor, welches skalierbare Sicherheit in Hinblick auf seine Effizienz bietet. Es ermöglicht die Vefolgung von Erpressungen und elektronischem Bankraub unter einem Off-line-Zahlungsprotokoll, wobei der Treuhänder nur während der Registrierung der Benutzer benötigt wird, weshalb sich das Protokoll neben der Verwendung als Internetzahlungssystem auch zur Implementierung auf einer elektronischen Geldbörse eignet.

1. Einleitung

Ein anonymes digitales Zahlungssystem wurde erstmals 1982 von David Chaum präsentiert [6]. Seither wurden zahlreiche Systeme vorgestellt (vgl. [17]). Leider lassen sich anonyme Zahlungssysteme von Kriminellen für ein *perfektes Verbrechen* mißbrauchen [22], das dadurch charakterisiert ist, daß die Angreifer erstmals nicht mehr physikalisch präsent sein müssen. Als neue Risiken ergeben sich die *Erpressung* oder der *Raub* von digitalem Geld (z.B. durch Hacker), die *Geldwäsche* sowie die unbemerkte Verwendung modifizierter Protokolle, um illegal an digitales Geld zu gelangen.

---

[1] Der Beitrag wurde mit Unterstützung eines Stipendiums des Wissenschaftsausschusses der NATO über den DAAD ermöglicht.

Um diese Angriffe zu verhindern, muß das anonyme Zahlungssystem eine Möglichkeit der Deanonymisierung bieten, die eine Verfolgung von Münzen bei einem der oben genannten Angriffe mittels einer vertrauenswürdigen dritten Partei erlaubt. Die ersten Systeme, die Erpressung und Geldwäsche verhindern, wurden in [2, 23] präsentiert. Weitere Systeme mit dieser Eigenschaft finden sich in [4, 5, 9, 10, 12]. Alle diese Systeme benötigen die Mitwirkung eines Treuhänders während der Kontoeröffnung oder während jedes Abhebevorgangs. Die einzigen Systeme, die den Treuhänder nur zur Initialisierung benötigen wurden in [3, 8] vorgestellt. Sie bieten jedoch keinen Schutz vor elektronischem Bankraub sowie unerkannter Verwendung modifizierter Protokolle. Diese Angriffe werden lediglich in den Systemen von [9, 10] mittels eines On-Line-Zahlungsprotokolls verhindert. Aufgrund ihres hohen Kommunikationsaufwandes sind sie für den Einsatz in elektronischen Geldbörsen wenig geeignet.

Ergebnisse

Unser Zahlungssystem erlaubt die Deanonymisierung der Münzen bei allen oben erwähnten Angriffen durch den Treuhänder, unter Verwendung eines *Off-line-Zahlungsprotokolls*. Dieses wird durch die Registrierung pseudonymer Benutzerschlüsselpaare beim Treuhänder erreicht, die bei der Zahlung verwendet werden. Obwohl unterschiedliche Zahlungen unter dem gleichen Pseudonym verknüpfbar sind, ist der Schutz der Privatsphäre durch den Einsatz mehrerer Pseudonyme – bis hin zu einem neuen Pseudonym für jede Zahlung, um totale Unverknüpfbarkeit zu gewährleisten – durch den Benutzer frei skalierbar. *Vorteile* dieses Ansatzes sind ein einfacher, modularer Systementwurf, der leicht verständlich und analysierbar ist und sich effizent implementieren läßt. Verschiedene kryptographische Mechanismen interagieren sowenig wie möglich (im Gegensatz etwa zu den Systemen in [1, 3, 10]). Das System ist zudem sehr vielseitig, da es die Integration mehrfach ausgebbarer oder teilbarer Münzen unterstützt.

2. Das faire elektronische Zahlungssystem

Vereinfachend wird ein Zahlungssystem mit *einer* Bank, *einem* Kunden, *einem* Geschäft und *einem* Treuhänder beschrieben. Es kann problemlos auf mehrere Teilnehmer erweitert werden. Ebenso beschränken wir uns auf *einen* Münzwert sowie *einen* Gültigkeitszeitraum für abgehobene Münzen. Allgemein läßt sich eine Kombination dieser beiden Aspekte erreichen, indem die Bank mehrere Schlüsselpaare einsetzt oder eine Technik, wie in [1] beschrieben, verwendet wird.

Für die Sicherheits-, Vertrauens- und Angreifermodelle sowie mögliche Angriffe und die daraus resultierenden Sicherheitsanforderungen für elektronisches Geld sei auf [17] verwiesen, ebenso für einen Überblick über bekannte Systeme und deren Charakteristika.

2.1 Grundlegendes Prinzip

Die Grundlage der meisten anonymen Zahlungssysteme bildet das System von Chaum [6, 7]. Wir verwenden es ebenso mit folgenden Modifikationen: Ein Kunde mit Identität ID_K eröffnet bei der Bank ein Konto. Er läßt sich beim Treuhänder registrieren und erhält ein zertifiziertes pseudonymes Schlüsselpaar (PS_x, PS_y). Anschließend hebt er unter Verwendung eines blinden Signaturverfahrens der Bank eine digital signierte Münze (M, σ_B) ab. Dabei wird der öffentliche Teil PS_y seines Pseudonyms mittels einer Hashfunktion h in die Münze eingebettet. Die blinde digitale Signatur σ_B der Bank gewährleistet die *Unfälschbarkeit* und *Unverfolgbarkeit* der Münze. Beim Bezahlen gibt er $h(M, PS_y)$ zusammen mit σ_B an ein Geschäft weiter und beweist die Kenntnis des Urbilds PS_x von PS_y. Hierzu generiert er eine digitale Signatur σ_M auf eine zufällige Nachricht m mit PS_x als geheimen Schlüssel. Dies verhindert das erneute Ausgeben der Münze durch das Geschäft und erfüllt die *Gerechtigkeitsanforderung* (vgl. [17]). Das Geschäft reicht zur Gutschrift der Münze das Tupel $(M, PS_y, \sigma_B, \sigma_M)$ an die Bank ein. Abbildung 1 stellt die wesentlichen Schritte denen im System von Chaum gegenüber.

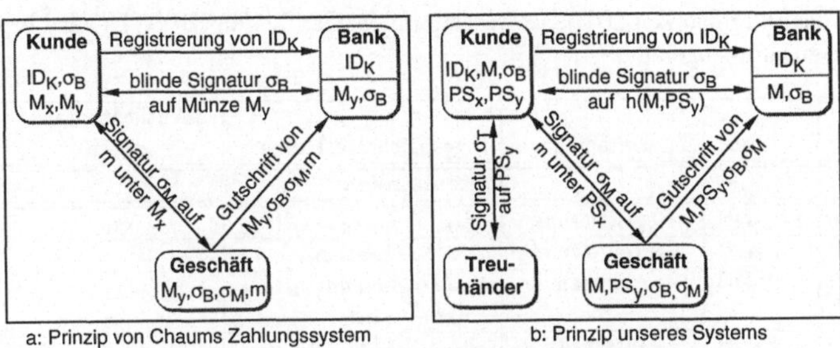

Abbildung 1: Gegenüberstellung der Grundkonzepte

2.2 Kryptographische Hilfsmittel

Wir benötigten sechs kryptographische Hilfsmittel. Dieses sind

1. eine kollisionsresistente kryptographische Hashfunktion h,

2. drei (evtl. verschiedene) digitale Signaturverfahren $(\mathcal{G}_T, \mathcal{S}_T, \mathcal{V}_T)$, $(\mathcal{G}_K, \mathcal{S}_K, \mathcal{V}_K)$, $(\mathcal{G}_M, \mathcal{S}_M, \mathcal{V}_M)$ mit zugehörigen Sicherheitsparametern, Nachrichtenräumen und Schlüsselgenerierungsalgorithmen $\mathcal{G}_T, \mathcal{G}_K, \mathcal{G}_M$, Signieralgorithmen $\mathcal{S}_T, \mathcal{S}_K, \mathcal{S}_M$ und Verifizieralgorithmen $\mathcal{V}_T, \mathcal{V}_K, \mathcal{V}_M$, die von den Treuhändern, Kunden und zur Signierung der Münzen bei der Zahlung eingesetzt werden,

3. ein blindes Signaturverfahren $(\mathcal{G}_B, \mathcal{S}_B, \mathcal{V}_B)$, der Bank zum Signieren digitaler Münzen, bei dem die Blindung/Aufdecken mit $blind(\cdot)/recover(\cdot)$ bezeichnet wird,

4. ein interaktives authentisches Schlüsselaustauschprotokoll $\mathcal{K}$ mit gegenseitiger Benutzerauthentifikation, das jeweils einen frischen Sitzungsschlüssel für die authentische Kommunikation zwischen A und B erzeugt. Wir bezeichnen ihn mit $K_{A,B} := \mathcal{K}(ID_A, ID_B)$ (für einen Überblick über geeignete Protokolle siehe [20]),

5. ein probabilistisches Verschlüsselungsverfahren $(\mathcal{E}_Z, \mathcal{D}_Z)$, $\mathcal{Z} \in \{T, G\}$ das für die vertrauliche Kommunikation eingesetzt wird. Das Chiffrat einer Nachricht m für $\mathcal{Z}$ wird als $e := \mathcal{E}_Z(m)$ bezeichnet und seine Entschlüsselung als $m := \mathcal{D}_Z(e)$,

6. ein symmetrisches Kryptosystem $(\mathcal{E}_K, \mathcal{D}_K)$, mit Sitzungsschlüssel K.

Diese Mechanismen sind wie üblich definiert (vgl. etwa [11]). Die drei digitalen Signaturverfahren werden von den Teilnehmern selbst gewählt. Dabei sollte $(\mathcal{S}_T, \mathcal{V}_T)$ ein existentiell unfälschbares Signaturverfahren sein, um die unerkannte Verwendung eines modifizierten Regisistrierungsprotokolls zu verhindern [10]. Die Schlüsselgenerierungsalgorithmen $\mathcal{G}_Z$, für $\mathcal{Z} \in \{B, T, K, M\}$, erzeugen asymmetrische Schlüsselpaare (x_Z, y_Z). Die Signaturgenerierung für eine Nachricht m durch Instanz $\mathcal{Z}$ wird als $\sigma_Z := \mathcal{S}(x_Z, m)$ und die Verifikation dieser Signatur als $\mathcal{V}_Z(y_Z, m, \sigma_Z) \in \{true, false\}$ beschrieben. Kunde K und Geschäft G erhalten die Kontonummern acc_K bzw. acc_G. $Ind(x)$ ist ein Zeiger auf einen gespeicherten Wert x.

2.3 Datenbanken

Der Inhalt der Datenbanken (DB), schwarzen und weißen Listen (BL,WL) wird in folgender Tabelle zusammengefaßt:

Besitzer	Bez.	Zugriff bei/nach	gespeicherte Daten	Authentizität durch	Vertraulichkeit
Datenbanken					
Treuhänder	PS_y-DB	Registrierung	$ID_K, PS_y, \sigma_K, Ind(x_T)$	σ_K	ja
Bank	K-DB	Kontoeröffnung	ID_K, acc_K	–	nein
	A-DB	Abheben	$ID_K, \tilde{M}, Ind(x_B),$ $\mathcal{E}_T(h(PS_y, M))$	$\tilde{\sigma}_B$	nein
	E-DB	Gutschrift	$M, PS_y, ID_G, \sigma_T, \sigma_M$	σ_T, σ_M	nein
Kunde	M-DB	Abheben und	$M, \sigma_B, Ind(PS_x)$	σ_B	nein
	PS_x-DB	Zahlung	$PS_x, (PS_y), \sigma_T$	σ_T	ja
Geschäft	Z-DB	Zahlung	$M, PS_y, \sigma_B, \sigma_T, \sigma_M$	σ_B, σ_M	nein
Widerrufungslisten					
Geschäft	K-BL	Kundenerpress.	ungültige PS_y	$\mathcal{S}_T(x_T, PS_y)$	nein
	M-WL	Bank-	gültige $h(PS_y, M)$	$\mathcal{S}_T(x_T, h(PS_y, M))$	nein
	B-BL	raub	ungültige y_B	$\mathcal{S}_T(x_T, y_B)$	nein
	PS-WL	Treuhänder-	gültige PS_y	$\mathcal{S}_T(\tilde{x}_T, PS_y)$	nein
	T-BL	erpressung	ungültige y_T	$\mathcal{S}_{CA}(x_{CA}, y_T)$	nein

Abbildung 2: Datenbanken und Widerrufungslisten

2.4 Protokolle

Kontoeröffnung

1. Der Kunde K mit Identität ID_K weist sich gegenüber der Bank aus und erhält ein Konto mit Nummer acc_K.

2. Die Bank speichert (ID_K, acc_K) in ihrer K-DB.

Registrierung

1. Um sich gegenüber dem Treuhänder zu identifizieren und einen authentischen Sitzungsschlüssel $K_{K,T}$ zu erhalten, nehmen der Kunde und Treuhänder am authentischen Schlüsselaustauschprotokoll $\mathcal{K}$ teil und erhalten $K_{K,T} := \mathcal{K}(ID_K, ID_T)$. Um

im folgenden die Vertraulichkeit der Kommunikation zu gewährleisten, wird $K_{K,T}$ zur Verschlüsselung benutzt.

2. Der Kunde erzeugt ein pseudonymes Schlüsselpaar $(PS_x, PS_y) := \mathcal{G}_M$. Dann berechnet er $\sigma_K := \mathcal{S}_K(x_K, (ID_K, PS_y))$ und sendet (ID_K, PS_y, σ_K) an den Treuhänder.

3. Der Treuhänder überprüft $\mathcal{V}_K(y_K, (ID_K, PS_y), \sigma_K) \overset{?}{=} true$ und berechnet $\sigma_T := \mathcal{S}_T(x_T, PS_y)$. Er speichert $(ID_K, PS_y, \sigma_K, Ind(x_T))$ in PS_y-DB und überträgt σ_T an den Kunden K.

4. Der Benutzer überprüft $\mathcal{V}_T(y_T, PS_y, \sigma_T) \overset{?}{=} true$ und speichert alle Werte.

Diese Schritte können mehrfach parallel ausgeführt werden, um mehrere pseudonyme Schlüsselpaare (PS_x, PS_y) zu erhalten. Für eine effiziente Implementierung auf einer elektronischen Geldbörse läßt sich das Protokoll derart modifizieren, daß der Treuhänder das Tupel $(PS_x, PS_y) := \mathcal{G}_M$ selbst erzeugt, PS_y signiert und anschließend alle Werte vertraulich an den Kunden überträgt. Der Treuhänder speichert PS_x als geheimen Schlüssel für eine spätere Verwendung in einem symmetrischen Kryptosystem.

Abheben

1. Um sich gegenüber der Bank zu authentifizieren und einen authentischen Sitzungsschlüssel $K_{K,B}$ zu erhalten, nehmen der Kunde und die Bank am Schlüsselaustauschprotokoll $\mathcal{K}$ teil und erhalten $K_{K,B} := \mathcal{K}(ID_K, ID_B)$. $K_{K,B}$ wird zur Verschlüsselung der folgenden Kommunikation verwendet.

2. Der Kunde K erzeugt eine zufällige Münze M (möglicherweise mit Redundanz) und berechnet $\tilde{M} := blind(h(M, PS_y))$ und $e_T := \mathcal{E}_T(h(PS_y, M))^1$. Er überträgt diese Werte zusammen mit seiner Identität ID_K an die Bank. Nach einer Erpressung der Bank, sendet diese alle gespeicherten Werte e_T an den Treuhänder (siehe unten).

3. Die Bank berechnet $\tilde{\sigma}_B := \mathcal{S}_B(x_B, \tilde{M})$ und sendet es K. Sie belastet den Gegenwert der Münze M dem Kundenkonto acc_K und speichert $(ID_K, Ind(x_B), \tilde{M}, e_T)$ in ihrer A-DB.

4. Der Kunde berechnet $\sigma_B := recover(\tilde{\sigma}_B)$ und überprüft $\mathcal{V}_B(y_B, h(M, PS_y), \sigma_B) \overset{?}{=} true$. Sofern die Überprüfung fehlschlägt, bittet der Kunde die Bank, ihm die blinde Signatur nochmals zu senden. Da angenommen wird, daß die Bank dem Kunden kein Geld stiehlt, wird dieser Vorgang wiederholt, bis die Verifikation von σ_B korrekt ist. Er speichert $(M, \sigma_B, ind(PS_y))$ als Münze.

Bezahlen

1. Das Geschäft sendet dem Kunde eine unikate Nachricht m.

2. Der Kunde erzeugt $\sigma_M := \mathcal{S}_M(PS_x, (M, ID_G, m), PS_y)$ und sendet das Zahlungstranskript $(M, PS_y, \sigma_B, \sigma_T, \sigma_M)$ an das Geschäft. Im Fall eines Internetzahlungssystems wird σ_M als $e_G := \mathcal{E}_G(\sigma_M)$ verschlüsselt, um Abhören zu verhindern.

3a. Sofern keine *Erpressungen* bekannt wurden, überprüft das Geschäft die Signaturen $\mathcal{V}_T(y_T, PS_y, \sigma_T) \overset{?}{=} true$, $\mathcal{V}_B(y_B, h(M, PS_y), \sigma_B) \overset{?}{=} true$ und

---

[1] Im Falle der Geldbörse wird vereinfacht $e_T := \mathcal{E}_{PS_x}(h(PS_y, M))$ berechnet, da der Schlüssel PS_x dem Treuhänder bekannt ist.

$\mathcal{V}_M(PS_y, (M, ID_G, m), \sigma_M) \overset{?}{=} true$. Es speichert das Zahlungstranskript zusammen mit der Nachricht m in seiner Z-DB.

3b. Nach einer *Benutzererpressung* erhält das Geschäft eine aktualisierte K-BL vom Treuhänder. Sofern $PS_y \in$ K-BL, weist er die Münze zurück. Ansonsten akzeptiert er sie wie unter 3a. beschrieben.

3c. Nach einer *Bankerpressung* erhält das Geschäft aktualisierte Listen B-BL und M-WL vom Treuhänder. Sofern ($y_B \in$ B-BL und $h(PS_y, M) \notin$ M-WL) weist er die Münze zurück. Ansonsten akzeptiert er sie wie unter 3a. beschrieben.

3d. Nach einer *Treuhändererpressung* erhält das Geschäft aktualisierte Listen T-BL und PS-WL vom Treuhänder. Sofern σ_T unter dem öffentlichen Schlüssel $y_T \in$ T-BL signiert wurde und $PS_y \notin$ PS-WL, weist er die Münze zurück. Ansonsten akzeptiert er sie wie unter 3a. beschrieben.

Schritt 1 kann ausgelassen werden, sofern der Kunde jedesmal eine frische, unikate Nachricht m, z.B. als Funktion der Systemzeit *time*, des Vertragsinhaltes *contract* usw. erzeugt. Im Fall einer teilbaren Münze muß m außerdem den auszugebenen Anteil der Münze M enthalten (vgl. Abschnitt 2.5).

Gutschrift

1. Das Geschäft schickt das Tupel $(M, PS_y, \sigma_B, \sigma_T)$ an die Bank.

2. Die Bank überprüft $\mathcal{V}_T(y_T, PS_y, \sigma_T) \overset{?}{=} true$, $\mathcal{V}_B(y_B, h(M, PS_y), \sigma_B) \overset{?}{=} true$ und verifiziert, ob M bereits unter Angabe des selben Pseudonyms PS_y gutgeschrieben wurde. Ist dies der Fall, so findet die Bank ein Tupel (M, σ'_M) in seiner E-DB und sendet σ'_M als Beweis an das Geschäft. Sofern $\sigma_M = \sigma'_M$ wird das Geschäft der mehrfachen Gutschrift der Münze M beschuldigt und kann sie daher nicht gutschreiben lassen. Sofern $\sigma_M \neq \sigma'_M$ ist oder die Münze M noch nicht gutgeschrieben wurde, schickt die Bank dem Geschäft hierfür eine signierte Bestätigung.

3. Das Geschäft übermittelt $(ID_G, acc_G, m, \sigma_M)$ an die Bank.

4. Die Bank verifiziert $\mathcal{V}_M(PS_y, (M, ID_G, m), \sigma_M) \overset{?}{=} true$ und vergewissert sich, ob M unter PS_y mehrfach ausgegeben wurde. Ist dies der Fall, so initiiert sie die unten beschriebene Benutzerverfolgung mit dem Treuhänder, um den betrügerischen Benutzer zu identifizieren. Als Beweis sendet sie dem Treuhänder die Zahlungstranskripte $(M, PS_y, m_1, \sigma_B, \sigma_T, \sigma_{M,1}), \ldots, (M, PS_y, m_k, \sigma_B, \sigma_T, \sigma_{M,k})$.

5. Sofern kein Betrug vorliegt, speichert die Bank das Zahlungstranskript in seiner E-DB und schreibt den Gegenwert von M dem Konto acc_G gut.

Wird angenommen, daß die Bank das Geschäft nicht betrügt, so kann die Signatur σ_M bereits im 1.Schritt übermittelt werden und Schritt 3. entfällt.

Verfolgung des mehrfachen Ausgebens und Kundenverfolgung

1. Um zu beweisen, daß die Münze M mehrfach ausgegeben wurde, sendet die Bank die Zahlungstranskripte $(M, PS_y, m_1, \sigma_B, \sigma_T, \sigma_{M,1}), \ldots, (M, PS_y, m_k, \sigma_B, \sigma_T, \sigma_{M,k})$ an den Treuhänder.

2. Der Treuhänder überprüft die Signaturen aller Transkripte. Sind sie korrekt, so sucht er in PS_y-DB das Tupel (ID_K, PS_y, σ_K) und sendet (ID_K, σ_K) der Bank.

3. Die Bank überprüft σ_K und identifiziert den Kunden.

Münzverfolgung

Die Münzverfolgung dient zur Verhinderung von Erpressungen oder Diebstahl von Münzen. In unserem System sind solche Angriffe nur möglich, wenn gleichzeitig das pseudonyme Schlüsselpaar (PS_x, PS_y) erpreßt bzw. entwendet wird. In diesem Fall greift die *Erpresserverfolgung* als Gegenmaßnahme.

Erpresserverfolgung

Es werden die Maßnhamen der angegriffenen Teilnehmer beschrieben. Die Konsequenzen hieraus für die Geschäfte beim Bezahlen wurden bereits in den Schritten 3b-d. des Zahlungsprotokolls erwähnt.

1. *Erpressung des geheimen pseudonymen Schlüssels PS_x:*
   Der Kunde meldet den Angriff dem Treuhänder, der in seiner Datenbank PS_y-DB nach dem Tupel (PS_y, ID_K) sucht. Sofern ID_K nicht der Kundenidentität entspricht, verweigert er die Bearbeitung des Angriffes, um Falschmeldungen durch unehrliche Benutzer vorzubeugen. Sonst setzt er PS_y auf die K-BL und verteilt diese schwarze Liste sofort an die Geschäfte. Er stellt dem Kunden die Signatur $\mathcal{S}_T(x_T, (ID_K, PS_y))$ aus, die es ihm erlaubt, unverwendete Münzen, die er unter dem Pseudonym PS_y abgehoben hat, bei der Bank in frische Münzen umzutauschen, nachdem er sich als deren legaler Besitzer ausgewiesen hat.

2. *Erpressung des geheimen Bankschlüssels x_B:*
   Um die Fälschung digitaler Münzen nach einer Erpressung von x_B zu verhindern, sendet die Bank die verschlüsselten Werte $E_T(h(PS_y, M_i))$ aller legal abgehobener Münzen M_i in seiner A-DB an den Treuhänder. Dieser entschlüsselt die Werte $h(PS_y, M_i)$ und setzt sie auf die weiße Liste M-WL. Außerdem wird der zu x_B gehörige öffentliche Schlüssel y_B auf die schwarze Liste B-BL gesetzt und beide Listen sofort an die Geschäfte verteilt. Die Bank erzeugt ein frisches Schlüsselpaar $(\tilde{x}_B, \tilde{y}_B) := \mathcal{G}_B$ für die künftige Ausstellung von Münzen.

3. *Erpressung des geheimen Treuhänderschlüssels x_T:*
   Nach einer Erpressung von x_T setzt der Treuhänder alle Pseudonyme PS_y, die er legal unter x_T signiert hat auf die weiße Liste PS-WL und gleichzeitig seinen öffentlichen Schlüssel y_T auf die schwarze Liste T-BL. Beide Listen werden sofort unter allen Geschäften verteilt. Der Treuhänder erzeugt ein frisches Schlüsselpaar $(\tilde{x}_T, \tilde{y}_T) := \mathcal{G}_T$ für die künftigte Zertifizierung von Benutzerpseudonymen.

4. *Verwendung eines modifizierten Abuchungsprotokolls:*
   Die unbemerkte Verwendung eines modifizierten Abbuchungsprotokolls ist unmöglich, da der Kunde sich eingangs mittes des authentischen Schlüsselaustauschprotokolls $\mathcal{K}$ gegenüber der Bank authentifiziert und damit jede abgehobene Münze automatisch seinem Konto belastet wird. Sofern ein betrügerischer Kunde die Verwendung eines modifiziertes Abbuchungsprotokolls ohne vorherige Identifikation erpreßt, weiß die Bank hiervon und ergreift die gleichen Gegenmaßnahmen wie bei einer Erpressung ihres geheimen Schlüssels x_B.

5. *Verwendung eines modifizierten Registrierungsprotokolls:*
   Die unbemerkte Verwendung eines modifizierten Registrierungsprotokolls mit dem Ziel dabei ein pseudonymes Schlüsselpaar (PS_x, PS_y) zu erhalten, zu dem der

Treuhänder die Identität des Kunden nicht ermitteln kann, ist aufgrund der Verwendung eines existentiell unfälschbaren Signaturverfahrens $(\mathcal{S}_T, \mathcal{V}_T)$ unmöglich [10]. Daher bekommt der Treuhänder den Angriff mit und ergreift die gleichen Gegenmaßnahmen wie bei einer Erpressung seines geheimen Schlüssels x_T.

2.5 Vielseitigkeit des Zahlungssystems

Das k-fache Ausgeben oder allgemeiner die *Teilbarkeit* der Münzen wird wie bei den Systemen in [10, 12] erreicht, indem der ausgegebene Anteil M_f der Münze M bei der Zahlung in die Nachricht m eingebettet und signiert wird. Die Verfolgung des mehrfachen Ausgebens wird entsprechend modifiziert. Das System unterstützt ebenfalls die in [10] vorgestellte *challenge semantics*. Weiterhin können für alle Protokollschritte *Quittungen* ausgestellt werden [7], um das gegenseitige Vertrauen der Teilnehmer zu minimieren.

2.6 Skalierbarkeit der kryptographischen Hilfsmittel

Die Auswahl konkreter kryptographischer Hilfsmittel erlaubt die Skalierbarkeit des Verfahrens hinsichtlich Sicherheit und Effizienz. Es werden unterschiedliche Mechanismen diskutiert, die die Implementierung auf der Kundenseite beeinflussen.

Wahl des blinden Signaturverfahrens $(\mathcal{S}_B, \mathcal{V}_B)$: Es existiert ein Dualismus zwischen der effizienten Blindung der Nachricht sowie der Speicherung von σ_B und der beweisbaren Sicherheit des Verfahrens. Das *effizienteste* Verfahren hinsichtlich der Blindung ist das blinde RSA-Signaturverfahren [6] mit kleinem öffentlichen Exponenten. Aufgrund der Nachrichtenrückgewinnung wird hier die signierte Münze nicht zusätzlich gespeichert. Das *effizienteste* Verfahren hinsichtlich der Signaturlänge ist das blinde Schnorr-Signaturverfahren [14]. Als *beweisbar komplexitätstheoretisch sicheres* Signaturverfahren kann das blinde Okamoto-Signaturverfahren gewählt werden [19]. Weiterhin könnte ein restriktiv blindes Signaturverfahren [1] eingesetzt werden, bei dem die Bank das mehrfache Ausgeben alleine verfolgen kann.

Wahl der Benutzer- und Münzsignaturverfahren $(\mathcal{S}_K, \mathcal{V}_K)$ ***und*** $(\mathcal{S}_M, \mathcal{V}_M)$: Es existiert ein Dualismus zwischen der effizienten Signaturgenerierung und der beweisbaren Sicherheit der Verfahren. CLE [24] ist eine *originelles, effizientes* Verfahren, das sich gut für eine Implementierung auf Chipkarten eignet. Eine *effiziente* und zugleich *beweisbar sichere* Alternative stellt das Schnorr-Signaturverfahren dar [21, 18]. Daneben kann jede beliebige Variante aus der Familie der ElGamal-Verfahren (vgl. [15]) oder das RSA-Verfahren verwendet werden.

Münze M: Es existiert ein Dualismus zwischen *effizienten*, mehrfach ausgebbaren oder teilbaren Münzen und dem Schutz der *Privatsphäre* der Kunden, da die einzelnen Teile solcher Münzen verknüpfbar sind.

3. Zusammenfassung

Es wurde ein effizientes faires Zahlungssystem vorgestellt. Es ist das erste System, das alle Arten von Erpresserangriffen unter einem Off-Line-Zahlungsprotokoll verhindert. Aufgrund seiner skalierbaren Sicherheit hinsichtlich seiner Effizienz, ist es möglich, sowohl eine sichere Realisierung für ein Internet-Zahlungssystem (Netzgeld) als auch eine effiziente Realisierung für eine elektronische Geldbörse (Kartengeld) abzuleiten.

Danksagung

Ich danke meinem Kollegen Guillaume Poupard für zahlreiche Diskussionen und Kritik am Zahlungssystem, die zu seiner Vereinfachung beigetragen haben.

Literatur

[1] S.Brands, „Untraceable Off-Line Cash in Wallets with Observers", LNCS 773, Advances in Cryptography – Crypto '93, Springer Verlag, (1994), S. 302 – 318.

[2] E.Brickell, P.Gemmell, D.Kravitz, „Trustee-based Tracing Extensions to Anonymous Cash and the Making of Anonymous Exchange", Proc. SODA, (1995), S. 457 – 466.

[3] J.Camenisch, U.Maurer, M.Stadler, „Digital Payment Systems with Passive Anonymity-Revoking Trustees", LNCS 1146, Proc. ESORICS'96, Springer Verlag, (1996), S. 31 – 43.

[4] J.Camenisch, J.-M.Piveteau, M.Stadler, „Faire anonyme Zahlungssysteme", Proc. GISI'95, Informatik aktuell, Springer Verlag, (1995), S. 254 – 265.

[5] J.Camenisch, J.-M.Piveteau, M.Stadler, „An efficient Fair Payment System", Proc. 3rd ACM Conf. on Computer and Communications Security, ACM Press, (1996), S. 88 – 94.

[6] D.Chaum, „Blind signatures for untraceable payments", Advances in Cryptography – Crypto '82, Plenum Press, (1983), S. 199 – 203.

[7] D.Chaum, „Privacy protected payments: Unconditional Payer and/or Payee Anonymity", Smart Card 2000: The future of IC Cards, North-Holland, (1989), S. 69 – 92.

[8] Y.Frankel, Y.Tsiounis, M.Yung, „ „Indirect Discourse Proofs": Achieving Efficient Fair Off-Line E-Cash", LNCS 1163, Advances in Cryptography – Asiacrypt'96, (1996), Springer Verlag, S. 286 – 300.

[9] E.Fujisaki, T.Okamoto, „Practical Escrow Cash System", LNCS 1189, Proc. 1996 Cambridge Workshop on Security Protocols, Springer Verlag, (1997), S. 33 – 48.

[10] M.Jakobsson, M.Yung, „Revokable and Versatile Electronic Money", Proc. 3rd ACM Conference on Computer and Communications Security, ACM Press, (1996), S. 76 – 87.

[11] A.Menezes, P.C.van Oorshot, S.Vanstone, „Handbook of Applied Cryptography", CRC Press, (1997), 780 Seiten.

[12] D.M'Raïhi, „Cost Effective Payment Schemes with Privacy Regulations", LNCS 1163, Advances in Cryptography – Asiacrypt '96, Springer Verlag, (1996), S. 266 – 275.

[13] National Institute of Standards and Technology, Federal Information Processing Standards Publication, FIPS Pub 180-1: Secure Hash Standard (SHA-1), 17.April, (1995), 14 Seiten.

[14] T.Okamoto, „Provable secure and practical identification schemes and corresponding signature schemes", LNCS 740, Advances in Cryptography – Crypto'92, Springer Verlag, (1993), S. 31 – 53.

[15] H.Petersen, „Digitale Signaturverfahren auf der Basis des diskreten Logarithmusproblems und ihre Anwendungen", Dissertation, TU Chemnitz-Zwickau, Shaker Verlag, (1996), 277 Seiten.

[16] H.Petersen, „Faires elektronisches Geld", Proc. 5. Deutscher IT-Sicherheitskongreß, Bonn, SecuMedia Verlag, (1997), S. 427 – 444.

[17] H.Petersen, „Anonymes elektronisches Geld - Der Einfluß der blinden Signatur", Datenschutz und Datensicherheit, Vol. 21, No. 7, (1997), S. 403 – 410.

[18] D.Pointcheval, J.Stern, „Security Proofs for Signatures", LNCS 1070, Advances in Cryptography – Eurocrypt'96, Springer Verlag, (1996), S. 387 – 398.

[19] D.Pointcheval, J.Stern, „Provably Secure Blind Signature scheme", LNCS 1163, Advances in Cryptography – Asiacrypt'96, Springer Verlag, (1996), S. 252 – 265.

[20] R.Rueppel, P.C.can Oorschot, „Modern key agreement techniques", Computer Communications, Vol. 17, Vol. 7, (1994), S. 458 – 465.

[21] C.P.Schnorr, „Efficient identification and signatures for smart cards", LNCS 435, Advances in Cryptography – Crypto '89, Springer Verlag, (1990), S. 239–251.

[22] S.von Solms, D.Naccache, „Blind signatures and perfect crimes", Computers & Security, Vol. 11 , (1992), S. 581 – 583.

[23] M.Stadler, J.-M.Piveteau, J.Camenisch, „Fair-Blind Signatures", LNCS 921, Advances in Cryptography – Eurocrypt '95, Springer Verlag, (1995), S. 209 – 219.

[24] J.Stern, „Designing identification schemes with keys of short size", LNCS 839, Advances in Cryptography – Crypto '94, Springer Verlag, (1995), S. 164 – 173.

A. Sicherheitsanalyse

Die Sicherheitsanalyse profitiert vom modularen Entwurf mittels bekannter kryptographischer Mechanismen. Dadurch ist die Analyse übersichtlich. Theoreme A.1–A.6 zeigen den Einfluß dieser Mechanismen auf die Sicherheit des Verfahrens, Theoreme A.7–A.8 die Notwendigkeit der Widerrufungslisten aus Abschnitt 2.3. Für eine Beschreibung der Angriffe und resultierenden Sicherheitsanforderungen sei auf [16, 17] verwiesen.

Definition A.1 *Ein Signaturverfahren* (S, V) *heißt* beweisbar komplexitätstheoretisch sicher, *sofern das existentielle Fälschen einer Signatur unter einem adaptiven Angriff äquivalent zur Lösung eines komplexitätstheoretisch harten Problems ist, für das keine polynomialen Lösungsalgorithmen bekannt sind (z.B. das diskrete Logarithmusproblem oder das Faktorisierungsproblem).*

Satz A.1 *Sei* (S_B, V_B) *ein beweisbar komplexitätstheoretisch sicheres blindes Signaturverfahren (im Sinne von [19]) und h ist eine kollisionsresistente Hashfunktion. Dann sind digitale Münzen unfälschbar und das System bietet Sicherheit vor* Münzfälschungen.

Beweis A.1: Da S_B sicher gegenüber existentiellem Fälschen ist, müssen alle Signaturen σ_B (blind) auf die Nachrichten $h(M, PS_y)$ von der Bank erzeugt worden sein. Da h kollisionsresistent ist, ist es unmöglich andere Werte M' oder PS'_y zu finden, die den gleichen Hashwert ergeben.□

Satz A.2 *Sei* (S_B, V_B) *ein beweisbar komplexitätstheoretisch sicheres blindes Signaturverfahren und* $(\mathcal{E}_T, \mathcal{D}_T)$ *ein starkes probabilistisches Kryptosystem. Dann sind digitale Münzen unverfolgbar und das System ist sicher gegen unautorisierte* Kundenverfolgung *durch die Bank.*

Beweis A.2: Da S_B ein perfekt blindes Signaturverfahren ist, ist eine signierte Münze (M, σ_B) unverfolgbar zur Sicht $(\tilde{M}, \tilde{\sigma}_B)$ der Bank beim Signieren. Weiterhin ergibt das Chiffrat e_T das mit der blinden Münze ausgeliefert wird, keinen Hinweis für die Bank, um die Münze später wiederzuerkennen, da es die gleiche Verteilung wie eine Zufallszahl hat. □

Satz A.3 *Sei* (S_M, V_M) *ein beweisbar komplexitätstheoretisch sicheres Signaturverfahren. Dann ist das System sicher gegen (1)* Imitation, *(2)* Erfindung des mehrfachen Ausgebens, *(3)* Kundenverfolgung mit Hilfe des Treuhänders *sowie (4)* Münzfälschung nach mehrfachem Ausgeben.

Beweis A.3: Da niemand ohne Kenntnis des geheimen Schlüssels PS_x eine digitale Signatur σ_M für das Tupel (M, ID_G, m) erzeugen kann, so daß $V_M(PS_y, \sigma_M, (M, ID_G, m)) = true$ ist, sind die Imitation sowie die Erfindung des mehrfachen Ausgebens unmöglich. Weiterhin hilft die Kenntnis digitaler Signaturen $\sigma_{M,1}, \ldots \sigma_{M,k}$ zur Münze der Bank nicht, um neue digitale Signaturen hieraus zu erzeugen. Dies verhindert die beiden letztgenannten Angriffe. □

Satz A.4 *Sei* (S_K, V_K) *ein beweisbar komplexitätstheoretisch sicheres Signaturverfahren. Dann bietet das System Sicherheit gegenüber* falsche Beschuldigungen *durch den Treuhänder.*

Beweis A.4: Da der Treuhänder eine gültige digitale Signatur $S_K(x_K, (ID_K, PS_y))$ zu einem gegebenen Wert PS_y und einer beliebigen Identität ID_K erzeugen müßte, um einen Kunden fälschlich zu beschuldigen, ist dieser Angriff unmöglich. □

Satz A.5 *Sei* (S_T, V_T) *ein beweisbar komplexitätstheoretisch sicheres Signaturverfahren. Dann kann die Verfolgung des mehrfachen Ausgebens sowie die Kundenverfolgung durchgeführt werden und das System ist sicher gegenüber* mehrfachem Ausgeben, Geldwäsche *und* unbemerkter Verwendung modifizierter Protokolle *gegenüber dem Treuhänder bei der Registrierung.*

Beweis A.5: Da S_T unfälschbar ist, muß der Treuhänder die Relation zwischen der Identität ID_K und dem signierten Pseudonym PS_y kennen. Bei der Kundenverfolgung sendet die Bank ein Tupel $(M, PS_y, m, \sigma_B, \sigma_T, \sigma_M)$ und der Treuhänder überprüft die Signaturen. Da bereits bewiesen wurde (s.o.), daß diese von den richtigen Teilnehmern erzeugt wurden, authentifiziert σ_B eine Münze $h(M, PS_y)$, weiterhin beweist σ_M, daß M unter dem Pseudonym PS_y abgehoben wurde und schließlich beweist σ_T, daß der Treuhänder ein Paar (PS_y, ID_K) kennt, um M der Identität ID_K zuzuordnen. Dadurch werden *mehrfaches Ausgeben* und *Geldwäsche* verhindert. Ferner ist die unbemerkte Verwendung eines modifizierten Registrierungsprotokolls mit dem Ziel, ein nicht registriertes, zertifiziertes Schlüsselpaar (PS_x, PS_y) zu erhalten, unmöglich, da der Angreifer sonst zwei gültige Signaturen nach nur einer Interaktion mit dem Treuhänder kennen müßte, was der existentiellen Unfälschbarkeit von S_T widerspricht.   □

Satz A.6 *Sei $\mathcal{K}$ ein sicheres authentisches Schüsselaustauschprotokoll und das symmetrische Kryptosystem $(\mathcal{E}_K, \mathcal{D}_K)$ ohne Kenntnis des symmetrischen Schlüssels K sicher. Damit bietet das System Sicherheit gegenüber dem* Abhören von Münzmaterial *sowie der Erfindung der mehrfachen Gutschrift gegenüber einem Geschäft.*

Beweis A.6: Die Kommunikation bei der Registrierung und dem Abheben von Münzen ist durch den authentischen Sitzungsschlüssel aus dem Protokoll $\mathcal{K}$ geschützt. Da dieses resistent gegenüber Abhören und Mittelsmann-Angriffen ist, vererbt sich die Eigenschaft auf die verschlüsselte Kommunikation. Da ferner, im Fall des Internetzahlungssystems, die Übertragung von σ_M zwischen Kunde und Geschäft mit y_G chiffriert wird, erfährt die Bank σ_M nicht vorzeitig und kann das Geschäft nicht unehrlicherweise der mehrfachen Gutschrift beschuldigen.   □

Satz A.7 *Sofern die Widerrufungslisten aus Abbildung 2 eingesetzt werden, erlaubt das System alle Arten der Erpresserverfolgung. Es ist damit sicher gegenüber der* Erpressung von Münzen *sowie jeglicher* Erpressung geheimer Parameter.

Beweis A.7: Die *Erpressung von Münzen vom Kunden*, die anschließend verwendet werden können, ist unmöglich, da der Zusammenhang zwischen M und PS_y durch σ_B gewahrt wird. Damit kann M nur eingelöst werden, sofern PS_x zur Signaturgenerierung bekannt ist. Die *Erpressung von Münzen vom Geschäft* ist unmöglich, da dessen Identität in die Signatur bei der Bezahlung eingeschlossen wird. Dies verhindert die Gutschrift auf andere Bankkonten. Im Fall einer *Erpressung* des geheimen Benutzerschlüssels PS_x, wird PS_y sofort auf eine schwarze Liste gesetzt. Damit werden alle Münzen, die unter diesem Pseudonym eingelöst werden, von den Geschäften zurückgewiesen und können von ihren legalen Besitzern umgetauscht werden (s.u.). Im Fall einer *Erpressung* des geheimen Bankschlüssels x_B werden legal abgehobene Münzen unter diesem Schlüssel in einer weißen Liste veröffentlicht, so daß alle nachträglich mit diesem Schlüssel ausgestellten Münzen nicht mehr akzeptiert werden. Im Fall einer *Erpressung* des geheimen Schlüssels des Treuhänders tritt der selbe Mechanismus in Kraft, wobei x_B durch x_T und Münzen durch Pseudonyme zu ersetzen sind.   □

Satz A.8 *Sofern die weißen Listen aus Abbildung 2 eingesetzt werden, geht keinem Benutzer durch eine Erpressung Geld verloren, das er noch nicht ausgeben hat.*

Beweis A.8: Die weiße Liste M-WL ermöglicht, alle legal abgehobenen Münzen nach einer Erpressung des geheimen Bankschlüssels weiterhin einlösen zu können, ebenso erlaubt die PS-WL, alle Münzen, die unter einem legal erhaltenen Pseudonym abgehoben wurden, weiterhin auszugeben. Sofern dem Benutzer ein Pseudonym PS_x gestohlen wird, kann er noch nicht ausgegebene Münzen mittels $S_T(x_T, (ID_K, PS_y))$ bei der Bank umtauschen.   □

Satz A.9 *Sofern die Computer der Benutzer den Diebstahl geheimer Information sofort entdecken, gewährt das System Sicherheit nach dem* Diebstahl geheimer Schlüssel.

Beweis A.9: Sofern ein Teilnehmer von einem Diebstahl Kenntnis hat, kann er die gleichen Gegenmaßnahmen wie im Fall einer Erpressung ergreifen.   □

Die elektronische Geldbörse

Susanne Hennecke[1], Petra Wohlmacher[2]

[1] Mannesmann Arcor AG & Co., Eschborn
[2] Universität Klagenfurt, Institut für Informatik

Kurzfassung: Der Beitrag gibt eine Übersicht zum Thema elektronische Geldbörse. Er beschreibt den grundlegenden Aufbau des Geldbörsensystems in Deutschland (GeldKarte). Geldbörsenkarten werden unterschieden nach kontogebundenen und kontoungebundenen Kundenkarten, Händlerkarten und Einreicherkarten. Für die Sicherheit des Geldbörsensystems wird ein umfassendes Schlüsselmanagement eingesetzt, mit Zertifikaten läßt sich die Authentizität und Integrität von Transaktionen feststellen. Jedoch ist die Anonymität des Kunden im Gegensatz zum Bezahlen mit Bargeld nicht gewährleistet. Der Erfolg der elektronischen Geldbörse muß letztendlich abgewartet werden.

1. Einleitung

Elektronische Geldbörsen in Form von Prozessorchipkarten werden in naher Zukunft das Bezahlen mit Münzgeld im Bereich der Automaten durch bargeldloses Bezahlen ablösen. Auch im Einzelhandel soll sich die elektronische Geldbörse zur Bezahlung kleinerer Geldbeträge etablieren. Der Erfolg einer solchen Debitkarte hängt von ihrer Entsprechung zu den wichtigsten Eigenschaften des Bargeldes ab:

- Die Karte muß branchenübergreifend als allgemeines Zahlungsmittel anerkannt sein.
- Das Auffüllen der Karte mit Geldeinheiten (Laden) und der Bezahlvorgang müssen einfach ablaufen.
- Die Anonymität der Zahlungen muß gewährleistet sein.
- Das auf der elektronischen Geldbörse gespeicherte Geld muß an andere Geldbörsen direkt weitergegeben werden können (Purse-to-Purse-Transaktion).
- Die Fälschungssicherheit von Geldeinheiten muß hoch sein.

2. Systemarchitektur der elektronischen Geldbörse in Deutschland

Grundlage für die elektronische Geldbörse in Deutschland (GeldKarte) ist die aus vier Teilen bestehende CEN Norm prEN 1546 [pr_1546]. Sie beschreibt das Gesamtsystem mit seinen Komponenten und Beziehungen, die Sicherheitsarchitektur mit Abwehrmaßnahmen möglicher Angriffe, Datenelemente und Protokolle sowie die Zustände der Geräte.

Elektronische Geldbörsen werden in zwei verschiedenen Ausführungen angeboten: Als kontobezogene Karte, bei der der Ladebetrag vom Kundenkonto abgebucht oder gegen eine andere Kreditkarte verrechnet wird, und als kontoungebundene Karte, bei der der Ladebetrag bar eingezahlt werden muß. Beide Kartenarten sind branchenunabhängig.

Grundlage der kontogebundenen Karte ist die herkömmliche ec-Karte, die seit dem letzten Quartal 1996 um einen Mikrochip ergänzt wurde. Auf dem Chip sind zwei getrennte Dateiverzeichnisse für die electronic cash- und die electronic purse-

Anwendung gespeichert, so daß mit dieser Hybridkarte die gewohnten Zahlungsmöglichkeiten weiterhin nutzbar sind.

Die kontounabhängige Karte wird auch als reine, weiße oder anonyme Karte bezeichnet. Sie trägt auf der Vorderseite den Namen des ausgebenden Kreditinstituts und ist auf der Rückseite frei gestaltbar. Mit einem Firmen-Logo versehen, können diese Karten vom Händler als Kundenkarten ausgegeben werden. Auch spezielle Anwendungen wie z.B. Rabattierungssysteme, Tickets oder Ausweise lassen sich mit dieser Karte realisieren - die GeldKarte wurde so konzipiert, daß auch nachträglich Zusatzfunktionen implementiert werden können. In Zukunft wird dies zu völlig neuen Kooperationsmöglichkeiten zwischen Handel und Kreditanstalten führen.

Systemarchitektur der kontogebundenen GeldKarte (ec-Karte)

Die Gesamtarchitektur der kontogebundenen GeldKarte ist in Abbildung 1 dargestellt.

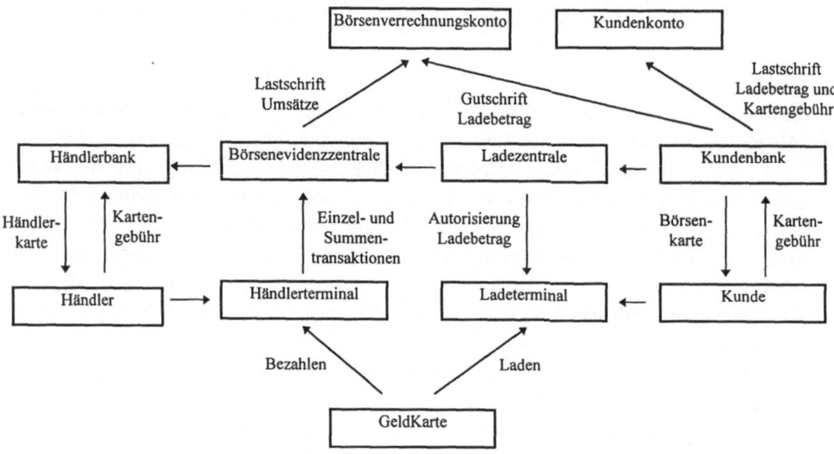

Abbildung 1: Gesamtarchitektur der kontogebundenen GeldKarte

Beim Aufladen der GeldKarte an den Ladeterminals wird der Ladebetrag vom Kundenkonto abgebucht: Die Ladezentrale, an die das Ladeterminal online angeschlossen ist, prüft mittels einer Online-Autorisierung die Verfügungsmöglichkeit des Karteninhabers (Kontodeckung) und gibt die GeldKarte zum Aufladen frei. Diese Information wird an die sogenannte Börsenevidenzzentrale weitergegeben, der Ladebetrag wird von der Kundenbank an ein Börsenverrechnungskonto transferiert.

Beim Bezahlen mit der GeldKarte fließt der elektronische Geldstrom zum Händlerterminal. Fast alle Zahlungen werden offline getätigt. Der Händler reicht Einzel- oder Summendatensätze an die Börsenevidenzzentrale weiter, die dann die Gutschrift auf das Konto des Händlers veranlaßt und den entsprechenden Betrag vom Börsenverrechnungskonto abbucht.

Der Strom des reellen Geldes und der Ware bzw. Dienstleistung ist dem Strom des elektronischen Geldes entgegengerichtet.

3. Technische Voraussetzungen

Um an dem System der GeldKarte teilnehmen zu können, ist auf der Händlerseite eine Anmeldung für die Servicebereitstellung der GeldKarte und eine entsprechende technische Ausrüstung erforderlich. Jeder Händler beantragt bei seiner Hausbank eine Händlerkarte, mit der er sich gegenüber dem Gesamtsystem authentifizieren kann. Die Händlerkarte gibt es in Form von Plug-in-Karten oder virtuell als Software, die in ein Sicherheitsmodul des Händlerterminals geladen wird.

Zur Einreichung der Zahlungsdatensätze stehen dem Händler zwei Möglichkeiten zur Verfügung. Zum einen kann er eine Einreicherkarte verwenden, auf der die Datensätze gespeichert werden. Bei Aktivierung der Kassenschnittfunktion des Terminals werden die Zahlungsdaten nicht in einer Einreicherdatei, sondern auf der Einreicherkarte gespeichert. Die Karte wird dann zur Händlerbank gebracht, die die Kommunikation mit der Evidenzzentrale übernimmt. Zur Zeit haben jedoch nur ca. 75 Zahlungen auf der Karte Platz.

Zum anderen kann der Händler den elektronischen Weg der Einreichung wählen. Hierfür benötigt er einen Netzanschluß (Telefon) und ein Modem, das direkt am Terminal oder an einen PC angeschlossen wird. Bei erfolgtem Kassenschnitt wird vom Terminal eine Einreicherdatei erzeugt, die direkt an die Evidenzzentrale übermittelt wird. Um Zahlungsdaten mehrerer Terminals zusammenzufassen und den online-Betrieb auf eine einzige Telefonleitung zu konzentrieren, benötigt der Händler einen Clustercontroller.

Um elektronische Geldbörsen laden zu können, sind auf der Bankseite Ladeterminals aufgestellt. Bislang stehen diese Terminals aus sicherheits- und einführungstechnischen Gründen in den Foyers der Banken. Ladeterminals sollen in Zukunft aber auch im Selbstbedienungsbereich, beispielsweise im Supermarkt, aufgestellt werden.

Dem Kundenberater der Bank muß ein Sonderfunktionsterminal zur Verfügung gestellt werden, mit dem folgende Funktionen ausgeführt werden können:

- Entladen der Karte mit Kontogutschrift oder Übertragung auf eine neue Karte; dies ist bei einer Kontoschließung oder einer defekten Karte zwingend erforderlich.
- Prüfung der Kartenechtheit.
- Prüfung einer defekten Karte; die Funktionsstörung soll geklärt oder sogar behoben werden können.
- Zurücksetzen des Fehlbedienungszählers für die PIN.
- Bei Unstimmigkeiten müssen die Protokolle der letzten 15 Zahlungen und 3 Aufladungen ausgelesen und ausgedruckt werden können.
- Aufladen der Karte für Kunden, die mit der Bedienung des Ladeterminals nicht vertraut sind.
- Aufladen der Karte gegen Kreditkarte oder Bargeld.

Damit der Kunde sein Recht auf Einsicht in seine elektronische Geldbörse wahrnehmen kann, gibt es sogenannte Taschenkartenleser, die an einem Display den aktuellen Ladestand und die letzten drei Ladungen und 15 Zahlungen anzeigen können.

4. Kosten des Systems

GeldKarte, Lade- und Bezahlvorgang sind für den Kunden zur Zeit kostenfrei.

Für den Händler dagegen entsteht zunächst eine Pauschalgebühr (Bank, ca. 40,- DM) für die Servicebereitstellung (Händlerkarte). Für die Einreichung von Börsenzahlungen fallen 0,3 % vom Gesamteinreichungsbetrag an, mindestens aber 0,02 DM für jede Einzeltransaktion (Autorisierungsgebühren der Kreditwirtschaft, zum Vergleich: 0,15 DM bei electronic cash). Hinzu kommen die Gebühren des Netzbetreibers (in der Regel ISDN-Anschluß) und die Gebühren für die Servicebereitstellung (wie Terminal, Modem). Bei nur einem aufgestellten Terminal und geschätzten Telefonkosten von 150,- DM pro Jahr (1 Anruf pro Tag, 6 Tage pro Woche) entstehen für das erste Jahr folgende Gesamtkosten (Grundlage der Berechnung sind Nettopreise der Fa. TeleCash, TS = Transaktionssumme/Monat):

1. 1.260,- DM bei Eigeninstallation, Analog-Modem und <= 1.000,- DM TS.

2. 1.890,- DM bei vor Ort Aufstellung, ISDN-Modem und <= 1.000,- DM TS.

3. 2.470,- DM bei Wartungsvertrag vor Ort, ISDN-Modem und <= 10.000,- DM TS.

5. Sicherheitsaspekte

Chipkartengestützte Zahlungssysteme müssen den Sicherheitsanforderungen der deutschen Kreditwirtschaft genügen [Schnit197]. Um diese Anforderungen zu erfüllen, werden bei der GeldKarte die folgenden Sicherheitsmaßnahmen durchgeführt [Schnit297]:

* Authentifikation des Benutzers gegenüber seiner GeldKarte mittels PIN-Prüfung. Die PIN ist mindestens vier, höchstens zwölf Ziffern lang, drei Fehlversuche sind erlaubt.

* Authentifikation der an der Kommunikation beteiligten Systemkomponenten mittels kryptographischer Verfahren. Zur Authentifizierung zwischen Karten und Terminals wird ein Challenge&Response-Verfahren basierend auf einem abgeleiteten Schlüsselkonzept eingesetzt. Die kartenindividuellen Schlüssel der GeldKarte werden von den Master Keys der Händlerkarten, Ladeterminals bzw. Bankensonderfunktionsterminals wie folgt abgeleitet: Aus dem Konkatenat eines öffentlich bekannten Initialwertes und den Kartenidentifikationsdaten der jeweiligen Geld-Karte wird mittels einer Hashfunktion [I_10118-2] ein 16 byte langer Hashwert berechnet, der dann mit dem Triple-DES-Verfahren im ECB-Mode, parametrisiert durch einen 16 byte langen Master Key, zu einem 16 byte langen, kartenindividuellen Schlüssel entschlüsselt wird.

  Das Einspielen nichtauthentischer Nachrichten wird durch Überprüfung von Sequenznummern erkannt.

* Nachweis der Integrität von Nachrichten. Kommando- und Antwortnachrichten können durch einen Message Authentication Code (MAC) auf Integrität geprüft werden: Mit einem DES-basierten Verfahren [A_X3.92, A_X9.19], parametrisiert durch einen 8 bzw. 16 byte langen, abgeleiteten Schlüssel, wird ein 8 byte langer einfacher bzw. Retail CFB-MAC berechnet. Der MAC wird zusammen mit der Nachricht versendet.

  Die Karten besitzen einen Schlüssel zur Erzeugung von 8 byte langen (Retail-) CBC-MACs, die als Einzeltransaktions- oder Summenzertifikate zur Einreichung bei der Evidenz-Zentrale dienen.

* Geheimhaltung von PINs und kryptographischen Schlüsseln. PINs und kryptographische Schlüssel müssen gegen unautorisiertes Auslesen und Verändern geschützt werden. Die eingesetzten Prozessorchipkarten und Sicherheitsmodule bieten hierfür geeignete Maßnahmen in Hard- und Software. Außerhalb gesicherter Bereiche er-

folgt die Übertragung von PINs und Schlüsseln DES- oder Tripel-DES-verschlüsselt im CBC-Mode.

- Sicheres Schlüsselmanagement. Kryptographische Schlüssel unterliegen bei ihrer Erzeugung, Verteilung, Verwaltung, ihrem Wechsel und Austausch strengen Sicherheitsauflagen. Beispielsweise werden alle Schlüssel mit Transportschlüsseln Triple-DES-verschlüsselt an ihren Empfänger übermittelt. Zusätzlich erhält der Empfänger einen Hashwert, der mit einer durch den Schlüssel parametrisierten Hashfunktion aus dem Schlüssel berechnet wurde. Damit ist der Schlüssel auf Integrität prüfbar. Jeder Schlüssel erhält zudem einen eindeutigen Schlüsselnamen und eine eindeutige Sequenznummer, die - mit einem Retail-CBC-MAC versehen - ebenfalls an den Empfänger übermittelt werden. Die Transportschlüssel werden mit Hilfe eines Pseudo-Zufallszahlengenerators in einem Sicherheitsmodul bei einem Trust-Center erzeugt und in zwei getrennten, 16 byte langen Hälften auf Papier bei zwei verschiedenen Sicherheitsbeauftragten in Tresorfächern aufbewahrt. Zur Integritätssicherung werden die Hashwerte der beiden Teilhälften hinzugefügt.

- Protokollierung. Verstöße gegen die Sicherheit und Daten, mit denen Abläufe rekonstruiert werden können, müssen protokolliert sowie authentisch und integer gespeichert werden.

- Verwendung von Sicherheitsmodulen. Sicherheitsrelevante Vorgänge wie kryptographische Berechnungen werden nur in Sicherheitsmodulen durchgeführt, sicherheitsrelevante Daten wie kryptographische Schlüssel nur in Sicherheitsmodulen gespeichert. Sicherheitsmodule müssen gegen Auslesen und Verändern geschützt sein.

- Organisatorische Maßnahmen bei der Herstellung und Personalisierung von Sicherheitsmodulen. Die Sicherheitsmodule müssen in einer sicheren Umgebung autorisiert hergestellt und personalisiert werden. Ihr Lebensweg muß aufgezeichnet werden.

- Sicherung der Abläufe. Die Abläufe, die für Transaktionen notwendig sind, und ihre sicherheitsrelevanten Daten dürfen nicht unerkannt manipuliert werden. Zwischen Karten und Terminals kann Secure Messaging [I_7816-4] gefahren werden. Die Börsenkarte verwendet zur Absicherung der Abbuchungs- und Rückbuchungstransaktionen einen von zehn 8 byte langen, kartenindividuellen Schlüssel. Zur Absicherung des Ladens der elektronischen Geldbörse und Entladens der kontobezogenen elektronischen Geldbörse setzt sie einen 16 byte langen, kartenindividuellen Schlüssel ein.
  Die Kommunikation zwischen Börsenkarte und Ladeterminals beim Laden gegen andere Zahlungsmittel wird durch einen von zehn 8 byte langen, kartenindividuellen Schlüssel gesichert.

- Sichere Abschottung aller auf einer Karte gespeicherten Anwendungen. Anwendungen, die neben der GeldKartenanwendung auf der Chipkarte gespeichert sind, dürfen die Sicherheit der GeldKartenanwendung nicht gefährden. Dies kann von den Chipkartenbetriebssystemen der eingesetzten Prozessorchipkarten gewährleistet werden.
  Alle Karten besitzen zur Sicherung von Administrationsfunktionen und PINs beim Transport vom Terminal zur Karte sowie für eine externe Authentifizierung zur Si-

cherung von Karten- und Applikationsinformation vor unberechtigtem Lesen und zur Authentifikation von Chipkartendaten jeweils eigene Schlüssel.

- Sicherstellung der Vertraulichkeit mittels etablierter Verschlüsselungsverfahren. Es dürfen nur solche Verschlüsselungsverfahren eingesetzt werden, von deren Sicherheit man sich durch öffentlich geführte Diskussionen überzeugt hat. Die Sicherheit der Verfahren darf nicht in der Geheimhaltung des Algorithmus, sondern nur in der Geheimhaltung der kryptographischen (geheimen) Schlüssel beruhen. Nachrichten werden mittels DES oder Triple-DES im CBC-Mode verschlüsselt.
- Eindeutige Repräsentation. Sender und Empfänger sicherheitsrelevanter Nachrichten müssen eindeutig identifizierbar sein.
- Vertrauenswürdiges Personal. Personen, die für die Durchführung der oben genannten sicherheitsrelevanten Vorgänge verantwortlich sind, müssen als vertrauenswürdig benannt sein.

6. Fazit

Elektronische Geldbörsen besitzen - trotz der Kosten - gegenüber der herkömmlichen Zahlungsweise entscheidende Vorteile, die zur Zeit jedoch noch hauptsächlich auf der Händlerseite zu finden sind. Hierzu zählen die Zeitersparnis bei Zahlungsabwicklungen, die Minimierung der Kleingeldhaltung, die Vermeidung von Kassendifferenzen, die Reduzierung der Kassenbestandszählung bei Kassenabschluß, die Senkung des Beraubungsrisikos bei Geldtransporten und Einwurf von Geldbomben, sowie die Abnahme von aufgebrochenen Automaten. Fraglich ist allerdings, ob für den kleinen Einzelhandel wie Kiosk die Teilnahme am elektronischen Geldbörsensystem letztendlich rentabel ist.

Die GeldKarte mit Kontobezug verstößt jedoch zur Zeit gegen eine der wichtigsten Eigenschaften des Bargeldes: die Anonymität. Durch die in der Evidenzzentrale in Schattenkonten geführten auflaufenden Datensätze können regelrechte Kundenprofile erstellt werden: die Ableitung von Kaufgewohnheiten und die Identifizierung einer Einzelperson sind möglich - die Sicherheit des Systems steht versus der Anonymität des Kunden.

Literatur

[A_X3.92]   American National Standard X3.92 - 1981: Data Encryption Algorithm.

[A_X9.19]   American National Standard X9.19 - 1986: Financial Institution Retail Message Authentication.

[I_10118-2] ISO/IEC 10118-2: Information Technology - Security techniques - Hash Functions, Part 2: Hash-functions using an n-bit block cipher algorithm, 1992

[I_7816-4]   ISO/IEC 7816-4: Identification Cards - Integrated circuit(s) cards with contacts - Part 4: Inter-industry commands for interchange, 1995

[pr_1546]   prEN 1546: Identification card systems - Inter-sector electronic purse - Part 1-4

[Schnit197] Schnittstellenspezifikation für die ec-Karte mit Chip, Kriterien für die Bewertung von chipkartengestützten Zahlungssystemen. Version 2.2, 22.01.1997

[Schnit297] Schnittstellenspezifikation für die ec-Karte mit Chip, Key-Management. Version 2.2, 22.01.1997

Sicherheitsproblematik in verteilten, digitalen Videoanwendungen und Präsentation eines technischen Lösungsansatzes zur transparenten Verschlüsselung von MPEG-2 Video

Jana Dittmann
Tel.: +49 - 6151 - 869 845
e-mail: jana.dittmann@darmstadt.gmd.de

Arnd Steinmetz
Tel.: +49 - 6151 - 869 862
e-mail: arnd.steinmetz@darmstadt.gmd.de

GMD (German National Research Center for Information Technology)
IPSI (Integrated Publication and Information System Institute)
Dolivostr. 15, D-64293 Darmstadt, Deutschland

Zusammenfassung

Die Sicherheitsproblematik ist zu einem entscheidenden Thema für die Verbreitung neuer Informationstechnologien geworden. Dieser Artikel zeigt Sicherheitsanforderungen an verteilte, digitale Videoanwendungen auf, um mit einer gezielten Gestaltung der verteilten Videoanwendung die Akzeptanzprobleme der Videoproduzenten gegenüber der neuen Technik zu beseitigen und die digitalen Marktmöglichkeiten lukrativer zu gestalten. Aus den einzelnen Sicherheitsaspekten wird die transparente Verschlüsselung herausgegriffen und ein technischer Lösungsansatz auf Basis von MPEG-2 repräsentiert und diskutiert.

1 Einleitung

Video- und Kinoproduktionen werden heute hauptsächlich auf einem digitalen Rechner bearbeitet und fertig gestellt. Der sich ständig verstärkende Trend zur weltweiten Vernetzung der einzelnen Rechner erfordert neue Konzepte, sowohl für verteilte Videoproduktionssysteme zur Unterstützung dieser Arbeitsweisen als auch für Mechanismen zum Schutz der dafür notwendigen Kommunikation und natürlich der daraus entstandenen Werke auf dem digitalen Markt. Es geht dabei nicht vordringlich um Wirtschaftstransaktionen, sondern um die Sicherung der Kommunikation sowie der entstehenden und entstandenen Werke. Dieser Artikel zeigt Sicherheitsanforderungen an digitale Videosysteme und digitales Videomaterial am Beispiel des digitalen Videoeditier- und schnittsystems (DiVidEd) der GMD Darmstadt, Institut IPSI auf, die notwendig sind, um bei den Anwendern eines solchen Systems Akzeptanz zu erreichen, die digitalen Produktionsmöglichkeiten und den digitalen Markt zur Verbreitung der Werke zu nutzen. Dazu werden bestehende Sicherheitsanforderungen besprochen und ein Lösungsansatz für transparente Verschlüsselungsmöglichkeiten für *Try&Buy* Transaktionen auf Basis des Videodatenformates MPEG-2 präsentiert.

2 System DiVidEd

Das DiVidEd Konzept ist ein verteiltes, digitales Videoeditier- und -schnittsystem. Es geht über die Funktionalität des herkömmlichen Videoeditings (der Postproduktion, dem Filmschnitt oder einfach Schnitt), hinaus und sichert die Konsistenz zwischen Skript, Drehbuch und abschließendem Schnitt. Da die gesamte Videoproduktion an örtlich verschiedenen Stellen stattfindet, ist das Werkzeug als verteilte Applikation realisiert, und reiht sich in eine verteilte Videoproduktionsumgebung (Distributed Video Production) ein, die alle Produktionsstufen und das gesamte Datenaufkommen über Netzwerk verbindet. Das Editiersystem besteht aus einer Server- und Clientkomponente. Der Editing Server führt die eigentliche Editierfuntionalität aus, während der Client eine Benutzeroberfläche bietet. Innerhalb der verteilten Videoproduktion wird eine Videodatenbasis und eine Metadatenbasis aufgebaut, auf welche das System DiVidEd zugreift, um Daten abzurufen, zu editieren und zu speichern, [Steinmetz1994].

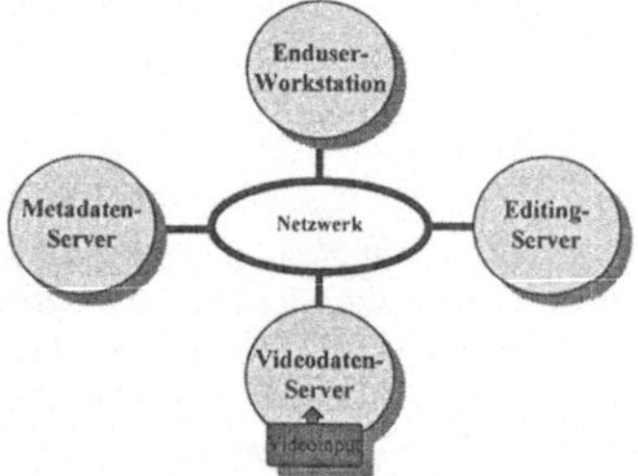

Abbildung - *Distributed Videoproduction*

Nach der Systemspezifikation muß das System sowohl in einem homogenen als auch heterogenen Netzwerk lauffähig sein, so daß das System vom Benutzer angefordert werden kann, wo immer er sich auch befindet und somit durch ein Netzwerk auf jeder Workstation gestartet werden kann, [Steinmetz1996]. Das System ist nicht nur als alleiniges Werkzeug zur Videoproduktion zu verstehen, sondern kann **auch als verteilte Videodatenbank** mit Video Data Base Servern, auf welchen Videos angeboten werden, angesehen werden. Das bedeutet, daß das System von vielen Anwendern genutzt werden kann, sei es zur eigentlichen Videoproduktion oder zur Abfrage von Videomaterial oder Metadaten zum Video in jeder Form. DiVidEd kann somit einerseits als eine offene kommunikationstechnische Infrastruktur gesehen werden, da die Dienste, die angeboten werden frei zugänglich sind. Andererseits ist ein offener, technisch gestützter Informations- und Datenaustausch von Videomaterial, Metadaten usw. Gewährleistet.

Sicherheitsfragen spielen bei der Offenheit von kommunikationstechnischen Infrastrukturen und bei offener Kommunikation eine wesentliche Rolle und werden im folgenden Kapitel erläutert.

3 Sicherheitsanforderungen an verteilte digitale Videoanwendungen

In die neuen Technologien im Videobereich werden große Hoffnungen gesetzt. Man sieht erhebliche **Chancen**, den Produktionsprozeß zu rationalisieren, effektiver und effizienter zu gestalten. Gleichzeitig bestehen jedoch auch **Sicherheitsbedenken** gegenüber der neuen Informationstechnik. Der Videoproduktionsprozeß ist ein Produktionsprozeß, der ein Endprodukt erzeugt, welches ein individuelles Werk (meist im Sinne einer geistigen Schöpfung) darstellt. Diese persönlichen Schöpfungen des Autors oder der Produzenten sind Erzeugnisse, die durch ihren Inhalt oder ihre Form etwas Neues und Eigentümliches darstellen. Es besteht somit das natürliche Interesse der Anwender, daß ihr Eigentum, sei es in Form eines digital gespeicherten Videos oder der gespeicherten Ideen, nicht offen für alle zugänglich ist. Der weltweite Ausbau digitaler Netze, dessen überaus dynamische Entwicklung weiter anhält, eröffnet weitläufige Angriffspunkte für das Abhören, Lesen, Manipulieren oder Löschen von Daten seitens Unbefugter. Es ergeben sich somit auch **erhebliche Risiken** für die Nutzung der Informationssysteme, **die von direkten finanziellen Schäden, rechtlichen Schäden bis hin zu Imageverlusten reichen**. Die erforderliche Täterlogistik ist nicht aufwendig und frei verkäuflich, ein PC mit Modem reicht häufig aus. Das Fehlen übergeordneter Kontrollinstanzen, der weitgehend freie Zugang und die häufig nur geringen Sicherheitsstandards bei Betreibern und Nutzern begünstigen den Mißbrauch. Zudem werden Mißbrauchsfälle von den Geschädigten nicht immer oder erst spät erkannt, die Schäden allerdings sind erheblich. Schließlich ist auch noch zu berücksichtigen, daß die Ermittlungs- und Beweisführung bei solcher High-Tech-Kriminalität Schwierigkeiten mit sich bringt, die sich bei internationaler Tatbegehung noch verstärken. Wirkungsvoll sind deshalb präventive Konzepte, die von Anfang an einen hohen Schutz bieten. Dies betrifft sowohl die **lokal gespeicherten Daten**, als auch die **übertragenen Daten**. Die Datenschutzinteressen der Anwender des Systems DiVidEd ergeben sich vor allem aus den Mißbrauchsgefährdungen in der offenen und verteilten Kommunikationsumgebung. Die Kommunikation erfolgt über jede Art von Kommunikationsnetz und Datenfernübertragung, wodurch die Übertragungswege nicht sicher zu kontrollieren sind und Unbefugte Zugriff auf die Daten erhalten können. Grundsätzlich müssen sich Schutzmaßnahmen sowohl gegen **externen wie internen Mißbrauch** richten, d.h. gegen unberechtigte und berechtigte Netz- und Systemteilnehmer sowie Systembetreiber.

Ein berechtigtes Interesse der Anwender einer verteilten Videoproduktionsumgebung ist, ihre Daten und ihre Kommunikation vor unberechtigtem Zugriff seitens anderer Benutzergruppen und Netzteilnehmer zu schützen und abzusichern. Dieses Bedürfnis zur Absicherung kann gerade in der Videoproduktion beobachtet werden, da hier ein hohes schöpferisches Potential eingebracht wird. Vertrauliche Daten müssen **vertraulich** bleiben. Sind fertige Videos produziert und stehen sie in der Videodatenbank zur Verfügung, kann es bei reinen Abfragefunktionen von Interesse sein, allen Anwendern des Systems diese Videodaten mit **urheberrechtlicher Kennung** zur Verfügung zu stellen, um das Produkt am Markt zu plazieren und entsprechende Urheber- und Lizenzrechte zu sichern. Hier spielen auch wirtschaftliche Interessen hinein, so daß Videos, für die seitens des Nachfragers noch keine Gebühr entrichtet wurde oder noch kein Preis und die entsprechenden Konditionen erfüllt sind, das Videomaterial nur im beschränkten Umfang zugänglich ist. zusammengefaßt also: *Try&Buy*. Der **Zugriffschutz** erfordert einerseits bei jeder Anfrage, den Zugreifer **sicher zu identifizieren** und in Abhängigkeit von dieser **Authentifizierung** den Zugriff zu gestatten oder zu verweigern. Andererseits muß auch die umgekehrte Richtung möglich sein. Der Nutzer will sicher sein, daß eine bestimmte Information tatsächlich vom gewünschten Partner oder Anbieter stammt und ihm nicht jemand anderes gefälschte Daten unterschiebt. Sollte jemand die übertragen Daten unterwegs manipulieren - unabhängig ob jemand die Orginaldaten entziffern kann oder nicht - muß der Empfänger diese **Verletzung der Integrität** sofort feststellen können. Eine verteilte, digitale Videoanwendung sollte deshalb das vollständige Potential an Sicherheitsaspekten umfassen, um die Akzeptanz der Nutzer zu erhöhen.

Sicherheitsaspekt	**Kurzbeschreibung**
Zugriffsschutz	Kontrolle des Systemzuganges und Zugriffsbeschränkungen auf Systemfunktionen und Datenbestände
Authentizität	Nachweis der Identität des Urhebers/Autors und der Echtheit des Datenmaterials, Authentikation ist die Verifizierung der Echtheit
Vertraulichkeit	Verhindert, daß unberechtigte Dritte auf Daten zugreifen können
Integrität	Erbringt den Nachweis, daß die Daten unverändert vorliegen
Nachweisbarkeit	Prüfung der Authentizität und Integrität der Daten auch von berechtigten Dritten, Gewährleistung der Verbindlichkeit der Kommunikation
Transparente Verschlüsselungsmöglichkeiten	Datenmaterial soll in verminderter Qualität zur Verfügung gestellt werden
Urheberrechte	Schutz des geistigen Eigentums, Copyrightmarkierungen zur Eigentümerkennzeichnung und Benutzerkennzeichnung, Identifizierung von Verbreitern illegaler Kopien
Persönlichkeitsschutz	Schutz der Privatsphäre: Unbeobachtbarkeit, Anonymität, Pseudonymität und Unverkettbarkeit zur Abwehr von unerlaubten Verkehrsflußanalysen und Analysen der Benutzung von Dienstprogrammen und anderer Ressourcen

Tabelle - Sicherheitsaspekte

Zu diesen Aspekten sind Mechanismen und Verfahren während den einzelnen Videoproduktionsstufen für die Kommunikation, den Datenaustausch und die Datenhaltung auf der Video- und Metadatenbank nötig, [Ditt1997]. Es sind verschiedene Sicherheitsmodule zu erstellen, die ein abgestuftes Sicherheitsniveau je nach Anwenderwünschen bieten. Betrachtet man die große Vielfalt der Anwendungsfunktionalität, so wird die Menge der unterschiedlichen Anforderungen an die Kommunikation augenscheinlich. Hier sind einerseits die Übertragung von Echtzeitvideodaten gefordert, bei der schnelle und damit oft schwache Algorithmen für die Sicherheit gefragt sind, andererseits kann man sich bei einer asynchronen Übertragung von Daten mehr auf den Sicherheitsaspekt konzentrieren. Die Sicherheitsmodule sollten zu diesem Zweck dynamisch konfigurierbar sein. Sie sind in die Funktionalität zu integrieren, so daß sie einen Bestandteil des Systems bilden. Die Sicherheitsmodule müssen auf alle vorkommenden Datenformate, wie zum Beispiel Text, Graphik und Video anwendbar sein.

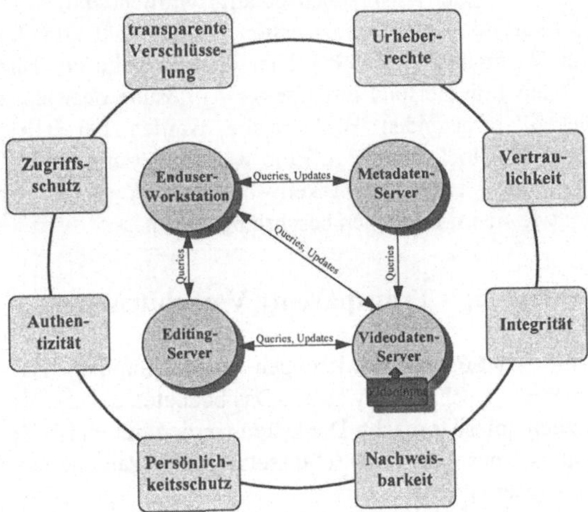

Abbildung - DiVidEd-Sicherheitskomponenten

Die Verarbeitungsrisiken einer globalisierten Kommunikation lassen sich nur mit Hilfe einer genauso global wirksamen Datensicherheitstechnologie eindämmen. Für die einzelnen Sicherheitsmodule wurden Konzepte entwickelt, so daß ein abgestuftes Sicherheitsniveau erreicht werden kann.

4 Transparente Verschlüsselungsmöglichkeiten

In diesem Artikel wird speziell eine Möglichkeit zur gezielten Begrenzung von Les- und Nutzbarkeit von MPEG Video präsentiert, die als transparenten Verschlüsselung bezeichnet wird. Im Rahmen der Sicherheitskonzeption für das System DiVidEd wurde dazu ein Verfahren auf Basis von MPEG-2 entwickelt, [Ditt1997].
Innerhalb einer verteilten Videoproduktionsumgebung werden neben den reinen Funktionalitäten der Videoproduktion selbst auch Videodaten zur Verfügung gestellt. Für die Anbieter von Videomaterial kann es durchaus von Interesse sein, nicht alle Daten geheim zu halten, für Unbefugte unkenntlich zu machen oder zu schützen. Diese Problematik entsteht unter anderem bei sogenannten *Try&Buy* Transaktionen, bei denen dem Kunden einerseits die Daten in einer Weise zur Verfügung gestellt werden sollen, die eine Bewertung ihres Inhalts und ihrer Verwendbarkeit ermöglichen, andererseits aber noch nicht in vollständiger Qualität anbieten, ehe gewisse Konditionen und Preise oder Absprachen getätigt wurden sind. Ein Verfahren TIE - Tool for Image Encryption stellt einen ersten Ansatzpunkt zum Verschlüsseln von Rasterbildern zur Verfügung [TIE1995]. Hier können Teile des Bildes verschlüsselt werden, welche dann zum Beispiel in niedriger Auflösung oder als schwarzer Kasten erscheinen. Die zu ersetzenden Bildbereiche werden bei TIE in sogenannten Application Extension Blocks abgelegt. Eine weitere Lösung für MPEG-2 Video zur Begrenzung der Les- und Nutzbarkeit bietet sich durch eine transparente Verschlüsselung, wie sie im folgenden beschrieben wird.

4.1 Lösungsvorschlag - Transparente Verschlüsselung

Eine Möglichkeit *Try&Buy* Anforderungen nachzukommen ist, daß die Daten sozusagen transparent verschlüsselt werden. Das bedeutet, daß die Verschlüsselung die Daten nicht gänzlich unlesbar macht. Die Daten werden mit einem Rauschen versehen. Erst durch Kenntnis eines geheimen Schlüssels nach Bezahlung des Lieferanten wird dieses Rauschen wieder entfernt.
MPEG-2 bietet einen interessanten Ansatz, dieses Problem zu lösen. **Die Idee**, die im Rahmen dieser Arbeit ausgearbeitet wird, ist, die Skalierbarkeit (Scalability) des Datenfomates für die transparente Verschlüsselung zu nutzen und Mechanismen zu entwickeln, wie die Skalierbarkeit erweitert werden kann, um transparente Verschlüsselungsmöglichkeiten zu erhalten. Dazu wird zuerst die Funktionalität der Skalierbarkeit vorgestellt und aufbauend auf den Erkenntnissen aus der Norm [ISO/IEC-13818-2] ein SNR-skalierbarer En- bzw. Decoder entwickelt. Aufbauend auf diesen Erkenntnissen wird dann gezeigt, wie solch ein spezifischer En- bzw. Decoder für eine transparente Verschlüsselung genutzt werden kann.

Die Skalierbarkeit ist eigentlich dazu gedacht, das Ausfallverhalten einer Übertragung oder die Qualität der Übertragung zu beeinflussen. Dazu wird aus den im Standard zur Verfügung stehenden Skalierbarkeitsmodi, die SNR-Skalierbarkeit herausgegriffen. Zur Erkennung, um welchen Modus es sich handelt existiert die sogenannte **sequence_scalable_extension**, welche ein Feld **scalable_mod** enthält [ISO/IEC-13818-2].

Die folgende Tabelle zeigt die möglichen Initialisierungen und deren Bedeutung:

scalable_mode	Meaning
sequence_scalable_extension() not present	
00	data partitioning
01	spatial scalability
10	SNR scalability
11	temporal scalability

Tabelle - Definitionen des Feldes scalable_mode [ISO/IEC13818-2]

Im folgenden sollen die Möglichkeiten der SNR-Skalierbarkeit näher betrachtet werden. Auf die anderen Modi wird nicht eingegangen. Es sind aber auch bei diesen Modi mögliche Ansätze für eine transparente Verschlüsselung zu finden. Desweiteren bietet der MPEG-2 Standard die Kombination von Skaliermodi an, so daß ein mehrstufiges Ausfallverhalten oder eventuell eine mehrstufige Qualitätsabstufung möglich ist, die für die transparente Verschlüsselung verwendet werden kann. Auch auf diese Möglichkeiten sei hier nur hingewiesen.

SNR Skalierbarkeit ist ein Werkzeug, daß dazu gedacht ist, Videoapplikationen die in Verbindung mit der Telekommunikation oder Videodiensten mit verschiedenen Qualitätsebenen stehen, zu unterstützen. SNR steht für Signal to Noise Ratio, der Rauschskalierbarkeit, die zwei Videoschichten der selben räumlichen Auflösung aber unterschiedlicher Videoqualität von einer einzigen Videoquelle generiert. Es entsteht ein Base Layer mit einer grundlegenden Videoqualität und einen Enhancement Layer zur Verbesserung der Videoqualität des Base Layers (Lower Layer). Wenn der Enhancement Layer zum Base Layer hinzugefügt wird, wird eine höhere Qualitätsreproduktion des Eingabestroms erreicht.
Diese Technik wird gewählt, um einen Layer niedrigerer Qualität zu haben, der genutzt wird, wenn der Enhancement Layer nicht verfügbar ist. Daneben läßt dieser Mechanismus einen guten Fehlerschutz vor Totalausfall zu, falls der Fehler hauptsächlich im Enhancement Layer auftritt. Im Fehlerfall im Enhancement Layer kann ausschließlich der Base Layer genutzt werden. Besonders bei sich ständig wiederholenden Fehlern, einem zeitweisen Verlust oder gar bei permanentem Verlust des Enhancement Layers ist das Verbergen des korrigierenden Layers sehr effektiv, [Reimers1995]. Ist der Enhancement Layer permanent unverfügbar, wird nur der Lower Layer decodiert.
Ausgehend von der generischen Spezifikation der SNR-Skalierbarkeit im Standard [ISO/IEC13818-2] kann man folgenden **konkreten Ablauf im Encoder** entwickeln:
Der Videodatenstrom wird vom Encoder eingelesen und es erfolgt wie im non_scalable Mode die Diskrete Cosinus Transformation (DCT). Die DCT wandelt ein 8x8 Pixelbild vom Ortsbereich in den Ortsfrequenzbereich. Dabei geht die Darstellung

bei einer Quantisierung des Eingangssignals von 8 Bit im Ortsbereich in eine Darstellung von 11 Bit in den DCT-Koeffizienten über. Dadurch wird zuerst eine Erhöhung der erforderlichen Bitzahl bewirkt. Der Vorteil der DCT liegt jedoch darin, daß viele der Ortsfrequenzkoeffizienten sehr klein oder Null sind, was für die Datenreduktion ausgenutzt wird.

Nach der Transformation in den Frequenzbereich wird eine Quantisierung der Koeffizienten der DCT vorgenommen mit dem Ziel einer Datenreduktion. Mit Hilfe eines psycho-physiologischen Experiments wurden Koeffizienten ermittelt, die für die Quantisierung benutzt werden. Somit werden Frequenzbereiche, die das menschliche Auge gut wahrnehmen kann, genauer quantisiert als weniger kritische Bereiche. Man kann tendenziell feststellen, daß je höher die Ortsfrequenzen werden, desto gröber die Quantisierung der DCT-Koeffizienten. Dies trägt auch der Tatsache Rechnung, daß das menschliche Auge feine Details nur mit einem geringeren Dynamikumfang wahrnehmen kann. Im MPEG-Standard existieren dazu entsprechende mögliche Quantisierungstabellen, die den DCT-Koeffizienten entsprechende Quantisierungsstufenhöhen zuordnen. Die Quantisierung wird nach folgender Formel vorgenommen [ISO/IEC13818-2]:

-      f_x, f_y: *spatial frequencies*

-      $G_Q(f_x, f_y)$:   *quantized DCT coefficients*

-      $G(f_x, f_y)$:   *unquantized DCT coefficients*

-      $Q(f_x, f_y)$:   *quantizer step size*

$$G_Q(f_x, f_y) = round \; \frac{G(f_x, f_y)}{Q(f_x, f_y)}$$

Nach dieser Gleichung werden die quantisierten Koeffizienten durch Division der DCT-Koeffizienten durch die gewünschte Quantisierungsstufenhöhe und anschließender Rundung auf ganze Zahlen berechnet. Der Divisor ist im allgemeinen Fall für jede Ortsfrequenz (f_x, f_y) unterschiedlich, damit eine Anpassung an den Dynamikumfang des menschlichen Sehsinns erreicht wird. Man erkennt: **je größer die Quantisierungsstufenhöhe** ist, **desto gröber erfolgt die Quantisierung** und **desto schlechter wird die Qualität des Bildes** für das menschliche Auge. Dieses Wissen macht sich nun die SNR-Skalierbarkeit zu nutze, um zwei Bitströme mit unterschiedlicher Qualität zu erzeugen.

<u>Die Quantisierung im SNR-Skaliermodus kann nun wie folgt gestaltet werden:</u>

Zur Erzeugung des Base Layers wird die Quantisierung der DCT-Koeffizienten zuerst grob durchgeführt. Grob bedeutet, daß statt der vorgeschlagenen Quantisierungsstufenhöhen, ein höherer Wert verwendet wird. Die berechneten Quantisierungskoeffizienten werden dann wie gewöhnlich VLC (<u>V</u>ariable <u>L</u>ength <u>C</u>oding) codiert, d.h. der Redundanzreduktion unterworfen und anschließend mit den erforderlichen Steuerinformationen versehen. Dieser Bitstrom bildet den Lower oder Base Layer. Durch den verwendeten groben Quantisierer wird somit ein Datenstrom erzeugt, der durch den eingetretenen Quantisierungverlust eine verminderte visuelle Qualität aufweist. Dieser Quantisierungsverlust muß nun gemessen und im

Enhancement Layer codiert werden, um die volle Qualität bei Bedarf zur Verfügung zu stellen. Dazu wird nach der Quantisierung mit einer höheren Quantisierungsstufenhöhe eine inverse Quantisierung des Lower Layers vorgenommen und die erhaltenen dequantisierten DCT-Koeffizienten werden mit den Original-DCT-Koeffizienten verglichen. Die gemessene Differenz wird dann über einen feineren Quantisierer, der eine Quantisierungsstufenhöhe von größer gleich 1 hat, verarbeitet und bildet dann nach der separaten Redundanzreduktion den Enhancement Layer.

Ein Beispiel soll diesen Vorgang nun verdeutlichen:

Zuerst sei der normale Vorgang beschrieben nach [ISO/IEC13818-2]:

1. Der DCT-Koeffizient sei für (f_x, f_y)=(1,3)=**36**.
2. Die Quantisierungsstufenhöhe beträgt für (f_x, f_y): $Q(1,3)$=**24**.
3. Dann ergibt sich ein Quantisierungskoeffizient von *round(36/24)*=**2**.
4. Es ist zu erkennen, daß für den DCT-Bereich von (35,60) der gleiche Quantisierungskoeffizient entsteht. Dieser Bereich wird um so größer, desto höher die Quantisierungsstufenhöhe gewählt wird: bei der Quantisierungsstufenhöhe von 48 ergibt sich ein Quantisierungskoeffizient von 1, der ein Intervall von (23,72) abdeckt, wodurch ein deutlich größerer Qualitätsverlust verursacht wird.

Im Skaliermodus wird wie folgt vorgegangen:

5. Quantisiert man nun **im SNR-Skaliermodus mit der groben Quantisierungsstufenhöhe von 48,** ergibt sich der Quantisierungskoeffizient von 1 für den Lower Layer, der Encoder führt dann eine inverse Quantisierung Q' aus, erhält als dequantisierten DCT-Koeffizienten 48, vergleicht diesen mit dem Original-DCT-Koeffizienten von 36 und berechnet die Differenz von -12. Diese Differenz geht in den Enhancement Layer ein und stellt den Qualitätsverlust, der durch die grobe Quantisierung entstanden ist, dar.

Aus der generischen Spezifikation zur SNR-Skalierbarkeit kann nun folgendes Blockschaltbild eines SNR-skalierbaren Encoders entwickelt werden:

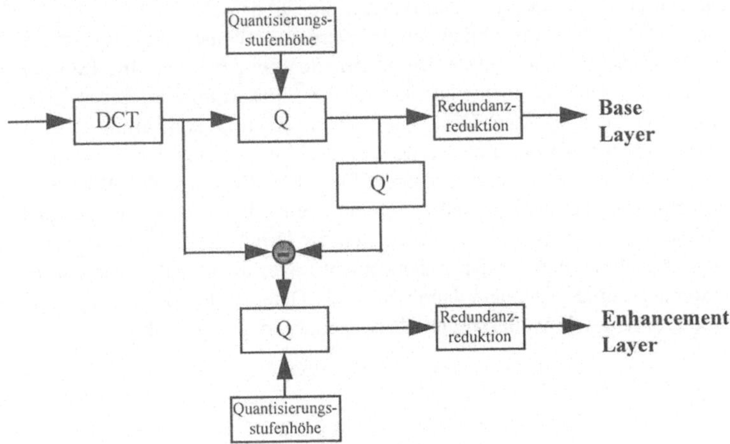

Abbildung - SNR-skalierbarer Encoder

Die Quantisierungsstufenhöhen sind so zu wählen, daß für den Lower Layer grober quantisiert wird, d.h. höhere Quantisierungsstufen verwendet werden. Für den Enhancement Layer können kleiner, feinere Quantisierungsstufen benutzt werden, das Minimum ist 1, wobei dann der Qualitätsverlust komplett durch den Enhancement Layer kompensiert wird.

Die SNR Skalierbarkeit definiert nun im **Decoder** nach ISO/IEC13818-2 einen Mechanismus, der die codierten DCT-Koeffizienten des Lower Layer verbessert, indem der Enhancement Layer, der den Qualitätsverlust ausgleicht, benutzt wird.

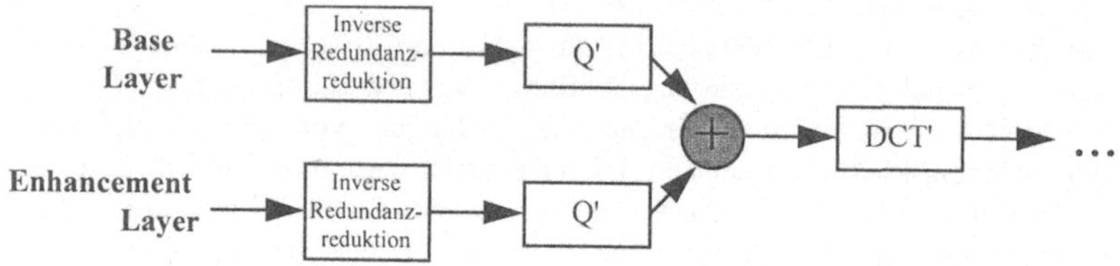

Abbildung - Illustration des Decodiervorganges für die SNR Skalierbarkeit [ISO/System]

Wie in der Abbildung illustriert, werden die Daten von zwei Bitströmen nach dem inversen Quantisierungsprozeß, Q', kombiniert, indem die DCT Koeffizienten addiert werden. Der Decodiervorgang der beider Schichten ist bis zur Kombination völlig unabhängig voneinander. Den Enhancement Layer vollständig zu decodieren, ist jedoch, ohne parallel den Lower Layer zu decodieren, nicht möglich. Um die beiden zusammengehörigen Layer identifizieren zu können, existieren vom Encoder vergebene Layer-Ids, wobei der Lower Layer eine um eins kleinere Id erhält als der zugehörige Enhancement Layer. Es existieren einige semantische und syntaktische Anforderungen an die beiden Layerstrukturen, die für den Decodiervorgang benötigt werden und vom Encoder hinzugefügt werden müssen. Es sei in diesem Zusammenhang lediglich auf diese Besonderheiten hingewiesen. Im ISO-Standard [ISO/IEC13818-2] sind sie unter dem Kapitel 7.8 SNR Scalability beschrieben. Der Ablauf der Redundanzreduktion (VLC) und der inversen Quantisierung läuft wie im non_scalable Modus ab. Von Interesse sei hier das Vorgehen bei der Kombination der beiden Bitströme. Die korrespondierenden DCT-Koeffizienten der Blöcke in den beiden Schichten werden addiert und bilden ein neues Signal. Aus dem Beispiel wäre das die Addition von 48 und -12, wodurch sich der Original-DCT-Koeffizient von 36 bildet, unter Annahme, daß für den Enhancement Layer die Quantisierungsstufenhöhe von 1 verwendet wurde. Nachdem dann wieder ein Datenstrom erzeugt wurde, verläuft der Decodiervorgang wie im herkömmlichen Decoder im non_scalable Modus.

Das Vorgehen zur Berechnung des zusammengesetzten Datenstroms ist prinzipiell für Luminanz- und Chrominanzblöcke anzuwenden. Der Vollständigkeit halber sei erwähnt, daß das Vorgehen für Chrominanzblöcke nach folgender Formel erfolgt, wenn das chrominanz_simulcast Feld auf 1 gesetzt ist [ISO/IEC13818-2]:

$$G'\,(0,0) = G'_{lower}\,(0,0) + G'_{enhance}\,(0,0)$$

$$G'\,(f_x,\,f_y) = G'_{enhance}\,(f_x,\,f_y), \quad \textit{für alle } f_x,\,f_y \textit{ außer } f_x = f_y = 0$$

Das bedeutet allerdings, daß auch im Encoder eine entsprechend veränderte Codierung vorgenommen werden muß. Dort wird dann nicht die Differenz zwischen den einzelnen Koeffizienten gebildet, sondern die Differenz bezogen auf den Koeffizienten (0,0).

Im Decoder wird für die Chrominanzblöcke dann der DC-Koeffizient (Gleichanteil) des Base Layers als Voraussage des DC-Koeffizienten im Block des Enhancement Layers benutzt, während die AC-Koeffizienten (Wechselanteil) des Base Layers abgelegt werden und die AC-Koeffizienten des Enhancement Layers das Ausgangssignal bilden.

Ergebnis: Ein Ansatzpunkt zur transparenten Verschlüsselung

SNR-skalierbares Codieren erstellt zwei Schichten mit der selben räumlichen Auflösung aber unterschiedlicher Bildqualität. Abhängig davon, ob ein oder beide Layer decodiert werden, ist im decodierten Bild eine hohe oder weniger gute Qualität zu erreichen. Diese Möglichkeit bietet auch einen guten **Ansatzpunkt für die transparente Verschlüsselung** und gute Möglichkeiten für *Try&Buy* Transaktionen. Die Idee ist nun, das Videomaterial mit dem entwickelten SNR-skalierbaren Encoder zu codieren und zwei Layer mit unterschiedlicher Bildqualität zu erzeugen. Den potentiellen Kunden wird dann lediglich der Base Layer zur Verfügung zu gestellt. Das kann auf unterschiedlicher Art und Weise erfolgen:

Entweder man codiert den Videostrom im SNR-Skalierbarkeits Modus und verschickt lediglich den Lower Layer, **oder** man verschickt beide entstandenen Layer, wobei dann der Enhancement Layer aber symmetrisch verschlüsselt ist. Für die erste Methode muß der Enhancement Layer nach dem Codiervorgang extrahiert werden. Wenn der Kunde die volle Qualität erhalten möchte, muß der komplette Enhancement Layer im Nachhinein zugeschickt werden. Die zweite Methode muß im SNR-Skaliermodus codieren, danach den Enhancement Layer verschlüsseln und dem Decoder den Enhancement Layer sozusagen verschweigen, so daß dort lediglich auf den Base Layer zugegriffen wird. Dies kann erreicht werden, in dem man den Decodiermodus von SNR-skalierbar auf den normalen (non-scalable) Modus setzt. Will der Kunde die volle Qualität, wird ihm der Schlüssel zugesandt und er kann den Enhancement Layer entschlüsseln, der Skaliermodus wird wieder auf SNR-skalierbar gesetzt und die gewünschte Qualität wird erreicht. Für diese Methode ist natürlich eine separates Werkzeug für die Ent- und Verschlüsselung notwendig, das auch dem Kunden zur Verfügung gestellt werden muß, wenn er entschlüsseln möchte. Es besteht somit die

Möglichkeit auf verschiedene Qualitätsstufen umzuschalten bei Kenntnis des geheimen Schlüssels. Bekannte symmetrische Verschlüsselungsverfahren, die zum Einsatz kommen, sind DES und IDEA.

4.2 Bewertung des Lösungsansatzes

Der Lösungsansatz für eine transparente Verschlüsselung auf der Basis der SNR-Skalierbarkeit wurde mit einem symmetrischen Verschlüsselungsverfahren vorge-schlagen. Der DES als ein symmetrischer Verschlüsselungsalgorithmus, ist zwar sehr schnell und sehr einfach zu implementieren, weist jedoch in verschiedenen Schlüsselbereichen Schwächen auf. Somit ist bei der Auswahl des Schlüssels auf dessen Sicherheit zu achten. Außerdem ist die Sicherheit abhängig von der Länge des verwendeten Schlüssels, [Schneier1994], [Stallings1995].

Es ist auf eine gezielte Schlüsselauswahl und auf die Geheimhaltung der verwendeten Schlüssel zu achten Ein asymmetrisches Verfahren hätte in seiner Anwendung mit öffentlichen und geheimen Schlüsseln bei der transparenten Verschlüsselung wenig Sinn, da der Zugriff nur gestattet sein darf, wenn der Schlüssel bekannt ist. Der Schlüssel zum Entschlüsseln muß also nur für berechtigte Nutzer zugänglich sein und sonst ebenfalls geheim bleiben. Das Schlüsselverteilproblem kann allerdings mit einem asymmetrischen Verfahren erfolgen, indem mit dem öffentlichen Schlüssel verschlüsselt wird und nur die Person mit dem geheimen Schlüssel entschlüsseln kann. Desweiteren ist es von Wichtigkeit, daß jeder Kunde eine transparente Kopie bekommt, die mit einem anderen Schlüssel chiffriert worden ist. Ansonsten könnten sich die Kunden die Schlüssel austauschen und erhielten dann legale Kopien.

Um die Sicherheit des Verfahrens zu beurteilen, sind ebenfalls Überlegungen hinsichtlich der Möglichkeit zu untersuchen, lediglich aus der Existenz des Lower Layer einen Enhancement Layer zu konstruieren, der den Datenstrom auf die volle Bildqualität bringt. Einem Angreifer sind dazu die Daten des Lower Layers verfügbar und die Quantisierungsstufenhöhen der En- und Decoder. Schaut man sich den Beispielvorgang aus dem Kapitel Mechanismen für transparente Verschlüsselungsmöglichkeiten an, so ist folgendes zu erkennen:

- Die **Quantisierungsstufenhöhe** für den Lower Layer beträgt **48** und für den Enhancement Layer **1**.
- Es steht der **Quantisierungskoeffizient** von **1** für den Lower Layer zur Verfügung.
- Der Angreifer führt dann eine inverse Quantisierung auf dem Quantisierungskoeffizienten mit der Quantisierungsstufenhöhe 48 aus und erhält als dequantisierten DCT-Koeffizienten 48. Da ihm nicht der Original-Koeffizient (von 36) vorliegt, kann er nun alle Möglichkeiten, die bei einer Quantisierungsstufenhöhe von 48, den Quantisierungskoeffizienten von 1 ergeben, daß ist gerade das Intervall von (23,72), untersuchen. Er bildet dazu alle möglichen Differenzen von 48 und den Intervallwerten. Die Differenzen sind mögliche Werte für den Enhancement Layer. Um herauszufinden, welche Differenz die richtige ist, muß er für jeden

Differenzwert eine vollständige Decodierung vornehmen und anschließend über das subjektive Empfinden entscheiden, welcher Differenzwert der richtige ist.

Diesen Vorgang muß der Angreifer für jedes einzelne Bild durchführen, was in der Praxis zu einem sehr hohen Aufwand führt, es sei denn, es existiert ein automatisiertes Beurteilungsverfahren, das erkennt, welcher Differenzwert für den Enhancement Layer die beste Qualität erzeugt. Der Aufwand ist dann immer noch abhängig von der Anzahl der Bilder im Videodatenstrom. Da die Festlegung der Bildqualität subjektiv und empirisch ist, existiert bisher kein automatisches Verfahren und es wird schwierig bleiben ein solches zu entwickeln.

Zu erwähnen ist noch, daß durch den SNR-Skalierungsmechanismus, der sehr komplex ist, mögliche Effizienzeinbußen beim Codier- als auch Decodiervorgang zu erwarten sind.

5 Fazit

Die Sicherheitsproblematik ist zu einem entscheidenden Thema für die Verbereitung neuer Informationstechnologien geworden. Im Rahmen dieser Arbeit wurde eine Analyse der Anwenderinteressen im Hinblick auf IT-Sicherheit an verteilte digitale Videoanwendungen durchgeführt. Aufbauend auf diesen Kenntnissen wurde ein neuer technischer Ansatz zur transparenten Darstellungsmöglichkeit von MPEG-2 entwickelt und bewertet. Dieser Ansatz soll Akzeptanzprobleme beseitigen, die sich bei der Verbreitung von digitalem Videomaterial wie etwa bei *Try&Buy* Transaktionen ergeben. Nur eine breite Akzeptanz der Videoproduzenten gegenüber der neuen Technik schafft die Voraussetzung für die zunehmende Nutzung der digitalen Marktmöglichkeiten. Die Produzenten können das Videomaterial in verminderter Qualität anbieten und erst nach Erfüllung bestimmter Konditionen seitens der Kunden das Material sozusagen freischalten, so daß eine größere Kontrolle bei der Verbereitung geschaffen wird. Der vorgestellte Ansatz bietet auch eine verbesserte Freischalt-Logistik. So kann die vollständige Qualität entweder durch Zusendung eines ergänzenden Videofilms, der nur mit dem bereits zugesendeten Material zusammenpaßt, erfolgen oder der Produzent kann dem Kunden den entsprechenden Schlüssels zusenden, mit dem er das Datenmaterial freischaltet. Probleme, die sich in Zusammenhang mit der Schlüsselverteilung und -zuordung ergeben werden an dieser Stelle nicht betrachtet, bilden aber eine wesentliche Gundlage für das vorgestellte Verfahren.

Literatur

[Barker1991] Barker, W.G.: Introduction to the analysis of the Data Encryption Standard (DES), Aegean Park Press, Laguna Hills, California, 1991

[Biham1993] Biham, Eli and Shamir, Adi: Differential Cryptoanalysis of the Data Encryption Standard, 1993, Springer-Verlag New York

[Ditt1997] Dittmann, J., Steinmetz, A.: Konzeption von Sicherheitsmechanismen für das Projekt DiVidEd, GMD-Studie Nr. 312, März 1997

[ISO/IEC13818-2] Internationaler Standard ISO/IEC 13818, Part2: Video, 1994

[ISO/IECSystem] ISO/IEC JTC1/SC29/WG11N0701, MPEG94, March 1994

[ISO/IECVideo] ISO/IEC JTC1/SC29/WG11 N0702rev Incorporating N702 Delta of 24 March 25 March 1994

[Lai1990] Lai,X. und Massey, J.: Markov Ciphers and Differential Cryptoanalysis, Veröffentlichung, Eurocrypt'91, 1991, Springer Verlag

[Matsui1994] Matsui, Mitsuru: The First Experimental Cryptoanalysis of the Data Encryption Standard, Advances in Cryptology - CRYPTO'94, 1994, Springer-Verlag

[Reimers1995] Ulrich Reimers: Digitale Fernsehtechnik: Datenkompression und Übertragung für DVB, Springer Verlag 1995

[Rivest1978] Rivest, Shamir, Adleman: Method for Obtaining Digital Signatures and Public Key Cryptosystems, Comm. ACM, Vol.21, Nr.2, 1987, 120-126

[Schneier1994] Schneier, Bruce: Applied Cryptography: Protocols, algorithms and source code in C, John Willey & Sons, 1994, New York

[Stallings1995] Stallings, W.: Sicherheit im Datennetz von William Stallings, 1995, Prentice Hall Verlag GmbH

[Steinmetz1994]: Steinmetz, Arnd und Hemmje, Matthias: Konzeption eines digitalen Videoeditiersystems auf Basis des internationalen Standards ISO/IEC 111782 (MPEG-1), 1994, GMD-Studie Nr. 245

[Steinmetz1996] Steinmetz, Arnd: DiVidEd A Distributed Video Production System, work in Progress, Proceedings of Visual96 Information Systems, February 1996

[TIE1995] Frauenhofer-Institut für Graphische Datenverarbeitung: Informationsblatt TIE, 1995 [http://www.igd.fhg.de/www/igd-a8/projects/tie/tie.htm]

Systematisierung und Modellierung von Mixen[1]

Elke Franz, Anja Jerichow, Andreas Pfitzmann

Technische Universität Dresden, D-01062 Dresden, {jerichow, efl, pfitza}@inf.tu-dresden.de

Zusammenfassung: Ausgehend von der ursprünglichen Definition der Mixe von D. Chaum und fortführenden Arbeiten von A. Pfitzmann et al. werden Erweiterungen dieses Konzepts betrachtet. Daraus wird ein mögliches allgemeines Modell abgeleitet, welches verschiedene Anwendungen des Mix-Konzepts berücksichtigt.

In einem ablauforientierten Modell werden die möglichen Konfigurationen eines Mixes vorgestellt. Anhand der allgemeinen Anforderungen sowie der durch die Betrachtung von Angriffen hinzukommenden Forderungen wird dieses Modell modifiziert. Das abgeleitete Modell zeigt die unter den getroffenen Annahmen sicheren Konfigurationen.

1 Schutz der Kommunikationsbeziehung durch Mixe

1.1 Beschreibung des Mix-Konzepts

Das Konzept der Mixe wurde erstmals in [Chau_81] beschrieben. Mixe dienen dem Schutz der Kommunikationsbeziehung. Die Verkettbarkeit zwischen Sendern und Empfängern von Nachrichten wird verhindert (Abbildung 1-1).

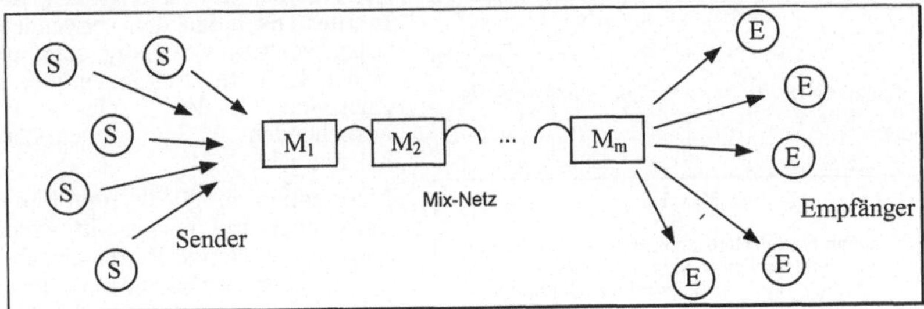

Abbildung 1-1: Schutz der Kommunikationsbeziehung durch Mixe

Ein Mix M_j sammelt Nachrichten, kodiert sie um, verändert ihre Reihenfolge durch Umsortieren (z.B. durch eine Sortierung nach alphabetischer Ordnung) und gibt die Nachrichten im Schub aus. Der nächste Mix, der diese Nachrichten erhält, verfährt genauso. Auf diese Weise kann der Weg der Nachrichten nicht durch das Mix-Netz verfolgt werden.

Durch Umkodierung wird erreicht, daß kein Außenstehender ein- und ausgehende Nachrichten anhand ihres äußeren Erscheinungsbildes einander zuordnen kann. Die Umkodierung innerhalb des Mixes erfolgt i.allg. mittels asymmetrischer Kryptographie. Das heißt: Ein Mix M_j hat einen öffentlich bekannten Schlüssel c_j und einen dazugehörigen privaten Schlüssel d_j. Die mit dem öffentlichen Schlüssel kodierten Nachrichten $c_j(N_i)$ gleicher Länge werden von M_j empfangen. Mittels seines privaten Schlüssels kann M_j dekodieren: $N_i := d_j(c_j(N_i))$. Die im Mix vorliegende Nachricht N_i wird zusammen mit den anderen in dieser Zeit angekommenen und bearbeiteten Nachrichten ausgegeben.

Bei deterministischer Umkodierung ist folgender Angriff möglich: Ein Mix kann überbrückt werden, indem ein Angreifer eine vom Mix ausgegebene Nachricht erneut mit dem öffentlichen Schlüssel des Mixes verschlüsselt. Das Ergebnis ist die zugehörige Eingangsnachricht. Durch Vergleich mit den zum Ausgabeschub gehörenden Ein-

---

[1] Teile der Arbeit wurden gefördert von der Deutschen Forschungsgemeinschaft (DFG).

gangsnachrichten erfolgt die Zuordnung der entsprechenden Ausgangsnachricht. Dieser Angriff kann durch **indeterministische Umkodierung** verhindert werden, indem der eigentlichen Nachricht ein zufälliger Teil z beigefügt wird: $c_j(N_i,z)$. Dieser kann vom Mix nach der Umkodierung einfach entfernt werden.

Trotz dieser Erweiterung erzeugt die Umkodierung aus gleichen Eingabenachrichten immer die gleichen Ausgabenachrichten. Um replay-Angriffe zu verhindern, darf jede Nachricht nur genau einmal vom Mix bearbeitet werden. Innerhalb eines Schubes wären ansonsten Verkettungen anhand von Nachrichtenhäufigkeiten möglich. Werden Nachrichten verteilt über mehrere Schübe wiederholt, so kann ein Mix durch das Bilden von Durchschnittsmengen bzw. Differenzen zwischen den jeweiligen Ein- und Ausgabenachrichten der entsprechenden Schübe überbrückt werden. **Wiederholungen von Nachrichten** sowohl innerhalb eines Schubes als auch verteilt über mehrere Schübe **sind zu ignorieren**.

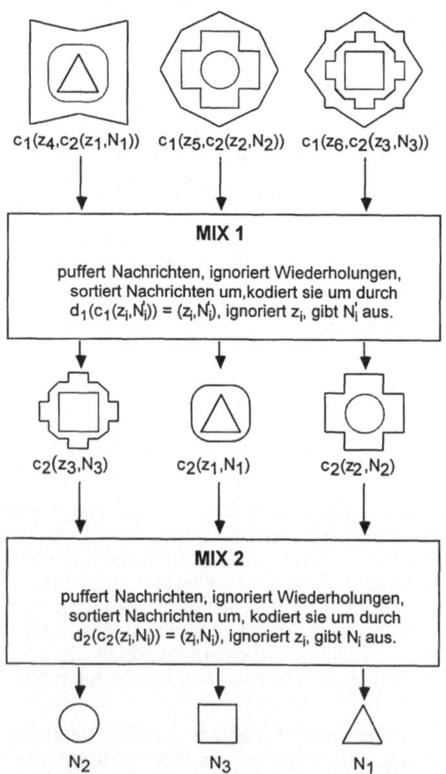

Abbildung 1-2: Mix-Prinzip für Senderanonymität nach [PfPW_88] (bei Kaskade entfallen Adressen)

Ein einzelner Mix weiß, von wem er Nachrichten erhält und an wen er sie weitergibt. Es sind also mindestens zwei Mixe nötig, damit einem einzelnen davon nicht Sender und Empfänger zugleich bekannt sind. Um Unverkettbarkeit zu erreichen, muß **mindestens einer** der verwendeten Mixe **vertrauenswürdig** sein, d.h. er darf keine Informationen über die Zuordnung der ein- und ausgehenden Nachrichten bekanntgeben.

Mixe sollten möglichst **unabhängig entworfen und hergestellt** werden sowie unabhängige Betreiber haben. Somit haben weder einzelne Personen noch Organisationen die Möglichkeit, den durch die Mixe erreichten Schutz der Kommunikationsbeziehung allein aufzuheben.

Diese Anforderungen wurden bereits in [Chau_81] gestellt. Mit weiteren Aspekten beschäftigen sich [Pfit_89] und [PfPW1_89].

Für die Wahl der Reihenfolge der durchlaufenen Mixe gibt es prinzipiell zwei Varianten: Entweder es wird von allen Nachrichten eine feste Reihenfolge durchlaufen, die Kaskade, oder die Reihenfolge der Mixe ist beliebig zu wählen.

In den Mixen kann durch verschiedene Umkodierungsschemata Sender- bzw. Empfängeranonymität sowie gegenseitige Anonymität erreicht werden (siehe 1.2 und 1.3). Bei diesen einfachen Schemata ändert sich jedoch die Länge der Nachrichten. Wird keine Kaskade verwendet, lassen sich daraus für einen Angreifer Informationen über den Weg der Nachricht gewinnen. Hier ist die Nutzung längentreuer Umkodierungsschemata nötig (siehe 1.5).

1.2 Senderanonymität

Der Sender verschlüsselt seine Nachricht N sowie Informationen über die Adresse A_{i+1} des jeweils nächsten Mixes M_{i+1} nacheinander mit den öffentlich bekannten Chiffrierschlüsseln c_j der Mixe, die auf dem Weg zum Empfänger (M_{m+1}) durchlaufen werden (Abbildung 1-2):

$$N_{m+1} \quad = \quad c_{m+1}(N)$$
$$N_i \quad = \quad c_i(z_i, A_{i+1}, N_{i+1}) \qquad \text{für } i = 1,...,m$$

Jeder Mix M_j wendet seinen geheimen Dechiffrierschlüssel d_j auf die Nachricht an und erhält so die nächste Darstellung der Nachricht und ggf. die Adresse des nächsten Mixes.

Hat der Sender mit dem ersten Mix einen geheimen Schlüssel ausgetauscht, so kann die entsprechende Verschlüsselung auch symmetrisch erfolgen [Pfit_89].

Der Empfänger kann bei diesem *direkten* Umkodierungsschema nicht nachvollziehen, wer ihm die Nachricht zugeschickt hat. Der Sender jedoch kennt das Aussehen der Nachricht in allen Punkten des Netzes einschließlich auf dem Weg vom letzten Mix zum Empfänger. Er kann infolgedessen den Empfänger beobachten.

1.3 Empfängeranonymität

Will der Empfänger B einer Nachricht anonym bleiben, so müssen zwei Verfahren kombiniert werden (Abbildung 1-3). Erstens muß B eine Adresse generieren, unter der er erreichbar ist, die sog. *anonyme Rückadresse* (RA).

$$R_{m+1} \quad = \quad e$$
$$R_j \quad = \quad c_j(k_j, A_{j+1}, R_{j+1}) \qquad \text{für } j = 1,...,m$$

Dabei ist e ein Kennzeichen für B. Die k_j sind symmetrische Schlüssel, die bei Anwendung der RA($:=R_1$) von den Mixen für die Verschlüsselung der Nachricht genutzt werden. Der Index $m+1$ bezieht sich dabei auf den Empfänger B. A_j steht für die

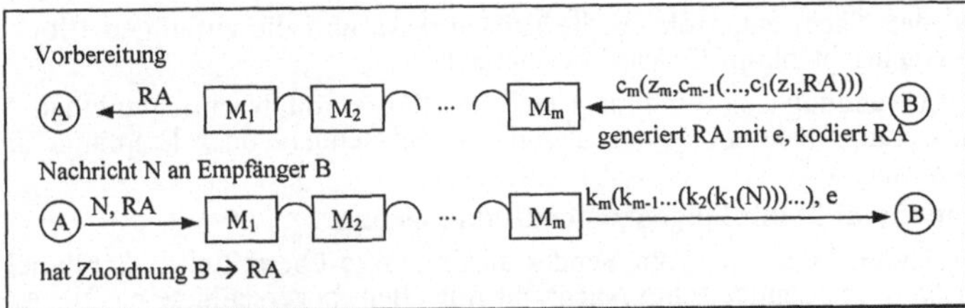

Abbildung 1-3: Empfängeranonymitätsschema

Adresse des Mixes j, c_j für seinen öffentlichen Schlüssel.

Zweitens sendet der Empfänger diese Rückadresse unter Wahrung seiner Anonymität an seinen Kommunikationspartner A gemäß dem Senderanonymitätsschema für Mixe. Die bei A hinterlegte RA wird beim Senden einer Nachricht an B an den ersten Mix geschickt, welcher die Nachricht verarbeitet und weiterleitet.

Der Empfänger kann mit diesem *indirekten* (oder *hybriden*) Umkodierungsschema seine Anonymität wahren, doch die Anonymität des Senders der Rückantwort wird nicht gewährt.

1.4 Gegenseitige Anonymität

Zur Wahrung gegenseitiger Anonymität können Sender- und Empfängeranonymitätsschema kombiniert werden. Auf dem Weg zwischen den Kommunikationspartnern gibt es einen Mix, welcher als Umlenkpunkt fungiert: Der Empfänger bildet eine anonyme Rückadresse, welche bei diesem Mix hinterlegt wird. Der Sender schickt seine Nachricht unter Benutzung des Senderanonymitätsschemas an diesen Mix.

Der Empfänger kennt den Nachrichtenweg vor dem Umlenk-Mix nicht, und der Sender verliert die Kontrolle über die Nachricht ab diesem Mix.

1.5 Längentreue Kodierung

Da jeder Mix aus der weiterzuleitenden Nachricht den Teil entfernt, in dem seine Adresse enthalten ist, zumindest aber die Zufallszahl, die er nicht ausgeben darf, erfolgt eine Verkürzung der Nachricht auf ihrem Weg durch die Mixe. Sollen keine Rückschlüsse aufgrund der Verkürzung der Nachrichten erfolgen, muß die Umkodierung *längentreu* erfolgen. Hierzu werden von jedem Mix zufällige Bitketten in die Nachricht eingefügt, so daß die Länge erhalten bleibt. Ein längentreues Umkodierungsschema ist immer indirekt.

Jede Nachricht hat die feste Länge von b Blöcken, in welchen eine anonyme Rückadresse, die zufälligen Bitfolgen und die eigentliche Nachricht enthalten sind. Die anonyme Rückadresse wird für dieses Verfahren folgendermaßen gebildet (die eckigen Klammern bedeuten Blockgrenzen):

$$R_{m+1} \quad = \quad [e]$$

$$R_j \quad = \quad [c_j(k_j,A_{j+1})], k_j\,(R_{j+1}) \quad \text{für } j = 1,...,m$$

Daraus ergibt sich, daß im ersten Block der Nachricht der vom empfangenden Mix M_j zu verwendende Schlüssel k_j sowie die Adresse A_{j+1} des nachfolgenden Mixes M_{j+1} enthalten ist. Dieser erste Block wird vom Mix M_j entfernt und der erhaltene Schlüssel k_j auf die restliche Nachricht angewandt. Vor den Blöcken, welche die eigentliche Nachricht enthalten, wird vom Mix M_j ein Block zufälligen Inhalts eingefügt und dadurch die Länge der Nachricht erhalten. Das Ergebnis dieser Operationen wird an den nachfolgenden Mix M_{j+1} geschickt, welcher wieder genauso verfährt.

Die Gesamtanzahl zu durchlaufender Mixe und die Nachrichtenlänge ist allen Mixen bekannt. Daraus können sie die entsprechende Stelle zum Einfügen ermitteln.

Der Teil der Nachricht, welcher die Rückadresse und die zufälligen Blöcke enthält, wird als Nachrichtenkopf (header) bezeichnet.

Ob die Anwendung des Schlüssels auf die Blöcke mit Nachrichteninhalt Ver- oder Entschlüsselung bedeutet, hängt davon ab, ob Sender- oder Empfängeranonymität erreicht werden soll.

Senderanonymität mit längentreuer Umkodierung:

Der Empfänger kennt die vom Sender an die Mixe übermittelten Schlüssel k_j nicht, folglich muß der Sender seine Nachricht mit allen k_j verschlüsseln. Die Anwendung des jeweiligen Schlüssels durch die einzelnen Mixe entspricht einer Entschlüsselung. Für die erste vom Sender anzuwendende Verschlüsselung muß dieser also den öffentlichen Schlüssel des Empfängers kennen, damit dieser durch Entschlüsselung mit seinem geheimen Schlüssel die Nachricht lesen kann.

Die Nachricht, die vom Sender gebildet werden muß, sieht folgendermaßen aus:

$$N_1 \quad = \quad H_1\,I \qquad \textit{mit} \quad H_1 \quad = \quad R_1$$

$$I_1 \quad = \quad k_1(k_2(...k_m(c_{m+1}(I))))$$

Dabei steht I für den Nachrichteninhalt. Statt der asymmetrischen Verschlüsselung im letzten Schritt könnte auch symmetrisch verschlüsselt werden, falls zwischen Sender und Empfänger Schlüssel ausgetauscht wurden. Das Kennzeichen e kann dann zur Zuordnung des richtigen Schlüssels verwendet werden.

Empfängeranonymität mit längentreuer Umkodierung:

Hierbei kennt der Sender die an die Mixe verteilten Schlüssel nicht. In seinem Rückadreßteil R ist der von ihm anzuwendende Schlüssel k_0 enthalten, mit welchem er seine Nachricht verschlüsselt und an den ersten Mix sendet.

Der Sender bildet also aus (k_0, A_1, R_1) die Nachricht

$$N_1 \;=\; H_1\, I_1 \qquad mit \;\; H_1 \;=\; R_1$$
$$I_1 \;=\; k_0(I)$$

Die Anwendung der Schlüssel durch die Mixe entspricht hierbei also einer Verschlüsselung. Der Empfänger kann durch das eindeutige Kennzeichen e die entsprechenden Schlüssel zuordnen und die Nachricht entschlüsseln.

2 Funktionen eines einzelnen Mixes

2.1 Sicherheit und Effizienz: Ableitung notwendiger Teilschritte

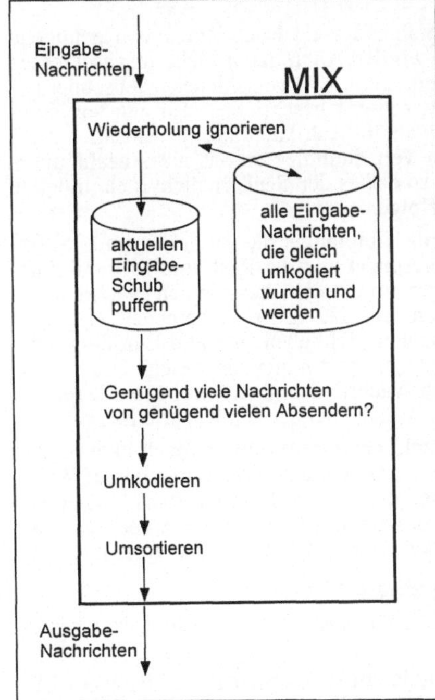

Abbildung 2-1: Grundfunktionen eines Mixes aus [Pfit_93]

Nach der Definition des Mixes in Kapitel 1.1 ergeben sich zunächst die folgenden Schritte:

- Sammeln von genügend vielen Nachrichten,
- Ignorieren von Wiederholungen,
- Umkodieren der Nachrichten,
- Umsortieren und Ausgeben der Nachrichten im Schub.

Abbildung 2-1 zeigt das Modell eines Mixes, welcher weitere notwendige Funktionen enthält. Die folgenden Funktionen könnte ein Mix zusätzlich zu den oben genannten Schritten noch ausführen:

- Erzeugen von dummies,
- Test auf genügend viele verschiedene Absender,
- Behandlung von Latenzzeiten.

Der Mix muß erst eine bestimmte Anzahl Nachrichten sammeln, ehe er sie weiterverarbeiten kann. Bei geringem Nachrichtenaufkommen können dadurch für die Nutzer große Wartezeiten entstehen. Die Festlegung von maximalen Wartezeiten erscheint daher sinnvoll. Ist die geforderte Anzahl von Nachrichten innerhalb der festgelegten Zeit noch nicht eingetroffen (Nachrichten könnten ja auch auf dem Weg durch das Kommunikationsnetz verloren gehen), so füllt der Mix seinen Nachrichtenpuffer mit bedeutungslosen Nachrichten, den **dummies**, auf. Die dummies sollen von den nachfolgenden Mixen ebenso wie alle anderen Nachrichten behandelt werden, sie sollen sich nicht von ihnen unterscheiden. Das Aussortieren der dummies könnte der letzte Mix vor dem Empfänger einer Nachricht übernehmen, wenn für ihn die dummies entsprechend gekennzeichnet sind.

Die Notwendigkeit einer weiteren Funktion ergibt sich aus dem **(n-1)-Angriff** auf einen Mix. Dabei kennt der Angreifer n-1 von n Nachrichten und kann somit die n-te, welche ihm unbekannt ist, leicht verfolgen. An die Nachrichten, die pro Schub umkodiert werden, wird darum die Forderung erhoben, daß sie von **genügend vielen verschiedenen Absendern** stammen müssen. Um die Forderung überprüfen zu können, müssen die **Absender verifizierbar** sein, was durch die Verwendung digitaler Signaturen erreichbar ist. Um digitale Signaturen leisten zu können, welche die Zuordnung zu einer Identität gewährleisten, sind ein asymmetrisches Kryptosystem und eine Zertifizierungsinstanz notwendig. Mit dem geheimen Schlüssel werden die Nachrichten verschlüsselt. Das Ergebnis dieser Umkodierung ist die Signatur, welche der Nachricht mitgegeben wird. Mit Hilfe des öffentlich bekannten Schlüssels kann jeder testen, ob die Nachricht wirklich von dem angegebenen Sender stammt. Um die Richtigkeit der Zuordnung von Schlüssel zu Schlüsselinhaber zu gewährleisten, ist die Zertifizierungsinstanz notwendig.

Die Überprüfung der Absender erfolgt vor dem ersten Mix (die anderen Mixe können an dieser Stelle mitprüfen). Ohne digitale Signaturen können die Absender jedoch nicht mit Sicherheit identifiziert werden - das Fälschen von e-mail-Absendern stellt kein ernsthaftes Problem dar. Nutzer können sich außerdem verschiedene Nutzernamen geben lassen.

Selbst wenn abgeprüft werden kann, ob genügend viele Nachrichten von genügend vielen verschiedenen Absendern gesammelt wurden, kann damit nicht ausgeschlossen werden, daß Angreifer zusammenarbeiten und damit trotz verschiedener Absender n-1 von n Nachrichten bekannt sind. Während der zuerst beschriebene Einsatz von dummies hauptsächlich der Garantie der Dienstqualität dient, kann hier ihr Einsatz aus Sicherheitsgründen erfolgen. Das Hinzufügen von dummies erhöht die Anzahl unbekannter Nachrichten pro Schub automatisch, so daß es den Sendern nicht mehr möglich ist, eine einzige unbekannte Nachricht zu verfolgen.

Dummies bieten jedoch keinen Schutz, falls die Empfänger die Angreifer sind. Im Extremfall könnte nur eine richtige Nachricht gesendet und der Rest vom Mix mit dummies aufgefüllt werden. Ist der Angreifer der Empfänger dieser einen Nachricht, so kann er den Sender bestimmen, da das Senden beobachtbar ist. Entsprechendes gilt für das Enthaltensein mehrerer richtiger Nachrichten: Schließen sich alle Empfänger bedeutungsvoller Nachrichten zusammen, können sie bestimmen, an welche Gruppe von Empfängern die in Frage kommenden Sender Nachrichten geschickt haben. Dummies bieten also vor allem Schutz gegen Angreifer, welche Sender von Nachrichten sind.

Um die Unsicherheit über den Weg einer Nachricht zu erhöhen, ist noch eine weitere Variante vorstellbar: von den Nutzern gewählte minimale (und/oder maximale) Wartezeiten der Nachrichten im Mix. Diese Verzögerungszeiten (**Latenzzeiten**) können von den Nutzern für jeden Mix, welchen die Nachricht passiert, gewählt werden, unabhängig von der Betriebsart des Puffers (Batch, Pool oder Mischform, siehe unten).

2.2   Beschreibung der Teilschritte in einem Mix

In Abbildung 2-2 wird ein allgemeines Modell für einen Mix vorgestellt, welches die soeben diskutierten Funktionen enthält.

Wie bereits erwähnt, kann die Erzeugung von dummies aus Sicherheits- bzw. aus Effizienzgründen notwendig sein. Da das Umkodieren bei Verwendung asymmetrischer Kryptographie eine aufwendige Funktion ist, wird die dummy-Erzeugung erst im letzten Schritt, dem Umsortieren, durchgeführt.

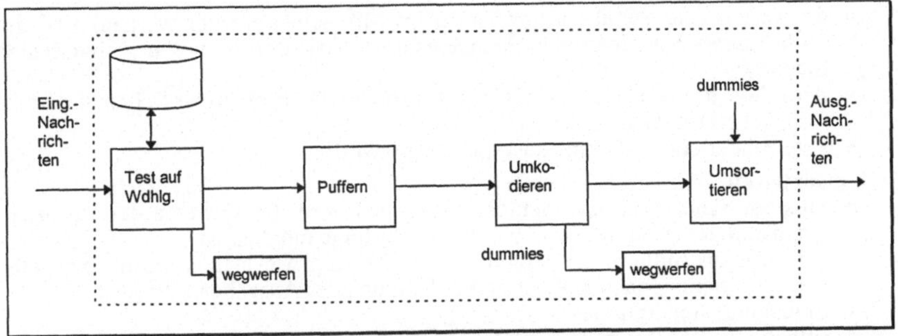

Abbildung 2-2: Allgemeines Modell eines Mixes

Nicht alle im Modell dargestellten Funktionen müssen in einem Mix auf einmal enthalten sein. Es gibt verschiedene Arten, sie zu kombinieren, so daß Unverkettbarkeit gewährleistet wird. Im folgenden sind die Aufgaben der Teilschritte zusammengefaßt.

1) Wiederholungen ignorieren
- um replay-Angriffe zu verhindern; mögliche Varianten:
  – Speicherung bereits gesendeter Nachrichten in einer Datenbank;
  – Vergabe von Zeitstempeln für die Nachrichten
  – Verhinderung von Wiederholungsangriffen durch Verwendung einer synchronen Stromchiffre (hybride Verschlüsselung)
- Wiederholungen werden vom Mix weggeworfen

2) Puffern
- Sammeln der Nachrichten, bevor diese weiterverarbeitet werden; zwei Varianten der Organisation des Puffers:
  – *Batchbetrieb*: m Speicherplätze für die Nachrichten vorgesehen; wenn diese Anzahl erreicht ist, wird der Puffer geleert,
  – *Poolbetrieb*: m Speicherplätze vorgesehen, trifft $(m+1)$-te Nachricht ein, wird aus Pool zufällig eine Nachricht ausgewählt und weiterverarbeitet
  – Mischformen: Pool von m Nachrichten, aus denen jedoch bei Eintreffen der $(m+1)$-ten Nachricht nicht nur eine, sondern eine Anzahl n $(n<m)$ Nachrichten zufällig ausgewählt und weitergeschickt wird
- Prüfen der Absender (Forderung: es müssen genügend viele Nachrichten von genügend vielen verschiedenen Absendern vorhanden sein, ehe weitergearbeitet wird; zur Verhinderung von $(n-1)$-Angriffen *weniger* Angreifer, Absender müssen verifizierbar sein)
- Behandlung von Latenzzeiten: Möglichkeit der Vergabe von Zeitschranken für die Weiterverarbeitung bzw. Verzögerung der Nachrichten:
  – Vergabe der Latenzzeiten durch die Sender der Nachrichten
  – Verzögerung um zufällige Dauer durch Mix ebenfalls vorstellbar

3) Umkodieren
- Speicherung von Zustandsinformationen (z.B. Datensätze der Teilnehmer bei verbindungsorientierter Übertragung, zur Verwaltung von Paßwörtern/Accounts (Remailer), Informationen zur Schlüsselerzeugung...)
- Verändern des Aussehens der Nachricht, außerdem Erhalten der Folgeadresse, an die gesendet wird (falls Mix-Netz, bei fester Kaskade keine Adresse nötig)
- je nach Verfahren und Position des Mixes symmetrische (synchrone Stromchiffre) oder asymmetrische Verschlüsselung

- der verwendete Schlüssel kann auch im Mix selbst erzeugt werden, z.B. mit Hilfe eines Pseudozufallszahlengenerators aus einem vorher hinterlegten Startwert
- Auffüllen der Nachrichten mit Zufallszahlen (für längentreue Schemata (siehe 1.5), bei NDM (siehe 2.3.5))
- vom Mix erkannte dummies werden weggeworfen

4) Umsortieren

- falls ein Schub von Nachrichten ausgegeben wird: Umsortieren der Nachrichten, damit keine Zuordnung über die Reihenfolge möglich ist
- Umsortieren sollte keine Möglichkeiten für Angriffe bieten, darum beispielsweise Sortierung nach alphabetischer Ordnung der Ausgabenachrichten
- Erzeugung von dummies
- enthält auch die Ausgabe der Nachrichten

Die Zuordnung von Aufgaben zu den einzelnen Teilschritten könnte an einigen Stellen auch anders erfolgen. So könnte die Speicherung von Zustandsinformationen auch Teil des Puffferns sein, da dort ohnehin Speicherung vorgesehen ist. Die Zuordnung zur Umkodierung betont, daß oftmals die Informationen entweder erst nach der Umkodierung vom Mix gelesen und gespeichert werden können oder daß gespeicherte Informationen zur Umkodierung benötigt werden, so z.B. zur Schlüsselgewinnung.

Das Senden bleibt beobachtbar. Daher kann von den Teilnehmern gefordert werden, daß sie zu jeder Zeit (immer zu festgelegten Takten) Nachrichten senden müssen, egal, ob es sich dabei um richtige oder um bedeutungslose Nachrichten handelt. Die zu einem Zeitpunkt sendenden Teilnehmer bilden jeweils eine Anonymitätsgruppe. Senden die Teilnehmer bedeutungslose Nachrichten, so kann die Erzeugung von dummies im Mix evtl. weggelassen werden.

Aus diesem allgemeinen Modell lassen sich viele Möglichkeiten konstruieren. Im folgenden werden zunächst existierende Varianten der Mixe vorgestellt. Anschließend werden die möglichen Kombinationen, welche man aus dem allgemeinen Modell ableiten kann, im Hinblick auf Sicherheitsanforderungen betrachtet.

2.3 Anwendungen des Mix-Konzepts - Beispiele für andere Modelle

2.3.1 Telefon-Mixe (PfPW1_89)

Diese Variante der Mixe gewährleistet den Schutz von der Kommunikationsbeziehung auf der Basis von ISDN. Die asymmetrische Verschlüsselung innerhalb der Mixe ist zeitaufwendig. DieVerwendungvon Mix-Kanälen verringert die Verzögerung durch die Mixe. Dabei wird die hybride Verschlüsselung zur Erhöhung der Geschwindigkeit der Übertragung genutzt, insbesondere der Verringerung der Verzögerungszeit.

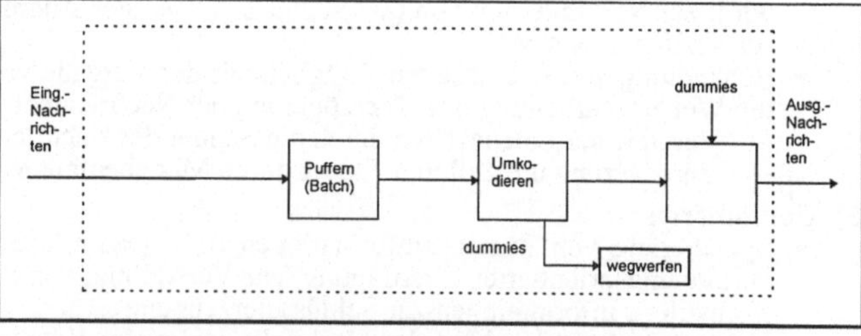

Abbildung 2-3: Funktionalität eines Telefon-Mixes (Senden der Nutzdaten)

Die Übertragung von Nachrichten wird aufgeteilt in die beiden Schritte:

1. Übertragung einer Kanalaufbaunachricht (asymmetrische Verschlüsselung) und
2. Übertragung des eigentlichen Nachrichteninhalts (symmetrische Verschlüsselung).

Bei der Übertragung des ersten Teils merkt sich jeder Mix die Zuordnung vom Eingabe- zum Ausgabeschub. Damit wird die Umsortierung der später gesendeten Nachrichten festgelegt. Die für die symmetrische Verschlüsselung der Nachrichten benötigten Schlüssel werden beim Kanalaufbau in den Mixen hinterlegt.

Baut sich der Sender einen Kanal auf, den Sendekanal, so kann er gegenüber dem Empfänger seine Anonymität wahren. Entsprechendes gilt für den Empfänger, welcher sich einen Empfangskanal aufbauen kann. Um gegenseitige Anonymität zu wahren, werden beide Kanäle nicht über die Teilnehmeridentitäten, sondern nur über ein sogenanntes Kanalkennzeichen verknüpft. Dieses müssen die Kommunikationspartner vor dem eigentlichen Kanalaufbau ausmachen. Dazu wird nach dem üblichen Schema eine Kanalwunschnachricht gesendet.

In Abbildung 2-3 ist das Modell eines solchen Mixes während der Übertragung der eigentlichen Nachrichten dargestellt. Diese werden mit einer synchronen Stromchiffre verschlüsselt. Durch die fortlaufende Schlüsselung sind Wiederholungsangriffe ausgeschlossen und der Test auf Wiederholungen entfällt.

Da Nachrichten eines Schubes gleich lang sein müssen, müßten gleichzeitig aufgebaute Kanäle auch gleichzeitig wieder abgebaut werden. Um zu verhindern, daß alle Kanäle auf das Ende des längsten gleichzeitig aufgebauten Kanals warten müssen, werden Zeitscheiben eingeführt: Der Auf- und Abbau von Zeitscheibenkanälen wird fortlaufend zu festen Zeittakten durchgeführt. Somit dürfen die Kanäle (die durch eine Abfolge von Zeitscheibenkanälen gebildet werden) beliebig lang sein, und alle zu den Zeittakten auf- bzw. abbauenden Teilnehmer bilden jeweils eine Anonymitätsgruppe.

2.3.2 Verbindungsorientierte Übertragung

[FFJM_97] beschreibt ein Verfahren zum Schutz der Aufenthaltsinformation in Mobilkommunikationsnetzen. Mixe werden hier mit erweiterter Funktionalität genutzt.

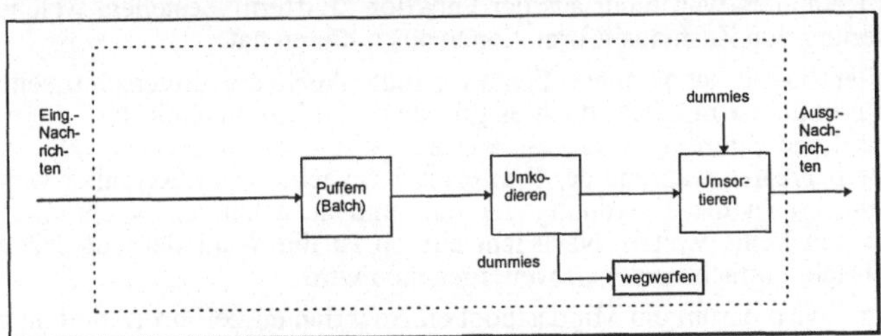

Abbildung 2-4: Mix für verbindungsorientierte Übertragung: Datensätze der Teilnehmer werden gespeichert

Um die Teilnehmer lokalisieren zu können, werden in existierenden Netzen Aufenthaltsinformationen in Registern gespeichert. Die Speicherung erfolgt hierarchisch aus Effizienzgründen.

Ankommende Verbindungswünsche werden durch die Hierarchieebenen zu den Teilnehmern weitergeleitet. Da jedoch alle Hierarchieebenen einem Betreiber gehören, hat dieser die globale Sicht auf die Teilnehmerdaten.

Damit trotz der Speicherung der Aufenthaltsort der Teilnehmer geschützt bleibt, werden zum Weiterleiten der Informationen Mixe vorgeschlagen, welche die Speicherung der Aufenthaltsinformationen in den einzelnen Hierarchieebenen übernehmen (Abbildung 2-4). Von den Mixen wird gefordert, daß sie nicht heimlich miteinander kooperieren. Da in zukünftigen Systemen die Netze verschiedener Betreiber integriert werden sollen, ist die Realisierung dieser Forderung vorstellbar. Damit Verbindungswünsche anonym weitergeleitet werden können, erfolgt die Speicherung der Teilnehmerdaten pseudonym.

Der Test auf Wiederholungen kann durch die Verwendung einer synchronen Stromchiffre entfallen.

2.3.3 Remailer

Eine Übersicht über bestehende Remailer- und Mixmaster-Implementierungen ist in [Scha_95] enthalten. Besitzt ein Remailer bzw. Mixmaster auch nicht die Funktionalität eines Mixes, so ist doch auch er zum Schutz der Kommunikationsbeziehung vorgesehen. Dabei existieren mehrere Varianten, welche verschiedene Sicherheit bieten.

Eine einfache Variante wurde in Finnland geschaffen (heute nicht mehr aktiv). Dieser Remailer führt einfach eine Ersetzung des Absenders mit seinem Absender durch. Der Empfänger ist dabei aus der e-mail des Senders erkennbar. Vor dem Remailer sind also beide Kommunikationspartner in der Nachricht erkennbar. Der Remailer kann leicht überbrückt werden, wenn das Netz auf Sender- und auf Empfängerseite überprüft werden kann, da keine Umkodierung stattfindet. Dieses System hält also nur sehr eingeschränkten Angriffen stand.

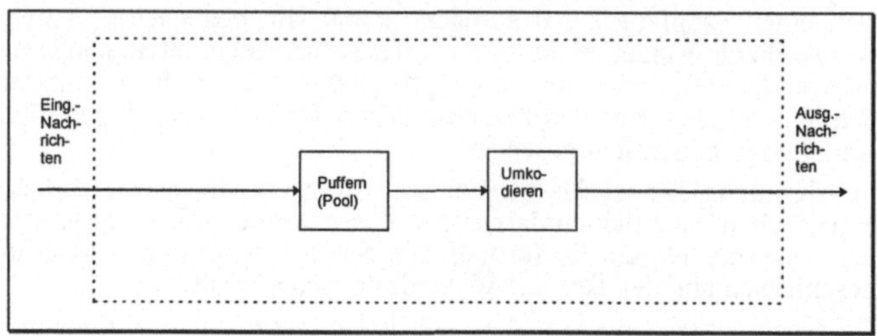

Abbildung 2-5: Remailer mit Behandlung von Latenzzeiten sowie Mindestpool

Um Rückantworten zu ermöglichen, wird jedem Nutzer ein Account erteilt. Mit Hilfe eines Paßwortes kann verhindert werden, daß Absender gefälscht werden. Allerdings wird das Paßwort im Klartext an den Remailer übertragen.

Das Modell dieses Remailers würde nur aus der Funktion „Puffern" bestehen, welche hier nur die Speicherung von Zustandsinformationen auszuführen hat.

Cypherpunk-Remailer bieten schon mehr Funktionalität. Auch die unverschlüsselte Weiterleitung vom e-mails ist möglich, doch es gibt auch die Möglichkeit der Umkodierung. Um die zeitliche Zuordnung von e-mails zu verhindern, können die Nutzer Latenzzeiten angeben. Treffen während der Wartezeit der Nachricht im Remailer weitere Nachrichten ein, so ist die Zuordnung der ein- und ausgehenden Nachrichten erschwert. Trifft jedoch keine weitere Nachricht ein, so ist die Wahl der Latenzzeit nutzlos, da die Nachricht einfach verzögert weitergegeben wird.

In weiteren Varianten wird darum ein Mindestpool an Nachrichten geführt (Abbildung 2-5). Nach wie vor besteht jedoch die Möglichkeit, replay-Angriffe durchzuführen oder die Nachrichten anhand ihrer Länge zuzuordnen.

2.3.4 Mixmaster

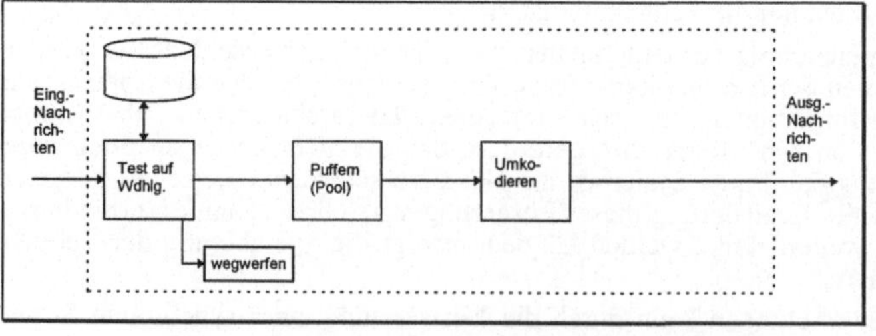

Abbildung 2-6: Funktionalität eines Mixmasters (Einzelbetrieb)

Die Mixmaster orientieren sich eng an der in [Chau_81] vorgestellten Variante. Die Nachrichten müssen ein festes Format haben, die Mixmaster führen Umkodierung durch und testen Wiederholungen anhand der Paket-ID. Arbeitet der Mixmaster im Einzelbetrieb, werden die Nachrichten in einem Pool gesammelt (Abbildung 2-6). Beim Stapelbetrieb werden sie zu festen Zeittakten ausgelesen, weiterverarbeitet und ggf. mit dummies aufgefüllt.

Ohne zusätzliche Sicherheitsmaßnahmen (wie Absenderprüfung) bietet der Pool mehr Sicherheit als ein Batch gleicher Größe.

Aus der Arbeitsweise des Pools ergibt sich, daß die Verweilzeit einer Nachricht im Mix ungewiß ist. Der Angreifer kann also im Gegensatz zum Batch nicht wissen, wann eine bestimmte Nachricht den Pool wieder verläßt. Um nun diese eine Nachricht dadurch zu verfolgen, daß sie die einzige unbekannte ist, müßte der Angreifer über längere Zeit den Pool nur mit seinen Nachrichten füllen und darüber hinaus in der Lage sein, nach dem Senden der interessierenden Nachricht jedes weitere Senden anderer Absender zu unterbinden, bis die Nachricht den Mix verläßt.

Für einen Angreifer, der dem Mix legal viele Nachrichten zuschicken, aber nicht unerlaubterweise die Kommunikation der anderen Teilnehmer stören kann, ist ein solcher Angriff sicher schwer durchzuführen. Kann der Angreifer jedoch das Senden anderer Teilnehmer verhindern, erscheint die gebotene Sicherheit nicht mehr ausreichend.

Ein anderes Konzept für Remailer wird in [GüTs_96] diskutiert. Die Nachrichten werden zufällig im Mix verzögert. Die Ausgabe erfolgt zu festen Zeittakten. Um einen $(n\text{-}1)$-Angriff durch Verzögerung von Nachrichten zu verhindern, werden hier Umwege zwischen den Mixen gewählt.

2.3.5 Non-Disclosure Method (NDM)

Bei der in [FaKK_96] vorgestellten Methode existieren eine Anzahl von unabhängigen SAs (security agents, Abbildung 2-7). Diese übernehmen die Rolle der Mixe. Die SAs benutzen für die Umkodierung ein asymmetrisches Kryptosystem (RSA). Aus den SAs wählt der Nutzer sich einen beliebigen aus. Ein SA sammelt keine Nachrichten im Batch oder Pool. Da auch Wiederholungen nicht abgewiesen werden, sind replay-Angriffe möglich.

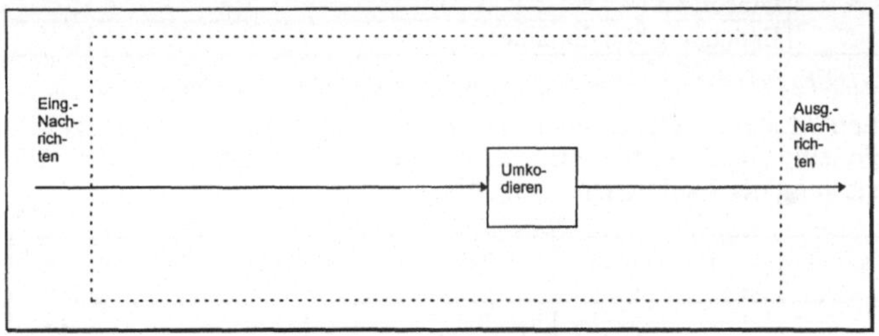

Abbildung 2-7: Security agent der NDM

Das Angreifermodell, welches ansonsten für Mixe angenommen wird, wurde für die Methode modifiziert. Ein Angreifer, welcher alle Kommunikation beobachten kann, wäre in der Lage, den Weg der Nachrichten zu verfolgen. Es wird jedoch davon ausgegangen, daß es schwierig ist, die Kommunikation zu beobachten. Dem Angreifer soll nicht bekannt sein, welche SAs gewählt werden. Er wird also nur mit einer gewissen Wahrscheinlichkeit die richtigen Stellen abhören, falls er nicht die gesamte Kommunikation im Netz beobachten kann. Die Möglichkeit, sämtliche Kommunikation zu beobachten, wird aber als zu aufwendig eingestuft. Gelingt es ihm jedoch, eine Nachricht an nur zwei Stellen abzuhören, können dazwischenliegende SAs durch replay-Angriffe überbrückt werden.

Um replay-Angriffe zu verhindern, werden noch zufällige Bitfolgen eingefügt. Der Sender fügt nur die erste Folge ein, die der erste SA entfernt. Nach dem Umkodieren erzeugt dieser eine neue Bitfolge und verschlüsselt sie zusammen mit der Nachricht mit dem öffentlich bekannten Schlüssel seines Nachfolgers. Dieser verfährt genauso.

Mit dieser Erweiterung wird eine Verbindungsverschlüsselung zwischen den einzelnen SAs erreicht. Eine Verbindungsverschlüsselung liefert jedoch immer nur Schutz gegen Angreifer auf der Verbindungsstrecke.

3 Kombinationen der Mix-Funktionen

3.1 Allgemeines ablauforientiertes Modell

Wie die Anwendungsbeispiele gezeigt haben, sind verschiedene Arbeitsweisen der einzelnen Teilschritte innerhalb eines Mixes möglich. Die Teilschritte selbst können auf verschiedene Arten miteinander kombiniert werden. Es ergibt sich theoretisch eine Vielzahl von Möglichkeiten. Um diese zu ordnen, soll ein Modell angegeben werden, welches die möglichen Kombinationen darstellt. Dazu werden zunächst die Varianten der Teilschritte in Tabelle 3-1 zusammengefaßt.

Teilschritt	Varianten	Abk.
Test auf Wiederho- lungen	Speicherung (Nachr., Hashwerte,...) in Datenbank	DB
	Verwendung von Zeitstempeln	Zst
	kein Test erforderlich, falls synchrone Stromchiffre	sSc
Puffern	Prüfung: genügend viele verschiedene Absender	Abs
	Behandlung von Latenzzeiten	Lz
	Batch	B
	Pool	P
	Mischformen	M
Umkodieren	Speicherung von Zustandsinformationen	Sp
	asymmetrisch	Ua
	symmetrisch (synchrone Stromchiffre)	Us
Umsortieren	Umsortieren	S
	Erzeugung von dummies	d

Tabelle 3-1: Übersicht über verschiedene Varianten zu den einzelnen Funktionen

In einem allgemeinen Modell, welches auch die vorgestellten Anwendungsbeispiele abdeckt, sind die einzelnen Teilschritte optional. Abbildung 3-1 zeigt eine solche *ablauforientierte* Darstellung der Funktionen eines Mixes.

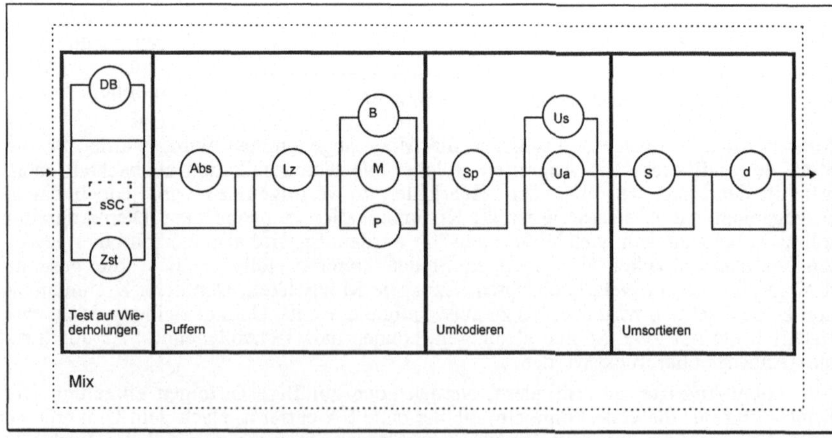

Abbildung 3-1: Ablauforientierte Darstellung der Teilschritte innerhalb eines Mixes (Abk. gemäß Tab. 3.1)

Nicht alle Systemparameter sind in einem solchen Modell darstellbar. So werden die Eingabenachrichten nicht näher charakterisiert. Die Teilnehmer können entweder nur bei Bedarf oder immer zu festen Zeittakten senden (dann mit dummy traffic).

Die Batch- bzw. Poolgröße könnte fest oder variabel sein. I.allg. wird von einer festen Größe ausgegangen.

Weiterhin kann der Mix entweder ereignis- oder zeitorientiert betrieben werden. Im ersten Fall wird der Puffer nur geleert, wenn genügend Nachrichten gesammelt wurden. Im zweiten Fall erfolgt das Auslesen zu festen Zeittakten. Die Verwendung von dummies ist dann unbedingt erforderlich. Doch auch beim ersten Fall sollte durch die Möglichkeit, dummies zu erzeugen, die Dienstqualität des Mixes gesichert werden.

Auch über die Ausgabe der Nachrichten werden keine Aussagen getroffen. Die einzelnen Nachrichten können entweder direkt zugestellt oder innerhalb einer Gruppe verteilt werden.

3.2 Anwendungsbeispiele unter dem vorgestellten allgemeinen Modell

Bei der Nachrichtenübermittlung durch die Telefon-Mixe (3-2 b)) werden die Nachrichten auch umsortiert. Die Umsortierung ist jedoch fest (entsprechend der Zuordnung von Eingabe- zu Ausgabekanälen bei der Signalisierung). Da die Umsortierung bei der Nachrichtenübermittlung keinen Aufwand mehr erfordert, wird sie in der Abbildung nicht als Funktion angegeben.

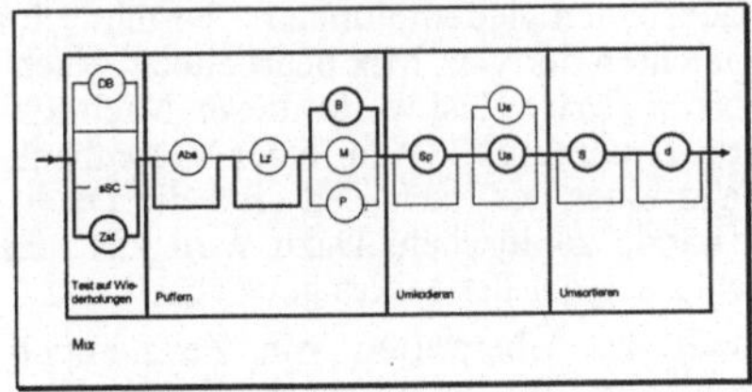

a) Telefon-Mix (Signalisierung)

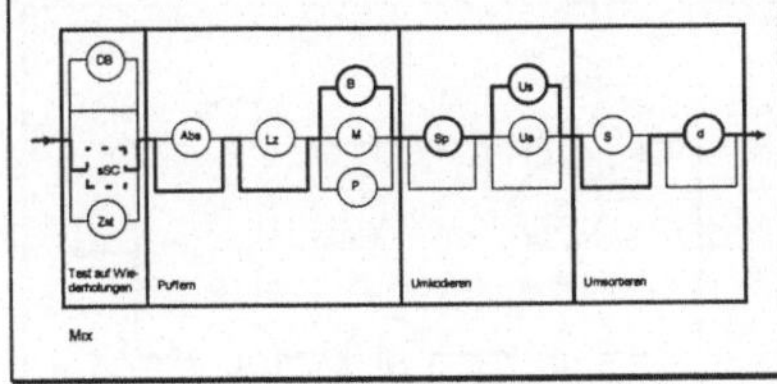

b) Telefon-Mix (Nachrichtenübermittlung)

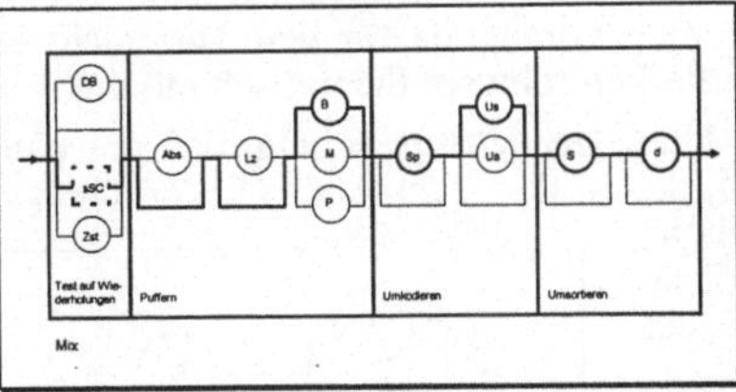

c) Verbindungsorientierte Übertragung (nach Aufbau)

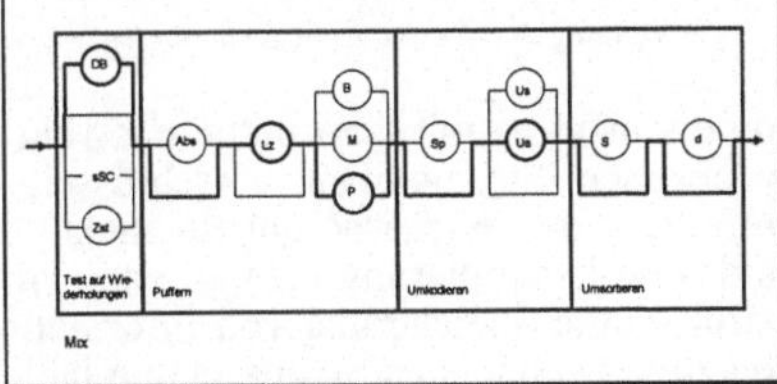

d) Cypherpunk-Remailer

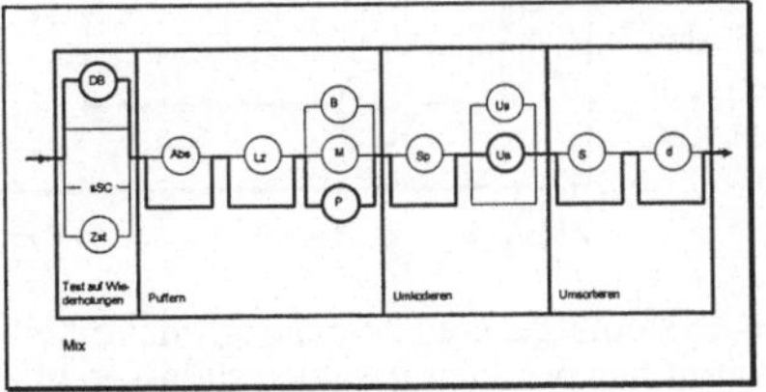

e) Mixmaster

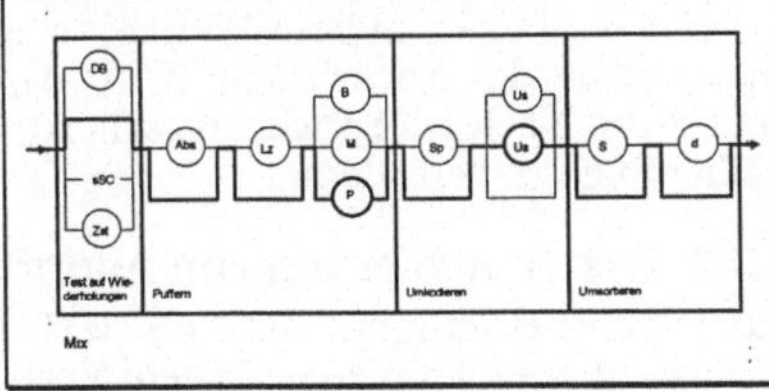

f) NDM-security agent

Abbildung 3-2: Kombinationsmöglichkeiten der Mixfunktionen (Darstellung der Anwendungsbeispiele von 2.3)

Beim NDM-security agent (3-2 f)) werden die Nachrichten nicht gesammelt. Die Zuordnung zum Pool erfolgte willkürlich, ebenso hätte die Betriebsart Batch gewählt werden können. Bei beiden Betriebsarten ist die Größe des Puffers gleich Eins.

In allen Darstellungen wurde davon ausgegangen, daß die Absender nicht verifizierbar sind. Darum wurde keine Absenderprüfung einbezogen. Nur bei den Telefon-Mixen wird die Forderung nach genügend vielen verschiedenen Absendern erfüllt, da jeder Teilnehmer ständig einen Sende- und einen Empfangskanal unterhält.

Das vorgestellte Modell enthält durch die Betrachtung sicherheitseingeschränkter Verfahren auch Wege, die eine unsichere Betriebsweise des Mixes zulassen. Im folgenden werden notwendige Einschränkungen sowie mögliche Kombinationen der Funktionen des Puffers diskutiert. Ziel ist ein modifiziertes Modell, welches nur Wege enthält, die unter bestimmten Voraussetzungen Unverkettbarkeit gewährleisten.

3.3 Notwendige Modifikationen und Angreifermodelle

3.3.1 Einschränkungen der Wege aufgrund allgemeiner Anforderungen

Aus den notwendigen Teilschritten eines Mixes ergeben sich bereits einige Einschränkungen des allgemeinen Modells. Daraus ergeben sich die im folgenden erläuterten Umsortierungen des ablauforientierten Modells. Sich beeinflussende Funktionen werden darin zusammengefaßt.

Test auf Wiederholungen, Umkodieren: Eine Möglichkeit der Verhinderung von Nachrichtenwiederholungen besteht im Speichern der vom Mix bearbeiteten Nachrichten (bzw. Hashwerte dieser Nachrichten, Nummern, ...) in einer Datenbank. Nach einer gewissen Zeit sind die Datenbestände zu löschen. Dazu wird z.B. das Schlüsselpaar des Mixes gewechselt.

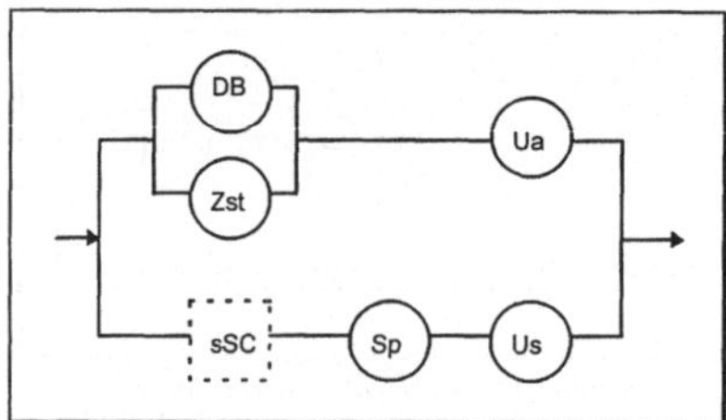

Abbildung 3-3: Test auf Wiederholungen

Auch das Überprüfen von Zeitstempeln verhindert Wiederholungen. Beide Varianten setzen keine symmetrische Umkodierung voraus, da für den Test nicht erforderlich (obwohl theoretisch möglich).

Anders sieht es mit dem unteren Zweig aus. Unter der Voraussetzung, daß mit einer synchronen Stromchiffre verschlüsselt wird, kann der Test auf Wiederholungen entfallen, da Wiederholungen nicht „möglich" sind. Die Umkodierung erfolgt also symmetrisch. Dafür ist die Speicherung von Informationen über den zu verwendenden Schlüssel unerläßlich.

Organisation des Puffers, Umsortieren:

Arbeitet der Puffer im Pool-Betrieb, so entfällt das Umsortieren der Nachrichten. Wird dagegen ein Schub von Nachrichten weitergeschickt, so ist die Reihenfolge zu verändern.

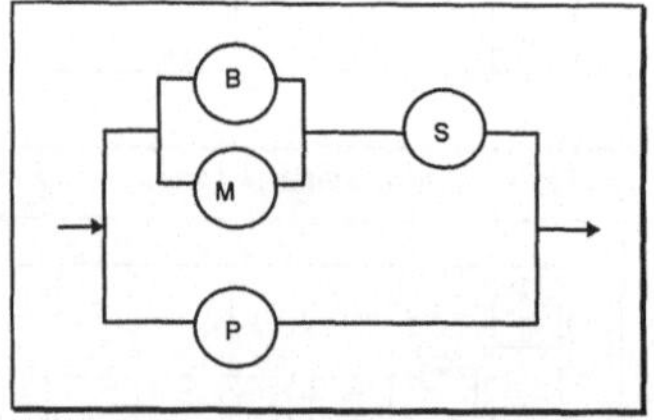

Abbildung 3-4: Organisation des Puffers

3.3.2 Systematisierung von Angriffen

Für Mixe ist es aufgrund der Komplexität des Systems generell schwierig, Sicherheitsbedingungen zu definieren. Andere Verfahren mit nur wenigen Parametern lassen klare Definitionen der Sicherheit zu. So kann beispielsweise die Vernam-Chiffre als informationstheoretisch sicher eindeutig beschrieben werden.

In den zugrunde gelegten Mix-Konzepten [Chau_81, Pfit_89] wird davon ausgegangen, daß Angreifer sämtliche Kommunikation im Netz abhören können. Es wird gefordert, daß mindestens ein Mix vertrauenswürdig ist.

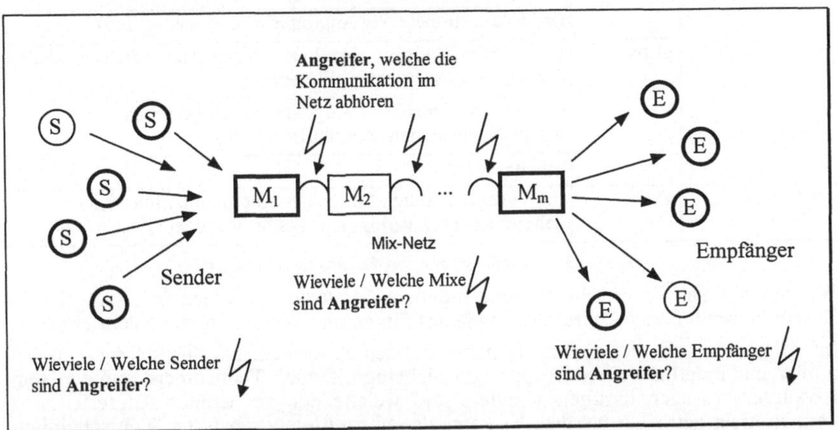

Abbildung 3-5: Mögliche Angreifer beim Mixkonzept

Bei der Betrachtung der Sicherheit ist eine Vielzahl von **Einflußgrößen** zu beachten:

- Es ist zu unterscheiden, **wer** angreift. Das könnten sowohl Outsider (Angreifer auf dem Netz) als auch Insider (Mixe oder Teilnehmer, also Sender oder Empfänger von Nachrichten) sein (Abbildung 3-5).
- Die **Stärke** der Angreifer muß beschrieben werden.
- Entscheidend sind auch **Systemparameter**, wie z.B.
  - die Organisation des Puffers (Batch, Pool oder Mischformen),
  - die Puffergröße oder
  - Aussagen über die Charakteristik des Ein- und Ausgabestroms von Nachrichten oder über den gewählten Weg durch das Mixnetz (Benutzung einer Kaskade oder Benutzung von Mixen in freier Reihenfolge).

In einem allgemeinen Angreifermodell werden Annahmen über die einzelnen Parameter getroffen. Die Verfahren fordern bestimmte Systemparameter als Voraussetzung. Anhand dieser Voraussetzungen kann die erreichte Sicherheit des Verfahrens unter einem eingegrenzten Angreifermodell betrachtet werden. Die ersten beiden Punkte - mögliche Angreifer und deren Stärke - sind unabhängig von einem speziellen Verfahren.

Die Stärke von Angreifern kann nach [Pfit_95] folgendermaßen charakterisiert werden: **Beobachtende** Angreifer führen im System nur erlaubte Handlungen und auch alle vorgeschriebenen aus, während **verändernde** Angreifer auch unerlaubte Handlungen im System ausführen oder vorgeschriebene unterlassen. Darüber hinaus kann noch unterschieden werden zwischen **passiven** und **aktiven** Angreifern. Passive Angreifer greifen nicht in den normalen Betrieb des Systems ein, sie beobachten nur die Aktionen des Angegriffenen. Aktive Angreifer dagegen führen schon vor dem eigentlichen Angriff gezielte Aktionen aus, mit denen sie den Angegriffenen zum vom Angreifer beeinflußten Einsatz seines kryptographischen Systems bringen.

In Tabelle 3-2 werden die bisher beschriebenen Angriffe aufgrund dieser Definitionen eingeteilt. Weitere aktive Angriffe werden in [PfPW1_89] beschrieben.

beobachtend	**passiv**	Abhören der Kommunikation auf dem Netz
		(n-1)-Angriff eines einzelnen Angreifers (wartet auf günstige Zusammensetzung des Puffers)
		(n-1)-Angriff mehrerer Angreifer (Zusammenschluß)
	aktiv	(n-1)-Angriff eines einzelnen Angreifers durch gezieltes Senden von Nachrichten (Fluten)
		(n-1)-Angriff mehrerer Angreifer durch gezieltes Senden von Nachrichten (Fluten, Zusammenschluß)
		replay-Angriffe
verändernd	**aktiv**	(n-1)-Angriffe durch Stören der Kommunikation anderer Teilnehmer (Komm. verhindern; Nachr. verzögern)

Tabelle 3-2: Einteilung der bisher beschriebenen Angriffe

Wie aus der Tabelle ersichtlich, sind mögliche Strategien für einen (n-1)-Angriff der Zusammenschluß von Angreifern sowie das Fluten und Verzögern von Nachrichten.

Im folgenden soll dieser Angriff näher betrachtet werden. Die Teilnehmer können immer nur innerhalb einer Gruppe von nichtangreifenden Teilnehmern anonym sein. Absolute Aussagen darüber, wieviele und welche der Teilnehmer Angreifer sind, können nicht getroffen werden. Es kann also nur mit einer gewissen Wahrscheinlichkeit Unverkettbarkeit erreicht werden. Die Funktionen des Mixes sollen so kombiniert werden, daß unter bestimmten Voraussetzungen dieser Angriff - ob nun durch einen einzelnen oder durch mehrere Angreifer - verhindert wird.

3.3.3 Modifikation des Modells zur Verhinderung von (n-1)-Angriffen

Übergeordnetes Kriterium der folgenden Angriffe ist die *Charakteristik des Eingabestroms* (Annahme A, B). Unter entsprechenden Annahmen werden Angreifer verschiedener *Stärke* betrachtet. Der Einfluß der *Puffergröße* stellt ein weiteres Unterscheidungsmerkmal dar. Die Verifizierbarkeit von Absendern - eine zur Verhinderung des Zusammenschlusses von Angreifern wichtige Funktion - wird in den folgenden Modellen mit „vAbs" gekennzeichnet.

Annahme A: Die Teilnehmer senden immer zu festen Takten, ggf. dummies.
Das bedeutet, daß sich zu allen Takten genügend viele Nachrichten von allen Teilnehmer im Puffer befinden. Entsprechend dem Angreifermodell lassen sich verschiedene Modelle konstruieren.

A.1) beobachtende Angreifer
A.1.a) passive Angreifer
   Im Puffer befinden sich genügend viele Nachrichten von genügend vielen verschiedenen Absendern. Die Angreifer könnten versuchen, sich zusammenzuschließen, um gemeinsam einen (n-1)-Angriff durchzuführen. Zwei Fälle sind zu unterscheiden:

- *Der Puffer ist groß genug, um den Zusammenschluß der Angreifer zu verhindern.*
  In diesem Fall sind keine Maßnahmen erforderlich.

- *Die Puffergröße ist nicht ausreichend groß.*
  Ein (n-1)-Angriff eines einzelnen Angreifers ist nach wie vor nicht möglich. Der Erfolg eines Angriffs durch Zusammenschließen von Angreifern muß jedoch durch weitere Maßnahmen (Erzeugung von dummies, Latenzzeiten) erschwert werden (Abbildung 3-6).

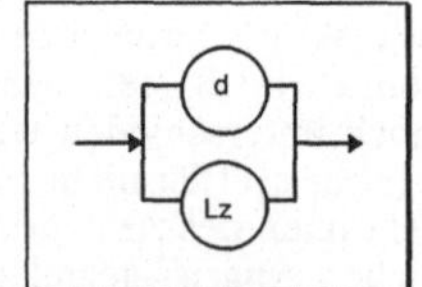

*Abbildung 3-6:
Maßnahmen bei beobachtenden, passiven Angreifern bei kleiner Puffergröße*

A.1.b) aktive Angreifer (senden Mix viele Nachrichten zu)
Es befinden sich immer noch genug Nachrichten im Puffer, aber es ist nicht gewährleistet, daß sie auch von genügend vielen verschiedenen Absendern stammen. Wieder sind zwei Fälle zu unterscheiden:

- *Der Puffer ist groß genug.*
  Da alle Teilnehmer ständig senden, wird es dem Angreifer nicht gelingen, in einen großem Puffer für n Nachrichten n-1 eigene zu plazieren. Besondere Schutzmaßnahmen sind also nicht erforderlich.

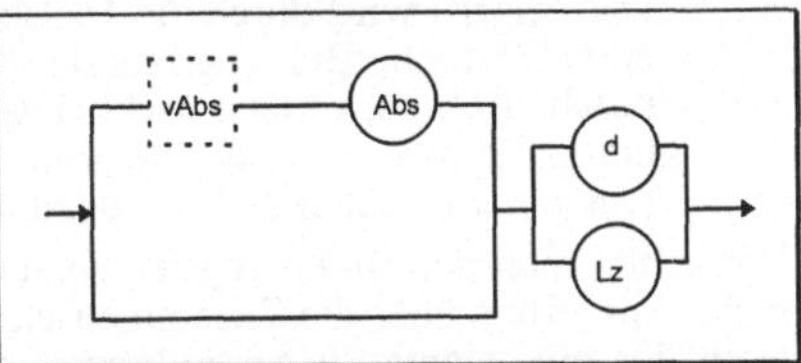

- *Der Puffer ist nicht groß genug.*
  In diesem Fall sind die Absender zu prüfen, falls diese verifizier-

Abbildung 3-7: Maßnahmen gegen beobachtende, aktive Angreifer bei kleiner Puffergröße

bar sind. Da der Puffer nicht ausreichend groß ist, könnten sich auch Angreifer zusammenschließen. Weitere Maßnahmen sind also unabhängig von der Verifizierbarkeit der Absender erforderlich (Abbildung 3-7).

A.2) verändernde Angreifer (die hier aufgeführten Angriffe sind aktiv)
Verändernde Angreifer können die Kommunikation der anderen Teilnehmer mit dem Mix stören. Die Wahrscheinlichkeit, daß sich im Mix dann genügend Nachrichten eines Angreifers für einen (n-1)-Angriff befinden, erhöht sich damit. Ebenso wird der Zusammenschluß von Angreifern erleichtert. Die Angreifer können außerdem Nachrichten mit gefälschten Absendern an den Mix schicken, falls Absender nicht verifizierbar sind. Auch hier sind die beiden Fälle zu unterscheiden:

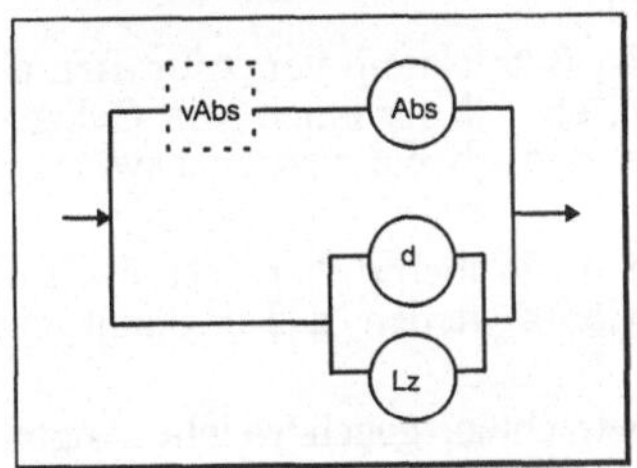

Abbildung 3-8: Maßnahmen gegen verändernde, aktive Angriffe bei großem Puffer

- *Der Puffer ist groß genug.*
  Im Gegensatz zu bisherigen Fällen sind auch in diesem Fall Schutzmaßnahmen erforderlich. Falls die Absender verifizierbar sind, ist eine Absenderprüfung vorzunehmen. Als Alternative dazu können dummies erzeugt bzw. Latenzzeiten verwendet werden (Abbildung 3-8).

- *Der Puffer ist nicht groß genug.*
  Die Absender sind ggf. zu prüfen. Darüber hinaus sind aber auf alle Fälle weitere Maßnahmen zu ergreifen (wie Abbildung 3-7).

Annahme B: Die Teilnehmer senden nur bei Bedarf.
Bei dieser Annahme kann nicht davon ausgegangen werden, daß sich normalerweise genügend viele Nachrichten von genügend vielen verschiedenen Absendern im Puffer befinden. Die Prüfung der Absender ist von noch größerer Bedeutung.

B.1) beobachtende Angreifer
B.1.a) passive Angreifer
Ist der Puffer ausreichend groß, sind keine Maßnahmen erforderlich. Ist er jedoch klein genug, daß (n-1) bekannte Nachrichten zusammenkommen könnten, so sind ggf. die Absender zu prüfen oder alternativ dummies zu erzeugen bzw. Latenzzeiten zu behandeln. Daraus ergibt sich ein Modell wie in Abbildung 3-8.

B.1.b) aktive Angreifer
Senden die Angreifer dem Mix viele Nachrichten zu, ohne daß die anderen Teilnehmer auch ständig senden, so sind unabhängig von der Puffergröße Maß-

nahmen erforderlich. Die Absender sind, falls möglich, zu prüfen. Außerdem ist der Erfolg eines Angriffs durch Zusammenschluß von Angreifern zu verhindern. Das Modell entspricht dem in Abbildung 3-7 gezeigten.

B.2) verändernde Angreifer (wiederum aktiv)
Der Angriff wird durch die Tatsache, daß die Teilnehmer nur bei Bedarf senden, erleichtert. Die Behinderung der Kommunikation ist nun nur für die Teilnehmer durchzuführen, welche wirklich senden wollen. Die Absender müssen überprüft und außerdem entweder dummies erzeugt oder Latenzzeiten behandelt werden. Auch hier ergibt sich das Modell aus Abbildung 3-7.

Betrachtet man den $(n\text{-}1)$-Angriff, so zeigt sich, daß das Vergrößern der Ungewißheit eines Angreifers über die Zusammensetzung des Puffers die wesentliche Aufgabe darstellt. Da eine sinnvolle Absenderprüfung nur bei verifizierbaren Absendern möglich ist, hat die Erzeugung von dummies bzw. die Verwendung von Latenzzeiten große Bedeutung. Daraus ergeben sich jedoch neue Probleme (siehe 2.1 sowie 3.3.4)

3.3.4 Diskussion weiterer Angriffsmöglichkeiten

Aus den unter 3.3.2 genannten Einflußgrößen für die Betrachtung von Sicherheit wurde hier nur ein Teil herausgegriffen. Es wurde nur ein Angriff näher untersucht, in welchem die Teilnehmer als Angreifer auftreten.

Obwohl $m\text{-}1$ der verwendeten m Mixe laut Angreifermodell korrupt sein dürfen, wurden die Mixe hier nicht explizit als Angreifer behandelt. Wollen sich die Teilnehmer gegen Angriffe der Mixe absichern, müssen sie die Möglichkeit haben, die Mixe zu kontrollieren. Kontrollierbar sind nur Operationen, welche deterministisch ablaufen. Haben die Mixe Gelegenheit, indeterministisch zu arbeiten, also etwa durch die Erzeugung von dummies, ergeben sich Möglichkeiten für nicht oder schwer entdeckbare Angriffe der Mixe. Die zum Schutz vor $(n\text{-}1)$-Angriffen vorgeschlagenen Maßnahmen sind demzufolge nicht ohne weiteres einzusetzen.

Daraus lassen sich weitere Ansatzpunkte für die Betrachtung von Angriffen ableiten. Auch für andere Funktionen ist zu untersuchen, ob sich für einen Mix Gelegenheiten zu unkontrollierbaren Manipulationen ergeben und ob ein vertrauenswürdiger Mix trotzdem noch Unverkettbarkeit garantiert.

So ist auch der Poolbetrieb für die Teilnehmer nicht überprüfbar. Verzögert der Mix einzelne Nachrichten, kann ihm nicht nachgewiesen werden, daß er damit einen Angriff versucht hat.

Es ist für getroffene Maßnahmen immer zu betrachten, gegen welche Angreifer Sicherheit erreicht werden kann und wessen Kontrolle erschwert wird. Es ist anzunehmen, daß es nicht möglich ist, „allgemeine Sicherheit" zu erreichen.

Die Angriffe auf einen einzelnen Mix sind auf Angriffe auf die gesamte Mixfolge auszuweiten. Die beiden Varianten Kaskade und freie Folge sind zu betrachten. Die Kaskade bietet gute Möglichkeiten der Kontrolle für die Teilnehmer. In einer freien Folge dagegen ist die Kontrolle sehr schwer - jeder Mix kann sowohl Nachrichten von Teilnehmern als auch weitergeleitete Nachrichten anderer Mixe erhalten. Für einen Angreifer erhöht sich in einem Netz die Komplexität seiner Angriffsversuche, aber die Angriffe sind schwerer zu entdecken. Die verschiedenen Optionen zur Gestaltung eines Mixes müssen auch für die Folge von Mixen betrachtet werden. Besonders bei freien Folgen kann sich von Mix zu Mix die Konfiguration ändern.

Die Möglichkeit der Kontrolle bietet den Teilnehmern den besten Schutz gegen Angriffe der Mixe. Es ist jedoch auch zu betrachten, ob die für Sicherheit notwendigen Kontrollmaßnahme überhaupt effizient genug durchführbar sind.

3.4 Modifiziertes Modell

Nach den Diskussionen in 3.1 und 3.2 wird ein modifiziertes Modell vorgestellt (Abbildung 3-9). *Ein* allgemeines „sicheres" Modell wird nicht angegeben. Durch die Komplexität der Einflußgrößen gilt die Sicherheit immer nur unter bestimmten Annahmen.

Das Modell teilt die Funktionen in drei Abschnitte auf:
1) realisiert den Test auf Wiederholungen sowie Umkodierung,
2) stellt die notwendigen Kombinationen von Organisationsformen des Puffers und der Funktion Umsortieren dar und in
3) werden weitere Funktionen kombiniert, um $(n-1)$-Angriffe abzuwehren.

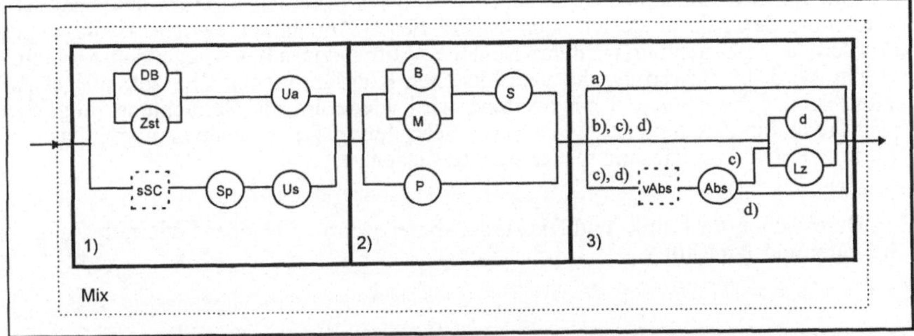

Abbildung 3-9: Modifiziertes Modell eines Mixes

Über die Sicherheit des Modells können folgende Aussagen getroffen werden: Durch die ersten beiden Teile wird hinreichende Sicherheit gegenüber beobachtenden, passiven Angreifern garantiert. Die Kombinationen in 1) gewähren außerdem hinreichende Sicherheit gegen replay-Angriffe beobachtender, aktiver Angreifer.

Der Einsatz von dummies und Latenzzeiten erhöht zwar die Unsicherheit angreifender Teilnehmer über den Inhalt des Puffers, stellt jedoch indeterministische Operationen dar. Über die erreichte Sicherheit im dritten Teil können darum keine allgemeinen Aussagen getroffen werden (siehe auch 2.1, 3.3.4). Unter der Annahme, daß nur die Teilnehmer angreifen, stellen die Wege a) bis d) notwendige Kombinationen dieser Funktionen dar, welche Maßnahmen gegen folgende Angreifer bieten:

Falls alle Teilnehmer ständig senden:
- beobachtende, passive / aktive Angreifer (ausreichend großer Puffer) (a),
- beobachtende, passive Angreifer (kein ausreichend großer Puffer) (b),
- beobachtende / verändernde, aktive Angreifer (kein ausr. großer Puffer) (c),
- verändernde, aktive Angreifer (ausreichend großer Puffer) (d).

Falls die Teilnehmer nur bei Bedarf senden:
- beobachtende, passive Angreifer (ausreichend großer Puffer) (a),
- beobachtende / verändernde, aktive Angreifer (kein ausr. großer Puffer) (c),
- beobachtende, passive Angreifer (kein ausreichend großer Puffer) (d).

4 Zusammenfassung und Ausblick

Ausgehend von Definitionen des Mixes in [Chau_81, Pfit_89, PfPW1_89] wurden Erweiterungen dieses Konzepts aus Sicherheits- und Effizienzgründen betrachtet. Daraus wurde ein allgemeines Modell abgeleitet, welches auf bekannte, teilweise Sicherheit bietende Anwendungsbeispiele angewendet wurde.

Aus der Systematisierung der verschiedenen aufgezeigten Varianten der einzelnen Funktionen eines Mixes wurde ein mögliches ablauforientiertes Modell entworfen.

Dieses Modell deckt die betrachteten Anwendungsbeispiele ab, enthält damit jedoch auch „unsichere" Konfigurationen.

Aus den Forderungen an einen Mix ergeben sich einige notwendige Einschränkungen der Wege durch dieses Modell. Die Betrachtung von Angriffen führt zu weiteren Kombinationsmöglichkeiten.

Um Aussagen über Angreifermodelle für Mixe zu machen, wurden mögliche Einflußgrößen auf die erreichte Sicherheit zusammengetragen. Am Beispiel des $(n\text{-}1)$-Angriffs wurden die Funktionen eines Mixes unter verschiedenen Annahmen über solche Einflußgrößen, speziell Stärke der Angreifer und Systemparameter, kombiniert.

Ein „einfaches" allgemeines Modell für einen „sicheren" Mix läßt sich aufgrund der Komplexität der Annahmen nicht angeben. Die vorgestellten Modelle sind mögliche Sichten auf den Mix unter den oben beschriebenen Annahmen. Weitere Bedingungen erfordern die Betrachtung weiterer Modelle. Aufbauend auf vorliegenden Modellierungen, wird die Bewertung der verschiedenen Konfigurationen Ziel zukünftiger Arbeiten sein. Schwerpunkt der Bewertung wird wiederum die Betrachtung möglicher Angriffe und in Zusammenhang damit die Betrachtung der erreichten Anonymität sein. Die skizzierten Angriffe sind weiter zu untersuchen.

Ein Dankeschön für Kritik- und Diskussionsbereitschaft an Hannes Federrath, Andreas Graubner und Jan Müller.

5   Literatur

Chau_81   D. Chaum: Untraceable Electronic Mail, Return Addresses, and Digital Pseudonyms; Communications of the ACM 24/2 (1981) 84-88.

FaKK_96   Andreas Fasbender, Dogan Kesdogan, Olaf Kubitz: Analysis of Security and Privacy in Mobile IP. 4th International Conference on Telecommunication Systems, Modelling and Analysis, Nashville, March 21-24, 1996.

FFJM_97   Hannes Federrath, Elke Franz, Anja Jerichow, Jan Müller, Andreas Pfitzmann: Ein Vertraulichkeit gewährendes Erreichbarkeitsverfahren (Schutz des Aufenthaltsortes in künftigen Mobilkommunikationssystemen), GI Fachtagung Kommunikation in Verteilten Systemen (KiVS) 97, 17.-22.2.97 in Braunschweig.

GüTs_96   C. Gülcü, G. Tsudik: Mixing Email with BABEL; Proc. Symposium on Networking and Distributed System Security, San Diego, IEEE Comput. Soc. Press, 1996, pp 2-16.

Pfit_89   A. Pfitzmann: Diensteintegrierende Kommunikationsnetze mit teilnehmerüberprüfbarem Datenschutz; Universität Karlsruhe, Fakultät für Informatik, Dissertation, Feb. 1989, IFB 234, Springer-Verlag, Berlin 1990.

Pfit_93   A. Pfitzmann: Technischer Datenschutz in öffentlichen Funknetzen; Datenschutz und Datensicherung, DuD 17/8 (1993), 451-463.

Pfit_95   A. Pfitzmann: Datensicherheit und Kryptographie; Vorlesungsskript TU Dresden, WS 1995/96.

PfPW1_89   Andreas Pfitzmann, Birgit Pfitzmann, Michael Waidner: Telefon-MIXe: Schutz der Vermittlungsdaten für zwei 64-kbit/s-Duplexkanäle über den $(2 \cdot 64 + 16)$-kbit/s-Teilnehmeranschluß; Datenschutz und Datensicherung DuD /12 (1989) 605-622.

PfPW_88   Andreas Pfitzmann, Birgit Pfitzmann, Michael Waidner: Datenschutz garantierende offene Kommunikationsnetze; Informatik-Spektrum 11/3 (1988) 118-142.

Scha_95   Steffen Schaarschmidt: Entwurf und Implementierung eines anonymen Remailers mit MIX-Funktionalität; Großer Beleg, TU Dresden, Institut für Theoretische Informatik, September 1995.

Unbeobachtbarkeit in Kommunikationsnetzen

Hannes Federrath, Anja Jerichow, Jan Müller, Andreas Pfitzmann
Technische Universität Dresden, Institut für Theoretische Informatik, 01062 Dresden
{federrath@, jerichow@, jm4@, pfitza@}inf.tu-dresden.de

Zusammenfassung

Es wird ein Verfahren zum Schutz der Vermittlungsdaten vorgestellt, das es den Nutzern von Kommunikationsnetzen ermöglicht, diese Netze unbeobachtbar und anonym zu nutzen. Dieses Verfahren der getakteten Mix-Kanäle [PfPW_89] basiert auf dem Mix-Netz, einer Methode für anonyme Kommunikation von D. Chaum. Die Realisierung dieses Konzeptes wird für die Kommunikationsnetze ISDN und GSM angegeben. Durch die beschriebenen Modifikationen ist eine anonyme und unbeobachtbare Netznutzung ohne Einbußen bei der Datenrate des Nutzkanals trotz Echtzeitanforderungen möglich. Die Effizienz des beschriebenen Verfahrens wird untersucht.

1. Motivation

Das schnelle Wachstum der Kommunikationsnetze erfordert eine ständige Auseinandersetzung mit neuen Anforderungen an die Technik. Die Gewährleistung von Datenschutz für die Nutzer der Systeme rückt immer mehr in den Mittelpunkt. Sensible, durch die Netznutzung entstehende, personenbezogene Daten sollen nicht mißbraucht werden können. Dazu gehört nicht nur der Schutz der Nutzdaten (z.B. Bild, Ton und Text), der durch Ende-zu-Ende Verschlüsselung gewährleistet werden kann, sondern auch der Schutz der Vermittlungsdaten (z.B. Adressen der Kommunikationspartner, Beginn, Dauer und Dienstart). Es wird gezeigt, daß sich Unbeobachtbarkeit und Anonymität von Teilnehmern bei der Signalisierung von Verbindungswünschen, Location Updates und anschließender Kommunikation bereits mit der zur Verfügung stehenden Bandbreite bestehender Kommunikationsnetze realisieren läßt.

Die bekanntesten Techniken für anonyme Kommunikation sind Mixe [Chau_81] und DC-Netze [Cha8_85, Chau_88]. Da das DC-Netz es erfordert, daß jeder Netzabschluß im physikalischen Sinne mindestens halb soviel sendet, wie alle Netzabschlüsse zusammen im logischen Sinne senden, ist dieses Konzept auf den existierenden schmalbandigen Signalisierungskanälen nicht realisierbar. Unter der gesetzten Zielstellung, vorhandene Netze zu modifizieren und nicht neue Netze zu entwerfen, kommt (entsprechend [PfPW_89]) nur das Mix-Netz in Frage.

Kapitel 2 stellt das Prinzip der getakteten Mix-Kanäle in allgemeingültiger Form vor. Anschließend wird in Kapitel 3 und 4 die Realisierung von anonymer Signalisierung am Beispiel von ISDN und GSM beschrieben. Die dafür notwendigen Modifikationen der Netzstrukturen, Signalisierungsprotokolle und Signalisierungsnachrichten werden in dieser Arbeit angegeben. In Kapitel 5 erfolgen Effizienzuntersuchungen zum beschriebenen Verfahren für ISDN und GSM.

2. Einführung der getakteten Mix-Kanäle

Die Kenntnis des Grundprinzips eines Mix-Netzes [Chau_81], mit den Verbesserungen aus [PfPW_89] zur Vermeidung der Angriffsmöglichkeiten aus [PfPf_89] wird in dieser Arbeit vorausgesetzt. Annahmen zum genutzten Mix-Netz:

- Es werden Mix-Kaskaden im Schubbetrieb genutzt. Diese können am einfachsten innerhalb der Vermittlungsstellen (Vst) des Kommunikationsnetzes implementiert werden.

- Alle Teilnehmer, die an dieselbe Vst angeschlossen sind, bilden dadurch eine Anonymitätsgruppe. Teilnehmeraktionen lassen sich nur einer Anonymitätsgruppe zuordnen, keinem einzelnen Teilnehmer.

- Für alle Mix-Nachrichten, die auf Teilnehmern exklusiv zugeordneten Kanälen übertragen werden, wird Dummy Traffic eingeführt, damit das physische Senden und Empfangen unbeobachtbar ist.

Das Mix-Netz eignet sich in dieser Form besonders für die unbeobachtbare Übertragung einzelner Nachrichten. Da für Telefonie-Dienste aber unbeobachtbare Kommunikationskanäle mit strengen Echtzeitanforderungen benötigt werden, muß das Mix-Konzept modifiziert werden. Analog der *hybriden Verschlüsselung* sollen im folgenden symmetrische unbeobachtbare Kanäle (Mix-Kanäle) durch eine asymmetrische Setup-Nachricht aufgebaut werden (siehe Abb. 2.1). Vollduplex-Kanäle werden im folgenden als zwei gegenlaufende Simplex-Kanäle betrachtet. Jeder Teilnehmer muß sowohl als Sender S als auch als Empfänger E in einer Kommunikationsbeziehung unbeobachtbar sein. Aus diesem Grund muß jeder Teilnehmer vor einer Kommunikation für sich einen unbeobachtbaren Sendekanal (ZS-Setup) und einen unbeobachtbaren Empfangskanal (ZE-Setup) aufbauen. Dadurch wird nicht nur die Unbeobachtbarkeit gegenüber Dritten gewährleistet, es bleiben auch beide Kommunikationspartner voreinander unerkannt.

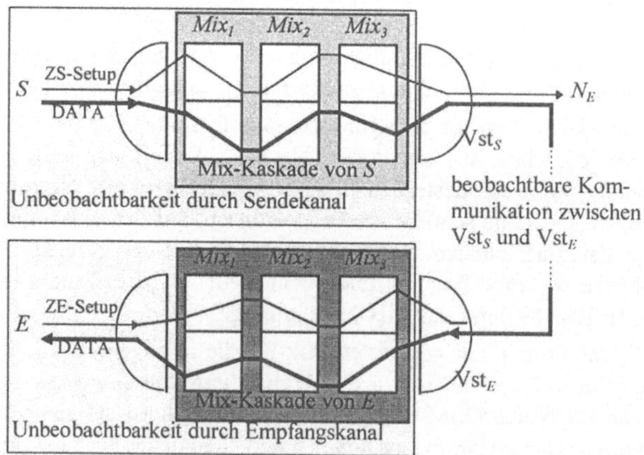

Abb. 2.1: Prinzip der Mix-Kanäle am Beispiel eines Simplex-Kanals von S nach E

Eine *Kanalaufbaunachricht* wird mit den öffentlichen Schlüsseln c_j der Mixe M_j der Kaskade der nächsten Vst asymmetrisch verschlüsselt. Mit dieser Nachricht wird jedem Mix M_j ein vom Teilnehmer gewählter symmetrischer Schlüssel k_{xyj} (x Sender/Empfänger, y Sende-/Empfangskanal, j Nummer des Mixes) für die folgende Signalisierung und Kommunikation übergeben. Außerdem speichert jeder Mix die Position dieser Nachricht in seinem Eingabeschub zusammen mit der Position dieser Nachricht im Ausgabeschub (nach der Umsortierung). Diese Zuordnung bleibt dann, ebenso wie der symmetrische Schlüssel, während der Kommunikation gleich.

Das bis zu dieser Stelle beschriebene Verfahren kann allerdings in dieser Form noch nicht eingesetzt werden. Forderungen an die Kanalaufbaunachrichten:

- Damit die Aktionen der Teilnehmer nicht beobachtbar sind, müssen Kanalaufbaunachrichten gleichzeitig gesendet werden und gleich viele Kanäle belegt werden.

- Außerdem müssen gleichzeitig aufgebaute Kanäle auch gleichzeitig wieder abgebaut werden, da die Nachrichten eines Schubes gleich lang sein müssen. Dadurch wird eine effektive Nutzung der zwei Nutzkanäle (ISDN) auch innerhalb einer kleineren Anonymitätsgruppe unmöglich.

Deshalb wird ein Systemtakt t_i (auch Zeitscheibe genannt) eingeführt. Der Systemtakt ermöglicht den Teilnehmern, unbeobachtbare Verbindungen, also unbeobachtbare Ende-zu-Ende-Kanäle von beliebiger Dauer zu unterhalten, die trotzdem von jedem Kommunikationspartner regelmäßig abgebaut werden können. Durch die Einführung der Zeitscheiben werden beliebig lang andauernde Verbindungen auf mehrere kurze (ca. 1s) und unmittelbar aufeinanderfolgende Mix-Kanäle, im folgenden *Zeitscheibenkanäle* genannt, aufgeteilt. Zu Beginn jeder Zeitscheibe hat jeder Teilnehmer die Möglichkeit, eine neue Verbindung aufzubauen, oder eine bestehende abzubrechen. Dieser Systemtakt erfolgt synchron für alle Teilnehmer des gesamten Netzes. Die Zeitscheibenkanäle werden im folgenden Zeitscheiben-Sende-(ZS)-Kanal und Zeitscheiben-Empfangs-(ZE)-Kanal genannt.

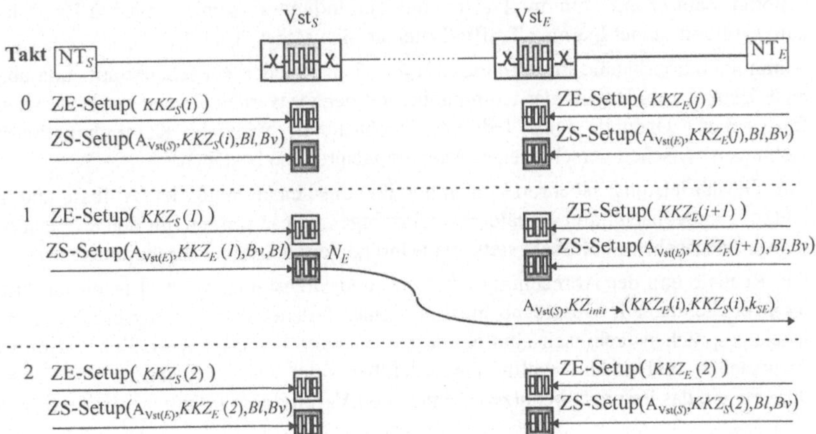

Abb. 2.2: Vereinfachter Beispielablauf eines Verbindungsaufbaus

Abb. 2.2 zeigt den vereinfachten Ablauf eines Verbindungsaufbaus. Um Unbeobachtbarkeit zu gewährleisten, muß jeder Netzabschluß in jeder Zeitscheibe für jeden Nutzkanal eine ZE- und eine ZS-Kanalaufbaunachricht signalisieren und Informationen auf den belegten Nutzkanälen übermitteln (notfalls jeweils Dummy Traffic).[1]

Mit dem Einsatz dieses Verfahrens muß außerdem die Übermittlung der Verbindungswunschnachricht N_E verändert werden, damit dadurch keine Beobachtungsmöglichkeiten entstehen. N_E ist so in die ZS-Setup-Nachricht codiert, daß sie erst auf der Netzseite der Vst im „Klartext" vorliegt. Anschließend kann diese Nachricht, wie bisher, zur Vst des Empfängers signalisiert werden. Dort muß sie allerdings an alle Teilnehmer einer Anonymitätsgruppe zuverlässig verteilt werden, damit das Empfangen der Verbindungswunschnachricht keine Beobachtungsmöglichkeiten bietet (Takt 1). Diese Nachricht kann nur der „gewünschte" Empfänger korrekt entschlüsseln. Die darin enthaltenen Informationen benötigt der Empfänger zum Aufbau seiner unbeobachtbaren Sende- und Empfangskanäle, z.B. Angaben (Kanalkennzeichen (KKZ)) zur Verkettung der unbeobachtbaren Kanäle beider Teilnehmer. Für jeden Sendekanal wird die Zieladresse $A_{Vst(x)}$ und das KKZ_x des dortigen Empfangskanals beim Kanalaufbau in der Netzseite der Vst gespeichert, für einen Empfangskanal wird nur das aktuelle KKZ gespeichert. Für diese $KKZs$ wird zu Beginn ein Initialwert KZ_{init} ausgetauscht, aus dem die KKZ der beiden Empfangskanäle berechnet werden. Diese werden mit einem Pseudozufallsgenerator in jedem Takt „weitergeschaltet".

Damit nicht beobachtbar ist, wann ein Teilnehmer einen Verbindungswunsch hat, muß auch für die Verbindungswunschnachricht Dummy Traffic signalisiert werden. Dieser sollte für die Netzseite einer Vst (vorher jedoch auf keinen Fall) als Dummy Traffic erkennbar sein (erste Bv/Bl-Kennzeichnung in ZS-Setup), damit solche Verbindungswünsche nicht verteilt werden müssen. Außerdem sollte eine analoge Kennzeichnung in der ZS-Setup-Nachricht für die übertragenen Nutzdaten erfolgen (zweite Bv/Bl-Kennzeichnung in ZS-Setup). Während eines Verbindungsaufbaus wird das Fernnetz dann nicht unnötig belastet. Wenn ein Teilnehmer einen Kommunikationskanal nicht benötigt, baut er mit Dummy Traffic eine Verbindung zu seinem eigenen Empfangskanal auf und sendet Dummy Traffic-Daten an sich selbst (Takt 0).

Kommunikationskanäle im Fernnetz werden nicht in jeder Zeitscheibe auf- und abgebaut. Einmal zum Beginn der Kommunikation belegt, werden sie erst am Ende wieder freigegeben. Dadurch wird lediglich beobachtbar, wieviele Kommunikationsbeziehungen zwischen verschiedenen Anonymitätsgruppen bestehen.

Zur Gewährleistung isochroner Kommunikationsdienste muß der Aufbau und die Nutzung unbeobachtbarer Kanäle parallel erfolgen. Diese werden mit den ausgetauschten symmetrischen Schlüsseln stets erst während der nächsten Zeitscheibe genutzt.

Die Realisierung der Abrechnung der Netz- und Dienstnutzung wird in dieser Arbeit nicht betrachtet. Für Gespräche innerhalb eines Ortsnetzes ist eine Abrechnung von Einzelgesprächen technisch sinnlos, da sie im Netz keinerlei zusätzlichen Aufwand verursachen und zudem völlig unbeobachtbar sind. Zur Abrechnung von Verbindungen, die das Fernnetz benutzen, können den Verbindungsaufbaunachrichten für den

---

[1] siehe Mix-Grundprinzip.

letzten Mix Gebührenmarken beigelegt werden. Einige ausführlichere Anmerkungen zur Abrechnung bei unbeobachtbarer Netznutzung können in [PfPW_89] nachgelesen werden.

3. Realisierung getakteter Mix-Kanäle in ISDN

Um für die Teilnehmer anonyme Signalisierung von Verbindungswunsch-, Kanalaufbau- und Kanalabbaunachrichten im ISDN gewährleisten zu können, muß die Netzstruktur, wie in Abb. 3.1 beschrieben, modifiziert werden:

- In den Ortsvermittlungsstellen (LE: Local Exchange) sind Mix-Kaskaden nötig.

- Ein Teil der Mix-Kaskade liegt in der Signalisier- und der andere Teil liegt in der Nutzkanalebene.

Die Nutzerseite einer LE erhält in jedem Takt des Kommunikationssystems von jedem angeschlossenen Teilnehmer (NT: Network Termination) sowohl Nachrichten auf dem Signalisierungskanal (D-Kanal) mit 16kbit/s, als auch auf den 2 Datenkanälen (B-Kanäle) mit je 64kbit/s. Diese Nachrichten sind ggf. bedeutungslos (Dummy Traffic). Für die Nutzerseite einer LE ist das allerdings nicht erkennbar. Sie kann keine Rückschlüsse auf eine bestehende Kommunikationsbeziehung bestimmter Teilnehmer ziehen.

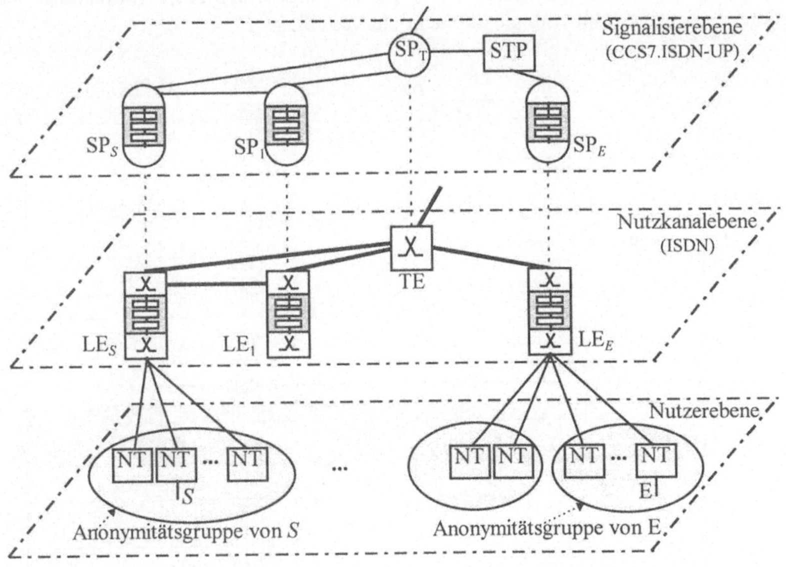

Abb. 3.1: Modifizierte Netzstruktur von ISDN

Die Netzseite der LE kann beobachten, wohin Nachrichten einer Anonymitätsgruppe signalisiert werden bzw. zu welchen LEs Kommunikationsbeziehungen aus der Anonymitätsgruppe heraus bestehen. Sie erhält aber auch nach Zusammenarbeit mit der Nutzerseite der LE keine Information darüber, von welchem Teilnehmer diese Nachrichten stammen. Erkennbar ist aber, wie lange wird von welcher LE mit welcher LE kommuniziert und wann bzw. zu welcher LE werden Verbindungswünsche signalisiert.

Neben der Netzstruktur von ISDN sind folgende Modifikationen des Signalisierungsprotokolls zum Verbindungsaufbau notwendig (siehe Abb. 3.2):

- Modifikation des Digital Subscriber Signalling System No. 1 (DSS1)-Protokolls zwischen den NTs und der LE.

- Einführung eines Mix-Protokolls innerhalb der Mix-Kaskade.

- Modifikation des Common Channel Signalling System No. 7 (CCS7)-ISDN-User Part (UP)-Protokolls zwischen den LEs.

Zum Aufbau einer unbeobachtbaren Kommunikationsverbindung initiiert der Sender S den Aufbau eines unbeobachtbaren ZE-Kanals durch die entsprechende DSS1-Nachricht über den Signalisierungskanal. Außerdem baut er einen ZS-Kanal für das unbeobachtbare Senden von Informationen auf. Die Netzseite der LE des Senders erhält mit der ZS-Setup-Nachricht die im CCS7-Netz zu sendende Verbindungswunschnachricht und die Adresse $A_{LE(E)}$ und sendet eine Initial Address Message (IAM) ab. Mit der im CCS7-Netz zum Empfänger gerouteten IAM wird bereits ein Kommunikationskanal bis zur LE_S belegt und geschaltet. Die mit der IAM im LE_E eintreffende Nachricht an den Empfänger E wird an alle angeschlossenen Teilnehmer der LE mittels einer DSS1-incoming call-Nachricht verteilt.

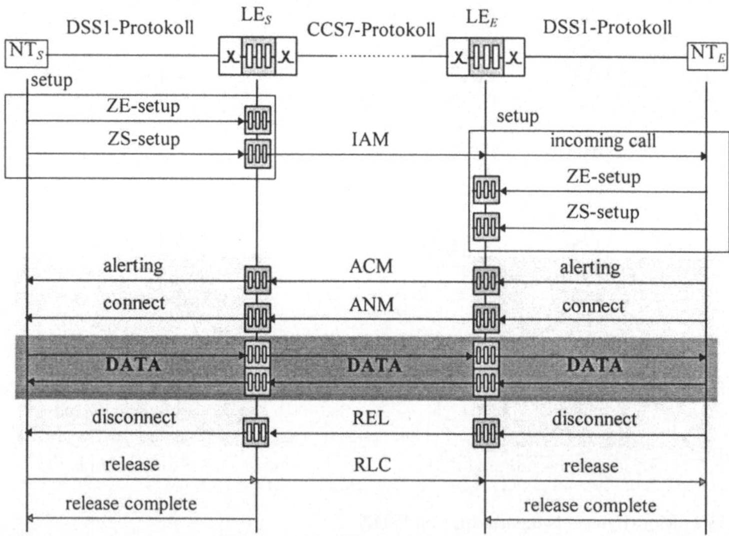

Abb. 3.2: Modifiziertes Protokoll des Verbindungsaufbaus in ISDN

Der gewünschte E kann als einziger die empfangene Verbindungswunschnachricht korrekt entschlüsseln. In dieser Nachricht erhält E alle notwendigen Informationen zum Aufbau der Verbindung. Dieser muß jetzt selbst aktiv werden und einen entsprechenden ZE-Kanal sowie ZS-Kanal aufbauen. Dadurch wird die Verbindung durchgängig geschaltet. Um Bandbreite auf dem schmalbandigen Signalisierungskanal zu sparen, wird der Zustand des Klingelns des Empfängertelefons mit der Alerting-Nachricht bereits im Datenkanal übertragen. Sobald der Empfänger abhebt, wird dies mit der Connect-Nachricht dem Sender über den Datenkanal mitgeteilt. Im Fernnetz werden die Signalisierungsnachrichten weiterhin im Signalisierungskanal übertragen. Die in Abb. 3.2 hellgrau hinterlegten Flächen gehören also bereits zum Datenkanal, wobei die eigentliche Datenübertragung in der Abbildung dunkelgrau dargestellt ist. Außerdem verdeutlichen die hellgrau geschriebenen Nachrichten den ursprünglichen Protokollablauf.

Durch Änderung des Protokolls ändert sich auch der Inhalt der Signalisierungsnachrichten. In dieser Arbeit sollen die notwendigen Modifikationen nur am Beispiel der ZS-Setup-Nachricht des Senders für eine Mix-Kaskade mit 3 Mixen angegeben werden.[2]

$$N_{\text{ZS-Setup}}(S) \quad := k_{S1}(t_i, \; c_2(t_i, \; k_{SS2}, \; c_3(t_i, \; k_{SS3}, \; \text{ALE}(E), \; KKZ_E(i), \; Bv/Bl,$$
$$Bv/Bl, \qquad\qquad\qquad\qquad\qquad\qquad\qquad\qquad\qquad\qquad N_E)))$$

$$\text{mit} \quad N_E \quad := c_E \, (\text{call\_setup\_msg}, \; t_i, \; \text{ALE}(S), \; KZ_{init})$$

Zur Vereinfachung hat jeder Teilnehmer mit dem ersten Mix bereits einen symmetrischen Schlüssel k_{S1} fest vereinbart, da diesem Mix ohnehin bekannt ist, wer sendet. Durch diese Modifikationen ändert sich die Länge der Signalisierungsnachrichten. Hinzu kommt, daß in jeder Zeitscheibe jeder Teilnehmer signalisieren und Daten senden muß (notfalls Dummy Traffic). Dadurch ändert sich die Performance im ISDN beträchtlich. In Kapitel 5 werden die notwendigen Effizienzuntersuchungen durchgeführt, und es wird nachgewiesen, daß sich dieses Verfahren ohne Einbußen bei der Datenrate auf dem Nutzkanal realisieren läßt.

4. Realisierung getakteter Mix-Kanäle in GSM

Um Anonymität und Unbeobachtbarkeit der Teilnehmer bei der Signalisierung auch für ein Mobilkommunikationssystem, wie GSM, gewährleisten zu können, ist nicht nur der Verbindungswunsch und der Kanalauf- bzw. -abbau zu schützen, sondern auch der Aufenthaltsort eines Mobilteilnehmers. In GSM sind für die Realisierung des Verfahrens folgende Annahmen nötig (siehe Abb 4.1):

- Eine Erweiterung der Ortsvermittlungsstellen (MSC: Mobile Switching Center) und deren Signalisierungspunkte um Mix-Funktionalität ist notwendig.

- Die Größe der Anonymitätsgruppe eines Mobilteilnehmers ist dynamisch. Sie besteht aus allen Mobilteilnehmern, die sich zur Zeit in einem Location Area (LA) aufhalten, das von demselben MSC verwaltet wird.

- Die Mixe im MSC schützen den Aufenthaltsort des Teilnehmers innerhalb eines MSC-Bereiches, d.h. ein MSC sollte möglichst viele verschiedene LAs verwalten.

---

[2] In [Müll_97] sind die Modifikationen der Nachrichteninhalte mit den notwendigen Änderungen der Protokolle im DSS1 und CCS7 detailliert beschrieben.

- Durch diese Mixe werden außerdem bestehende Kommunikationsbeziehungen geschützt.

- Zusätzlich muß in GSM der Signalisierungspunkt des HLRs um Mix-Funktionalität erweitert werden. Dadurch wird bei einem Location Update geschützt, aus welchem MSC-Bereich Änderungen vorgenommen werden. Bei einem Verbindungswunsch wird verborgen, in welchen MSC-Bereich der Verbindungswunsch zum Empfänger geroutet wird. Eine zweite Anonymitätsgruppe, bezüglich der Aufenthaltsorte, besteht demzufolge aus allen MSCs, die einem HLR zugeordnet sind.[3]

An dieser Stelle soll jedoch auf die stets existierende Möglichkeit der Peilung und Ortung einer sendenden Mobilstation hingewiesen werden, die auch durch DS/SS (siehe [FeTh_95]) nicht ganz verhindert werden kann.

---

[3] Die Beziehung zwischen einem MSC und dessen VLR ist dagegen zur Zeit nicht durch Mixe zu schützen, da es sich um eine 1:1 Beziehung handelt. Sollten jedoch zukünftig mehrere MSCs zu einem VLR zugeordnet werden, sollte auch diese Kommunikationsbeziehung analog der MSC-HLR-Beziehung durch Mixe im SP des VLRs geschützt werden.

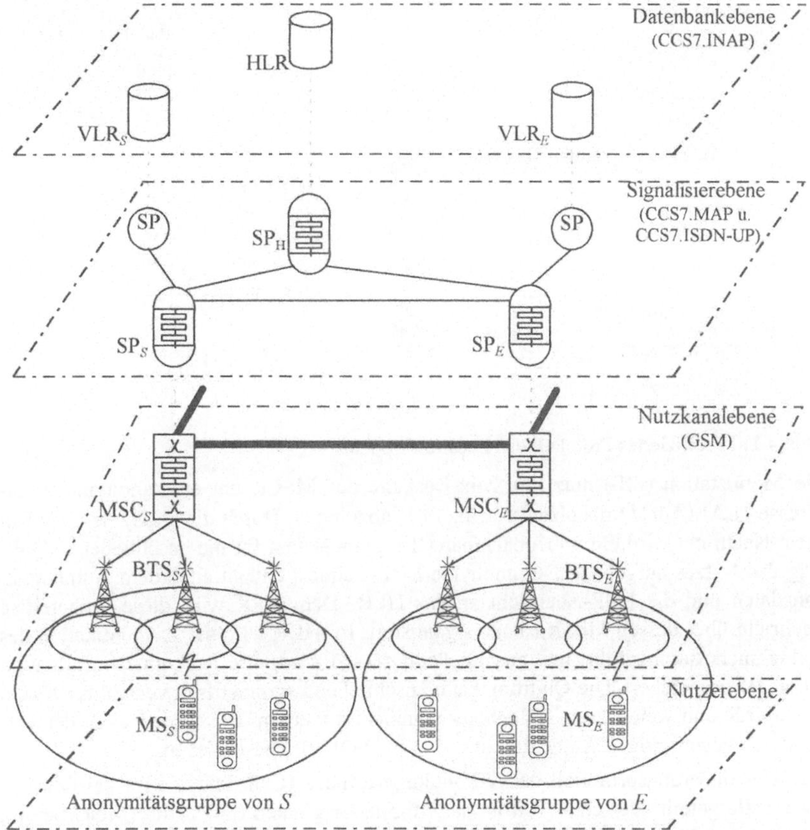

Abb. 4.1: Modifizierte Netzstruktur von GSM

Die Verfahren des Location Managements und des Verbindungsaufbaus werden in den folgenden Abschnitten diskutiert, ebenso verschiedene Angriffsmöglichkeiten und das erreichbare Wissen eines Angreifers. Einschränkungen bezüglich der Unbeobachtbarkeit von Mobilteilnehmern wird es im Mobilfunk allerdings stets durch die geringe Bandbreite auf der Funkschnittstelle geben. Außerdem kann man von akkugetriebenen Mobilstationen keine Aussendung von Dummy Traffic erwarten. Aus diesen Gründen läßt sich bei Mobilfunksystemen nie der gleiche Schutz erreichen, der in Festnetzen, wie ISDN, möglich ist.

Nach den Modifikationen der GSM-Netzstruktur ist das Signalisierungsprotokoll des Location Updates (LUP) ebenfalls zu modifizieren. Optimierungskriterium ist diesbezüglich die Anzahl und die Länge der Nachrichten auf der Funkschnittstelle (siehe Abb. 4.2). Auch diese Nachrichten müssen aus den in Kapitel 2 genannten Gründen getaktet werden. Allerdings kann dieser LUP-Takt deutlich länger (ca. 10s) sein, da LUPs nicht so häufig notwendig sind.

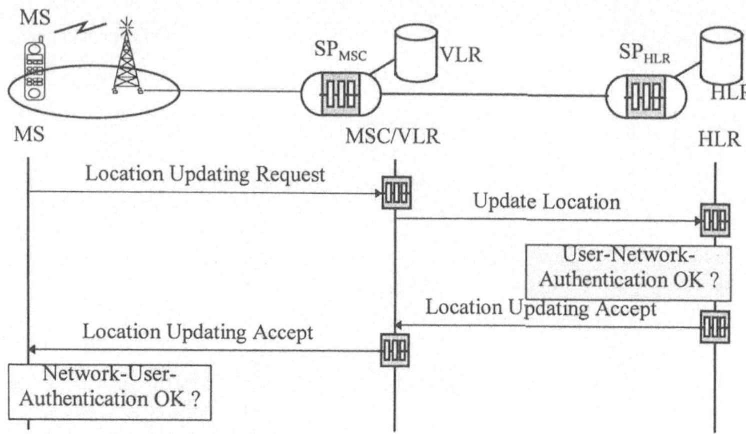

Abb. 4.2: Modifiziertes Protokoll des Location Updates

Die Mobilstation (MS) nutzt die Mix-Kaskade des MSCs, um eine anonyme Rückadresse $\{LAI,IAdr\}^4$ unbeobachtbar im VLR abzulegen. Durch die anonyme LUP-Request-Nachricht wird ein unbeobachtbarer Empfangskanal für die anschließende Quittung des Netzes aufgebaut. Der anonyme LUP-Request enthält außerdem Authentikationsdaten und die LUP-Nachricht an das HLR. Dem HLR wird diese vorbereitete Nachricht über dessen Mix-Kaskade signalisiert. Im HLR wird die Authentikation des Teilnehmers durchgeführt und anschließend dessen verdeckte Aufenthaltsinformation $\{VLR,P\}$ gespeichert. Die Quittung kann anschließend symmetrisch verschlüsselt zum MSC/VLR und weiter zum Teilnehmer signalisiert werden. Die Einzelheiten der verbesserten gegenseitigen Authentikation sind in [Müll_97] beschrieben.

Das Signalisierungsprotokoll des Verbindungsaufbaus (Call Setup) wird bei GSM in zwei Teile geteilt. Der erste Teil ist das Mobile Originated Call Setup (MOC-Setup), ein von der MS ausgehender Ruf, und der zweite Teil ist das Mobile Terminated Call Setup (MTC-Setup), ein die MS erreichender Ruf.

---

4   Die anonyme Rückadresse [LAI,IAdr} wird im VLR abgelegt und schreibt die Nutzung der Mix-Kaskade des MSCs bei der Übermittlung eines Verbindungswunsches vor. Diese Rückadresse wird in der Mix-Kaskade stufenweise entschlüsselt und gibt am Ende unverkettbar den Aufenthaltsort LAI des Teilnehmers und dessen dortige Kennung IAdr für eine Signalisierung bekannt.

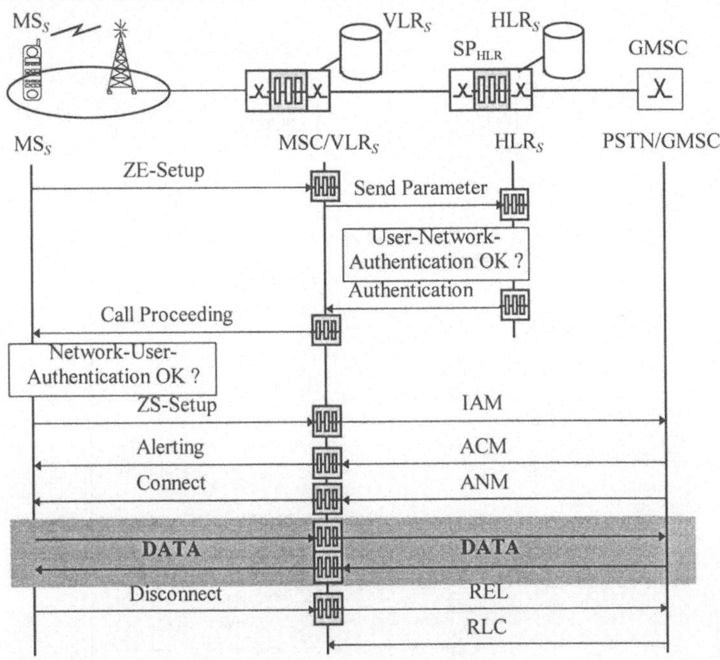

Abb. 4.3: Modifiziertes Protokoll des Mobile Originated Call Setup

Im folgenden sollen nur die Änderungen des MOC-Setup-Protokolls aus Abb. 4.3 beschrieben werden. Auch hier ist eine gegenseitige Authentikation vorgesehen (siehe [Müll_97]). Zuerst wird von S ein Signalisierungskanal (SDCCH) angefordert und ein ZE-Kanal über die Mix-Kaskade des MSC/VLR aufgebaut. Anschließend baut S den ZS-Kanal mit den Informationen über den Verbindungswunsch auf. Das MSC/VLR sendet darauf die IAM an das HLR$_E$. Dort wird die anonyme Rückadresse $\{VLR,P\}_E$ ermittelt und die IAM mit dieser Rückadresse über die Mix-Kaskade des HLRs gesendet und durch die in $\{VLR,P\}_E$ enthaltenen Informationen zum VLR$_E$ geroutet. Dabei kann noch kein Nutzkanal reserviert werden, da die Mix-Kaskade des HLRs nicht auf dem direkten Weg zum gewünschten VLR liegt und die spätere Nutzkanalkommunikation einen kürzestmöglichen Weg im Fernnetz nehmen soll. Nun muß der Sender warten, bis der Empfänger eine ACM auf dem direkten Weg sendet, wie anschließend im MTC beschrieben. Mit dieser Nachricht wird der bidirektionale Nutzkanal vollständig geschaltet. Im Netz wird bekannt, welches MSC/VLR mit welchem MSC/VLR kommuniziert. Die beteiligten Teilnehmer bleiben innerhalb der verwalteten Aufenthaltsgebiete unbeobachtbar. Während der Aufbauphase einer Verbindung bzw. wenn diese abgewiesen wird, erfährt ein Angreifer nichts über den ungefähren Aufenthaltsort (MSC/VLR-Bereich) von E. Bekannt ist im Netz lediglich, daß ein Teilnehmer aus dem MSC/VLR-Bereich von S einen der Teilnehmer erreichen möchte, die im HLR$_E$ eingetragen sind.

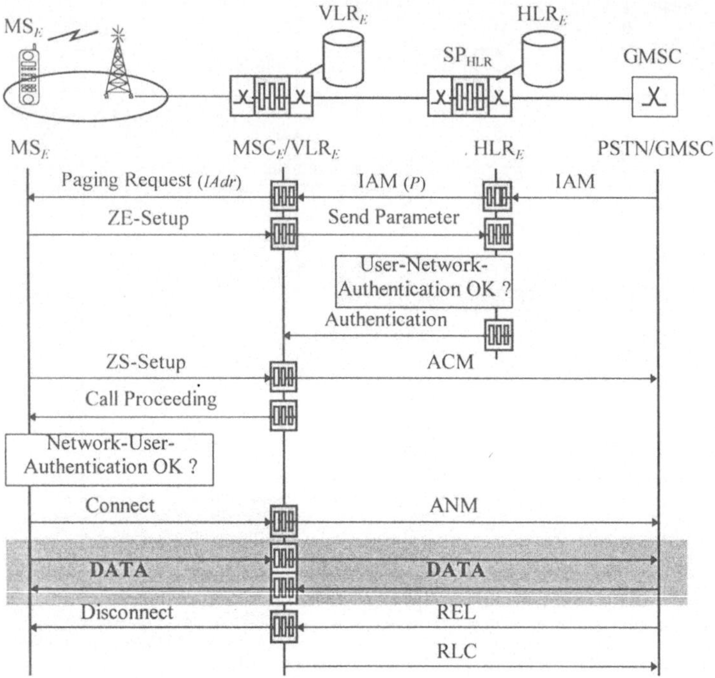

Abb. 4.4: Modifiziertes Protokoll des Mobile Terminated Call Setup

Abb. 4.4 zeigt das modifizierte Gegenstück des Verbindungsaufbaus, das MTC-Setup. Vom HLR_E wird die IAM mit der anonymen Rückadresse $\{VLR,P\}_E$ zum VLR_E weitergeleitet. Dort ist E mit dem Pseudonym P angemeldet und hat eine weitere anonyme Rückadresse hinterlegt, welche an die Mix-Kaskade im MSC des Empfängers zusammen mit dem Paging-Request signalisiert wird. Der letzte Mix dieser Kaskade erhält die Location Area Identification (LAI) des LAs, in dem der Paging-Request auszusenden ist. Der Empfänger erkennt an der signalisierten impliziten Adresse (*IAdr*), ob ein Verbindungswunsch für ihn ist. Ist dies der Fall, baut er, wie der Sender auch, den entsprechenden unbeobachtbaren ZE- und ZS-Kanal einschließlich Authentikation auf. Er sendet anschließend die ACM direkt zum ZE-Kanal im MSC/VLR des Senders und schaltet damit den Nutzkanal auch im Fernnetz auf dem direkten Weg.

Der Aufwand an Dummy Traffic, den eine anonyme Signalisierung erfordert, ist in GSM beträchtlich. Hauptproblem ist der Bandbreiteengpaß auf der Funkschnittstelle und der Einsatz von Dummy Traffic bei den Mobilstationen. Diese können sich nur an dem Dummy Traffic zum Schutz aller Teilnehmer eines MSC/VLR-Bereiches beteiligen, wenn sie nicht mit Akku betrieben werden. Den fehlenden Dummy Traffic aus einem LA gleicht die empfangende BTS aus. Dadurch entsteht im MSC/VLR kein zusätzliches Wissen, die BTS allerdings hat genauere Informationen über die Teilnehmer einer Anonymitätsgruppe und als Angreifer damit bessere Chancen zur Deanonymisierung einzelner Teilnehmer.

Auch für GSM soll beispielhaft nur die Modifikation einer komplexen Signalisierungsnachricht mit 3 Mixen angegeben werden (ausführlicher in [Müll_97]).

$$N_{\text{LUP-Request}} := c_1(t_{Li}, k_{SE1}, c_2(t_{Li}, k_{SE2}, c_3(t_{Li}, k_{SE3}, KKZ_S(Li), N_{\text{VLR})))$$

$$N_{\text{VLR}} := c_{MSC/VLR}(\text{lup\_msg}, t_{Li}, P, \{LAI, IAdr\}, Bv/Bl, N_{\text{HLR}})$$

$$N_{\text{HLR}} := c_{H1}(t_{Li}, k_{H1}, c_{H2}(t_{Li}, k_{H2}, c_{H3}(t_{Li}, k_{H3}, KKZ_{HLR}(Li), N_{\text{HLR}}^*)))$$

$$N_{\text{HLR}}^* := k_{HLRS}(\text{lup\_msg}, s_{MS}(t_{Li}, RAND_1), IMSI, RAND_2, Bv/Bl, t_{Li}, \{VLR, P\})$$

$$\{VLR, P\} := c_{H3}(k_{H3}, c_{H2}(k_{H2}, c_{H1}(k_{H1}, A_{VLR}, P)))$$

$$\{LAI, IAdr\} := c_3(k_{S3}, c_2(k_{S2}, c_1(k_{S1}, LAI, IAdr)))$$

Die Länge dieser LUP-Request-Nachricht erhöht sich bei der Realisierung in GSM am meisten, da der Teilnehmer diese Nachricht für die Benutzung von zwei Mix-Kaskaden vorbereiten und zwei anonyme Rückadressen für Verbindungswünsche in dieser Nachricht übertragen muß. Dennoch läßt sich das Verfahren mit geringen Modifikationen bei der Kanalvergabe und -nutzung (siehe Kapitel 5) effizient realisieren.

5. Performance-Betrachtungen

5.1 Voraussetzungen der Performanceanalyse

Festlegungen bezüglich der eingesetzten Kryptosysteme:

- Einsatz einer symmetrischen Stromchiffre,
- Einsatz einer asymmetrischen Blockchiffre,
- Einsatz von hybrider Verschlüsselung.[5]

Abb. 5.1 faßt die benötigten Parameter von ISDN und GSM zusammen. Diese wurden aus [BGGH_95] für ISDN und [GSM_04.08, GSM_09.02 und GSM_09.10] für GSM entnommen.

Parametername	Abk.	Länge	Parametername	Abk.	Länge
allgemeine Parameter:			Nutzkanalidentifikator	CIC	12 Bit
Zeitscheibennummer	t	30 Bit	Kanalkennzeichen	KKZ	28 Bit
Initialwert des Kanalkennzeichens	KZ_{init}	128 Bit	Block des asym. Systems	b_{asym}	660 Bit
Schlüssel des symmetrischen Systems	s_{sym}	128 Bit	digitale Signatur	$s_{MS}(N)$	200 Bit
bedeutungsvoll/-los Kennzeichen	bv/bl	1 Bit	Anzahl der MIXe	m	10 Stück
ISDN-Parameter:			Adresse eines Local Exchanges	A_{LE}	14 Bit
GSM-Parameter:			Implizite Adresse	$IAdr$	128 Bit
International Mobile Subscriber Identity	IMSI	60 Bit	Intern. Mobile Equipment Identity	IMEI	64 Bit
Temporary Mobile Subscriber Identity	TMSI	32 Bit	Mobile Subscriber ISDN-Number	MSISDN	60 Bit
Location Area Identifikation	LAI	48 Bit	Pseudonym	P	128 Bit

Abb. 5.1: Länge der Parameter in den Kommunikationsnetzen

Abb. 5.2 zeigt die Länge der Signalisierungsnachrichten bisher, sowie die berechnete Länge der modifizierten Nachrichten mit 10 eingesetzten Mixen pro Kaskade für ISDN und GSM. Die bisherige Länge im DSS1 wurde aus [ETS_300_403-1] ermittelt, im

---

[5] Bei hybrider Verschlüsselung wird von der zu verschlüsselnden Nachricht stets ein möglichst großer Teil N^* des Klartextes N in den ersten, asymmetrisch verschlüsselten Block hineingezogen und außerdem ein symmetrischer Schlüssel k_{SE} ausgetauscht, mit dem der Rest N^{**} effizient symmetrisch verschlüsselt werden kann.

CCS7-ISDN-UP aus [ITU_Q.763], im RIL3[6] aus [GSM_04.08] und im CCS7 Mobile Application Part (MAP) aus [GSM_09.02]. Parameter mit einer variablen Länge innerhalb bestimmter Grenzen mit ihrem Mittelwert und solche mit variabler Länge wurden mit dem Doppelten der Mindestlänge angenommen. Außerdem wurden auch optionale Parameter zur Längenberechnung als vorhanden eingestuft, so daß die Performance in der Praxis stets besser sein sollte, als im in Abb. 5.2 dargestellt.

Signalisier-system	Nachrichten	ISDN		GSM	
		bisher	10 Mixe	bisher	10 Mixe
DSS1 bzw. RIL3	Location Updating Request	—	—	148 Bit	7578 Bit
	Location Updating Accept	—	—	124 Bit	824 Bit
	Setup	1084 Bit	—	853 Bit	—
	ZE -Setup	—	2322 Bit	—	3914 Bit
	ZS-Setup	—	2536 Bit	—	2738 Bit
	Incoming Call	—	1040 Bit	—	—
	Call Proceeding	296 Bit	—	195 Bit	948 Bit
	Paging Request	—	—	176 Bit	1420 Bit
	Alerting	296 Bit	458 Bit	280 Bit	356 Bit
	Connect	404 Bit	530 Bit	448 Bit	356 Bit
	Disconnect	204 Bit	366 Bit	416 Bit	356 Bit
CCS7. ISDN-UP	IAM	976 Bit	1166 Bit	976 Bit	1594 Bit
	ACM	1076 Bit	470 Bit	1076 Bit	470 Bit
	ANM	1216 Bit	478 Bit	1216 Bit	478 Bit
	REL	912 Bit	374 Bit	912 Bit	374 Bit
CCS7. MAP	Send Parameter from HLR	—	—	64 Bit	2098 Bit
	Authentication Parameter	—	—	324 Bit	516 Bit
	Update Location	—	—	120 Bit	3863 Bit
	Location Updating Accept	—	—	54 Bit	324 Bit

Abb. 5.2: Länge der modifizierten Signalisierungsnachrichten im Vergleich

5.2 Minimale Dauer des Systemtaktes

Jeder Teilnehmer muß für jeden Systemtakt und jeden zur Verfügung stehenden Nutzkanal eine ZE- und eine ZS-Setup-Nachricht signalisieren. Da es sich bei dem Signalisierungskanal in ISDN und GSM um einen Vollduplexkanal handelt, hat die Verteilung der Verbindungswünsche keinen Einfluß auf die Größe der Zeitscheibe. Ein Teil des Signalisierungskanals wird zur Sicherung der Daten, für einige schmalbandige Dienste benötigt. Daher soll im folgenden stets von einer etwas geringeren Datenrate ausgegangen werden. Die minimale Zeitscheibenlänge z berechnet sich wie folgt:

$$z \geq \frac{Anz_{Nutzkanäle} \cdot \left(\left| N_{ZE-Setup} \right| + \left| N_{ZS-Setup} \right| \right)}{Signalisierungskapazität}$$

Bei einem Einsatz von 10 Mixen pro Kaskade erhält man für ISDN (2 Nutzkanäle, 12kbit/s) $z \geq 0{,}81$s und für GSM (1 Nutzkanal, 0,6kbit/s) $z \geq 11{,}09$s.

In GSM existieren zur Signalisierung von Verbindungswünschen und LUPs zwei Signalisierungskanäle (SDCCH mit 0,782kbit/s und SACCH mit 0,391kbit/s). Der

---

[6] Radio Interface Layer 3 Protokoll auf der Funkschnittstelle in GSM.

SDCCH soll mit 0,6kbit/s für den modifizierten Verbindungsaufbau und mit 0,18kbit/s für das modifizierte LUP genutzt werden. Der SACCH bleibt für die übrigen Signalisierungsaufgaben frei. Zur Verbesserung der Performance werden folgende Modifikationen in der Kanalvergabe und -nutzung vorgeschlagen: Der Signalisierungskanal FACCH wird bisher in GSM nur Nutzkanälen zugeordnet. Bei Aerly-TCH-Assignment wird der FACCH belegt, jedoch bisher nicht für die Signalisierung zum Verbindungsaufbau verwendet. Die Modifikation besteht darin, den ungenutzten FACCH zum Verbindungsaufbau zu verwenden. Der FACCH kann als Halfrate- (/H) mit 4,6kbit/s bzw. als Fullrate-Kanal (/F) mit 9,2kbit/s betrieben werden, wobei wiederum eine Aufteilung in Verbindungsaufbau und LUP erfolgt. Da der FACCH diese Aufgaben übernimmt, steht der SDCCH für andere Aufgaben zur Verfügung. Es wird folgende Aufteilung angenommen: FACCH/H 3,5kbit/s für Verbindungsaufbau und 1,1kbit/s für LUP bzw. FACCH/F 7,5kbit/s und 1,7kbit/s. Für letzteres beträgt die minimale Zeitscheibenlänge $z \geq 0{,}89s$.

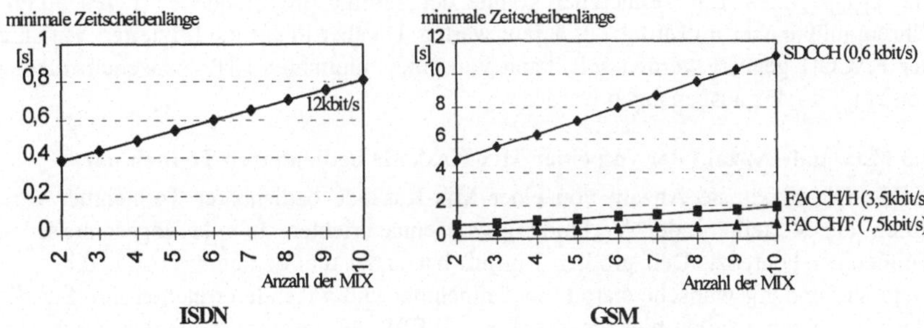

Abb. 5.3: Abhängigkeit der Zeitscheibenlänge von der Mix-Anzahl

In GSM muß zusätzlich die minimale Länge des LUP-Taktes z_{LUP} berechnet werden. Die LUP-Häufigkeit hängt von mehreren Faktoren ab, z.B. von der Durchschnittsgeschwindigkeit v der MS, dem Radius r der Funkzellen und der Anzahl der Funkzellen pro Location Area N_{LA}. In [FuBr_94] wird für die durchschnittliche Anzahl von LUPs λ_{LUP} pro MS während der Stoßzeiten

$$\lambda_{LUP} = \frac{v}{\pi \cdot r} \cdot \left(1 - \frac{N_{LA} - 1}{2 \cdot N_{LA}}\right)$$

angegeben. Es gilt $z_{LUP} \leq \dfrac{\left|N_{LUP-Request}\right|}{Signalisierungskapazität_{LUP}}$.

Abb. 5.4 links zeigt die durchschnittliche LUP-Rate in Abhängigkeit von dem Zellradius und der Anzahl der Funkzellen pro LA. Dabei wird von einer Durchschnittsgeschwindigkeit analog zu [FuBr_94] von 15 km/h ausgegangen.

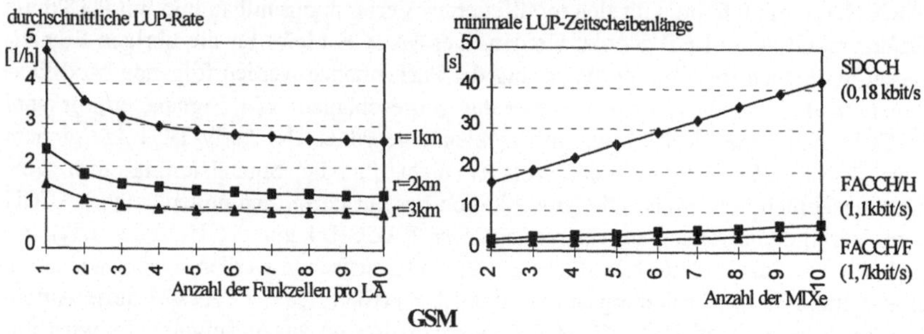

Abb. 5.4: Abhängigkeit der LUP-Zeitscheibenlänge von der Mix-Anzahl

Würde man nur den SDCCH mit 0,18 kbit/s für die LUP Signalisierung nutzen, so erhielte man für die minimale LUP-Zeitscheibe nach obenstehender Formel einen Wert von $z_{LUP} \leq 42,1s$. Ein Teilnehmer könnte demzufolge pro Stunde 85 LUPs durchführen, müßte aber im Mittel 21s darauf warten. Da aber in der modifizierten Variante der FACCH genutzt werden soll, kann von einer minimalen LUP-Zeitscheibenlänge von $z_{LUP} \leq 4,46s$ ausgegangen werden.

5.3 Maximale Anzahl der von einer Mix-Kaskade bedienbaren Teilnehmer

ISDN: Die maximale Anzahl von einer Mix-Kaskade bedienbarer Teilnehmer wird durch die Verteilung der Verbindungswunschnachrichten an alle angeschlossenen Teilnehmer begrenzt. Den größten Einfluß hat dabei die Verkehrsstatistik, d.h. wieviele Verbindungswünsche startet ein Teilnehmer und wieviele erreichen ihn. Für die Beispielrechnung sollen hier die Werte aus [PfPW_89] verwendet werden. Dort wird von einer maximalen Ankunftsrate von 12 Verbindungswünschen pro Teilnehmer und Stunde ($\lambda = 1/300$ /s) in Stoßzeiten ausgegangen. Dieser Wert ist sehr hoch gewählt, in der Literatur findet man beispielsweise auch durchschnittliche Werte von 2 pro Stunde (in [PoMG_95]) bis 0,8 pro Stunde (in [FuBr_94]. Außerdem ist für die Realisierung von anonymer Signalisierung nur die Rate ankommender Verbindungswünsche interessant.

Es können maximal $\mu := b/N$ Verbindungswunschnachrichten pro Sekunde verteilt werden. Für ISDN (b=12kbit/s, N=1040Bit) erhält man $\mu = 11,54$ /s. T_V ist die Zeit, die ein Verbindungswunsch im Mittel benötigt, um bei den Netzabschlüssen einzutreffen (nach [PfPW_89] aus [Klei_75]), und berechnet sich wie folgt:

$$T_V = \frac{2 \cdot \mu - n \cdot \lambda}{2 \cdot \mu \cdot (\mu - n \cdot \lambda)}. \text{ Daraus ergibt sich für n: } n = \frac{\mu}{\lambda} \cdot \frac{\mu \cdot T_v \cdot -1}{\mu \cdot T_v - 0,5}.$$

Für $T_V \leq 0,5s$ erhält man $n \leq 3133$ Teilnehmer pro Mix-Kaskade. Geht man von nur 5 Verbindungswünschen pro Teilnehmer und Stunde (λ=1/720 /s) in Stoßzeiten aus, könnte man bereits 7520 Teilnehmer an jede Mix-Kaskade anschließen.

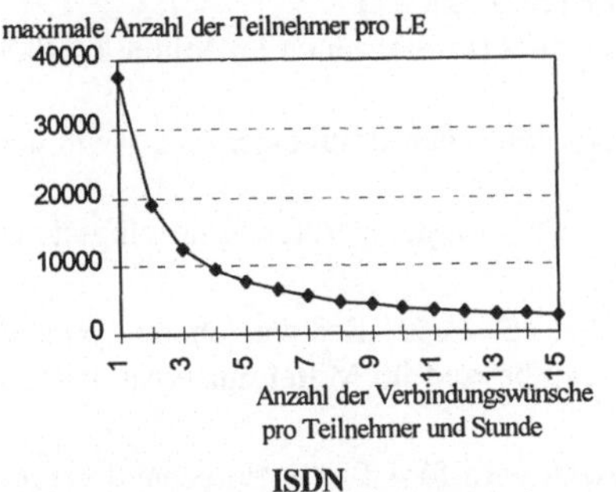

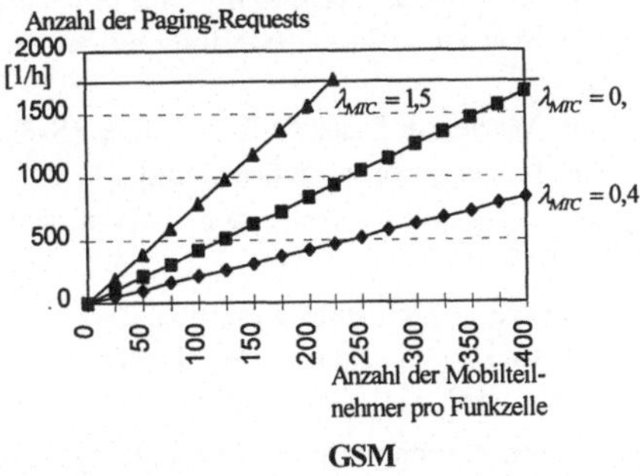

Abb. 5.5: Anzahl der bedienbaren Teilnehmer

GSM: In GSM wird die Aussendung eines Verbindungswunsches Paging genannt. Die Paging Request-Nachricht wird bei einem MTC von der BTS ausgesendet, um den Empfänger eines Verbindungswunsches davon zu unterrichten. Diese Nachricht ist 1420 Bit lang. Es wird der Signalisierungskanal PCH (0,782kbit/s simplex downlink) verwendet. Wenn davon 0,7 kbit/s genutzt werden können, dauert die Übermittlung einer Verbindungswunschnachricht 2,03s. Die durchschnittliche Anzahl notwendiger Paging-Nachrichten in einer Funkzelle hängt wiederum von verschiedenen Größen ab. In [FuBr_94] wird für die durchschnittliche Paging Request-Rate

$$\lambda_{PAGING} = N_{MS} \cdot \lambda_{MTC} \cdot \left(A_P \cdot \left(N_{LA} - 1 \right) + 1 + \frac{\left(A_F - 1 \right) \cdot P_{KF}}{1 + P_{KF} + P_{KA}} \right)$$

angegeben, wobei N_{MS} die durchschnittliche Anzahl der Mobilteilnehmer pro Funkzelle, λ_{MTC} die durchschnittliche Anzahl der ankommenden Verbindungswünsche, A_F die maximale Anzahl der Paging Request Versuche pro MTC in der besuchten Funkzelle (A_F=2), A_P die Anzahl der Paging Request Versuche in jeder anderen Funkzelle des besuchten LAs (A_P=2) und N_{LA} die Anzahl der Funkzellen pro LA ist (N_{LA}=3). P_{KF} ist die Wahrscheinlichkeit, daß kein Funkkontakt möglich ist (P_{KF}=0,4) und P_{KA} die Wahrscheinlichkeit, daß eine MS nicht antwortet (P_{KA}=0,3).

In Abb. 5.5 rechts wird die Anzahl der notwendigen Paging Requests innerhalb einer Funkzelle in Abhängigkeit von der Anzahl der Mobilteilnehmer pro Funkzelle und deren Verkehrscharakteristik λ_{MTC} gezeigt. Da die Übermittlung einer Paging Request-Nachricht 2,03s dauert, können in einer Stunde nicht mehr als 1773 Paging-Nachrichten signalisiert werden. Dadurch wird auch die maximale Anzahl der versorgbaren Mobilteilnehmern begrenzt, wobei deren Verkehrsstatistik und die Anzahl der Funkzellen pro LA sowie deren Durchmesser ebenfalls diese Größe beeinflussen.

5.4 Verbindungsaufbauzeit

ISDN: Die Berechnung der Verbindungsaufbauzeit für ISDN wurde aus [PfPW_89] entnommen. Sie setzt sich im wesentlichen zusammen aus:

- dem Warten auf die Übermittlung des Verbindungswunsches durch die Mix-Kaskade der Ortsvermittlungsstelle des Senders ($\leq z$),
- der Verzögerungszeit der Mix-Kaskade der Senderortsvermittlungsstelle ($m \cdot 0,01$ s),

- der Laufzeit des Verbindungswunsches im Fernnetz ($\leq$0,2 s),
- dem Warten auf die Verteilung an den Empfänger (in Stoßzeiten im Mittel $T_V\leq$0,5 s),
- dem Warten auf das Aufbauen des ZS-Kanals durch den Empfänger ($\leq z$, wenn der Empfänger sofort antwortet) und
- der Verzögerungszeit der Mix-Kaskade der Empfängerortsvermittlungsstelle ($m\cdot$0,01 s).

Daraus ergibt sich eine Zeit von $2\cdot(z+m\cdot0,01s)+T_V+0,2s$. Bei den obigen Werten ($z = 0,8s$) also ca. 2,5s, wobei die Verbindungsaufbauzeit im Mittel nur etwa halb so groß sein wird.

GSM: Die Berechnung der Verbindungsaufbauzeit wird für GSM entsprechend vorgenommen und setzt sich im wesentlichen zusammen aus:

- der gegenseitigen Authentikation von Sender und Netz ($\leq z$),
- dem Warten auf die Übermittlung des Verbindungswunsches durch die Mix-Kaskade des MSCs des Senders ($\leq z$),[7]
- der Verzögerungszeit der Mix-Kaskade des MSCs des Senders ($m\cdot$0,01 s),
- der Laufzeit des Verbindungswunsches im Fernnetz zum HLR des Empfängers ($\leq$0,2 s),
- dem Warten auf die Übermittlung des Verbindungswunsches durch die Mix-Kaskade des HLRs des Empfängers ($\leq z$),
- der Verzögerungszeit der Mix-Kaskade im HLR des Empfängers ($m\cdot$0,01 s),
- der Laufzeit des Verbindungswunsches im Fernnetz zum VLR der Empfängers ($\leq$0,2 s),
- dem Warten auf die Übermittlung des Verbindungswunsches durch die Mix-Kaskade des MSC/VLRs des Empfängers ($\leq z$),
- der Verzögerungszeit der Mix-Kaskade im MSC/VLR des Empfängers ($m\cdot$0,01 s),
- dem Warten auf die Verteilung an den Empfänger (in Stoßzeiten im Mittel $T_V\leq$5 s),
- der gegenseitigen Authentikation von Empfänger und Netz ($\leq z$),
- dem Warten auf das Aufbauen des ZS-Kanals durch den Empfänger ($\leq z$, wenn der Empfänger sofort antwortet) und
- der Verzögerungszeit der Mix-Kaskade des MSCs des Empfängers ($m\cdot$0,01 s).

Daraus ergibt sich eine Zeit von $4\cdot(z+m\cdot0,01)+2\cdot z+2\cdot0,2+T_V$. Bei den obigen Werten (z=0,9s) also ca. 11,2s, wobei die Verbindungsaufbauzeit im Mittel nur etwa halb so groß sein wird. Dieser Wert zeigt deutlich, daß die Bandbreite auf den Signalisierungskanälen in GSM erhöht werden müßte.

6. Ausblick

Bereits mit den Bandbreitemöglichkeiten der bestehenden Kommunikationsnetze läßt sich Unbeobachtbarkeit und Anonymität bei der Signalisierung von Verbindungswünschen und Location Updates realisieren. Im Nutzkanal entstehen keine Datenrateeinbußen. Für die Erklärung der Verfahren zur Erreichung von Teilnehmerunbeobachtbar-

---

[7] Der Aufbau von zugeordneten Signalisierungs- und Nutzkanälen auf der Funkschnittstelle gehört zu diesem Punkt.

keit und der beispielhaften technischen Realisierung wurden das Festnetz ISDN und das Mobilfunknetz GSM (D1, D2, E-plus) gewählt.

Im ISDN wurden, ausgehend von der Idee der Telefon-Mixe, die Netzstruktur und die Protokolle entsprechend modifiziert. In den Effizienzbetrachtungen wurde nachgewiesen, daß sich Teilnehmerunbeobachtbarkeit bei der Signalisierung von Verbindungswünschen mit vertretbarem Aufwand erreichen läßt. Es wurden allerdings nur die Basisabläufe des Verbindungsaufbaus untersucht und modifiziert. Zusatzdienste, wie Anrufweiterleitung, Anklopfen und Halten, wurden nicht analysiert. Es ist allerdings zu erwarten, daß sich ein großer Teil der Zusatzdienste auch trotz Unbeobachtbarkeit und Anonymität der Nutzer realisieren läßt.

In GSM mußten die Verfahren zum Schutz der Signalisierungsbeziehungen an die Teilnehmermobilität angepaßt werden. Auch für GSM wurden Modifikationen der Netzstruktur und der Verbindungsaufbau- bzw. LUP-Protokolle entwickelt und diskutiert, wobei durch die schmalbandigen Signalisierungskanäle nicht die Leistungsfähigkeit erreicht werden konnte wie im ISDN. Außerdem können sich akkubetriebene Mobilstationen nicht am notwendigen Dummy Traffic auf der Funkschnittstelle beteiligen. Dadurch verringert sich die erreichbare Unbeobachtbarkeit erheblich, da eine MS der BTS vertrauen muß, daß diese den notwendigen Dummy Traffic im Festnetz erzeugt. In der Leistungsbewertung wurde ermittelt, daß die Realisierung zu spürbaren Einbußen in der Dienstqualität führt. Die Wartezeiten können allerdings durch die beschriebene modifizierte Kanalvergabe und -verwendung deutlich reduziert werden, so daß sich auch diese Lösungen praktizieren lassen. Nicht untersucht wurde in GSM das Handover und die Zusatzdienste.

Die Methode zum Schutz der Signalisierungsbeziehungen wurde zu Beginn dieser Arbeit kurz allgemeingültig beschrieben. Eine Adaption auf andere Kommunikationsnetze, wie Asynchronous Transfer Mode (ATM) im Festnetzbereich oder Universal Mobile Telecommunication System (UMTS) im Mobilkommunikationsbereich ist deshalb möglich.

Die hier dargestellten Konzepte zeigen Möglichkeiten, die Privatsphäre der Nutzer und deren informationelle Selbstbestimmung zu gewährleisten. Dies sollte bei der rasanten Entwicklung neuer Kommunikationstechnologien und -dienste nicht vernachlässigt werden.

Literaturverzeichnis

BGGH_95   Gerhard Bandow et.al.: Zeichengabesystems – Eine neue Generation für ISDN und intelligente Netze; L.T.U.-Vertriebsgesellschaft, Bremen, 1995.

Chau_81   D. Chaum: Untraceable Electronic Mail, Return Addresses, and Digital Pseudonyms; Communications of the ACM 24/2 (1981) 84-88.

Chau_88   D. Chaum: The Dining Cryptographers Problem: Unconditional Sender and Recipient Untraceability; Journal of Cryptology 1/1 (1988) 65-75.

Cha8_85   D. Chaum: Security without Identification: Transaction Systems to make Big Brother Obsolete; Communications of the ACM 28/10 (1985) 1030-1044.

ETS_300_403-1 ETSI: ISDN, DSS1 protocol, Signalling network layer for circuit-mode basic call control, Part 1 Protocol specification; November 1995.

FeTh_95   Hannes Federrath, Jürgen Thees: Schutz der Vertraulichkeit der Aufenthaltsortes von Mobilfunkteilnehmern; Datenschutz und Datensicherung, DuD 6 (1995) 338-348.

FuBr_94   Woldemar F. Fuhrmann, Volker Brass: Performance Aspects of the GSM Radio Subsystem; Proceedings of the IEEE, Vol. 82, No. 9, September 1994.

GSM_04.08 ETSI: ETSI/TC GSM: 04.08 Mobile Radio Interface Layer 3 Specification; Version 4.10.1; February 1995.

GSM_09.02 ETSI: ETSI/TC GSM: 09.02 Mobile Aplication Part (MAP) specification; Version 4.9.1; February 1995.

GSM_09.10 ETSI: ETSI/TC GSM: Information element mapping between MS - BSS and BSS-MSC Signalling procedures and the MAP; Version 4.2.2; February 1995.

Klei_75   Leonard Kleinrock: Queueing Systems - Volume 1: Theory; Wiley, New York 1975.

Müll_97   Jan Müller: Anonyme Signalisierung in Kommunikationsnetzen; Diplomarbeit, TU Dresden, Institut für Theoretische Informatik, 1997.

PfPf_89   Andreas Pfitzmann, Birgit Pfitzmann: How to Break the Direct RSA-Implementation of Mixes; Eurocrypt '89, LNCS, Springer-Verlag, Berlin, 1989.

PfPW_89   Andreas Pfitzmann, Birgit Pfitzmann, Michael Waidner: Telefon-Mixe: Schutz der Vermittlungsdaten für zwei 64-kbit/s-Duplexkanäle über den (2o64+16)-kbit/s-Teilnehmeranschluß; Datenschutz und Datensicherung DuD, 12 (1989) 605-622.

PoMG_95   Gregory P. Pollini, Kathleen S. Meier-Hellstern, David J. Goodman: Signalling Traffic Volume Generated by Mobile and Personal Communications; IEEE Communications Magazine, 6 (1995) 60-65.

Über die Modellierung steganographischer Systeme[*]

J.Zöllner, H.Federrath**, A.Pfitzmann**, A.Westfeld**, G.Wicke**, G.Wolf**

Technische Universität Dresden, 01062 Dresden
**Institut für Betriebssysteme, Datenbanken und Rechnernetze*
***Institut für Theoretische Informatik*
{zoellner, federrath, pfitza, westfeld, wicke, g.wolf}@inf.tu-dresden.de

Zusammenfassung

Nach einer kurzen Einführung in die Steganographie und der Abgrenzung zu kryptographischen Systemen werden verschiedene Modellierungsmöglichkeiten für steganographische Systeme vorgestellt und hinsichtlich ihrer Allgemeingültigkeit diskutiert. Es wird ein allgemeines Modell für steganographische Konzelationssysteme abgeleitet. Weiterhin werden Bedingungen für sichere Steganographie formuliert.

1 Einführung

Sicherheitsanforderungen an Kommunikationssysteme werden meist durch eine Kombination von Vertraulichkeits-, Integritäts- und Verfügbarkeitseigenschaften beschrieben. Die Einhaltung bzw. Realisierung dieser Sicherheitseigenschaften wird durch Sicherheitsmechanismen garantiert. Mit Hilfe kryptographischer Systeme können vor allem die Eigenschaften Vertraulichkeit und Integrität gewährleistet werden. In diesem Kontext spielen vor allem Konzelationssysteme und Authentikationssysteme eine bedeutende Rolle. Konzelation beschreibt Funktionen zur Sicherung der *Vertraulichkeit des Inhaltes* geheimer Daten, während Authentikation die Richtigkeit (und Echtheit) von Information sicherstellen soll. Die zugrundeliegenden Mechanismen arbeiten mit der Verschlüsselung von Daten. Dieser Problematik widmet sich insbesondere die Kryptographie. Demgegenüber beschreibt Steganographie Funktionen zur Sicherung der *Vertraulichkeit der Existenz* von geheimen Daten.

[*] Wir danken den Teilnehmern der „Stegorunde" in Dresden, die nicht als Autoren auf diesem Papier erscheinen. Viele der hier dargestellten Ideen wurden in gemeinsamen Diskussionen erarbeitet und präzisiert. Diese Arbeit wurde finanziell unterstützt vom Bundesministerium für Bildung, Wissenschaft, Forschung und Technologie (BMBF) sowie der Gottlieb-Daimler- und Karl-Benz-Stiftung Ladenburg.

1.1 Steganographie - was ist das?

Bruce Schneier kennzeichnet Steganographie folgendermaßen [Schn_96, S. 10]: „Steganographie hat den Zweck, Nachrichten in anderen Nachrichten zu verstecken, um die bloße Existenz einer geheimen Botschaft zu verbergen". Als historische Beispiele nennt er „...unsichtbare Tinte, winzige Einstiche in ausgewählten Buchstaben, kleinste Unterschiede in handgeschriebenen Zeichen, handschriftliche Markierungen auf getippten Buchstaben...".

Steganographie ist also nicht neu, sie erfährt jedoch durch den Einsatz von Computer- und Multimediatechnik eine Renaissance. Besonders gilt dies für Grafik- und Audiodaten. In Abbildung 1 wird die Anwendung von Steganographie auf Grafikdaten veranschaulicht.

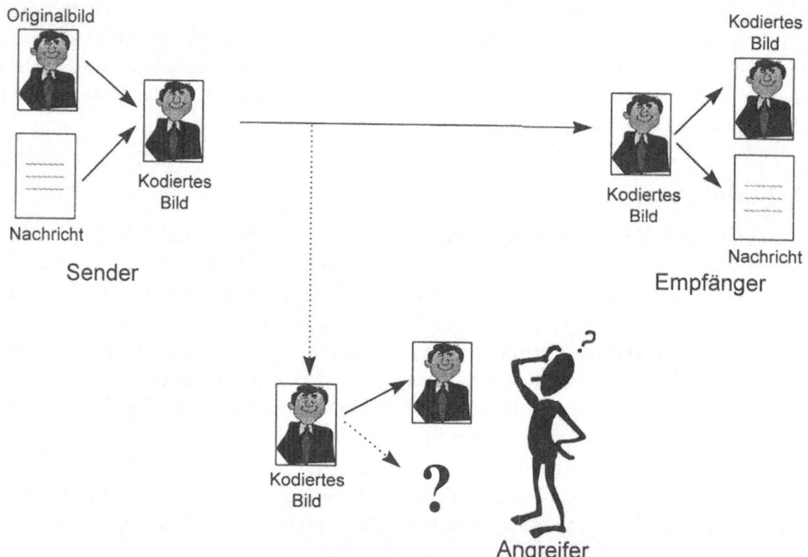

Abbildung 1: Steganographie am Beispiel von Grafikdaten

Auf der linken Seite ist zu sehen, daß der Sender der Nachricht diese in eine Grafikdatei einbettet. Die hier als „kodiertes Bild" bezeichnete modifizierte Grafik wird zum rechts symbolisierten Empfänger übertragen. Dabei wird sie durch den Angreifer (unten) abgehört. Der Empfänger kann die eingebettete Nachricht aus dem kodierten Bild extrahieren, der Angreifer nicht. Das gelingt selbstverständlich nur, wenn zwischen Sender und Empfänger ein gemeinsames „Geheimnis" existiert. Dies könnte der

Algorithmus zum Extrahieren sein oder bestimmte Algorithmenparameter, z.B. Schlüssel.

Wie sich bei Kryptosystemen schon oft zeigte, können Varianten, die sich auf die Geheimhaltung des Algorithmus zur Realisierung der Konzelation verlassen, in offenen Anwendungen nicht als langfristig sicher betrachtet werden. Vor allem bei großer Teilnehmerzahl ist die Geheimhaltung des Algorithmus schwierig. Es ist also notwendig, den (öffentlichen) Algorithmus mittels eines Schlüssels zu parametrisieren. Die Geheimhaltung dieses Schlüssels ist analog zu kryptographischen Systemen: sie ist essentiell für die sichere Anwendung des Systems.

Bei steganographischen Systemen kann man zwischen zwei Varianten mit unterschiedlichen Zielrichtungen unterscheiden:

1. Steganographie zum vertraulichen und versteckten Datenaustausch (siehe Abbildung 1) und

2. sogenanntes Watermarking, dessen Ziel es ist, mittels der eingebetteten Informationen Urheberschaft von digitalen Daten nachzuweisen.

Beide Varianten zielen darauf ab, die Originaldaten möglichst geringfügig zu verändern. Während jedoch bei 1. die Geheimhaltung der eingebetteten Daten oberste Priorität hat, kommt es bei den Watermarkingsystemen auf die robuste Anbringung des eingebetteten Kennzeichens an. Es darf durchaus sehr einfach nachweisbar sein, seine Entfernung soll jedoch ohne signifikante Beeinträchtigung der Originaldaten für alle außer den Eigentümer schwierig sein. Watermarking wird in diesem Papier nicht betrachtet.

1.2 Vertraulichkeit durch Geheimhaltungssysteme

Vertrauliche Nachrichtenübermittlung ist sowohl mit Steganographie als auch mit Kryptographie möglich. In beiden Fällen spielt Geheimhaltung eine bedeutende Rolle. Bei Konzelation im Sinne von Inhaltsdatenverschlüsselung soll eine geheime Botschaft vor einem Außenstehenden (Angreifer) vertraulich bleiben, während bei Steganographie sogar die Existenz einer solchen Botschaft verborgen bleiben soll.

Entsprechend lassen sich zwei Eigenschaften von Geheimhaltung definieren:

1. Konzelationseigenschaft: Ein Angreifer ist ohne Kenntnis eines Geheimnisses nicht in der Lage, an den Inhalt einer geheimen Botschaft zu gelangen.

2. Steganographische Eigenschaft: Ein Angreifer ist ohne Kenntnis eines Geheimnisses nicht in der Lage, in einer offenen Übermittlung (von Daten) die Existenz einer verborgenen Nachricht zu entdecken.

Durch diese Eigenschaften läßt sich beschreiben, wann ein Geheimhaltungssystem gebrochen ist. Bei Konzelationssystemen ist dies einfach: Ein Konzelationssystem ist gebrochen, wenn die Konzelationseigenschaft verletzt ist. Im Gegensatz dazu ist dieses Brechen bei einem steganographischen Konzelationssystem zweistufig:

1. Stufe: Ein steganographisches System ist gebrochen, wenn die steganographische Eigenschaft verletzt ist. Dies bedeutet jedoch nicht, daß der Angreifer den Inhalt der verborgenen Nachricht besitzt. Erst in einer zweiten Stufe kommt er in Kenntnis des Inhaltes:

2. Stufe: Zusätzlich gelingt es dem Angreifer, den Inhalt der verborgenen Nachricht aufzudecken. Dies entspricht der Verletzung der Konzelationseigenschaft.

Wir wollen annehmen, daß ein Brechen der 1. Stufe bereits genügt, um ein steganographisches System als unsicher zu bezeichnen. Ist das Brechen der 2. Stufe erfolgt, ist automatisch auch die 1. Stufe gebrochen, d.h. es ist nicht möglich, die Konzelationseigenschaft zu verletzen, ohne auch die steganographische Eigenschaft zu verletzen. Anders formuliert: Die steganographische Eigenschaft ist die weitergehende Eigenschaft eines Geheimhaltungssystems. Folglich muß jedes Stegosystem auch ein Konzelationssystem sein.

1.3 Was bezwecken Modelle?

Um das Phänomen Steganographie zu erfassen und Eigenschaften steganographischer Systeme beschreiben zu können, werden im folgenden Modelle aufgestellt. Als abstrakte Systeme beschreiben sie nicht nur die wesentlichen Objekte (Sender, Empfänger, Angreifer) und ihre Beziehungen untereinander (z.B. „unentdeckt kommunizieren"), sondern sie erheben den Anspruch, eine homomorphe Abbildung des betrachteten Ausschnitts der Welt der sicheren Kommunikation zu sein. In [Klir_89] findet man einen Überblick über die Systemmodellierung.

Dies geschieht mit dem Ziel zu zeigen, was einerseits notwendige, andererseits auch hinreichende Bedingungen für diese besondere Art der vertraulichen Kommunikation sind, so daß letztlich Aussagen darüber getroffen werden können, ob ein gegebenes steganographisches System dem allgemeinen Modell entspricht und die Modellannahmen, insbesondere zur Sicherheit, erfüllt sind. Ein weiterer Aspekt der Modellierung ist es, Steganographie in den größeren Zusammenhang der Geheimhaltung zu stellen.

Während der Modellierung eines allgemeinen steganographischen Systems wird besonderer Wert auf die Abgrenzung der Objekte gelegt. Das geschieht mit dem Ziel, die Außensicht auf das System mit der Sicht des Angreifers abzustimmen.

2 Modellierungsmöglichkeiten für steganographische Konzelationssysteme

2.1 Das Einbettungsmodell

Ausgangspunkt der folgenden Überlegungen zur Modellierung steganographischer Konzelationssysteme soll Abbildung 2 sein, eine leichte Erweiterung von [Pfit_96].

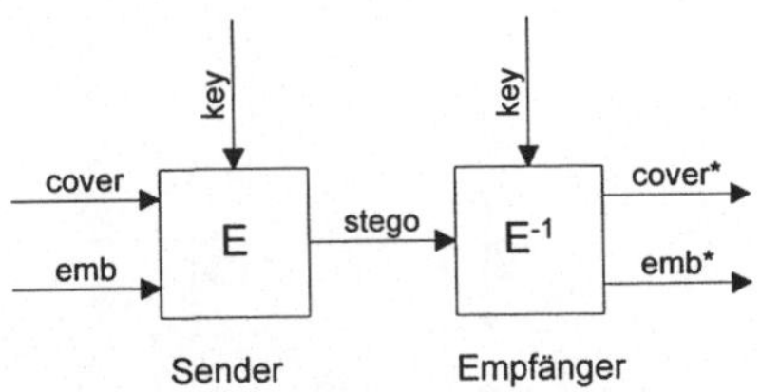

E: steganographische Funktion „Einbetten"
E^{-1}:   steganographische Funktion „Extrahieren"
cover: Hüllnachricht
emb: einzubettende Nutzdaten
stego: Hüllnachricht mit eingebetteten Daten
key: Parameter für E und E^{-1}
cover*: Hüllnachricht nach dem Extrahieren (meist: cover* = stego)
emb*: erhaltene Nutzdaten nach dem Extrahieren (Ziel: emb* = emb)

Abbildung 2: Das Einbettungsmodell

Die im Folgenden angeführten Modelle sind prinzipiell Erweiterungen dieses Modells oder beinhalten detailliertere Darstellungen des Schrittes E, wobei die Abgrenzung „gehörig zu Schritt E oder nicht" oft nicht leichtfällt. Im Schritt E wird die zu verbergende Nachricht *emb* in die Hüllnachricht eingebettet. Das Ergebnis dieser Operation ist das kodierte Bild *stego* (vgl. Abbildung 1). E^{-1} ist die Umkehroperation zu E, die *emb* wieder extrahiert. Es gilt also *stego* = E(*cover*, *emb*, *key*) und $emb^* = E^{-1}(stego,$ *key*). Dabei wird natürlich angestrebt, daß *emb* und emb^* identisch sind, da sonst das Einbetten nicht umkehrbar ist.

In diesem Modell wird vorausgesetzt, daß E beim Sender ausgeführt wird und E^{-1} beim Empfänger der Nachricht. Dazwischen findet eine Übertragung von *stego* zum Empfänger statt. Diese Annahmen sind auch für alle folgenden Modelle gültig.

Da es momentan noch keine echte asymmetrische Steganographie gibt (Ansätze sind in [Ande_96] und [HuPf_96] zu finden), befassen wir uns auch bei der Modellbildung ausschließlich mit symmetrischen Systemen. Aus diesem Grunde wird in Abbildung 2 im Gegensatz zu *cover* und *emb* auch keine Unterscheidung zwischen den in E und E^{-1} verwendeten Schlüsseln getroffen.

Anmerkung: Es ist ohne weiteres ein Stegosystem vorstellbar, das ohne *cover* auskommt, indem es sich ein in E zu verwendendes *cover* generiert bzw. alleine aus *emb* ein *stego* erzeugt (das dann implizit ein *cover* enthält). Daraus ergibt sich zwangsläufig die Frage: Braucht man für ein allgemeines Modell ein *cover*?

In einem solchen Modell (vgl. Abbildung 3) kann *stego* als *emb* plus eine bestimmte hinzugefügte Datenmenge (Redundanz) betrachtet werden. Diese hinzugefügten Daten müssen so gestaltet sein, daß das resultierende *stego* einem Angreifer plausibel erscheint. Ein Beispiel dafür sind z.B. Programme, die in fraktalen Bildern Daten unterbringen.

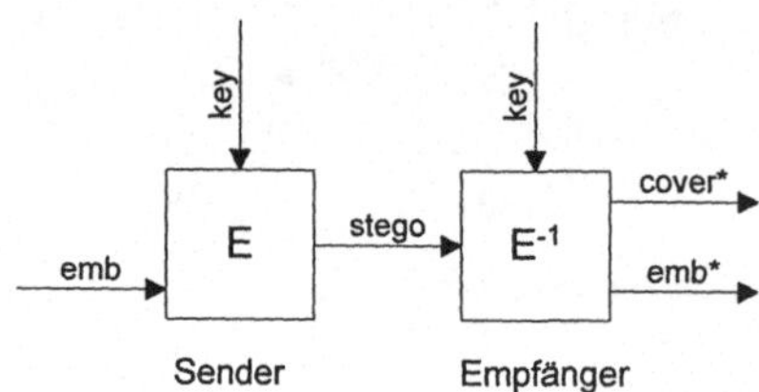

Abbildung 3: Modell eines Stegosystems ohne die Eingangsgröße cover

Dieses Modell setzt die Möglichkeit voraus, *cover* weitgehend frei generieren zu können. Ein allgemeines Modell muß aber alle Anwendungsfälle abdecken und kann daher nicht von dieser speziellen Freiheit ausgehen. Das in diesem Papier herausgearbeitete allgemeine Modell (vgl. Abschnitt 2.4) schließt auch den eben angeführten Spezialfall ein und zeigt damit seine umfassendere Anwendbarkeit.

2.2 Erweiterung um Kryptographie

Die Anwendung kryptographischer Systeme ist für die Verschlüsselung von *emb* interessant. Es gibt also die der eigentlichen Stegofunktion vor- bzw. nachgelagerten Schritte Verschlüsseln und Entschlüsseln (Abbildung 4).

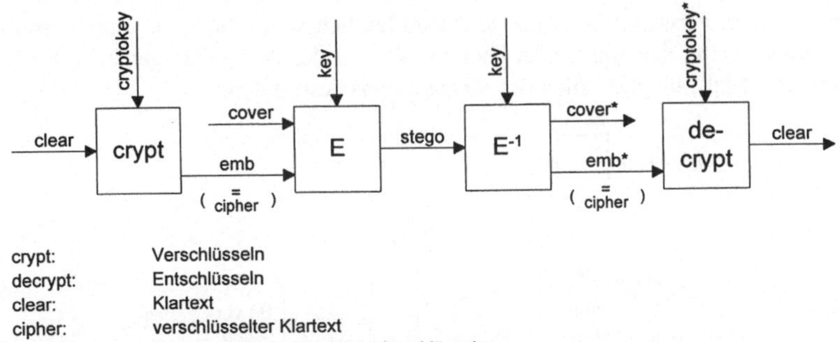

crypt:	Verschlüsseln
decrypt:	Entschlüsseln
clear:	Klartext
cipher:	verschlüsselter Klartext
cryptokey:	Parameter für die Funktion Verschlüsseln
cryptokey*:	Parameter für die Funktion Entschlüsseln

Abbildung 4: Erweiterung des Modells um Kryptographie

Der Einsatz eines asymmetrischen Kryptosystems bei crypt bzw. decrypt führt nicht zu einer asymmetrischen Steganographie, wie man vermuten könnte; ebensowenig wie ein mit asymmetrischer Kryptographie ausgeführter Austausch des (symmetrischen) Steganographie-Schlüssels *key* (z.B. Diffie-Hellman-Schlüsselaustausch).

So lassen sich mehrere vorgeschlagene Systeme, die asymmetrische Schlüssel verwenden, auf eine Erweiterung schlüsselloser symmetrischer steganographischer Funktionen um asymmetrische Verschlüsselung zurückführen.

Wie in Abschnitt 1.2 dargelegt wurde, ist das Ziel eines Stegosystems die Wahrung der steganographischen Eigenschaft. Mit der zusätzlichen Anwendung von Kryptographie wird nur die Wahrung der Konzelationseigenschaft eines Stegosystems unterstützt, die steganographische Eigenschaft jedoch nicht. Daher wird diese um kryptographische Verfahren erweiterte Variante nicht als allgemeingültiges Modell für steganographische Systeme betrachtet, wenngleich der Einsatz von Kryptographie durchaus sinnvoll sein kann.

2.3 Erweiterung um Vorverarbeitungsfunktionen

2.3.1 Analyse der Ein- und Ausgangsgrößen

Das in Abbildung 5 dargestellte Modell ist eine Erweiterung des allgemeinen Falles durch eine dem Schritt E vorgelagerte Analysephase. Während der Analysephase wird anhand des angenommenen Angreifermodells zu bewerten versucht, ob die vorliegende *cover/emb*-Kombination sichere Steganographie ermöglicht. Auf diese Art und Weise wird das System von den Eingangsgrößen unabhängiger und insgesamt sicherer, da es die Möglichkeit zur Zurückweisung unsicherer Datenkombinationen hat.

Hier zeigt sich ein wesentlicher Unterschied zwischen Steganographie und Kryptographie: während kryptographische Systeme als Schlüsseltext möglichst ideales weißes

Rauschen erzeugen sollen, muß *stego* anderen Anforderungen genügen. Insbesondere muß es dem verwendeten *cover* genügend ähnlich sein (abhängig vom Angreifermodell). Das heißt, daß Annahmen über das Wissen des Angreifers gemacht werden, die in hohem Maße die Sicherheit des Stegosystems bestimmen.

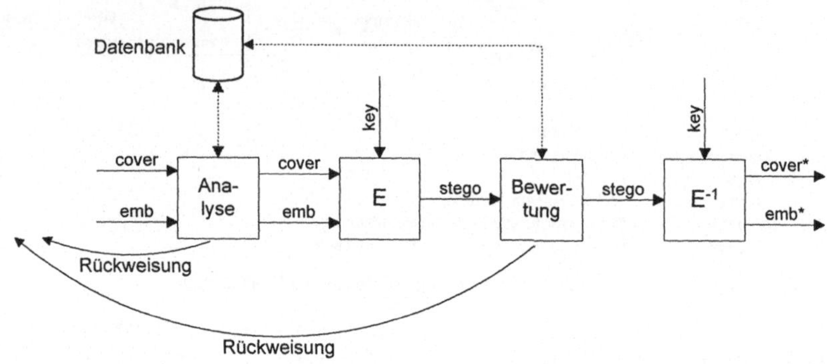

Abbildung 5: Analyse der Eingangsdaten und Ergebnisbewertung

Wie weiterhin in Abbildung 5 gezeigt, kann auch eine Datenbank zum System hinzugefügt werden. Diese Datenbank ist ein Hilfsmittel für den Analyseschritt. Sie kann zur Sicherstellung der einmaligen Verwendung eines bestimmten *cover* genutzt werden.

Eine weitere Rückweisung ist im Schritt „Bewertung" nach der Erzeugung von *stego* und damit vor dem Transfer zum Empfänger vorgesehen. Hier ist eine gute Überprüfung des Einbettungsergebnisses möglich, da man sämtliche bekannten, vom Angreifer verwendeten Methoden selbst anwenden kann. Die Bewertungskriterien ergeben sich aus dem Angreifermodell, also insbesondere aus dem Wissen über *cover*, das dem Angreifer zugebilligt wird, wie etwa bestimmte statistische Kenntnisse. Hält *stego* diesen Überprüfungen stand, kann es als sicher betrachtet werden. Auch dieser Schritt kann, wie in Abbildung 5 zu sehen ist, durch eine Datenbank unterstützt werden.

Die beiden Schritte Analyse und Bewertung bedingen einander nicht, d.h. es sind auch Systeme realisierbar, die nur mit einem der beiden auskommen.

2.3.2 Wann ist ein Stegosystem sicher?

Wenn dem Angreifer das verwendete *cover* und das daraus erzeugte *stego* bekannt sind, ist es für ihn trivial zu erkennen, daß Steganographie verwendet wurde, da er Unterschiede zwischen *stego* und *cover* feststellen kann. Aus diesem Grund darf ihm das konkrete *cover* nicht bekannt sein. Die dann für den Angreifer existierende Unsicherheit (Entropie) über *cover* wird vom Stegosystem ausgenutzt (siehe auch [KlPi_96]).

Ein allgemeines sicheres Stegosystem bei gleichzeitiger Kenntnis von *cover* und *stego* durch den Angreifer ist also nicht realisierbar.

Eine Ausnahme gilt es zu beachten: Im Fall *cover* $\equiv$ *stego*, d.h. *cover* enthält bereits steganographische Daten (auch als „selective steganography" bezeichnet), wäre theoretisch sichere Steganographie möglich. Voraussetzung ist jedoch, daß es dem Sender gelungen ist, in einem *echt zufälligen Prozeß* ein *cover* zu finden, das die gewünschten eingebetteten Daten bereits enthält. Die systematische Ausnutzung dieses Sachverhalts ist jedoch praktisch nicht möglich.

2.3.3 Auswahl aus breiten Eingabeströmen am Beispiel einer Videoaufnahme

Als ein interessanter Ansatz erweist sich die Idee, nicht mehr ein einzelnes *cover* zu betrachten, sondern eine große Anzahl von *covers*, die dann als *Source* bezeichnet wurden.

Ein Beispiel dafür ist die Aufnahme eines Videobildes mit einer Kamera (vgl. Abbildung 6), die durch viele Parameter beeinflußt wird. Einige Parameter sind nicht reproduzierbar (oder nur mit einer bestimmten Genauigkeit), woraus die nötige Unsicherheit (Entropie) über *cover* selbst bei vollständiger Kenntnis von *Source* für einen Angreifer erzielt wird. Aus *Source* wird eine Auswahl getroffen, in die ggf. noch zufällige Parameter eingehen, über die keiner (auch der Nutzer des Stegosystems nicht) hundertprozentige Aussagen treffen kann.

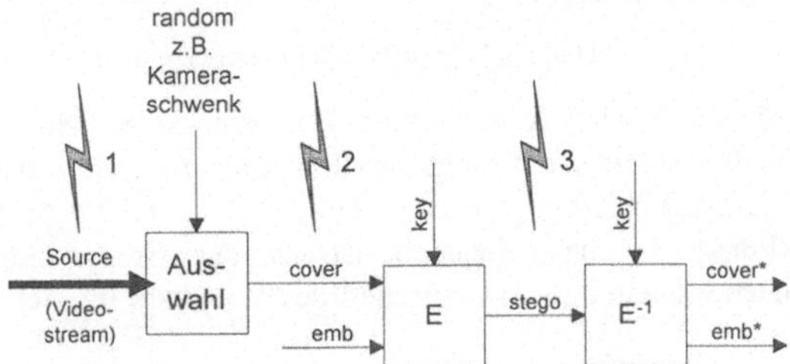

Source: Menge an Eingangsdaten
Auswahl: Auswahlprozeß
random: Zusammenfassung der in die Auswahl eingehenden Zufälligkeiten
1, 2, 3:   mögliche Angriffspunkte

Abbildung 6: Auswahl aus breiten Eingabeströmen (*Source*) am Beispiel Videoaufnahme

Der Vorteil eines solchen Verfahrens ist die extreme Erschwerung der Ermittlung des konkret verwendeten *cover* durch den Angreifer. Diese wird durch die eingehenden Zufallsgrößen noch verstärkt.

In Abbildung 6 bedeutet das konkret: beim Angriff an Stelle 1 ist Steganographie möglich (der Angreifer kennt nur *Source*); beim Angriff an Stelle 2 nicht (Angreifer kennt *cover*). Stego (Punkt 3) ist ihm in beiden Fällen bekannt. Selbstverständlich darf die Kenntnis von *stego* durch den Angreifer keine Schwächung des Stegosystems darstellen.

2.3.4 Vorgelagerte Quantisierung

Wie bereits in den Abschnitten 2.3.2 und 2.3.3 aufgeführt wurde, muß für den Angreifer eine bestimmte Ungenauigkeit in seinem Wissen über *cover* existieren, die z.B. auch durch Digitalisierung/Quantisierung eines analogen Signals erreicht werden kann. Der Angreifer darf Wissen über das analoge Signal haben und auch das analoge Signal *Source* selbst ganz genau kennen, und dennoch ist Steganographie möglich.

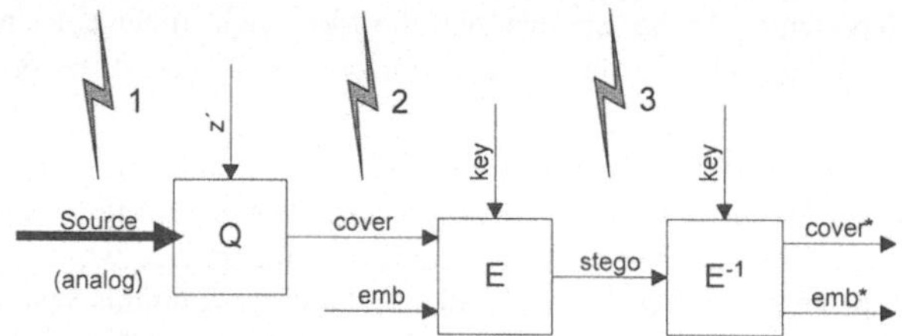

Abbildung 7: Modell mit Quantisierungsschritt

Ein Beispiel für das Modell in Abbildung 7 ist ein ISDN-Telefon mit in den Digital/Analog-Wandler integrierter Stegofunktion: Der Angreifer darf das eingehende Analogsignal (*Source*) beliebig genau kennen (Punkt 1) und die Ausgabe des Bausteins beobachten (Punkt 3), und dennoch ist es der Stegofunktion möglich, das (indeterministische) Quantisierungsrauschen des Wandlers für sichere Steganographie zu nutzen.

Einem Angriff an Punkt 2 entspricht es, wenn der Angreifer Zugriff auf die digitalisierten Daten vor dem Einbetten erhält. Das könnte z.B. der Fall sein, wenn Digitalisiereinheit und Stegofunktion auf getrennten Bausteinen untergebracht sind und die Verbindung zwischen ihnen abgehört werden kann.

2.4 Ergebnis der Diskussion

Die Hauptbedingung für sichere Steganographie ist, daß dem Angreifer *cover* nicht bekannt sein darf. Das Modell aus Abschnitt 2.1. beschreibt somit die steganographische Kernfunktion eines Stegosystems, die von genau dieser Annahme ausgeht. Will man analog zur Kryptographie davon ausgehen, daß dem Angreifer alle Ein- und Ausgangsgrößen außer dem Schlüssel bekannt sind, erweist sich dieses Modell als unzureichend, wenngleich es nichts von seiner Gültigkeit einbüßt.

Der Funktion Einbetten muß ein Schritt vorangestellt werden, der sicherstellt, daß dem Angreifer *cover* nicht beliebig genau bekannt ist (vgl. Abschnitt 2.3.2 ff.). Das führt uns zu folgender Darstellung:

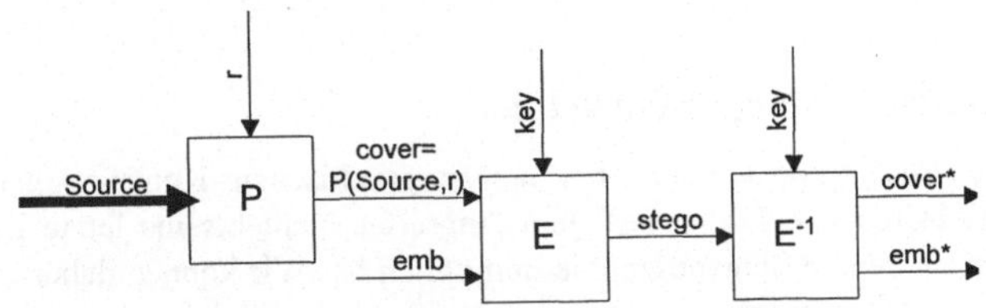

P:　　Preprocessing; Vorverarbeitungsfunktion
r:　　durch P entstehender zufälliger Anteil von cover (kann Eingangsgröße von P sein, muß aber nicht)

Abbildung 8: Modell mit Vorverarbeitungsfunktion

Die Funktion P kann z.B. ein Auswahlprozeß analog Abschnitt 2.3.3 oder die Digitalisierung eines Analogsignals (vgl. 2.3.4) sein.

Faßt man die Funktionen E und P zusammen, kommt man zum Modell eines symmetrischen steganographischen Konzelationssystems:

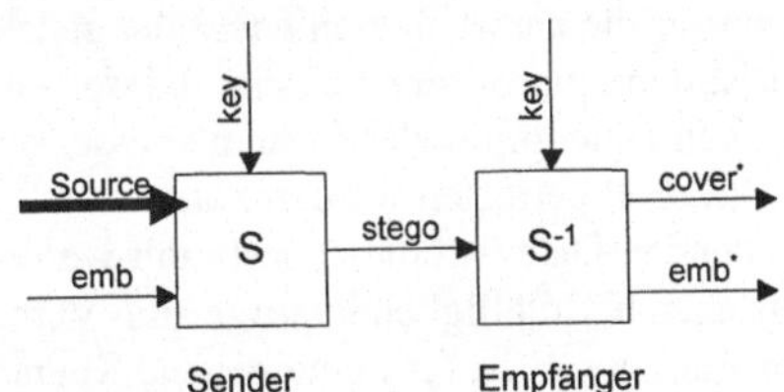

S:　　indeterministische steganographische Gesamtfunktion „Verbergen"
S⁻¹:　　Umkehroperation zu S

Abbildung 9: Modell eines symmetrischen steganographischen Konzelationssystems

Der Unterschied zur anfänglichen Betrachtungsweise besteht darin, daß jetzt nicht mehr nur die steganographische Kernfunktion, sondern das gesamte System betrachtet wird. Mittels der Eingangsgröße *Source* wird die Unbestimmtheit des Parameters *cover* der Einbettungsfunktion modelliert und nicht mehr nur einfach vorausgesetzt. Die Gesamtfunktion S ist damit indeterministisch, die Kernfunktion E dagegen deterministisch.

Mit dieser Modelldarstellung wird eine weitgehende Analogie zur Modellierung von Kryptosystemen dahingehend erreicht, daß das Kerckhoff-Prinzip gilt. Dieses besagt,

„... daß die Sicherheit eines Verschlüsselungsverfahrens nur von der Geheimhaltung des Schlüssels abhängen darf" [Schn_96, S. 6]. Ein Angreifer darf also alle Eingangsgrößen und Ausgangsgrößen eines Systems sowie die Algorithmen selbst kennen, nur den Schlüssel nicht. Das ist bei dem in Abbildung 9 dargestellten Modell der Fall, beim ursprünglichen Einbettungsmodell (Abbildung 2) dagegen nicht.

3 Zusammenfassung und Ausblick

Besondere Beachtung verdient bei steganographischen Konzelationssystemen die Hüllnachricht (*cover*). Diese darf dem Angreifer nicht bis ins letzte Detail bekannt sein, da sonst sichere Steganographie unmöglich ist. Wir können daher zwei Anforderungen an ein Stegosystem nennen:

1. Dem Angreifer bleibt der Schlüssel *key* unbekannt.

2. Dem Angreifer ist das *konkrete cover* unbekannt; er hat jedoch eine gewisse Menge an Wissen über *cover* (z.B. Verteilungen).

Die Geheimhaltung des Schlüssels *key* entspricht der von symmetrischen Kryptosystemen.

Punkt 2 kann zunächst nur als Voraussetzung für sichere Steganographie formuliert werden. Alternativ besteht jedoch die Möglichkeit, eine als *Source* bezeichnete Menge an Eingangsdaten zu betrachten, aus der das Stegosystem ein konkretes *cover* auswählt (wobei für einen Angreifer die auswählbaren *cover* die gleiche Wahrscheinlichkeit haben). Dem Angreifer ist dann nur *Source* bekannt. Diese Herangehensweise ist für die Modellierung und Implementierung realer Systeme besser geeignet, da Aussagen über die Preprocessing-Funktion P getroffen werden können und die Einbettungsfunktion E optimal auf die eingehenden Datenströme abgestimmt werden kann. E nutzt dabei die im Verlauf von P entstehende Zufälligkeit in *cover* zum Verbergen von Daten. Für eine gute Realisierung der Funktion E ist also eine genaue Kenntnis bezüglich der Indeterminiertheit von P notwendig.

Es gibt etliche offene Probleme und Fragestellungen auf dem Gebiet der Steganographie, die weiteren Arbeiten vorbehalten bleiben bzw. in diesem Papier nicht betrachtet werden. Einige davon sind:

– die Beschreibung und Bewertung der Sicherheit von Stegosystemen mit den Mitteln der Informationstheorie (dieser Punkt wir ausführlich in [KlPi_97] diskutiert),

– die möglichst umfassende Modellierung und Bewertung von Angriffen auf Stegosysteme,

– die Betrachtung von existierenden oder zu realisierenden Stegosystemen unter den beiden o.a. Gesichtspunkten sowie der Konformität zum erarbeiteten allgemeinen Modell für steganographische Konzelationssysteme.

Darüber hinaus hat Steganographie zum Urheberrechtsschutz seine Bedeutung, die in diesem Papier nicht betrachtet wurde. Bezüglich der hier vorgenommenen Modellbildung kann man davon ausgehen, daß existierende Watermarkingsysteme ebenfalls nach den hier vorgestellten Modellen arbeiten. In Zukunft ist jedoch zu erhoffen, daß auch „echte" asymmetrische Systeme entstehen werden.

4 Literatur

[Ande_96] Ross Anderson: Stretching the limits of Steganography. In: Proceedings: Information Hiding. Workshop, Cambridge, U.K., May/June, 1996, LNCS.

[HuPf_96] Michaela Huhn, Andreas Pfitzmann: Erste Gedanken zu Steganographie mit öffentlichen Schlüsseln. Internes Arbeitspapier TU Dresden, Dresden 1996.

[Klir_89] G. J. Klir: Inductive Systems Modelling: An Overview. In: M. S. Elzas, T. I. Ören, B. P. Zeigler (Hrsg.): Modelling And Simulation Methodology, Knowledge Systems' Paradigms. Elsevier Science Publishers B.V., North Holland 1989, 55-75.

[KlPi_97] Herbert Klimant, Rudi Piotraschke: Informationstheoretische Bewertung steganographischer Konzelationssysteme. eingereicht für: VIS '97, Freiburg/Brsg.

[Pfit_96] Birgit Pfitzmann: Information Hiding Terminology. In: Proceedings: Information Hiding. Workshop, Cambridge, U.K., May/June, 1996, LNCS.

[Schn_96] Bruce Schneier: Angewandte Kryptographie, Addison-Wesley, Bonn 1996.

Informationstheoretische Bewertung steganographischer Konzelationssysteme

Herbert Klimant, Rudi Piotraschke

Technische Universität Dresden, Institut für Theoretische Informatik, 01062 Dresden
{klimant, pi}@tcs.inf.tu-dresden.de

Zusammenfassung

Aufgrund einer zu erwartenden breiteren Anwendung der Steganographie in Kommunikationssystemen (Vertraulichkeit durch *Verbergen der Existenz* geheimer Daten) stellt sich die Frage nach der Sicherheit steganographischer Systeme. In der vorliegenden Arbeit wird diese Frage auf der Grundlage informationstheoretischer Modelle beantwortet. Es wird nachgewiesen, daß theoretische Sicherheit (bezüglich der Vertraulichkeit von Informationen) möglich ist, wenn eine Reihe von Bedingungen, in Form von Entropie–Schranken dargestellt, erfüllt wird.

1   Einführung

Die Gewährleistung der *Vertraulichkeit* gehört zu den wichtigsten Sicherheitsanforderungen in Kommunikationssystemen. Dazu dienen meistens kryptographische Systeme, die die Vertraulichkeit des *Inhalts* der Daten sicherstellen, wobei es allgemein bekannt sein kann, daß die betreffenden Daten geheime Informationen enthalten. Eine andere Möglichkeit zur Gewährleistung der Vertraulichkeit besteht darin, daß die *Existenz* geheimer Daten verborgen wird. Hierzu dienen steganographische Systeme /ZÖL 97/.

Die prinzipielle Aufgabe der Steganographie besteht darin, eine geheimzuhaltende Nachricht (embedded e) in einer anderen, umfangreicheren Nachricht (cover c) zu verstecken. Dies geschieht mit einer steganographischen Funktion f, wobei ein Schlüssel (key k) verwendet wird. Im Ergebnis entsteht ein Text mit dem verborgenen e (stegotext s), d.h. es gilt $s = f(c, e, k)$. Dabei seien s, c, e und k Elemente entsprechender Alphabete S, C, E und K.[1]

Als Beispiel stellen wir uns eine aus n Bytes bestehende Datei vor, in der m Bits ($m < n$) in den niederwertigsten Stellen „versteckt" werden sollen.

---

[1]Die Elemente dieser Alphabete stellen demnach Zeichenketten bzw. Datensätze dar.

Dann bedeuten:

C Menge der unterschiedlichen Bytefolgen,

c eine konkrete Bytefolge der Länge n,

E Menge der unterschiedlichen Bitfolgen,

e eine konkrete Bitfolge der Länge m, die in c versteckt werden soll,

K Menge der Schlüssel,

k ein konkreter Schlüssel, z.B. zur Bestimmung der Positionen, in denen e versteckt werden soll,

S Menge der unterschiedlichen Stegotexte $(S = C)$,

s ein konkreter Stegotext (Bytefolge der Länge n, die e enthält).

Mit dem Verbergen der *Existenz* einer Nachricht e (Steganographie) wird auch gleichzeitig der *Inhalt* der verborgenen Nachricht geheimgehalten (Konzelation). Wir sprechen deshalb von einem *steganographischem Konzelationssystem* (im folgenden kurz: *Stegosystem*). Es stellt sich dann die Frage, wie **sicher** ein solches Stegosystem ist, d.h., ob es überhaupt möglich ist bzw. unter welchen Bedingungen gewährleistet werden kann, daß die *Existenz eines eingebetteten e von einem potentiellen Angreifer nicht entdeckt* werden kann.

Die Untersuchung und Beantwortung dieser Frage ist Gegenstand der vorliegenden Arbeit. Dazu werden informationstheoretische Beschreibungsmittel (Entropie, bedingte Entropie, Verbundentropie, Transinformation) verwendet /KPS 96/.

Für ein Alphabet X ist die *Entropie* $H(X)$ ein Maß für die „Unbestimmtheit über X", d.h. die Unbestimmtheit über das Auftreten eines *konkreten* Elementes $x \in X$.[2] Als *Verbundentropie* $H(X, Y)$ wird die Entropie der *Vereinigung* der Alphabete X und Y bezeichnet. Während die *bedingte Entropie* $H(X|Y)$ die verbleibende Unbestimmtheit über X bei Kenntnis von Y angibt, stellt die *Transinformation* $H_T(X; Y)$ die Informationsmenge dar, die über X aus der Beobachtung von Y gewonnen werden kann.

Entropien können anschaulich durch *VENN-Diagramme* dargestellt werden: während man die Entropien der einzelnen Alphabete durch Kreisflächen darstellt, werden die Abhängigkeiten der Alphabete voneinander durch mehr oder weniger starke Überschneidungen der einzelnen Kreisflächen zum Ausdruck gebracht.

Ein potentieller *Angreifer* wird wie folgt charakterisiert:

– er *vermutet*, daß ein e in s versteckt ist,

– er *kennt* die steganographischen Funktionen,

– er *beherrscht* das notwendige Instrumentarium, um einen gezielten Angriff auf das Stegosystem durchzuführen.

Wenn trotz Einsatz unbegrenzter Mittel in unbegrenzter Zeit eine Vermutung bzw. Anfangshypothese über ein mögliches e nicht bestätigt werden kann, soll das System als „theoretisch sicher" gelten. Da diese Bewertung in der vorliegenden Arbeit auf informationstheoretischer Grundlage erfolgt, werden wir ein solches System als „**informationstheoretisch sicher**" bezeichnen.

---

[2]Dieser Zusammenhang ist zu beachten, wenn im folgenden teilweise nur auf die *Alphabete* Bezug genommen wird.

2 Modellfall (1)

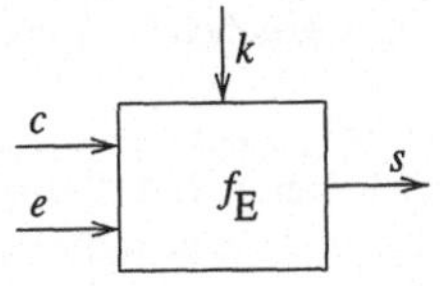

Annahmen:

f, s und c bzw. S und C sind allgemein bekannt (d.h. auch einem potentiellen Angreifer), während k nur die Nutzer kennen.

Bild 1

Das Stegosystem kann bei diesen Annahmen als „informationstheoretisch sicher" bezeichnet werden, wenn aus der Beobachtung von s und c bzw. von S und C keine Information über e bzw. E gewonnen wird. Dann muß die Transinformation zwischen E und (S, C) verschwinden:

$$H_T(E;(S,C)) = H(E) - H(E|(S,C)) = 0.$$

Daraus folgt die fundamentale Sicherheitsbedingung:

$$H(E|(S,C)) = H(E),$$

d.h., die Unbestimmtheit über E, also die Entropie $H(E)$, darf sich nach Kenntnis von S und C nicht verringern.

Das bedeutet: E muß **unabhängig** von S und C sein.

Im folgenden wollen wir überlegen, ob diese Bedingung erfüllbar ist.

Aus rein pragmatischer Sicht ergibt sich, daß ein Angreifer (bei den gegebenen Annahmen) durch den konkreten Vergleich von s und c ggf. Informationen über ein verstecktes e gewinnt. Dabei kann allgemein vorausgesetzt werden, daß nicht nur die Alphabete S und C, sondern auch ihre Entropien $H(S)$ und $H(C)$ gleich sind.

Bezüglich der bedingten Entropien bestehen jedoch signifikante Unterschiede, nämlich

- *ohne* eingebettete Information: $H(S|C) = H(C|S) = 0$,
- *mit* eingebetteter Information: $H(S|C) = H(C|S) > 0$.

Aufgrund des Zusammenhangs zwischen Unbestimmtheit und Information können wir nun sagen: Das Maß an Unbestimmtheit über S, wenn C bekannt ist (bzw. über C, wenn S bekannt ist), entspricht der Information über E, die aus der Beobachtung von S und C gewonnen werden kann.

Bei einer Einbettung von $e \in E$ erhalten wir demnach eine Transinformation

$$H_T(E;(S,C)) = H(E) - H(E|(S,C)) > 0.$$

Daraus folgt

$$H(E|(S,C)) < H(E),$$

d.h., die Sicherheitsbedingung des Stegosystems ist damit verletzt. Als *notwendige* und *hinreichende* Bedingung für *informationstheoretisch sichere* Steganographie gilt somit:

$$H(S|C) = H(C|S) = 0.$$

Diese kann nur erfüllt werden, wenn

$$s_i = c_i \quad \text{mit } s_i \in S,\, c_i \in C \quad \text{für alle } i$$

gilt.

Das steganografische Problem wird damit auf einen praktisch irrelevanten Spezialfall reduziert.[3] Wenn man diesen ausschließt, ergibt sich die *Schlußfolgerung*:

Die **Sicherheitsbedingung** $H(E|(S,C)) = H(E)$ ist bei den gegebenen Annahmen **nicht erfüllbar.**

Diese Feststellung gilt, wie bereits gesagt, für den Fall, daß einem potentiellen Angreifer sowohl der konkrete covertext als auch der konkrete stegotext bekannt sind. Dabei wird vorausgesetzt, daß beide Texte in gleicher Darstellung, z.B. als Bilddateien vorliegen. Bei diesem *„rauschfreien"* Fall kann ein Beobachter jede Veränderung an c durch direkten Vergleich mit s feststellen und damit auch e entdecken.

Ist dem Angreifer dagegen der covertext z.B. nur als Bild (Foto) zugänglich, die daraus erzeugte Bilddatei jedoch geheim, dann liegt ein vollkommen anderer Fall vor, der im nächsten Abschnitt untersucht wird.

Als Anmerkung sei hinzugefügt, daß diese Aussagen gemacht werden können, ohne den Einfluß des Schlüssels zu berücksichtigen.

3   Modellfall (2)

Nachdem im Abschn. 2 festgestellt wurde, daß keine informationstheoretisch sichere Steganographie möglich ist, wenn S und C bekannt sind, soll für die folgenden Untersuchungen die Bedingung gelten:

Bei Kenntnis von S verbleibt eine Restunbestimmtheit über C, d.h. $H(C|S) > 0$, so daß ein Angreifer nicht entscheiden kann, ob an einem konkreten c Veränderungen vorgenommen wurden. Dazu führen wir eine neue Größe $\overline{C}$ mit der Entropie $H(\overline{C})$ ein. Über dem Alphabet $\overline{C}$ werden konkrete Nachrichten $c \in \overline{C}$ für autorisierte Empfänger erzeugt.

Die nötige Unbestimmtheit über C kann dadurch erreicht werden, daß jedes konkrete c in einem *zufälligen* Prozeß erzeugt und geheim gehalten wird. Dies ist z.B. beim Scannen eines Bildes oder durch Quantisierung eines analogen Sprachsignals möglich. Bezogen auf das Beispiel im Abschn. 1, in dem für c eine Bytefolge der Länge n angenommen wurde, würde das eine Folge von Abtastwerten (mit je 8 Bit kodiert) des Sprachsignals in einem bestimmten Zeitabschnitt bedeuten. Ein potentieller Angreifer darf zwar das zu übertragende stochastische Sprachsignal, aber nicht seine konkrete Kodierung kennen. Aufgrund des Quantisierungsrauschens wäre das ursprüngliche Signal dann nur mit entsprechender Ungenauigkeit reproduzierbar. Solange sich die Veränderungen von c infolge des Einbettens von e in diesem Unschärfebereich bewegen, bleibt dem Angreifer die Existenz von e verborgen.

Für den Modellfall (2) gelten dann folgende *Annahmen*:

– f, S (bzw. s) und $\overline{C}$ sind allgemein bekannt,

– C und K (bzw. c und k) sind potentiellen Angreifern unbekannt.

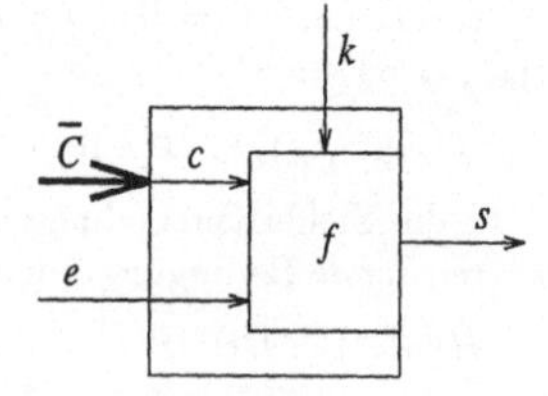

Bild 2

---

[3]Spezialfall $s \equiv c$ bedeutet: Es müßte ein c gefunden werden, welches das vorgesehene e enthält.

Es soll informationstheoretisch untersucht werden, ob unter diesen Annahmen sichere Steganographie möglich ist.

Unter dem Aspekt eines **Angreifers**, der die Existenz von $e \in E$ aufdecken will, lautet die *erste Grundbedingung*:

$$H(E|(S,\overline{C})) = H(E),$$

d.h., *die Unbestimmtheit über E darf durch Kenntnis von S und $\overline{C}$ nicht verringert werden*, bzw. mit anderen Worten: E muß **unabhängig** von S und $\overline{C}$ sein (im VENN–Diagramm, Bild 3, kommt diese Unabhängigkeit dadurch zum Ausdruck, daß $H(E)$ *außerhalb* von $H(S,\overline{C})$ liegt).

Wie kann diese Bedingung erfüllt werden?

Damit aus $s \in S$ keine Information über Veränderungen von $c \in C$, d.h. über eine mögliche Einbettung von $e \in E$ gewonnen werden kann, ist die o.g. Forderung bezüglich einer bleibenden Restunbestimmtheit $H(C|S) > 0$ zu erfüllen. Ihre *erforderliche* Größe ergibt sich aus dem Zusammenhang von bedingter Entropie und Transinformation (vgl. Abschn. 2):

$$H(C|S) \geq H_T(E;(S,C)) = H(E) - H(E|(S,C)).$$

Geht man davon aus, daß bei Kenntnis von S und C auch E vollständig bestimmt werden kann (was bezüglich der steganographischen Sicherheit den ungünstigsten Fall darstellt), dann ist

$$H(E|(S,C)) = 0.$$

Im VENN–Diagramm wird diese vollständige Abhängigkeit dadurch ausgedrückt, daß $H(E)$ *innerhalb* von $H(S,C)$ liegt.

Somit gilt

$$H(C|S) \geq H(E).$$

Diese Bedingung kann so interpretiert werden: Da die Transinformation maximal die Größe von $H(E)$ haben kann, muß die *erforderliche* Unbestimmtheit über C bei Kenntnis von S mindestens auch so groß sein, damit ein Angriff über S erfolglos bleibt.

Die gleichen Überlegungen treffen natürlich auch für einen möglichen Angriff über $\overline{C}$ zu. Deshalb soll gelten

$$H(C|\overline{C}) = H(C|S)$$

und

$$H(C|\overline{C}) \geq H(E).$$

Wir benötigen unter diesen Bedingungen demnach eine *Verbundentropie*

$$H_0 = H(\overline{C},C) = H(\overline{C}) + H(C|\overline{C}),$$

d.h., die Auswahl von c aus $\overline{C}$ muß mit einer Unbestimmtheit $H(C|\overline{C})$ verbunden sein. Wegen $C \subseteq \overline{C}$ gilt

$$H(\overline{C}) \geq H(C),$$
$$H(\overline{C}|C) \geq H(C|\overline{C}).$$

Deshalb kann die untere Schranke für die erforderliche Verbundentropie auch wie folgt dargestellt werden:

$$H_0 \geq H(C) + H(C|S)$$

bzw.

$$H_0 \geq H(C) + H(E).$$

Unter Berücksichtigung, daß $S \subseteq \overline{C}$ und $H(\overline{C}) \geq H(C)$ gelten, kann auch die sicherheits-relevante Schranke

$$H(\overline{C}|S) \geq H(E)$$

erfüllt werden. Alle hier dargestellten Entropiebeziehungen gehen auch unmittelbar aus dem VENN–Diagramm (Bild 3) hervor.

Die bisherigen Ableitungen erlauben die *Schlußfolgerung*:

Bei Einhaltung der unteren Schranken für $H(C|S)$ und $H(C,\overline{C})$ sind Angriffe über S und $\overline{C}$, um die Existenz von E aufzudecken, erfolglos, d. h., die *erste Grundbedingung* für ein sicheres Stegosystem kann damit als *erfüllbar* angesehen werden.

Nach der Analyse der Bedingungen bei einem Angriff auf E wollen wir nun noch die Stego–Schlüssel K in die folgenden Untersuchungen einbeziehen.

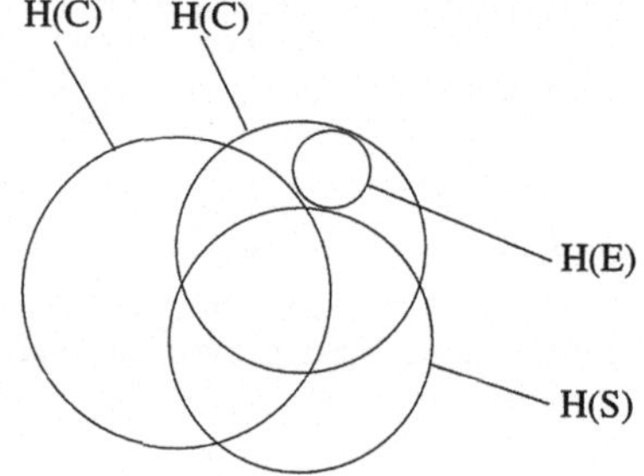

Bild 3

Dabei gehen wir von der Situation rechtmäßiger Nutzer aus und werden die daraus zu gewinnenden Erkenntnisse auf die Situation potentieller Angreifer übertragen.

Grundsätzlich gilt: *Wenn neben S auch K bekannt ist, muß auch E eindeutig bestimmt werden können*, d.h., die Transinformation H_T zwischen (S,K) und E muß in diesem Fall gleich $H(E)$ sein:

$$H_T(E;(S,K)) = H(E) - H(E|(S,K)) = H(E).$$

Daraus erhalten wir unmittelbar eine weitere *Grundbedingung* für ein Stegosystem (nämlich die aus der Sicht eines **Nutzers**):

$$H(E|(S,K)) = 0.$$

Mit dieser Bedingung wird eine grundsätzliche Forderung erfüllt, die jeder Nutzer an ein Stegosystem stellt.

Bezüglich des für diese Situation zutreffenden VENN–Diagramms besagt diese Bedingung, daß $H(E)$ innerhalb der Verbundentropie $H(S,K)$ liegen muß.

Wir untersuchen nun die Bedingungen, unter denen das Stegosystem auch bei einem *Angriff auf E, der über K erfolgt*, informationstheoretisch sicher ist.

Dabei gehen wir von der grundsätzlichen Forderung aus, daß ein potentieller Angreifer (der S und $\overline{C}$ kennt) keine Information über (K,E) gewinnen darf.[4]

---

[4]Obwohl diese Forderung eigentlich zu hoch ist, kann sie für den theoretischen Ansatz gewählt werden, denn im Ergebnis wird die angenommene Bedingung wieder abgeschwächt.

Diese Bedingung wird durch folgende Transinformation ausgedrückt:

$$H_T((K,E);(S,\overline{C})) \;=\; H(K,E) - H((K,E)|(S,\overline{C})) = 0$$
$$=\; H(K,E) - H(K|(S,\overline{C})) - H(E|(S,\overline{C},K)) = 0.$$

Unter Berücksichtigung, daß $H(E|(S,\overline{C},K)) = 0$ sein muß, ergibt sich daraus:

$$H(K|(S,\overline{C})) = H(K,E) \quad \text{bzw.}$$

$$H(K|(S,\overline{C})) = H(E) + H(K|E) \geq H(E).$$

Mit dieser Schranke für die erforderliche Schlüsselentropie haben wir die *zweite Grundbedingung* für informationstheoretisch sichere Stegosysteme gewonnen.

Das nebenstehende VENN–Diagramm (Bild 4) zeigt die Entropieverhältnisse unter dem Aspekt von möglichen Angriffen auf E und K, wobei die Nutzer–Bedingung $H(E|(S,K)) = 0$ berücksichtigt wird.[5]

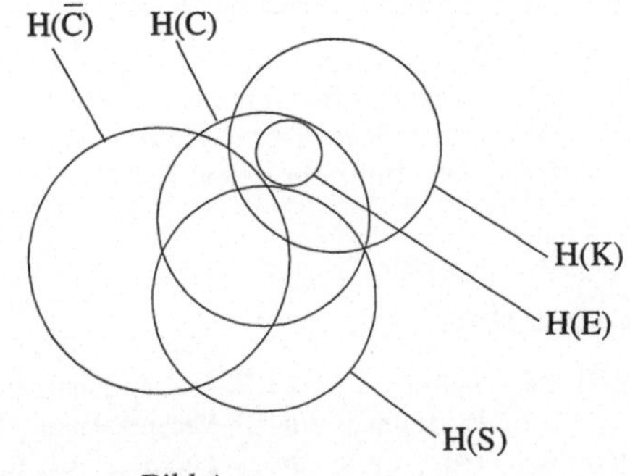

Bild 4

Es zeigt sich, daß informationstheoretische Sicherheit auch gewährleistet werden kann, ohne daß statistische Unabhängigkeit von (K,E) und $(S,\overline{C})$ gefordert werden muß. Wesentlich ist nur, daß die *bedingte* Schlüsselentropie $H(K|(S,\overline{C}))$ mindestens so groß wie $H(E)$ ist.

Ergebnisse:

Die durchgeführten Untersuchungen zeigen, daß ein **informationstheoretisch sicheres** Stegosystem möglich ist, wenn *zwei Grundbedingungen* (**Sicherheitsbedingungen**) erfüllt sind (s. auch Bild 4):

1. Durch Kenntnis von $\overline{C}$ und S bzw. s darf die Unbestimmtheit über E bzw. e nicht verringert werden:

   $$H(E|(S,\overline{C})) = H(E|S) = H(E).$$

   Zur Erfüllung dieser Bedingung müssen folgende Schranken eingehalten werden:
   $$H_0 = H(C,\overline{C}) \geq H(C) + H(E),$$
   $$H(\overline{C}|S) \geq H(E),$$
   $$H(C|S) \geq H(E).$$

2. Die bedingte Schlüsselentropie muß mindestens so groß wie $H(E)$ sein, damit ein Angriff auf E über K erfolglos bleibt:

   $$H(K|(S,\overline{C})) \geq H(E).$$

---

[5] Anmerkung: Das VENN–Diagramm ist so zu interpretieren, daß $H(E)$ innerhalb von $H(S,K)$ liegt (und *nicht* innerhalb von $H(K)$ allein).

Die *dritte Grundbedingung* (**Nutzerbedingung**) eines Stegosystems lautet:
Für den rechtmäßigen Empfänger (der den Schlüssel $k \in K$ besitzt) darf keine Unbestimmtheit über $e \in E$ existieren:

$$H(E|(S,\overline{C},K)) = H(E|(S,K)) = 0.$$

Aufgrund dieser Ergebnisse kann Modell (2) jetzt auch allgemein gemäß Bild 5 mit

$$s = f^*(\overline{c}, e, k)$$

dargestellt werden, wobei mit $\overline{c}$ ein cover bezeichnet wird, das gegenüber einem Angreifer die erforderliche Unbestimmtheit enthält.

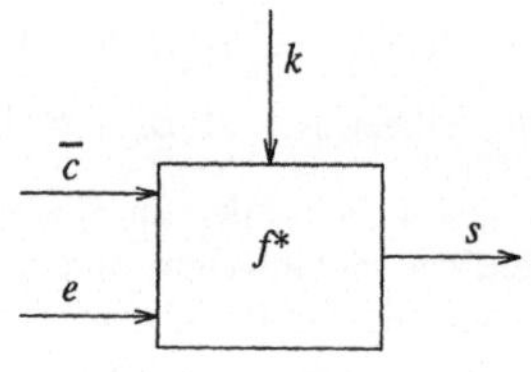

Bild 5

Literatur

[ZÖL 97]   Zöllner, J.; u.a.: Über die Modellierung steganographischer Systeme. Eingereicht zur GI–Veranstaltung VIS 97, 30.9. – 2.10.97, Freiburg/Brsg.

[KPS 96]   Klimant, H.; Piotraschke, R.; Schönfeld, D.: Informations– und Kodierungstheorie. B. G. Teubner Verlagsgesellschaft Stuttgart und Leipzig, 1996.

Common Criteria

Die neuen IT-Sicherheitskriterien - Eine Zwischenbilanz

Dr. Eckart Brauer

Bundesamt für Sicherheit in der Informationstechnik, PF 20 03 63, 53133 Bonn

- Eingeladener Vortrag -

Übersicht

Die Common Criteria for Information Technology Security Evaluation, Version 1.0 (Common Criteria, CC) sind Ende Januar 1996 fertiggestellt worden. Im Anschluß daran waren die Common Criteria Gegenstand der kritischen Begutachtung durch die internationale Fachöffentlichkeit. Gleichzeitig begannen Probeevaluationen von IT-Sicherheitsprodukten nach den CC. Daraus sind bis heute zahlreiche CC Observation Reports und verschiedene Einzelbeiträge zu Detailfragen hervorgegangen. Dies dient als Grundlage der Erarbeitung der CC-Version 2.0, die nicht vor Ende 1997 erwartet wird.

Parallel dazu sind die Common Criteria Gegenstand eines ISO-Normungsvorhabens. Gegenwärtig hat die CC-Version 1.0 den Status eines ISO Committee Drafts. Der vorliegende Beitrag liefert eine Zwischenbilanz zum Thema Common Criteria und wagt einen Blick in die Zukunft dieser neuen IT-Sicherheitskriterien.

1.   Einleitung

Der folgende Teil stellt die Entwicklung des Common-Criteria-Projektes bis zur Version 1.0 vor und geht auf mögliche Perspektiven der CC ein.

1.1   Hintergrund

Im Dezember 1992 lag ein erster Entwurf der Federal Criteria for Information Technology Security (Federal Criteria, FC) als geplanter Ersatz für die Trusted Computer System Evaluation Criteria (Orange Book, TCSEC) vor. Der notwendige Übergang vom Orange Book zu den Federal Criteria in den USA wurde durch das National Institute of Standards and Technology (NIST) und die National Security

Agency (NSA) zur Zusammenarbeit mit den Herausgebern anderer IT-Sicherheits-
kriterien genutzt. Als Ziel wurden weltweit gültige und international akzeptierte
IT-Sicherheitskriterien als Grundvoraussetzung für die gegenseitige Anerkennung von
Evaluationsergebnissen anvisiert.

Mit den Federal Criteria hatten sich die USA weitgehend den europäischen
Information Technology Security Evaluation Criteria (ITSEC) angenähert, so daß sich
eine enge Zusammenarbeit mit Frankreich, Großbritannien, den Niederlanden und
Deutschland anbot, die mit der Entwicklung der ITSEC befaßt waren. Auch Kanada
erkannte die Vorteile einer Zusammenarbeit, weil die Canadian Trusted Computer
Product Evaluation Criteria (CTCPEC) wie die Federal Criteria nur als Entwurf
vorlagen. Die neu zu entwickelnden Common Criteria for Information Technology
Security Evaluation (Common Criteria, CC) sollten die Weiterentwicklung und
Harmonisierung des amerikanischen Orange Book bzw. der Federal Criteria, der
kanadischen CTCPEC und der europäischen ITSEC werden.

Die USA, Kanada, Frankreich, Großbritannien, die Niederlande und Deutschland
vereinbarten, die in gemeinsamer Arbeit zu entwickelnden Common Criteria der
Internationalen Standardisierungsorganisation (ISO) für deren laufendes Normungs-
vorhaben Evaluationskriterien zur IT-Sicherheit zur Verfügung zu stellen und damit
einen weltweit gültigen IT-Sicherheitskriterienstandard zu etablieren.

1.2   Durchführung

Im Juni 1993 konstituierte sich das Common Criteria Editorial Board (CCEB), das die
Common Criteria erarbeiten sollte. Im CCEB waren neben NIST und NSA auch das
Communications Security Establishment (CSE) für Kanada sowie der Service Central
de la Sécurité des Systèmes d'Information (SCSSI) für Frankreich, die Communi-
cations-Electronics Security Group (CESG) für Großbritannien, die Netherlands
National Communications Security Agency (NLNCSA) für die Niederlande und das
Bundesamt für Sicherheit in der Informationstechnik (BSI) für Deutschland beteiligt.

Während das CCEB die technische Arbeit leistete, fielen die meisten politischen
Entscheidungen zu den Common Criteria auf der übergeordneten Ebene der Sponsors
bzw. Sponsor Representatives (SRs). Diese Rollen wurden von den Behörden- bzw.
Bereichsleitern der die einzelnen Länder vertretenden Institutionen wahrgenommen.

In Deutschland war die Entwicklung der Common Criteria durch den Nationalen
Arbeitskreis IT-Sicherheitskriterien (NAK) begleitet, in dem Vertreter von
akkreditierten Prüfstellen, von Wirtschaft und Wissenschaft sowie von Behörden und
IT-Anwendern durch das BSI über den Fortschritt im CC-Projekt informiert wurden
und eigene Kommentare und Vorschläge zur Entwicklung der CC einbringen konnten.
Die Zwischenversion 0.9 der Common Criteria, die das Konzept bereits klar erkennen
ließ und nur beispielhafte Sicherheitsanforderungen enthielt, wurde im Herbst 1994
verteilt. Es gingen weltweit mehrere tausend Einzelkommentare dazu ein. Auf Basis
dieser Kommentare wurden die CC überarbeitet und dann vervollständigt.

1.3   Heutiger Stand

Seit der erfolgreichen Fertigstellung der CC-Version 1.0 Anfang 1996 werden die Common Criteria für Probeevaluationen von IT-Sicherheitsprodukten verwendet. Daraus sind bis heute zahlreiche CC Observation Reports und verschiedene Einzelbeiträge zu Detailfragen hervorgegangen. Diese Anwendungserfahrungen werden vom Common Criteria Implementation Board (CCIB), dem Nachfolgegremium des CCEB, bei der Überarbeitung der CC-Version 1.0 zur Version 2.0 ausgewertet. Nach wie vor fallen politische Entscheidungen zu den Common Criteria auf der übergeordneten Ebene der Sponsors bzw. Sponsor Representatives.

Das Common-Criteria-Projekt umfaßt nach Fertigstellung der CC-Version 1.0 mehrere Teilprojekte. Neu hinzu gekommen sind z.B. das Common Evaluation Methodology Editorial Board (CEMEB), die Cryptographic Support Working Group (CSWG) und die Mutual Recognition Working Group (MRWG).

Das CEMEB befaßt sich mit der Entwicklung einer Evaluationsmethodik für die Common Criteria, vergleichbar mit dem Information Technology Security Evaluation Manual (ITSEM) für die ITSEC. Die CSWG befaßt sich mit der Frage, inwieweit die CC das Thema Kryptographie in den IT-Sicherheitsanforderungen aufgreifen sollen. Die CC-Version 1.0 wurde schon zusammen mit dem technischen Anhang Evaluation Criteria for Cryptography (Crypto Annex) ausgeliefert. Die MRWG befaßt sich mit der gegenseitigen Anerkennung von Evaluationsergebnissen. Die Ergebnisse der Teilprojekte werden größtenteils erst nach Fertigstellung der CC-Version 2.0 vorliegen.

Außerdem werden in allen an der Entwicklung der CC beteiligten Nationen Probeevaluationen nach den CC durchgeführt. Aus verschiedenen objektiven Gründen können die Probeevaluationen nicht international abgestimmt werden. Der Informationsaustausch innerhalb der Trial Evaluations Working Group (TrialEval) wird über das CCIB abgewickelt. Die Probeevaluationen liefern den hauptsächlichen Beitrag zur Anwendungserfahrung mit den CC und sind damit Grundlage der Überarbeitung der Version 1.0 zur Version 2.0.

1.4   Ausblick

Die Common Criteria sollen nach gründlicher Erprobung und Überarbeitung die europäischen ITSEC, das amerikanische Orange Book und die kanadischen CTCPEC ersetzen. Für Deutschland bedeutet dies, daß die Common Criteria ins Deutsche übersetzt und im Bundesanzeiger veröffentlicht sein müssen, ehe sie die ITSEC offiziell als Evaluationskriterien ablösen. Dies wird noch geraume Zeit in Anspruch nehmen. Die zahlreichen Gemeinsamkeiten zwischen ITSEC und CC stellen heute schon sicher, daß der Wert von ITSEC-Zertifikaten erhalten bleibt.

In etlichen Detailfragen wird sich die neue CC-Version 2.0 von der jetzt gültigen Version 1.0 unterscheiden. Es wird erwartet, daß eine Reihe von IT-Sicherheitsanforderungen und der zugehörigen Erläuterungen überarbeitet werden. Ebenso wird damit gerechnet, daß die Frage der Abhängigkeiten unter den einzelnen Komponenten neu geregelt wird. Die Schutzprofile (Protection Profile, PP) als eines der tragenden IT-Sicherheitskonzepte werden sicher auch nicht-CC-Anforderungen enthalten dürfen, was sich u.U. auf die gegenseitige Anerkennung von nicht streng CC-konformen Schutzprofilen auswirkt. Andere Verbesserungen werden die Prägnanz und Konsistenz der CC insgesamt erhöhen. Weitere Regelungen werden das Umfeld der Common Criteria betreffen, z.B. die Registrierung von Schutzprofilen und der zukünftige Status des Interim Protection Profile Registers (IPPR).

Das ISO-Normungsvorhaben, das die Teile 1 bis 3 der Common Criteria umfaßt, nimmt ebenfalls Einfluß auf die Gestaltung der Version 2.0. Das hat im wesentlichen zwei Gründe. Erstens haben CCIB und die Arbeitsgruppe ISO/IEC JTC 1 SC 27 WG 3 vor dem Hintergrund des gemeinsamen Liason-Statements teilweise unterschiedliche Vorstellungen über den Fortgang und die Zuständigkeiten innerhalb des CC-Projekts. Zweitens ist die relative Beständigkeit eines vollwertigen ISO-Standards Auslöser der Befürchtung, man könne nicht schnell genug auf neue Entwicklungen im IT-Sicherheitsbereich reagieren.

Dies führt regelmäßig zu unterschiedlichen Vorschlägen für Handhabung und Inhalt der Common Criteria. Teilweise wird auch die vollständige Umgestaltung der einzelnen Teile der CC in Erwägung gezogen. Im Interesse der Einhaltung des Zeitplans bis zur Fertigstellung der Version 2.0 sowie der Planungssicherheit bei der Anwendung der Common Criteria wäre die konsequente Verbesserung der CC-Version 1.0 zur Version 2.0 auf der Grundlage der bisher gesammelten Anwendungserfahrung wohl die beste Lösung. Eine größere Umgestaltung der CC wäre nicht unbedingt ein technisches Problem, sondern eher eine Frage nach der Zielrichtung der Common Criteria insgesamt.

Die weltweite gegenseitige Anerkennung von Evaluationsergebnissen war und ist das erklärte Ziel der Common Criteria. Je konkreter die Common Criteria und der zugehörige ISO-Standard werden, desto weniger Spielraum bleibt für deren Auslegung und desto einfacher wäre die gegenseitige Anerkennung von Evaluationsergebnissen. Auf der anderen Seite würde ein allgemein gehaltener ISO-Standard weniger Aktualisierungsaufwand im Falle neuer Entwicklungen im IT-Sicherheitsbereich verursachen. Je nach Blickwinkel kann also die ISO-Standardisierung der Common Criteria als Vorteil oder als Nachteil gesehen werden. Es bleibt abzuwarten, wie die Entwicklung weitergeht.

2.   Common Criteria Version 1.0

Der folgende Teil stellt die Common Criteria Version 1.0 vor.

2.1   Einführung und Allgemeines Modell - Teil 1

Der Teil 1 der Common Criteria gibt eine Einführung in die IT-Sicherheitsevaluation und stellt dazu das allgemeine Modell der Vorgehensweise der Common Criteria dar. Die einzelnen Zielgruppen der CC werden angesprochen und der Anwendungsbereich der CC wird abgesteckt. Die möglichen Evaluationsergebnisse werden definiert.

Die Common Criteria sind für IT-Anwender, Hersteller und Evaluatoren gleichermaßen interessant. Vor allem IT-Anwender sollen von den Evaluationsergebnissen profitieren. Der Anwendungsbereich der Common Criteria umfaßt die durch geeignet funktionale und vertrauenswürdige Informationstechnik (IT) herzustellende IT-Sicherheit. Die IT-Sicherheit im Sinne der CC hat die Bewahrung von schützenswerten Informationen vor Verlust der Vertraulichkeit, Integrität und Verfügbarkeit zum Inhalt.

Bei der Beschreibung des IT-Sicherheitsproblems werden aus Bedrohungen, Sicherheitsstrategien und Annahmen zum sicheren Betrieb die Sicherheitsziele und daraus die IT-Sicherheitsanforderungen an Funktionalität und Vertrauenswürdigkeit abgeleitet. Das allgemeine Modell berücksichtigt jedoch nicht nur funktionale und Vertrauenswürdigkeitsanforderungen an die Informationstechnik selbst, sondern geht auch auf Aspekte der unmittelbaren Umgebung ein.

2.1.1 Evaluationsansatz der Common Criteria

Der Evaluationsansatz der CC entspricht weitgehend dem der ITSEC. Die Funktionalität wird vom Hersteller beschrieben und vom Evaluator geprüft. Dabei wird der Teil 2 der CC zur Anwendung empfohlen. Der Hersteller legt weiterhin die Stufe der Vertrauenswürdigkeit fest, die durch die Evaluation bestätigt werden soll.

Die Evaluation der Funktionalität untersucht die Konsistenz zwischen vorgegebenem und tatsächlichem Verhalten des Evaluationsgegenstandes (EVG). Daneben werden Penetrationstests für die Schwachstellenbewertung und andere Prüfungen durchgeführt.

Die Evaluation der Vertrauenswürdigkeit besteht zum großen Teil aus der Untersuchung der Korrektheit der Entwurfsunterlagen des EVG. Dazu wird die Abbildung der in den Sicherheitsvorgaben auf hoher Ebene beschriebenen Funktionen auf die niedrigeren Ebenen der Spezifikation des EVG untersucht.

Die Anforderungen an die Wirksamkeit sind Teil der Vertrauenswürdigkeitsstufen, der Evaluation Assurance Levels (EAL). Die Wirksamkeit geht nicht separat in das Evaluationsergebnis ein.

Die CC unterscheiden nicht zwischen IT-Sicherheitsprodukten und -systemen. Es kann Hardware und Software nach den CC evaluiert werden, wobei der Schwerpunkt bei der Software liegt.

Verschiedene Bereiche, die die IT-Sicherheit unterstützen, lassen sich nicht mit Hilfe der Common Criteria bewerten. Dies betrifft insbesondere
- physikalische Aspekte, z.B. kompromittierende Abstrahlung,
- Spezialfälle der IT-Sicherheit, z.B. Stärke kryptographischer Algorithmen,
- juristische und verwaltungstechnische Vorgaben, z.B. Zulassung von IT-Sicherheitsprodukten für die Verschlußsachenbearbeitung sowie
- Systemakkreditierung, d.h. Interpretation von CC-Evaluationsergebnissen in einem umfassenden IT-Sicherheitskonzept.

2.1.2 Evaluationsergebnisse der Common Criteria

Die CC lassen frei definierte funktionale Sicherheitsanforderungen zu, d.h. die Nutzung des Teils 2 der CC zur Beschreibung der Funktionalität ist freigestellt. Die Evaluation nach den CC baut jedoch, ähnlich den ITSEC, immer auf eine Stufe der Vertrauenswürdigkeit auf (EAL). Die EAL können durch Vertrauenswürdigkeitsanforderungen aus dem Teil 3 der CC ergänzt werden.

Die Evaluationsergebnisse der Common Criteria orientieren sich daran, welche Rolle die Anforderungen der CC als Grundlage der Evaluation spielen. Eine Übersicht liefert die Tabelle 1 (siehe unten).

CC-Evaluationsergebnisse	Grundlage der Evaluation
konform zu Teil 2	funktionale Anforderungen aus Teil 2
Teil 2 erweitert (extended)	funktionale Anforderungen aus Teil 2 sowie frei definierte funktionale Anforderungen
konform zu Teil 3	EAL aus Teil 3
Teil 3 mit Zusatz (augmented)	EAL aus Teil 3 sowie weitere Vertrauenswürdigkeitsanforderungen aus Teil 3
Teil 3 erweitert (extended)	EAL aus Teil 3, weitere Vertrauenswürdigkeitsanforderungen aus Teil 3 sowie frei definierte Vertrauenswürdigkeitsanforderungen
konform zu einem Schutzprofil (Protection Profile)	Schutzprofil

Tabelle 1 - CC-Evaluationsergebnisse und Grundlagen der Evaluation.

2.1.3 Schutzprofil (Protection Profile)

Ein Schutzprofil (Protection Profile, PP) ist die Darstellung eines IT-Sicherheitsproblems und einer allgemeinen, CC-konformen Lösung. Aus einem Schutzprofil können verschiedene Sicherheitsvorgaben (Security Target, ST) einfach abgeleitet werden, so daß ein Schutzprofil auf viele Produkte anwendbar ist. Damit wird der umfassende Ansatz eines Schutzprofils erkennbar. Das Konzept der Schutzprofile stammt aus den Federal Criteria.

Ein Schutzprofil thematisiert das IT-Sicherheitsproblem, indem zunächst die relevanten Bedrohungen, Sicherheitsstrategien und Annahmen zum sicheren Betrieb dargestellt werden. Daraus werden dann IT- und nicht-IT-Sicherheitsziele abgeleitet. Aus diesen folgen zuletzt die IT-Sicherheitsanforderungen an Funktionalität und Vertrauenswürdigkeit sowie die Anforderungen an die Umgebung des Evaluationsgegenstandes. Der Teil 1 beschreibt die Struktur von Schutzprofilen.

Schutzprofile sind insbesondere für IT-Anwender interessant. Schutzprofile nach CC-Version 1.0 müssen einen EAL enthalten und dürfen ansonsten nur Sicherheitsanforderungen aus den Teilen 2 und 3 verwenden.

Ein Schutzprofil muß nach der Klasse Schutzprofil-Evaluation im Teil 3 der CC evaluiert sein, bevor es Bestandteil der Common Criteria wird. Damit wird sichergestellt, daß ein Schutzprofil in sich geschlossen, sinnvoll und widerspruchsfrei ist. Dies ist besonders für Hersteller im Vorfeld der Entwicklung eines IT-Sicherheitsprodukts interessant. Die Verwendung eines Schutzprofils als Grundlage einer Evaluation ist freigestellt.

2.1.4 Sicherheitsvorgaben (Security Target)

Die Sicherheitsvorgaben (Security Target, ST) sind stets Ausgangspunkt einer Produktevaluation. Sicherheitsvorgaben und Schutzprofile sind zueinander kompatibel, enthalten im Kern also die gleichen Informationen. Die Sicherheitsvorgaben enthalten darüber hinaus die notwendigen produktspezifischen Details und stellen eine Übersichtsspezifikation des Evaluationsgegenstandes bereit. Anhand der Klasse Sicherheitsvorgaben-Evaluation im Teil 3 der CC wird evaluiert, ob die Sicherheitsvorgaben in sich geschlossen, sinnvoll und widerspruchsfrei sind. Danach beginnt die eigentliche Produktevaluation.

Sicherheitsvorgaben nach CC-Version 1.0 müssen einen EAL enthalten und sind ansonsten bzgl. der Definition von Anforderungen frei. Die möglichen Evaluationsergebnisse können aus der Tabelle 1 (siehe oben) entnommen werden. Der Teil 1 beschreibt die Struktur von Sicherheitsvorgaben.

2.1.5 Gliederungsprinzip der Sicherheitsanforderungen

Die IT-Sicherheitsanforderungen in den CC gliedern sich nach dem Schema Klasse - Familie - Komponente - Element, wie Abbildung 1 zeigt (siehe unten).

Eine Klasse deckt ein ganzes Gebiet von IT-Sicherheitsanforderungen ab und ordnet
die einzelnen Teilgebiete unter einen Oberbegriff.

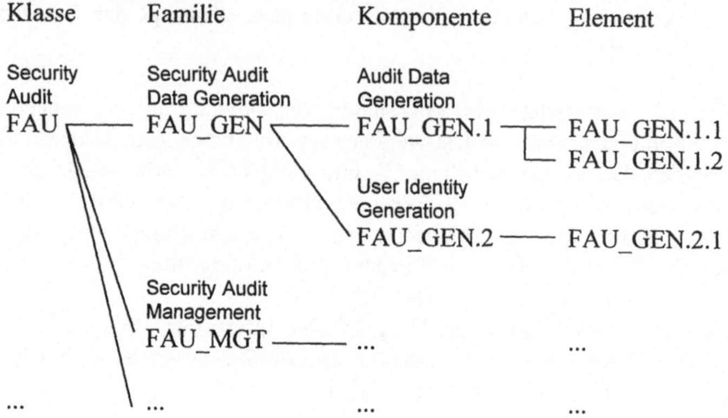

Abbildung 1 - Gliederungsprinzip der IT-Sicherheitsanforderungen in den CC

Die Familien einer Klasse decken ein jeweils anderes Teilgebiet von IT-Sicherheitsanforderungen ab, ordnen sich jedoch dem gemeinsamen Ziel ihrer Klasse unter.

Die Komponenten einer Familie decken stets das gleiche Teilgebiet von
IT-Sicherheitsanforderungen ab. Sie unterscheiden sich jedoch in der Breite und der
Schärfe der Anforderungen. Komponenten können verschieden hierarchisch
zueinander stehen. Der Teil 3 der CC enthält nur streng hierarchische Komponenten in
den Familien, d.h. hier gilt die Aussage, je höher die Einordnung, desto mehr
Vertrauenswürdigkeit ist bei der Erfüllung der Komponente gegeben.

Der Teil 2 der CC enthält streng hierarchische, teilhierarchische und nicht
hierarchische Komponenten in den Familien, d.h. hier wäre im Einzelfall zu prüfen, ob
durch eine höhere Einordnung mehr Funktionalität bereitgestellt wird. Mehr
Funktionalität bedeutet nicht unbedingt mehr Vertrauenswürdigkeit.

Die Komponenten sind die kleinste auswählbare Einheit von IT-Sicherheitsanforderungen für Schutzprofile und Sicherheitsvorgaben. Die Elemente einer
Komponente sind die kleinste in den CC identifizierte Einheit von IT-Sicherheitsanforderungen und werden bei einer Evaluation einzeln auf ihre Erfüllung durch einen
Evaluationsgegenstand geprüft.

2.2   Funktionale Sicherheitsanforderungen - Teil 2

Der Teil 2 der Common Criteria enthält einen umfangreichen Katalog von
funktionalen Anforderungen zur IT-Sicherheit. Der Teil 2 kommt insbesondere bei der
Erstellung von Sicherheitsvorgaben und Schutzprofilen zur Anwendung. Er ist aber
auch zur Auswahl von technischen IT-Sicherheitsmaßnahmen außerhalb von

CC-Anwendungen geeignet. Der Teil 2 wird zur Beschreibung der Funktionalität eines Evaluationsgegenstandes (EVG) empfohlen.

Die Funktionalität eines EVG dient der Abwehr von Bedrohungen in der Einsatzumgebung und der Umsetzung der Sicherheitsstrategien bzw. dem Erreichen der gesteckten Sicherheitsziele. Die enthaltenen funktionalen Sicherheitsanforderungen sind vielfach erprobt. Bei prinzipiell neuen, nicht durch den Teil 2 abgedeckten Bedrohungslagen, Sicherheitszielen usw. kann die Funktionalität eines EVG immer frei definiert werden.

2.2.1 Aufbau der funktionalen Sicherheitsanforderungen

Die funktionalen Sicherheitsanforderungen sind alphabetisch nach Klassen und Familien geordnet. Zu jeder Klasse, Familie und Komponente werden Zusatzinformationen bereitgestellt, die verschiedene Hinweise zur Anwendung geben. Ein Großteil dieser Hinweise ist im Anhang zum Teil 2 enthalten.

Viele Komponenten nehmen Einfluß auf die Protokollierung sicherheitsrelevanter Aktionen. Dadurch ergibt sich aus den gewählten Komponenten eine Liste von mindestens notwendigen Eintragungen in das Protokoll.

Viele Komponenten nehmen über die Abhängigkeiten Einfluß auf die Auswahl anderer Komponenten, indem sie das Vorhandensein einer bestimmten Funktionalität verlangen oder bestimmte Maßnahmen zur Vertrauenswürdigkeit voraussetzen. Daraus ergibt sich eine ausgewogene Auswahl von Komponenten, z.B. für ein Schutzprofil.

Viele Komponenten können über erlaubte Operationen an einen bestimmten EVG angepaßt werden. Dafür sind Auswahlmöglichkeiten, Zuweisungen und Verfeinerungen vorgesehen. Als Beispiel für eine Auswahlmöglichkeit sei die von den Komponenten verlangte Tiefe der Protokollierung 'minimum', 'basic' oder 'detailed' genannt.

Innerhalb einer Hierarchie von Komponenten werden neu hinzukommende oder veränderte Sicherheitsanforderungen fett dargestellt. Damit ist leicht zu erkennen, was sich beim Übergang zur nächste Komponente für einen Evaluationsgegenstand ändert.

2.2.2 Klassen funktionaler Sicherheitsanforderungen

Neben den erprobten Sicherheitsanforderungen, z.B. für die Identifizierung und Authentisierung, werden in den Common Criteria neue Klassen funktionaler Sicherheitsanforderungen vorgestellt, z.B. die Klasse Schutz der Privatsphäre, die funktionale Anforderungen für den Datenschutz bereitstellt.

Die gewachsene Bedeutung von Integrität und Verfügbarkeit für heutige, sichere IT-Anwendungen spiegelt sich in der Verteilung dieser Schwerpunkte innerhalb der

Klassen wider, z.B. in der Klasse Schutz der vertrauenswürdigen Sicherheitsfunktionen.

Die CC-Version 2.0 wird eventuell schon funktionale Anforderungen zur Kryptographie enthalten. Dies betrifft nicht die Krypto-Algorithmen selbst, sondern deren
Zusammenwirken mit den bereits bestehenden funktionalen Anforderungen. Damit
rückt die Kryptographie erstmals aus Anhängen, Ergänzungen und anderen
Einzelbeiträgen in den Hauptteil eines Kriterienwerks zur IT-Sicherheit vor, was ihrer
heutigen Bedeutung gerecht wird.

Der Umfang des Teils 2 der CC läßt die Behandlung der einzelnen Klassen
funktionaler Sicherheitsanforderungen in diesem Rahmen nicht zu.

2.3   Sicherheitsanforderungen zur Vertrauenswürdigkeit - Teil 3

Der Teil 3 der Common Criteria enthält die Anforderungen zur Vertrauenswürdigkeit
der IT-Sicherheitsfunktionalität und definiert die mit den CC erreichbaren Vertrauenswürdigkeitsstufen (EAL). Der Teil 3 kommt insbesondere bei der Erstellung von
Sicherheitsvorgaben und Schutzprofilen zur Anwendung, wenn es um die Auswahl
eines EAL und eventuell notwendiger Zusätze geht. Die gemachten Aussagen zum
Aufbau der funktionalen Sicherheitsanforderungen gelten sinngemäß auch für den
Teil 3 der Common Criteria (Abbildung 1, siehe oben).

2.3.1 Vertrauenswürdigkeitsstufen

Die Evaluation nach den Common Criteria baut immer auf eine Stufe der Vertrauenswürdigkeit auf (Evaluation Assurance Level). Die einzelnen Stufen der Vertrauenswürdigkeit sind streng hierarchisch definiert, d.h. beim Übergang zu einer höheren
Stufe wird die erreichte Vertrauenswürdigkeit größer.

Es gibt sieben EAL in der CC-Version 1.0 (EAL 1, ..., EAL 7). Die EAL der Common
Criteria entsprechen inhaltlich in etwa den E-Stufen der ITSEC. Der EAL 1 liegt
unterhalb von E1 der ITSEC und wurde entwickelt, um Herstellern von IT-Sicherheitsprodukten den Zugang zur Evaluation zu erleichtern. Damit ergibt sich eine
Vergleichbarkeit von CC EAL 2 mit ITSEC E1, CC EAL 3 mit ITSEC E2 usw.

Die Common Criteria definieren die EAL über eine Tabelle, in der alle zum jeweiligen
EAL gehörenden Komponenten aufgeführt werden. Dadurch wird der Zuwachs an
Vertrauenswürdigkeit beim Übergang zu einem höheren EAL sofort erkennbar. In den
alphabetisch nach Klassen und Familien geordneten Anforderungen zur Vertrauenswürdigkeit lassen sich Details leicht nachlesen.

Zusätzlich werden die EAL ähnlich den E-Stufen der ITSEC im Volltext wiedergegeben. Bei einmal ausgewähltem EAL stehen dadurch alle notwendigen
Informationen zur Vertrauenswürdigkeit auf wenigen Seiten zusammengefaßt bereit.

Alle ITSEC-Aspekte und -Phasen zur Vertrauenswürdigkeit lassen sich in den CC wiederfinden, was die Wertstabilität von ITSEC-Zertifikaten unterstreicht. Die Tabelle 2 gibt die Einzelheiten wieder (siehe unten).

ITSEC	Common Criteria
- Anforderungen	- Funktionale Spezifikation
- Architekturentwurf	- Entwurf auf hoher Ebene
- Feinentwurf	- Entwurf auf niedriger Ebene
- Implementierung	- Ausführung der Implementierung - Funktionale Tests
- Konfigurationskontrolle	- Automatisierung des Konfigurationsmanagements - Leistungsvermögen des Konfigurationsmanagements - Anwendungsbereich des Konfigurationsmanagements
- Programmiersprachen und Compiler	- Werkzeuge und Techniken
- Sicherheit beim Entwickler	- Sicherheit bei der Herstellung
- Benutzerdokumentation	- Benutzerhandbuch
- Systemverwalter-Dokumentation	- Systemverwalter-Handbuch
- Anlauf und Betrieb	- Installierung, Generierung und Anlauf
- Auslieferung und Konfiguration	- Auslieferung
- Eignung der Funktionalität - Zusammenwirken der Funktionalität	- Evaluation der Sicherheitsvorgaben
- Stärke der Mechanismen	- Stärke der EVG-Sicherheitsfunktionen
- Bewertung der Konstruktions-schwachstellen - Bewertung der operationellen Schwachstellen	- Schwachstellenanalyse
- Benutzerfreundlichkeit	- Mißbrauch

Tabelle 2 - Vergleich der Vertrauenswürdigkeitsaspekte in ITSEC und CC.

2.3.2 Sicherheitsanforderungen zur Vertrauenswürdigkeit

Der Umfang des Teils 3 der Common Criteria läßt die Behandlung der einzelnen Klassen von Vertrauenswürdigkeitsanforderungen in diesem Rahmen nicht zu. Hier sollen nur einige besondere Aspekte kurz erläutert werden.

Der Teil 3 der CC enthält die Klassen Schutzprofil-Evaluation und Sicherheits-vorgaben-Evaluation. Diese beiden Klassen nehmen eine Sonderrolle gegenüber den anderen Klassen von Vertrauenswürdigkeitsanforderungen ein, da sie mehr die Voraussetzungen für eine erfolgreiche Evaluation schaffen als sich mit den Spezifika eines Evaluationsgegenstandes selbst auseinanderzusetzen. Die Aufnahme dieser Klassen in die CC resultiert aus der Erkenntnis, daß viele Hersteller von IT-Sicherheitsprodukten Unterstützung bei der Formulierung von kriterienkonformen Sicherheitskonzepten erwarten.

Der Teil 3 der CC enthält die Klasse Schwachstellenbewertung, die auch den ITSEC-Aspekt Stärke der Mechanismen abbildet. Die Stärke von Sicherheitsmechanismen wird nur noch dann untersucht, wenn diese Angabe für den Evaluationsgegenstand sinnvoll ist. Die beiden ITSEC-Aspekte Eignung der Funktionalität und Zusammenwirken der Funktionalität werden bereits bei der Evaluation der Sicherheitsvorgaben abgedeckt.

Der Teil 3 der CC enthält verschiedene Anforderungen an die Vertrauenswürdigkeit, die erst nach dem Abschluß einer Evaluation auf ihre Erfüllung geprüft werden können, z.B. die Familie Fehlerbehebung aus der Klasse Lebenszyklus-Sicherung. Solche Anforderungen sind nicht in der Definition der Vertrauenswürdigkeitsstufen (EAL) enthalten, können aber als Zusatz zu einem EAL verwendet werden. Damit kann während der Evaluation wenigstens das zugrundeliegende Konzept geprüft werden, z.B. für die Fehlerbehebung.

2.4   Vordefinierte Schutzprofile - Teil 4

Der Teil 4 der Common Criteria enthält drei vordefinierte Schutzprofile und ist ein Beispiel für die Anwendung der Common Criteria Version 1.0 auf verschiedene Felder der IT-Sicherheit. Ein Schutzprofil (Protection Profile, PP) ist die Darstellung eines IT-Sicherheitsproblems und einer allgemeinen, CC-konformen Lösung. Durch das Einbringen der produktspezifischen Details und einer Übersichtsspezifikation des Evaluationsgegenstandes (EVG) werden aus dem allgemeinen Schutzprofil EVG-spezifische Sicherheitsvorgaben.

Der Teil 4 der Common Criteria enthält drei Schutzprofile, Commercial Security 1 (CS1), Commercial Security 3 (CS3) und Network / Transport Layer Packet Filter Firewall (PFFW). Ein noch zu etablierendes Registraturverfahren, möglicherweise auf der Basis des Interim Protection Profile Register (IPPR), soll die gleichbleibende Qualität und die weltweite Anerkennung der Schutzprofile gewährleisten. Registrierte Schutzprofile sind nach Maßgabe der CC-Version 1.0 Bestandteil der Common Criteria selbst.

Der Aufbau eines Schutzprofils widerspiegelt die Herangehensweise an ein IT-Sicherheitsproblem. Im Teil 1 der CC wird die Struktur und der Inhalt eines Schutzprofils definiert.

Das vorliegende IT-Sicherheitsproblem wird aus verschiedenen Blickrichtungen auf den EVG und die unmittelbare Umgebung betrachtet. Dies sind zunächst die relevanten Bedrohungen, denen der EVG während des Betriebs ausgesetzt ist. Dies sind weiterhin die organisatorischen Sicherheitsstrategien, denen der EVG entsprechen muß. Außerdem sind dies die Annahmen zum sicheren Betrieb, deren Berechtigung in der Verantwortung der Umgebung des EVG liegt.

Aus Bedrohungen, Sicherheitsstrategien und Annahmen zum sicheren Betrieb werden IT-Sicherheitsziele und nicht-IT-Sicherheitsziele abgeleitet. Dies ist eine Zusammenfassung dessen, was durch den EVG und seine unmittelbare Umgebung erreicht werden soll.

Aus den IT-Sicherheitszielen werden immer funktionale und Vertrauenswürdigkeitsanforderungen an den EVG abgeleitet. Ebenso können sich auch Anforderungen an die IT-Umgebung des EVG ergeben. Dies ist dann der Fall, wenn die Erfüllung aller sich ergebenden funktionalen und Vertrauenswürdigkeitsanforderungen durch den EVG selbst nicht möglich oder nicht angemessen ist.

Verschiedene von den CC geforderte Zusatz- und Hintergrundinformationen zum Schutzprofil bilden einen speziellen Erläuterungsteil (Rationale). Ein Rationale ist in erster Linie für die Evaluation des Schutzprofils interessant.

Der Umfang des Teils 4 der Common Criteria läßt nur die stichwortartige Behandlung der Schutzprofile zu.

2.4.1 Commercial Security 1 (CS1)

Der Teil 4 der CC enthält das Schutzprofil Commercial Security 1 (CS1) zum Thema Basic Controlled Access Protection. CS1 setzt das Schutzprofil FC CS1 der Federal Criteria in die Common Criteria um. Dieses Schutzprofil FC CS1 wiederum ist die Umsetzung der Klasse C2 des Orange Book. Aus der Sicht der Funktionalität gibt es damit auch eine Beziehung zwischen der ITSEC-Funktionalitätsklasse F-C2 und dem Schutzprofil CS1 der Common Criteria.

Das Schutzprofil CS1 behandelt das klassische Sicherheitsproblem, bei dem mehrere Benutzer gleichzeitig auf eine gemeinsame Ressource zugreifen. Typische Anwendungen für CS1 sind Betriebssysteme und Datenbanken. CS1 kompatible Produkte schützen vor zufälligen oder nicht systematischen Versuchen, die IT-Sicherheit zu kompromittieren. Dabei kommt die benutzerbestimmbare Zugriffskontrolle (DAC) zum Einsatz.

2.4.2 Commercial Security 3 (CS3)

Der Teil 4 der CC enthält das Schutzprofil Commercial Security 3 (CS3) zum Thema Role-Based Access Protection. CS3 setzt das Schutzprofil FC CS3 der Federal Criteria in die Common Criteria um.

Das Schutzprofil CS3 behandelt ebenso wie CS1 das klassische Sicherheitsproblem, bei dem mehrere Benutzer gleichzeitig auf eine gemeinsame Ressource zugreifen. Typische Anwendungen für CS3 sind Betriebssysteme und Datenbanken, aber auch Vorgangsbearbeitungssysteme. CS3 bietet starke Sicherheitsfunktionen und ist für Anwendungen geeignet, bei denen ein Ausfall der Sicherheitsfunktionen nicht toleriert werden kann.

Die wichtigste Eigenschaft ist die Unterscheidung von Rollen, die den Benutzern durch den Administrator zugewiesen werden. Die Rollen in CS3 bestehen aus der Verbindung von Benutzer, erlaubten Operationen und freigegebenen Datenobjekten. Dies kann auch zur Einschränkung der Rechte eines Administrators selbst verwendet werden. Daneben kommt die benutzerbestimmbare Zugriffskontrolle (DAC) zum Einsatz.

Das Schutzprofil CS3 ist in der Schlußphase der Erarbeitung der CC-Version 1.0 entstanden, so daß späte Änderungen besonders am Teil 2 der Common Criteria nicht mehr berücksichtigt werden konnten. Die intensive Prüfung der Konformität zum Schutzprofil FC CS3 steht noch aus.

2.4.3 Netzwerk/Transportschicht Paketfilter Firewall (PFFW)

Der Teil 4 der CC enthält das Schutzprofil Network / Transport Layer Packet Filter Firewall (PFFW). Das Schutzprofil PFFW wurde entwickelt, um die Konzepte der Common Criteria an einem für IT-Sicherheitskriterien neuen Anwendungsbereich zu erproben. Das Schutzprofil PFFW behandelt das Sicherheitsproblem, bei dem ein Computernetz über eine gesicherte Schnittstelle (Firewall) an ein anderes, unsicheres Netz angeschlossen wird.

Wichtigstes Merkmal des Schutzprofils ist die Beschränkung der Kontrolle auf die Paket-Ebene der verwendeten Protokolle. Dies hat die Konsequenz, daß erstens ankommende und abgehende Pakete nur anhand der in jedem einzelnen Paket verfügbaren Informationen identifiziert werden müssen (Konzept des maschinellen Benutzers ohne Authentisierung).

Zweitens findet keine Untersuchung des Inhalts aufeinander folgender Pakete statt. Daher können nach einem einmal erlaubten Verbindungsaufbau ohne weitere Kontrollen beliebige Informationen übertragen werden.

Das Schutzprofil PFFW stellt nur minimale Anforderungen an die Vertrauenswürdigkeit und soll insgesamt den Einstieg in die Evaluation von IT-Sicherheitsprodukten erleichtern.

3. Zusammenfassung

Die Common Criteria Version 1.0 ist das Ergebnis der Harmonisierung von Orange Book / Federal Criteria, CTCPEC und ITSEC. Das gesamte CC-Projekt wird von den USA, Kanada, Frankreich, Großbritannien, den Niederlanden und Deutschland getragen. In mehreren CC-Teilprojekten werden spezielle Aspekte der Common Criteria und des Umfeldes bearbeitet.

Die Common Criteria sind Gegenstand eines Normungsvorhabens bei der Internationalen Standardisierungsorganisation.

In Probeevaluationen von IT-Sicherheitsprodukten nach der CC-Version 1.0 wurden zahlreiche Anwendungserfahrungen gesammelt, die in die Erarbeitung der Version 2.0 einfließen. Die Version 2.0 wird nicht vor Ende 1997 fertiggestellt sein.

Die Common Criteria sollen in der Perspektive die europäischen ITSEC, das amerikanische Orange Book und die kanadischen CTCPEC ersetzen. Die zahlreichen Gemeinsamkeiten zwischen ITSEC und CC stellen heute schon sicher, daß der Wert von ITSEC-Zertifikaten erhalten bleibt.

Der Teil 1 der Common Criteria gibt eine Einführung in die IT-Sicherheitsevaluation und stellt dazu das allgemeine Modell der Vorgehensweise der Common Criteria dar. Der Teil 2 enthält einen umfangreichen Katalog von funktionalen Anforderungen zur IT-Sicherheit. Der Teil 3 der Common Criteria enthält die Anforderungen zur Vertrauenswürdigkeit der IT-Sicherheitsfunktionalität und definiert die mit den CC erreichbaren Vertrauenswürdigkeitsstufen. Der Teil 4 enthält drei vordefinierte Schutzprofile und ist ein Beispiel für die Anwendung der Common Criteria auf verschiedene Felder der IT-Sicherheit.

Die Zielgruppen der Common Criteria sind Hersteller, IT-Anwender und Evaluatoren. Hersteller und IT-Anwender können über Schutzprofile und Sicherheitsvorgaben maßgeschneiderte IT-Sicherheitskonzepte erstellen. Evaluatoren prüfen vor der eigentlichen Evaluation die Geschlossenheit, Sinnhaftigkeit und Widerspruchsfreiheit des IT-Sicherheitskonzepts und können sich dabei auf umfangreiche Hintergrundinformationen in den CC abstützen.

Literatur

Brauer

Akzeptierbare Sicherheit - Common Criteria : internationale Sicherheitskriterien, unix open, 1/96, pp.86-88

Brauer, van Essen

Common Criteria - Gemeinsame Kriterien für die Prüfung und Bewertung der Sicherheit von Informationstechnik, KES, 4/96, pp. 48-51

Brauer

Durchgecheckt - Common Criteria : Internationaler Standard für Sicherheitskriterien, c't, 8/96, pp. 64-65

Stevens

The Common Criteria and Distributed Network Security, PhD Thesis, Manchester Metropolitan University, 1997

Tober

Same Security for All ?, BYTE, 12/96, pp. 40 IS 7-10

CC   Common Criteria for Information Technology Security Evaluation, Version 1.0, Common Criteria Editorial Board, January 1996

CTCPEC   Canadian Trusted Computer Product Evaluation Criteria, Version 3.0, Can. System Security Centre, Comm. Security Establishment (CSE), January 1993

FC   Federal Criteria for Information Technology Security, Draft Version 1.0, jointly publ. by Nat. Inst. of Standards and Technology (NIST), Nat. Security Agency (NSA) and US Government, January 1993

ITSEC   Information Technology Security Evaluation Criteria, Version 1.2, ISBN 92-826-3003-X, Commission of the European Communities, June 1991

ITSEM   Information Technology Security Evaluation Manual, Version 1.0, ISBN 92-826-7087-2, Commission of the European Communities, September 1993

TCSEC   Trusted Computer System Evaluation Criteria, DoD 5200.28-STD, US Dept. of Defense, December 1985 (Orange Book)

Anhang

Das BSI gibt auf Anfrage Möglichkeiten zum Bezug der CC bekannt (Tel. 0228-9582-300; Fax 0228-9582-427; e-mail cc@bsi.de; Anschrift s.o.).

Combining Assessment Techniques from Security and Safety to Assure IT System Dependability - The SQUALE Approach

Alan Hawes[*], Angelika Steinacker[**]

[*]Admiral Management Services Limited
Kings Court, 91-93 High Street, Camberley, Surrey GU15 3RN UK
Email: Alan_Hawes@compuserve.com

[**]Industrieanlagen-Betriebsgesellschaft mbH, Dept CC33
Einsteinstr. 20, D-85521 Ottobrunn
Email: steinacker@iabg.de

Abstract. SQUALE stands for „Security, Safety and Quality Evaluation for Dependable Systems" and the goal of SQUALE project is to develop a scheme for the assessment of systems with a wide range of dependability requirements, using a common dependability framework, and combining techniques from current and emerging standards. We present the ideas of the project in order to promote discussion of the issues, and provide publicity for the new concepts which have been developed. To illustrate the application of the concepts an example is given which consists of the examination of a part of a railway signalling installation.

1   Introduction

With the ever increasing use of IT in the modern world, greater reliance is being placed on the dependability of IT systems. Quite often the replacement of more traditional approaches, with the automation which IT can provide, brings great advantages in the cost, effectiveness or convenience of systems, but these are sometimes gained at the cost of a reduction in the dependability of such systems. In this paper we address the issue of the dependability of IT systems, a concept concerned with the level of trust or confidence that can be placed in a system's correct and effective operation.

The basic problem of ensuring that IT systems are sufficiently dependable has been addressed in great depth by those concerned about the safety or security of those systems which have IT components. They have developed effective techniques for assessing systems, their design, implementation and operation, to ensure that a suitably high level of trust can be placed in them. Many current and emerging standards provide guidance on how particular aspects of a systems dependability (e.g. safety or confidentiality) can be assessed, for each of a wide range of application fields.

Increasingly, as such systems are taking over more complex functions, the users are insisting that the full range of aspects or attributes of dependability should be assured. So, for example, a system may need to preserve confidentiality and be highly available, or it may need to be safe in an environment of malicious interference.

The new approach taken by the SQUALE project is to develop a scheme for the assessment of systems with a wide range of dependability requirements, using a common dependability framework, and combining techniques from current and emerging standards. To do this, it has taken an existing generic approach to the issues of dependability, and extended this to develop a set of assessment techniques or Criteria [DPCRIT-D2.0], which can be instantiated for a range of application fields, and hence applied to a wide range of systems. The existing generic dependability framework has thus been extended by the project, from a theoretical concept, to a practically reality, which can be used in an industrial context.

The project has concentrated on the issue of assurance of dependability of systems, rather than defining ways in which they should be designed, implemented or operated. It has, as a consequence of this focus, produced a set of dependability Criteria, which are a set of techniques for the assessment of dependability attributes, which are used together to provide the required evidence of dependability. This is a different approach from that of the Common Criteria [CC 96], which has Protection Profiles, having a combination of defined security functions and assurance. SQUALE is based on the ITSEC [ITSEC 91] type of criteria, as the flexibility of defining assurance requirements, independently of functionality, is needed for the wide range of system types which are to be addressed by the SQUALE criteria.

We present some of the ideas from the SQUALE project, in a simplified form, in order to promote discussion of the issues, and provide publicity for the new concepts which have been developed. The concepts are the result of the work of all the project partners, as identified below, and not just the ideas of the authors. The project is partly funded by the European Commission, as part of its ACTS programme.

This paper provides some information on the SQUALE project, introduces the dependability concepts on which the approach is based, and describes the SQUALE Criteria which have been defined by the project. It then provides a worked example of the application of the Criteria, in order to illustrate the use of the dependability concepts and show how the Criteria can be used in practice.

2   The SQUALE Project

2.1   Background to the Project

The need for reliability, safety and security in IT systems is rapidly growing as those systems are now increasingly used for business and life critical applications. Failures of IT systems, may their origin be accidental, due to program or equipment fault or due to deliberate attack, become highly critical. So confidence is needed in the correctness and effectiveness of any safeguard measures implemented to avoid system failures or to minimise their impact.

The need for a generic concept combining reliability, safety and security aspects is not new. The definition of "dependability" which tries to cover all these aspects, was introduced some years ago, but until now, no generic framework to gain confidence in the correctness and effectiveness of systems has been introduced which covers all dependability aspects. The main objective of the SQUALE project is therefore to provide such a framework and test its applicability by applying it to systems with critical requirements in a range of dependability areas.

There are several standards and techniques which address the problem of selecting the appropriate measures to counter threats, and of achieving confidence in the correctness and effectiveness of those measures. The safety and security sectors have developed their own techniques and standards in this area. But due to the different viewpoints, and the fact that they have developed their standards separately from each other, the techniques used in the two sectors are not harmonised, and show substantial differences, although they are intended to address mainly the same problem. The reason for those differences arises from the different viewpoint of the two sectors.   The safety sector deals mainly with failures arising from faulty equipment or accidental human errors. In contrast, the security sector deals mainly with the prevention of deliberate attacks against the system.

In many cases it cannot be determined whether a failure was caused deliberately, by accident or by equipment fault. The consequences are in any case identical. Also the countermeasures proposed by both sectors are very similar. Therefore a distinction between countermeasures resulting from safety and security is not useful. A harmonised approach to gain confidence in the correctness and effectiveness of critical functions of a system is needed.

A problem that arises with a framework covering all dependability aspects is the possibility that requirements from different dependability areas may overlap or even be contradictory. Hence a simple combination of practices used in the safety domain and those used in the security domain is not guaranteed to result in a design satisfying both requirements. A harmonised approach is needed to address safety and security requirements simultaneously. But in the present state of the art it is very difficult to identify and define all these requirements, and to design and implement a system that can satisfy all of them with an appropriate level of confidence.

2.2   Outline of the Project

The SQUALE project has the objective to set up such a harmonised approach by analysing and combining existing methods from both fields, and defining a common approach taking the benefits of methods used in the individual sectors. To validate the usability of the new combined method it is to be applied to systems with critical requirements in both safety and security.

The objectives of SQUALE are therefore:

- the definition of a harmonised scheme to measure the confidence in the dependability of IT systems;
- the validation of the approach using systems with critical dependability requirements as demonstrators.

The SQUALE project has analysed the existing standards and practices in the safety and security areas and defined a combined harmonised approach to gain confidence in the correctness and effectiveness of systems with safety and security requirements. The project has defined draft dependability Criteria describing this harmonised approach [DPCRIT-D2.0] and will apply these Criteria to a demonstrator system.

The new Criteria are based on the many existing standards in the safety and security world, e.g. the ITSEC standard [ITSEC 91], which provides the basis for the Security Evaluation of IT systems and products throughout Europe and several other countries, and the draft IEC 1508 standard [IEC 1508], which is intended to set out a generic approach for all Safety Lifecycle activities for electrical/electronic/ programmable electronic systems. A mapping to existing standards (which all cover only specific dependability aspects) is needed and will be provided as part of the project. This will ensure that the Criteria and practices developed can (and will) be applied in practice.

The main results of the project will be:

- an overview and analysis of existing safety and security standards and practices
- generic dependability Criteria describing how confidence in the correctness and effectiveness of functions in systems with safety and security requirements can be gained
- the application of those Criteria to existing systems with safety and security requirements
- the results of the discussion of the mapping to existing standards with standardisation organisations.

The project is intended to develop an approach to increase the confidence in dependable systems significantly. The partners within the project combine long years of experience of development and assessment in the safety and security areas. The SQUALE partners are:

- Conception et Réalisation d'Application Automatisées-Décision Internationnal (CR2A-DI)
- Industrieanlagen-Betriebsgesellschaft mbH (IABG)
- Admiral Management Services Limited (Admiral)
- Laboratoire d'Analyse et d'Architecture des Systèmes (LAAS)
- Matra Transport (MATRA)

The duration of the project is two years and four months, and we have reached just over half way through the workplan. The initial research and analysis needed to produce the draft Criteria has been completed, and the remaining work will focus on the validation of the approach through use of the Criteria on demonstrator projects.

3   Dependability Concepts

We explain the dependability concepts used by SQUALE in this section of the paper. Some definitions are needed for the terms which are used to describe dependability issues. These definitions are not only of academic interest but useful and necessary for the understanding of the dependability problems of IT systems. The background of the terminology presented here originates in the Fault-Tolerant Computing community (as described in [LAP 95] for example) but has much wider applicability.

Dependability is defined as that property of a computer system such that reliance can justifiably be placed on the service it delivers. The service delivered by a system is its behaviour as it is perceived by its user(s); a user is another system (human or physical) which interacts with the system under consideration.

Dependability may be viewed according to different properties which can be grouped into three classes as Figure 1 shows.

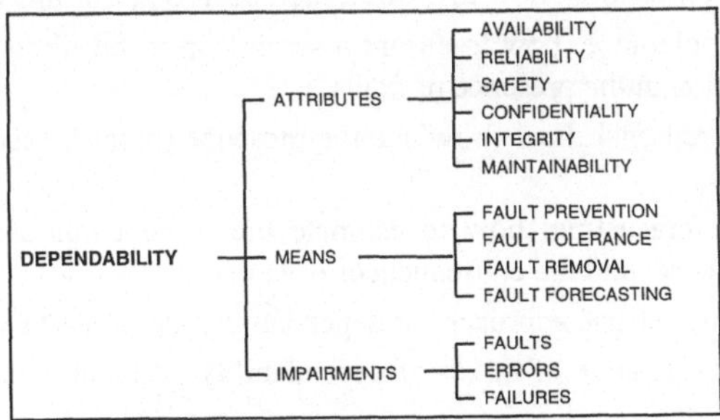

Figure 1: Dependability Classes

The different attributes are useful to understand the different needs for dependability for the system as the user sees it. The differentiation between fault, error and failure is helpful because it enables the user to see clearly that there is a separation between sources of failures and their effects, what effect impairments to systems can have, and what the logical relationships between them are. The classification of means helps the user to find methods which enhance the dependability of a system.

So the **attributes** of dependability are the following:

- the *readiness for usage* leads to **availability,**
- the *continuity of service* leads to **reliability,**
- the *non-occurrence of catastrophic consequences on the environment* leads to **safety,**
- the *non-occurrence of unauthorized disclosure of information* leads to **confidentiality,**
- the *non-occurrence of improper alterations of information* leads to **integrity,**
- the *ability to undergo repairs and evolution* leads to **maintainability.**

The system attribute of security is usually, in current standards, defined as a combination of confidentiality, availability and integrity, although the last two are restricted to consideration of unauthorised actions, rather than all causes of failure.

A system **failure** occurs when the delivered service deviates from fulfilling the system **function**, the latter being what the system *is aimed at.* An **error** is that part of the system state which is *liable to lead to subsequent failure*: an error affecting the service is an indication that a failure occurs or has occurred. The *adjudged or hypothesised cause* of an error is a **fault.**

The development of a dependable computing system calls for the *combined* utilisation of a set of **means** (methods and techniques) which can be classed into:

- fault prevention: how to prevent fault occurrence or introduction,
- fault tolerance: how to ensure a service up to fulfilling the system's function in the presence of faults,
- fault removal: how to reduce the presence (number, seriousness) of faults,
- fault forecasting: how to estimate the present number, the future incidence, and the consequences of faults.

The design of this generic approach for dependability has at least two justifications:

- the wide range of mixes of dependability attributes in real systems, and
- the wide range of system lifecycles.

The goal of the application of these Criteria is to ensure confidence in the dependability of an IT system. Not all IT systems have the same mix of attributes and not all IT systems have the same level of criticality for each attribute. So the approach should allow the user to assign the appropriate attributes with corresponding criticality levels to his system.

The Criteria should be usable in various application domains in both safety and security fields, using various lifecycles for product and system development. This means the dependability concepts must be compatible with any lifecycle. Therefore, the concepts and Criteria do not prescribe the lifecycle on which the development process relies. However, they make the assumption that any lifecycle is composed of construction, operational and decommissioning phases.

4   The SQUALE Criteria

On the SQUALE project, we have concentrated on defining the assessment activities that are needed to ensure confidence in the dependability of a system - these are defined as a set of Criteria that must be applied to the system to judge whether the required level of trust has been achieved.

The project has produced a set of Criteria, which is now in its second draft stage. The next stage of the project is to validate these Criteria with a number of demonstrators. The current Criteria are based on the knowledge and background of the partners in the development and assessment of dependable systems in the safety and security fields.

The SQUALE Criteria address the activities needed to gain a specific level of confidence that failures of critical systems will not lead to situations with disastrous consequences. They are intended to be used with a wide range of lifecycle or system development models. The purpose of the Criteria is the definition of assessment activities that are necessary to gain confidence that systems meet their dependability objectives.

It is the intention of the Criteria to define an approach which is compatible with existing Criteria from the safety or security sector. A high level of effort has been spent to define the Criteria in a way which allows the derivation of many individual criteria from the safety and security sector as special instantiations of these Criteria.

The Criteria require a basic set of activities that need to be performed for each dependable system. These activities are :

- The hazards identified in the hazard analysis of the system should be rated for their criticality. The assumptions made about the system and its environment during the rating process should be clearly described.

- Dependability Objectives should be derived describing the overall dependability goals that the system has to achieve. These objectives should completely address all hazards rated as unacceptable.

- These dependability objectives, as well as the results from the hazard analysis and rating process, should be used to define the specific dependability requirements for all parts of the system. These requirements should cover all dependability objectives.

- The (potential) consequences of a failure of an individual part of the system defines the criticality for this part of the system. A confidence level that reflects this criticality should be chosen for each part of the system. This confidence level impacts on the development, operation, maintenance and decommissioning of the system as well as on the assessment activities.

- A dependability target document should be produced, to form the basis for the dependability assessment activities. This document should identify when and how all of the required assessment activities are to be performed, in a way that is integrated with the project or system lifecycle.

- Assessment activities should be performed to achieve confidence that the system meets its dependability objectives. The individual assessment activities depend on the level of confidence needed for a specific part of the system as well as on the dependability objectives relevant to this part of the system.

- For highly critical systems, assessment activities should be defined that allow for continual monitoring that the objectives are achieved.

It has to be taken into account that the whole range of dependable systems can not be covered with a single fixed set of Criteria. The aspects that have to be addressed are diverse, and require assessment activities specific to the individual dependability attributes. For this reason the Criteria don't define a fixed set, but provide a toolkit which can be used to specify the assessment activities needed for a specific system. Rules are defined which describe how a specific assessment plan can be derived. These rules ensure that the set of assessment activities selected to achieve a specific confidence level are comparable for the same level.

The **dependability confidence level** for a system is a composite measure, with separate levels defined for each of the dependability attributes (safety, confidentiality, etc.). This allows for a profile of confidence levels related to each of the dependability attributes, to allow for the wide range of possible system requirements. A system may be critical relative to one attribute, such as integrity, but not have such high level availability requirements. Each attribute is rated on a scale of 1 to 4 , in a way similar to the Safety Integrity Levels (SILs) in the safety field. A system may thus have a dependability confidence profile, which reflects its criticality with respect to each dependability attribute.

The dependability confidence profile is used in the Criteria to select those activities that are needed to establish the required level of confidence in the system. It is also used to determine some of the characteristics of those assessment activities, according to the confidence level required - the levels of rigour, detail and independence that should be used when the assessment is performed.

5   Example of the Use of the Criteria

5.1   Scope of the example

In order to illustrate the application of the concepts described above, a simple example is given of the use of our ideas, based on one of the "demonstrators" which form a part of the project. Of course, space limitations prevent the presentation of a realistic example here, so the limited scope of the example should not be taken as indicating the full range of application of our techniques.

We will consider a fictional system which is part of a railway signaling installation, providing a connection from a central controller to trackside equipment via a communications link. The link carries "critical" messages, which are important to the safe, secure and efficient running of the railway.

In this example we will only consider the assessment that is needed during the design and development phases of the lifecycle, looking at the initial hazard analyses and design evaluations which will be required before the system goes into operation. The Criteria can be used to define assessments needed in all lifecycle phases for a system.

5.2   Description of the Example

An illustration of the system which will be the basis for our worked example for the use of the SQUALE Criteria is given below. It is a part of a system which transmits messages between a central control centre and equipment at the trackside, and hence to trains. The data transmitted has a range of functions, some of which are concerned with functions with dependability attributes, such as Automatic Train Operation and Automatic Train Protection.

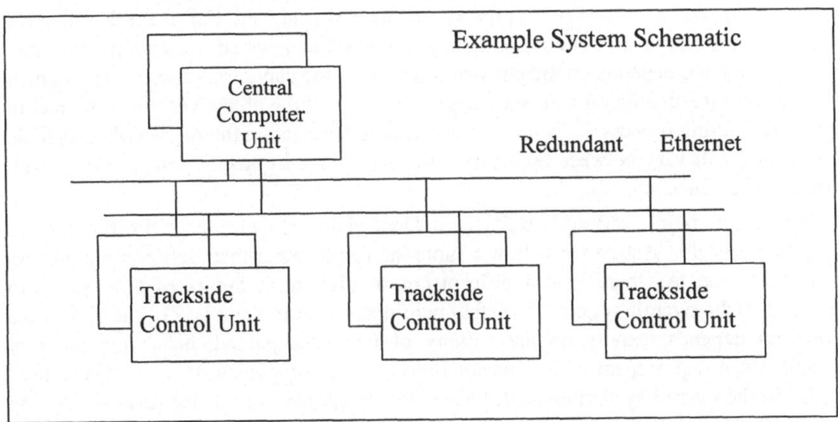

Figure 2: Illustration of the example

5.3   Dependability Requirements Definition

A Hazard Analysis of this system, considering hazardous conditions which could lead to unsafe or unacceptable operation of the railway, results in the following partial hazard list:

1. Loss of messages due to equipment malfunction
2. Loss of messages due to software error in development
3. Loss of messages due to malicious software
4. Loss of messages due to accidental damage to link
5. Loss of messages due to vandalism to link
6. Corruption of messages due to equipment malfunction
7. Corruption of messages due to software error in development
8. Corruption of messages due to malicious software
9. Corruption of messages due to accidental interference with link
10. Corruption of messages due to malicious interference with link.

These indicate that the system has a number of Dependability Objectives:

1. Detection of message loss
2. Recovery from message loss
3. Prevention of physical access to link
4. Detection of message corruption
5. Recovery from message corruption
6. Detection of malicious software and software errors.

All of these objectives will have associated with them a success rate of how effectively they can be met with the system as it is designed. For example objective six depends on the strength of configuration control and system validation processes; objective three depends on the physical strength of the cable trunking. An assessment is made of the likelihood and consequences of each of the objectives not being met in a hazard rating process. This process involves judgments on the degree of acceptable risk, and will vary between application fields (e.g. road transport, air traffic control, nuclear, defence, etc.).

The hazard rating activity results in a Dependability Confidence Profile for the system. For this system we will just quote the result rather than defining the process of derivation. As there is a significant contribution from failure of this system to safety of the overall system, the safety confidence requirement is S3. The rest of the system depends heavily on the validity of messages passed, hence the integrity confidence requirement is I3. Unavailability of this component does not pose a high risk in the overall system context, hence the availability confidence level is A2. The overall Dependability Confidence Profile is therefore:

S3, I3, A2.

This profile defines the level of confidence that is required in the correct and effective operation of the system in meeting its Dependability Objectives.

5.4   Dependability Target

In order to bring together all the requirements for assessment activities which will be needed to demonstrate achievement of the confidence level, a document, the Dependability Target, is produced. This will provide both a claim that this level has been achieved and also provide a basis for the necessary assessment work to support it.

For this system, the dependability target will contain:

- **Introduction**

  (introduction to the document)

- **Description of the Target of Dependability Assessment (TDA)**

  (as described above)

- **System and Environment Analysis**

  (as described above)

- **Hazards**

  (as described above)

- **Dependability Objectives**

  (as described above)

- **Dependability Policy**

  (a description of how the dependability objectives will be met using a combination of procedural, physical and technical measures)

- **Dependability Function Specification and Allocation**

  (a description of how the technical dependability countermeasures are implemented in the components of the system)

- **Required Confidence Level Profile**

  (for the above system this is: S3, I3, A2)

- **Specific Assessment and Validation Plan**

  (the Criteria provide a list of assessment methods related to specific dependability attributes and to specific phases of the system lifecycle. Within this section of the Dependability Target document the assessment methods that are appropriate for the TDA, with its specific dependability attributes, are selected. A (preliminary) work plan for the assessment of the example system is established in the next section.)

5.5   Assessment Workplan

Using the selection techniques given in the SQUALE Criteria, the techniques to be used to assess the system are chosen. For each assessment technique a degree of rigour, detail and independence of the assessment is defined, selected by using the dependability confidence profile. The precise meaning of the techniques and the levels chosen is defined in the Criteria. The table below shows the selection that has been made for the example system.

No	Assessment Activity	Rigour	Detail	Independence
1.	Preliminary Hazard Identification	high	high	high
2.	Hazard Rating	high	high	high
3.	Preliminary Hazard Analysis	medium	high	medium
4.	Static Analysis	medium	high	medium
5.	Behavioural Analysis	high	medium	medium
6.	Formal Methods and Proof-of-correctness	medium	low	medium
7.	Testing Techniques	medium	high	high
8.	Traceability Analysis	high	high	high
9.	Penetration Analysis	high	medium	medium
10.	Failure Modes, Effects and Criticality Analysis (FMECA)	high	medium	medium
11.	Hazard and Operability Studies (HAZOPS)	high	high	medium
12.	Common Cause Analysis	high	high	medium
13.	Process Quality Assessment	medium	high	medium

It should be noted that many of the activities listed in this table are not exactly those known in current standards under these names - the Squale Criteria have extended many of these activities to allow them to address the full range of dependability attributes. For example FMECA will include consideration of maliciously induced failures, and Penetration Analysis will address the impact of all failure types.

Each of the above activities needs to be planned to fit in with the design and development plans for the example system, in order to ensure that the inputs required (as defined in the Criteria) are available. It is also important to schedule the assessment activities in such a way that there is minimal impact on the development timescale (particularly the completion date), while ensuring that the assessment results are produced in a timely way to support the systems acceptance into service by a regulator or accreditor.

6   The Future

This paper has given an introduction to the SQUALE project, and the dependability concepts on which the project's approach is based. It has described the SQUALE Criteria which have been defined by the project, and presented a worked example of the application of the Criteria, in order to show how the Criteria can be used in practice.

There are a number of areas where the project still needs to do work, in order to extend the applicability and usefulness of the SQUALE approach. One is the issue of performing trade-offs between the requirements of different dependability attributes. The model used on the project allows attributes such as safety and availability to be defined on the same basis, and this should allow the impact of countermeasures for one attribute on all the other attributes to be assessed. This needs further work, which is planned for the later stages of the project.

Another issue which will be addressed is the mapping from the generic SQUALE Criteria to the needs of specific sectors - providing sector specific instantiations of the Criteria. One aspect of this is the mapping to existing or emerging standards for those sectors, and another is the issue of how these standards can be extended to address the full range of dependability issues.

The primary future work of the project remains, however, the validation of the Criteria through their use on a number of demonstrators. The major demonstrator planned is a newly developed railway signaling system, which has recently gone into service, and which has a range of dependability requirements. It is hoped that this demonstrator will enable us to address the technical correctness of the Criteria and their usability, and give an insight into the process of creating a sector specific version of the Criteria.

Bibliography

[CC 96]   Common Criteria for Information Technology Security Evaluation, Common Criteria Editorial Board, Version 1.0, 31. Jan 1996

[DPCRIT-   SQUALE - Definition of Draft Criteria for the Assessment of Dependable
D2.0]   Systems - Draft 2 -
               ACTS95/AC097, 15.01.1997

[IEC 1508]   Draft IEC 1508 - Functional safety: safety-related systems, Part 1 to 7, IEC CD, June 1995

[ITSEC 91]   Information Technology Security Evaluation Criteria (ITSEC), Harmonised Criteria of France, Germany, the Netherlands, the United Kingdom, Commission of the European Communities, 1991

[LAP 95]   J.C. Laprie, "Dependability - Its attributes, impairments, and means", in *Predicably Dependable Computing Systems*, B.Randell, J.C. Laprie, H. Kopetz, B. Littlewood, eds, Springer-Verlag, 1995, pp. 3-18.

The Extended Commercially Oriented Functionality Class for Network-based IT Systems

Alexander Herrigel[*] Roger French[†] Herrmann Siebert[‡] Helmut Stiegler[§] Haruki Tabuchi[**]

European Computer Manufactures Association (ECMA), Geneva, TC 36, http://www.ecma.ch

Abstract

This paper presents a new approach for security evaluation criteria of network-based IT systems. The Extended Commercial Oriented Functionality Class (E-COFC) addresses a minimum set of security functionalities for the commercial market to reduce technical complexity, and to allow the cost-and time effective application. The standard addresses today's commercial requirements with its different legal parties involved. In contrast to state-of-the art approaches such as the Common Criteria, the standard address the contractual relationships the business processes are based on. The E-COFC is considered as a baseline standard commercial enterprises can measure against.

1.   Introduction

With the rapid increase of networking and new methods in information technologies like the Internet, Electronic Data Interchange (EDI), Intelligent Networking (IN), Distributed Databases (X500), more and more mission critical business processes from an enterprise are directly depended from network-based IT systems. These network-based systems require a complex management and auditing, since they involve typically several different platforms with specific operating systems and different technologies such as DCE, Corba, Asynchronous Messaging, HTTP and many others. Figure 1 shows a typical IT infrastructure for Web-based services. These technologies exploit the TCP/IP protocol suite. Since this suite has been developed over the last period at different universities and companies for different systems, some of them have inherent security deficiencies. In addition, the main design objective at the beginning of the protocol suite development was availability and not security. Different security aspects are, therefore, of great significance for the enterprises business and its continuity. The enterprise systems and the processed information are subject to a number of threats from different sources. Effective protection of enterprise systems and information databases requires a structured management approach to security. This approach must affect different domains of the enterprise such as people, IT systems, communication networks, buildings, business and facility planing.

---

[*]r³ security engineering ag , Zurichstrasse 151, CH - 8607 Aathal, Switzerland, Email: herrigel@r3.ch .

[†]Digital Equipment Corporation, U.S.Email: french@zeke.ENET.dec.com .

[‡]EDP Consulting, Germany, Email: 100041.3255@compuServe.com .

[§]STI Consulting, Germany, Email: helmut.stiegler@muenchen.org .

[**]Fujitsu Ltd, Japan, Email: Tabuchi@Saint.NM.Fujitsu.co.jp .

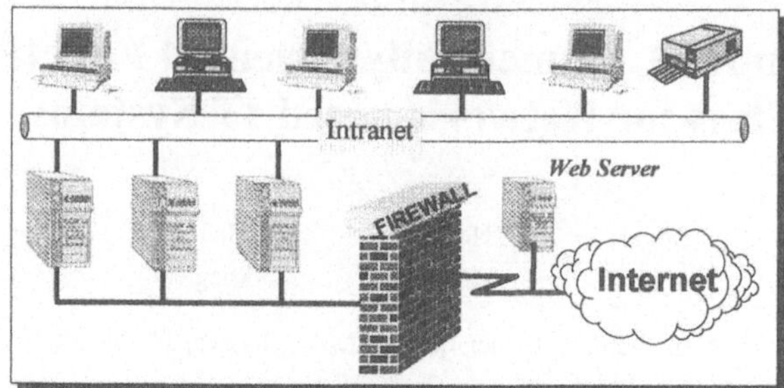

Figure 1: Web-based IT infrastructure.

The core part of any security policy is the risk assessment and auditing of the network-based infrastructure which constitutes the backbone for the mission critical business processes. Effective protection of these business processes affects the IT security strategy of the enterprise. IT security evaluation criteria have to be investigated for the selection of adequate systems to implement an appropriate protection.

2.   State-of-the-Art Approaches

A number of standards or quasi standards are available or under development, which can be applied to evaluate a specific assurance level of security for a hardware/operating system combination. One is the somewhat dated and US government/DoD-centric TCSEC, the "Orange Book" [1]. Others [2-5] like the Information Technology Security Evaluation Criteria (ITSEC) [2], or the Federal Criteria (FC) [4] have been developed, including the 800 page Common Criteria (CC) [5] which is a five government effort to provide a harmonized criteria for the U.S., Canada and the European Union. There is also the ISO 3 part criteria (ISO/IEC JTC/SC27/WG3) and several industry specific criteria. To prove the feasibility of the CC some evaluation trials are currently applied to check for possible inconsistencies. From a commercial perspective, the standards have the following main limitations:

The standards do not address different legal parties which might be involved in business process actions. A great number of business processes, however, are based on the Internet architecture which is supporting an IT environment for different business partners. With respect to the contractual relationships between different business partners, special requirements have to be satisfied for addressing the audit of legal contract issues during the execution of the business processes. This important commercial and legal aspect was not taken into consideration during the specification phase of the CC. Many standards have been specified under the control of governmental agencies. Some requirements, however, for governmental environments do not match with the requirements for the commercial market*. Some standards such as the CC have a very high technical complexity and are not concise. The auditor and management accep-

---

*Confidentiality, for example is often less important for a financial institution than for a government agency. In addition, repudiation aspects are important for a financial institution, but not for a military environment.

tance is, therefore, in a commercial environment very limited since time to market deployment overrules in many cases other objectives.

3.   The Extended Commercially Oriented Functionality Class

The Extended Commercially Oriented Functionality Class (E-COFC) is an ECMA standard which specifies security evaluation criteria for interconnected IT systems. The systems are interconnected through a communication network which is considered a priori as not secure. The systems may be located at different sites, cities or countries, and are connected through leased lines, public or private networks. The E-COFC applies to the security of data processing in a commercial business environment, independent of hardware and software platforms of the participating systems. Its functions are selected to satisfy the minimum security requirements for typical business applications on interconnected systems. The E-COFC is based on an IT Security Policy of a commercial enterprise taking typical environmental and organizational constraints into account. The identified minimal security requirements of this standard have to be supported by the TOE but not necessarily by each individual system. The support of the security enforcing functions within a system may be based on the Operating System (OS) or on the combination of the OS and secure hardware or software products.

3.1.   Identified Hierarchical Classes

With respect to the commercial requirements, the E-COFC is partitioned into the following three hierarchical classes of commercial security requirements:

1. The **Enterprise Business Class** (EB-class) (includes COFC [6] requirements).
2. The **Contract Business Class** (CB-class) (includes the EB-class and COFC requirements).
3. The **Public Business Class** (PB-class) (includes the CB-class, EB-class and COFC requirements).

Each subclass specifies for the imposed commercial environment the security requirements, the resulting threats and the identified security functionalities.

In practice, electronic business actions between business partners are not only based on secure data communication, but also on the provision of legal proof. Those commercial requirements are the foundation for the CB-and the PB-class, while the EB-class provides the necessary network communication security.

4.   The Enterprise Business Class

The following characteristics have been identified for the EB-Class. All users are employees of a single enterprise (legal entity). The usage of the IT systems which are part of the TOE is regulated by the employee contract. Only one legal party is responsible for all business actions. In case of outsourcing, the responsibility can be partly delegated to other legal parties on the basis of special contracts. The model describing the enterprise business is shown in the following figure. The exchange of business information is done on behalf of the involved system users. The security of the exchange is enabled by the security services provided by management of the enterprise. Conflict mediation is provided by management actions on the basis of the employee contracts. No specific business actions between different legal parties have to be investigated for this class.

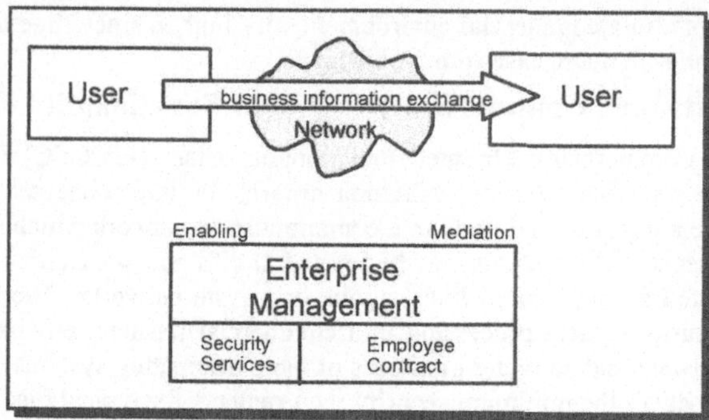

Fig. 2: The Enterprise Business Model.

The identified threads, the countermeasures, and the derived security functionalities are described in detail in the E-COFC standard [7].

5. The Contract Business Class

The CB-class is built on top of the EB-class. It adds to the network security requirements of the EB-class those requirements which are typical for business actions (business information exchange) between independent enterprises which belong to a closed user group. The enterprises agree in a contract on a defined mode of operation, the business conditions and the security rules, which are the foundations of their business actions. They establish a Notary, or Regulatory Board, which acts as impartial judge to mediate conflicts within the user group and acts also as trusted third party to handle security matters (for example key management, authentication, audit, etc.). All business partners who sign the contract form a closed user group. Users shall only get access to the systems if they belong to a business partner that has signed the contract. The business within the closed user group can be described with the following model:

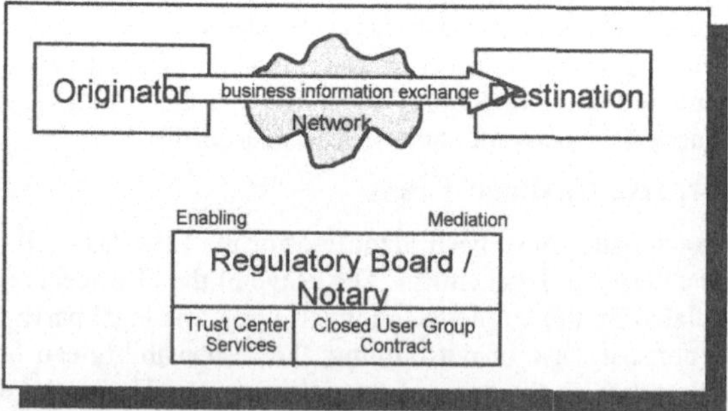

Figure 3: The Contract Business Model.

The business information is exchanged between the originator and the destination. These terms describe functional roles in the business process. The originator is a user or process in behalf of a user that sends business information (document, order, invoice etc.) to a destination. The destination is a user or process in behalf of a user that receives the business information and acts on it (e.g. processes an order). If the business information flow is reversed then the originator becomes the destination and the destination becomes the originator. The originator and the destination must be authorized according to the closed user group contract to perform the business actions. If the information is confidential, the originator and destination are persons (processes) that are authorized to read or use the exchanged information. The secure exchange of business information is enabled by trust center services like key generation, key distribution, key revocation, key certification and security logging. The trust center services are under control of the Regulatory Board.

The TOE has to establish a secure basis for various kinds of business. On the basis of the security functions provided by the underlying TOE it has to be accredited that the security requirements of all the business actions stated in the contract of the cooperating enterprises can be fulfilled by the enterprises.

The identified threads, the countermeasures, and the derived security functionalities are described in detail in the E-COFC standard [7].

6. The Public Business Class

The PB-class is built on top of the EB-class and the CB -class. It adds on those requirements which are typical for public business in an open environment (no closed user group). Public business typically covers areas like selling of goods, tickets and other merchandise, but also network-based information services and others. The terms Customer and Provider are used in this class. A provider system which is offering a dedicated service contains a set of business actions. For each business action the term Originator and Destination can be applied as outlined in the CB-class. The business is described by the following model:

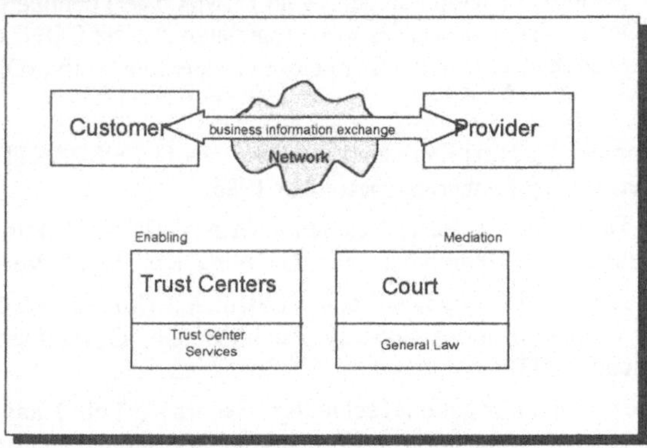

Figure 4: The Public Business Model.

The Public Business Class is characterized by business on the basis of pre-existing contracts that legally connect the provider and the customer for a set of pre-defined business actions. In contrast to the CB-class there is no regulatory board which resolves possible conflicts. Conflicts have to be resolved on the basis of the law. Trust centers are necessary to enable secure business. They provide as independent organizations the required services for the sequence of secure business actions. In contrast to the EB-and CB -class the business relationships may be beyond the contractual relationships. Different scenarios of contractual relations are possible. In the case of electronic advertising the contracts are provided by business and consumer law. The formal contractual relationship between the customer and the provider is either direct or through other business organizations such as credit card organisations.

The identified threads, the countermeasures, and the derived security functionalities are described in detail in the E-COFC standard [7].

7. Conclusions and Future Work

From our perspective, the E-COFC has the following advantages:

1. The standard is easy to understand since the specification of a minimum set of security functionalities reduces the technical complexity. The derived set provides a reasonable protection for commercial multi-user, network-based IT systems.

2. The standard is based on today's commercial requirements avoiding military and governmental bias. Major efforts have to be undertaken for the CC to come up with a concise security evaluation standard that fulfills the constraints for these three very different environments identified for electronic commerce.

3. The standard is independent of assurance criteria and evaluation methods. It is consistent with TCSEC, ITSEC and MSFR concepts. The COFC can, therefore, be applied with an appropriate assurance scale and methodology, which might be TCSEC, ITSEC, FC, CC or any scale of assurance criteria and evaluation methods provided by any other organization, if accepted by the user.

4. In contrast to all other approaches, the E-COFC addresses the security requirements for business processes with different legal parties.

We are currently planning to setup and deploy an ECMA based commercial protection profile registry. One of the first profiles to be registered are the COFC, the E-COFC, and the CC profiles such as CS1 and CS2 in close co-operation with ISO.

References

[1] "Trusted Computer Systems Evaluation Criteria", DoD 5200.28-STD, Department of Defense, United States of America, December 1985.

[2] "Information Technology Security Evaluation Criteria (ITSEC)-Harmonized Criteria of France, Germany, the Netherlands, and the United Kingdom ", Vers 1.2, 1991.

[3] "The Canadian Trusted Computer Product Evaluation Criteria", Canadian System Security Center, Communications Security Establishment, Government of Canada, Version 3.0e, January 1993.

[4] "Federal Criteria for Information Technology Security", Vol. 1 and Vol. 2, Dec. 1992, National Institute Of Standards and Technology & National Security Agency.

[5] "Common Criteria for Information Technology Security Evaluation", Version 1.0, CCEB.

[6] "Standard ECMA-205, Commercially Oriented Functionality Class for Security Evaluation (COFC) ", ECMA, December 1993.

[7] "Draft Standard ECMA-999, Security Functionalities of the E-COFC, ECMA, March 1997.

Online-Dienste im Internet – eine kombinierte Anforderungs- und Risikoanalyse

Ph. Kirsch, H. Weidner

Institut für Informatik, Universität Zürich
Winterthurerstrasse 190, CH-8057 Zürich, Switzerland
Tel. +41 1 257 43 34, Fax. +41 1 363 00 35
{kirsch, weidner}@ifi.unizh.ch

Zusammenfassung

Die Nutzung des Internet wird für immer mehr Unternehmungen wichtiger. Insbesondere kleinere und mittlere Unternehmungen (KMUs) können durch die Nutzung des Internet profitieren. Diesen Chancen stehen jedoch für KMUs nicht durchschaubare Risiken im Sicherheitsbereich entgegen. Übliche Verfahren des Sicherheitsmanagements sind für diese Firmen und für den Einsatzbereich Internet ungeeignet. In diesem Artikel wird daher ein Ansatz entworfen, der Anforderungen an die Nutzung und Sicherheit integriert behandelt. Die Vorteile sind eine Verkürzung der Entwicklungszeiten und die Reduzierung der Komplexität. Darüber hinaus muß bei der Nutzung von Online-Diensten die Sicherheit von Anfang an berücksichtigt werden. Dies schließt eine sequentielle Betrachtung von Anforderungen und Sicherheit aus.

1 Einführung

Eine wachsende Anzahl von Unternehmen betrachtet die Nutzung des Internet als wichtiges zukünftiges Geschäftsfeld und Marketinginstrument. Internet-Dienste sind hervorragend geeignet, Unternehmen in der Informationsbeschaffung, der Informationsgewinnung, bei Geschäftsbeziehungen (electronic Commerce) und in Ihrer Zusammenarbeit (Gruppenunterstützung), die auf den Austausch von Informationen basieren, zu unterstützen (vgl. [KaSi96, S. 16f]).

Neben diesen Möglichkeiten ist ein Anschluß an das Internet allerdings mit Risiken verbunden. Spätestens, wenn das Firmennetz als Intranet mit dem weltweiten Internet gekoppelt werden soll, entstehen oft unbeachtete Risiken für die Unternehmung.

Dieser Artikel führt in die Sicherheitskonzeption als wesentlicher Bestandteil des Sicherheitsmanagements ein (zu Sicherheitsmanagementaufgaben vgl. [KTW97]). Im Zentrum der Betrachtung stehen insbesondere kleinere und mittlere Unternehmen (KMUs), die mit einer vollständigen Durchführung von Sicherheitsmaßnahmen häufig überfordert sind; zum Einstieg in diese neue Themenstellung benötigen sie zunächst leicht durchführbare Konzepte, die im Lauf der Zeit professionalisiert werden können.

Viele Sicherheitskonzeptionen gehen von der nachträglichen Absicherung von Diensten bzw. Informationssystemen aus; d.h. Sicherheit wird als "Add On" zur primären Systemfunktion betrachtet (diese Ansätze finden sich z.B. in [BSI96] und [BFI95]). Durch diese organisatorische Trennung von Anforderungsanalyse und Bedrohungsanalyse sind sicherheitsrelevante Funktionen häufig unzureichend in Informationssysteme integriert. Ein typisches Beispiel ist das Zusammenspiel von E-Mail Systemen mit Verschlüsselungs- und Signatursystemen. Die Folge ist, daß Sicherheitsaspekte aufgrund mangelnder Integration in die Systeme vernachlässigt werden.

Die hier vorgeschlagene Konzeptionierung versucht diese Trennung, durch eine Integration von Anforderungs- und Risikoanalyse zu einem gemeinsamen Gestaltungsprozeß zu vermeiden. Dieses Verfahren besitzt drei wesentliche Vorteile gegenüber einem getrennten Vorgang:

- *Sicherheit von Anfang an*: Gerade bei der Nutzung von Internet-Diensten, ist die nachträgliche Absicherung des Systems nicht zu verantworten; Die Öffnung des Unternehmens muß von Anfang an abgesichert werden.

- *Schnelle Bereitstellung von Diensten*: Im Umfeld sich schnell entwickelnder Systeme, das Internet kann als solches bezeichnet werden, ist die Durchführung einer abgetrennten Sicherheitskonzeption sehr zeitaufwendig. Dies ist der Hauptgrund, weshalb das Sicherheitsmanagement vernachlässigt wird.

- *Integration von Sicherheitsmaßnahmen*: Sicherheitsmaßnahmen werden häufig aufgrund des Zusatzaufwandes bei der Arbeit umgangen (z.B. verschicken sicherer E-Mails); eine Integration könnte Sicherheitsmaßnahmen benutzungsfreundlicher gestalten.

Das vorgeschlagene Verfahren besitzt eine große Ähnlichkeit zum Rapid Product Development Ansatz im Softwareengineering (vgl. [HeIn88], [Bla94]). In Anlehnung an diesen Begriff bezeichnen wir diese Sicherheitskonzeption auch als *Rapid Secure Development* (RSD), d.h. als schnelle und sichere Implementierung von Informationssystemen. Implementierung bezieht sich dabei mehr auf die organisatorische Einführung von Informationssystemen (in unserem Szenario z.B. Internet-Dienste) und weniger auf einen Softwareentwurfsprozeß.

Die vorliegende Arbeit entstand innerhalb des SINUS Projekts (Sichere Nutzung von Online-Diensten) [KTW97] und enthält erste wesentliche Erkenntnisse bei der Internetanbindung einer Unternehmung. Das Projektziel ist die Entwicklung eines integrierten Sicherheitsmanagement-Ansatzes, der sowohl organisatorische als auch technische Aspekte sowie die dynamische Entwicklung bei der Verwendung von Internet-Diensten berücksichtigt.

2 Vorgehen — Rapid Secure Development

Bei der Anbindung eines Unternehmens an das Internet oder auch bei der Neugestaltung der Internetnutzung zu Unternehmenszwecken empfiehlt sich ein Vorgehen, das sowohl Nutzungspotentiale als auch Bedrohungspotentiale gleichermaßen zu einer sicheren Gestaltung von Internet-Diensten verschmelzen läßt.

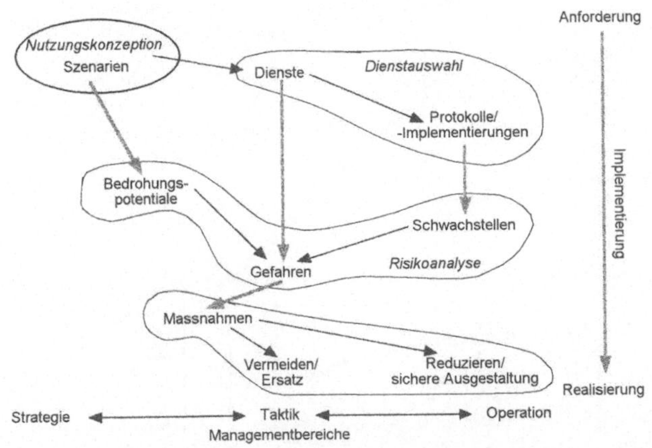

Abbildung 1 Nutzungs- und Sicherheitskonzeption

Im RSD-Ansatz lassen sich fünf Phasen unterscheiden; die Aufgaben und Probleme, die in den jeweiligen Phasen zu lösen sind, lassen sich durch fünf Fragen beschreiben (vgl. Abbildung 1):

- *Nutzungskonzeption (Szenarien)*: Welchen Nutzen möchte die Unternehmung aus dem Internet ziehen?

- *Dienstauswahl (Dienste, Protokolle)*: Welche Dienste sind für diese Anforderungen geeignet?

- *Risikoanalyse (Bedrohungen, Gefahren, Schwachstellen)*: Welche Risiken entstehen bei der Verwendung der Dienste?

- *Maßnahmen*: Welche Maßnahmen schützen vor diesen Risiken?

- *Realisierung (Ersatz, sichere Ausgestaltung)*: Wie können diese Dienste und Maßnahmen realisiert werden?

Die in Abbildung 1 dargestellten Schritte von den Szenarien über die Dienstauswahl zur Betrachtung der Risiken und Erstellung von Gegenmaßnahmen lassen sich unterschiedlichen Mangementbereichen und unterschiedlichen Entwurfsstadien zuordnen. Die Einkreisung zusammengehörender Begriffe dient zur Einordnung dieser Begriffe in die Kategorien Anforderungsanalyse, Internet-Dienste, Risikoanalyse und Maßnahmen.

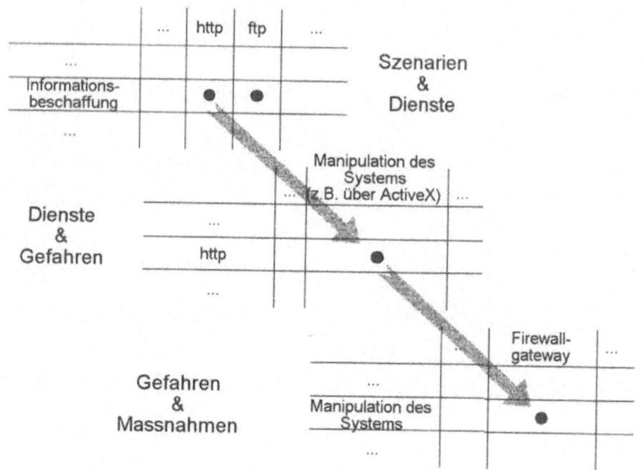

Abbildung 2 Beziehungstabellen von den Szenarien zu den Massnahmen

Die Generierung von Beziehungen zwischen den einzelnen angedeutete Teilbereichen
kann mittels folgender Tabellen unterstützt werden (vgl. Abbildung 2):

- *Szenarien - Dienste bzw. Protokolle*: Zuordnung der Nutzungskonzepte zu Internet-Diensten.

- *Dienste - Risiken*: Zuordnung der Risiken zu den Diensten unter Berücksichtigung der Dienstnutzung.

- *Risiken - Maßnahmen*: Zuordnung von Diensten und den mit diesen verbundenen Risiken zu Maßnahmen, die getroffen werden müssen.

Abbildung 2 illustriert dieses Vorgehen anhand eines Beispiels. Ausgehend von dem
Wunsch, sich Informationen über das Internet zu beschaffen (Szenario), kann ein geeigneter Dienst zur Erfüllung dieses Wunsches ausgewählt werden, z.B. Nutzung des
HTTP Protokolls als Basis des World Wide Web Dienstes. Durch diese Nutzung können Gefahren entstehen, z.B. das Einschleusen und Ausführen unerwünschter Programme. Gegen diese Gefahr läßt sich u.a. ein Firewall (Application Gateway) einsetzen.

Der in Abbildung 1 dargestellte Entwurfsprozeß sieht noch keinerlei Rückkopplung
vor. Diese treten allerdings in der Realität zu jeder Zeit auf. Die Sicherheitskonzeption
in Abbildung 3 trägt diesem Umstand Rechnung.

Der in Abbildung 3 dargestellte Prozeß erinnert stark an Entwurfsprozesse der Systementwicklung (z.B. bei [LaLo95, S. 270]). I.d.R. werden KMUs jedoch auf vorhandene Dienste und Informationssysteme zurückgreifen, um Ihre Internetnutzung sicher zu gestalten.

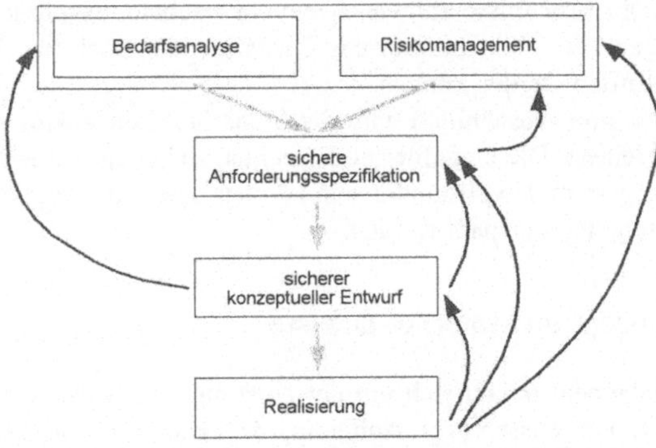

Abbildung 3 Sicherheitskonzeption mit Rückkopplung

Neben dem normalen Top-Down Vorgang entstehen gerade in der organisatorischen Gestaltung dieses für viele Unternehmungen recht jungen Betätigungsfeldes viele Rückkopplungen. Auch wenn Rückkopplungen in möglichst frühen Phasen vorgenommen werden sollten, ist es nicht auszuschließen, daß erst bei der Realisierung grundlegende Fehler oder Probleme in der Nutzungskonzeption oder der Risikobetrachtung gefunden werden können.

Ändert man die Darstellung in Abbildung 3 so um, daß aus diesen Rückkopplungen ein Kreislauf entsteht, erhält man ein Vorgehensmodell wie in Abbildung 4 dargestellt.

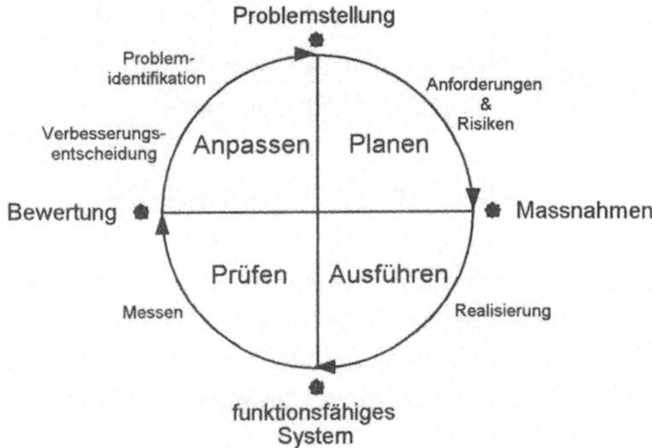

Abbildung 4 TQM Ansatz (PAPA Modell)

Das resultierende Vorgehensmodell (Abbildung 4) entspricht dem Total Quality Management (TQM) Ansatz (vgl. [Art93], [Var95], [OzAs90]). Ausgehend von der Anforderungsanalyse und der Bedrohungsanalyse (Planen) wird ein Internet-Dienst oder ein beliebiges Informationssystem sicher gestaltet (Ausführen) und anschließend bewertet (Prüfen). Die Ergebnisse führen zu einer Entscheidungsphase (Anpassen), in der Entscheidungen zur Verbesserung des Dienstes oder auch zur Neustellung der Ausgangslage getroffen werden können.

In den nächsten drei Abschnitten wird die Sicherheitskonzepzion nach dem RSD-Ansatz weiter verfeinert. Die eingeführten Beziehungstabellen werden genauer formuliert, auch wenn sie niemals vollständig sein können. Die Tabellen können bei der individuellen Übertragung angepaßt werden.

3 Nutzungskonzept und Internet-Dienste

Informationsmanagement befaßt sich mit der Nutzung und Nutzbarmachung von Daten, Informationen und Wissen (vgl. Abbildung 5). Daten und Informationen besitzen je nach Unternehmensebene (strategisch, taktisch, operativ) unterschiedliche Struktur und Wertschätzung.

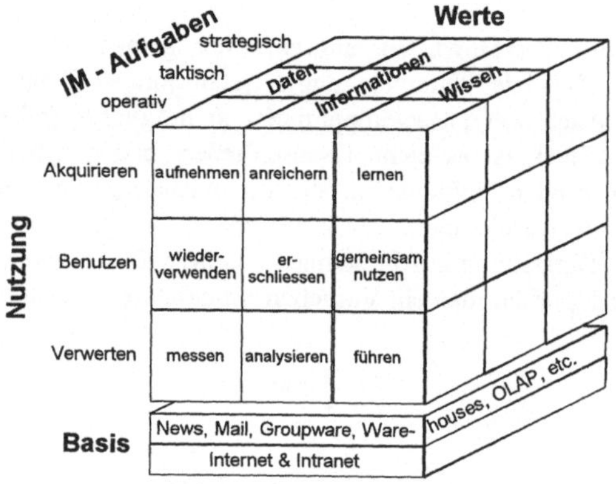

Abbildung 5 Nutzungspotentiale

Daten sind zunächst eine Ansammlung von syntaktischen Elementen, die bestimmten definierten Regeln gehorchen. Erst die Erfassung der Bedeutung einer Nachricht führt zu einer Semantik, die den Sinn bzw. die Bedeutung erschließt. Die zweckorientierte Verarbeitung der Information führt zu einem pragmatischem oder zweckorientiertem Wissen (vgl. [Bro92, S. 8ff]).

Dieser Prozeß läßt sich von der Akquisition bis zur Verwertung (Wissenserwerb) der Informationen durch die Nutzung von Internet-Diensten effizient unterstützen.

Diese prozessuale Sicht der Informationsverarbeitung führt unmittelbar auf vier grundlegende Nutzungsmöglichkeiten, die sich über das Internet besonders gut realisieren lassen (in Anlehnung an [Bor96, S. 71ff], vgl. [TSMB95, S. 10ff und S. 23ff]):

- *Informationsbeschaffung*: Bei der Informationsbeschaffung fließen Informationen von der Quelle zur Senke, z.B. besorgt sich ein Mitarbeiter Informationen über ein neues Produkt von einer informationsbereitstellenden Firma. Die Initiative geht vom Mitarbeiter aus.

- *Informationsbereitstellung*: Der gleiche Vorgang kann aus anderer Perspektive betrachtet werden; eine Firma möchte Informationen über ihr Produkt bereitstellen, damit Interessierte diese Informationen abrufen können.

- *Handel (electronic Commerce)*: Bei dieser Form der Kooperation erfolgt der Informationsfluß zum Zweck des Austauschs von Gütern und Werten in beide Richtungen. Es entsteht eine Kunden-Lieferanten Beziehung, die meist im Rahmen einer *Geschäftsbeziehung* stattfindet.

- *Zusammenarbeit (Group Support)*: Bei dieser Form der Kooperation erfolgt der Informationsfluß zum Zweck der Zusammenarbeit in beide Richtungen. Es steht eine gemeinsame Aufgabe (bei einem Team ein gemeinsames Ziel) im Vordergrund. Die Kooperation muß hier zusätzlich durch Koordination unterstützt werden. Auch die Zusammenarbeit geschieht zumeist im Rahmen einer *Geschäftsbeziehung*.

Diese Nutzungspotentiale konkretisieren sich zu Anforderungen. Diese Anforderungen lassen sich i.d.R. durch vorhandene Internet-Dienste realisieren. Z.B. bietet der Mail-Dienst die Funktionalität des Informationsaustausches (Dokumente beliebiger Art) in elektronischer Form zwischen genau definierten Partnern (z.B. von einer Person zu einer anderen).

Legende: ● geeignet — ○ nur für spezielle Zwecke, bzw. nicht überall verfügbar

Nutzungsszenario		Telnet Client	Telnet Server	FTP Client	FTP Server	E-Mail MTA	E-Mail MUA	WWW Client HTML	WWW Client PlugIns	WWW Client JavaScript	WWW Client Java	WWW Client ActiveX	WWW Server CGI	rlogin Client	rlogin Server	rsh Client	rsh Server	rcp Client	rcp Server	SSH Client	SSH Server
Informationsbeschaffung	intern	○		●		●	●	○		●	○		○		○		○		○		
	extern	○		●		●	●	○		●	○		○		○		○		○		
	öffentlich	○		●		●	●			●									○		
Informationsbereitstellung	intern		○	●	●		●	○		●	○	●									○
	extern		○	●	●		●	○		●	○	●									○
	öffentlich		○	●	●		●			●		●									○
elektronischer Handel	intern					●	●	●	○	●	○	●									
	extern					●	●	●	○	●	○	●									
	öffentlich					●	●	●		●		●									
Zusammenarbeit	intern	○	○			●	●	●	○	●	○	●								○	○
	extern	○	○			●	●	●	○	●	○	●								○	○

Abbildung 6 Nutzungsszenarien und Dienste

Abbildung 6 zeigt für typische Nutzungspotentiale die verwendbaren Internet-Dienste auf. Z.T. wird auf Besonderheiten oder Einschränkungen verwiesen; so kann z.B. die Informationsbeschaffung im Unternehmen über die Installation von WWW-Clients (sog. Browsern) geschehen, es ist jedoch auf die Verwendung von neueren Zusätzen zu HTML wie ActiveX, Java und JavaScript zu achten. Die Szenarien ergeben sich aus Betrachtung der Nutzungspotentiale und der Adressatengruppe. Aus Sicherheitsbetrachtung ist es ein wesentlicher Unterschied, ob Informationen im Unternehmen (intern), zwischen Unternehmen (extern) oder weltweit (öffentlich) fließen sollen.

Die Tabelle könnte noch um einige Dienste erweitert werden; ebenso ließen sich die Nutzungsmöglichkeiten noch näher ausführen. Insbesondere beinhaltet die gewählte Zuordnung bereits eine Eignungsbewertung, die im konkreten Fall anders ausfallen kann. Die Grundfunktion dieser Tabelle ist dennoch ersichtlich; sie soll einen Rahmen für eine begründete Verwendung der Dienste zur Erfüllung bestimmter Szenarien liefern. Besonderheiten bei der Dienstnutzung sind aus der Tabelle nicht ersichtlich und müssen zudem beachtet werden.

4 Risiken bei der Nutzung von Internet-Diensten

Legende: ● möglich — ○ eingeschränkt, bzw. indirekt

		Stören der Verfügbarkeit von Netzkomponenten	Ausführen von Prog. mit unerwünschter Wirkung (Systemanomalien)	Leugnung der Kommunikation	Leugnung des Empfangs	Manipulation der eigenen Daten	Spionage interner Informationen	Verkehrsflussanalyse	gefälschte Authetifizierung	Authentizität des Kommunikationspartners
Telnet	Client	●	●		●		●	●	●	●
	Server	●				●	●	●	●	●
FTP	Client	●	●		●		●	●	●	●
	Server	●	●			●	●	●	●	●
E-Mail	MTA	●		●	●		●	●	●	●
	MUA	●	●	●	●	●		●	●	●
WWW	HTML	●								
	PlugIns	●	●	●	●	●	●			
	JavaScript	●	●			●	○			
	Java	●	○	●	●	●	○		●	●
	ActiveX	●	●	●	●	●	●		●	●
	CGI	●	●				●	●	●	●
Berkeley r-Tools	Client	●	●			●	●	●	●	●
	Server	●					●	●	●	●
ssh	Client	●	○			●	●	●		
	Server	●					●	●		

Abbildung 7 Zuordnung von Bedrohungen und Diensten

Nachdem ein Nutzungskonzept erstellt wurde (vgl. Abschnitt 3), können nun Risiken für bestimmte Internet Dienste (bzw. deren Nutzungsintention) bestimmt werden. Abbildung 7 zeigt typische Bedrohungen, die mit der Nutzung bestimmter Dienste verknüpft sind (vgl. [ChZw96, S. 399], [Var93, S. 48ff], [Woj91, S. 30ff]). Auch hier ist die Darstellung sicherlich nicht vollständig; ein sinnvolles Verfahren zur Vervollständigung ist die Betrachtung der Grundbedrohungen: Verlust der Verbindlichkeit, Vertraulichkeit, Verfügbarkeit und Integrität (vgl. [Ker91, S. 47], [Var95, S. 67]).

In der Abbildung 7 sind bewußt einige Bedrohungen nicht mit aufgenommen worden; dies sind Bedrohungen, die nicht in direktem Zusammenhang mit der Nutzung von Online Diensten stehen (z.B. die Zerstörung von IT Anlagen) oder solche Bedrohungen, die nicht von der Informationssicherheit betrachtet werden (wie z.B. Feuerschäden).

Die in Abbildung 7 angegebenen Bedrohungen müssen einer Risikoanalyse unterzogen werden, in der Bedrohungen, Schwachstellen, die Werte der Informationen und die Eintrittswahrscheinlichkeit der Bedrohungen analysiert werden. Dies ist in Abbildung 1 angedeutet. Die Markierung von relevanten Bedrohungen bei der Nutzung der Dienste wie sie in Abbildung 7 angegeben wurde beinhaltet durch eine erste Wertung bereits einen Teil der Risikoanalyse. Die Zuordnung von Bedrohung zu Dienst kann sicherlich auch anders bewertet werten.

Nachdem für einen bestimmten Dienst die relevanten Bedrohungen (Risiken) ermittelt wurden, können nun Maßnahmen zur Risikominimierung ergriffen werden.

5 Risiken und Maßnahmen

Grundsätzlich gibt es mehrere Möglichkeiten, Risiken zu minimieren (vgl. [Sch-B92]):

- **Vermeidung**: Ein Dienst wird nicht eingesetzt oder durch einen alternativen aber sicheren Dienst ersetzt.

- **Reduzierung**: Trotz eines Risikos bei der Nutzung eines Dienstes kann das Risiko reduziert werden, indem zusätzliche Sicherheitsvorkehrungen getroffen werden.

- **Begrenzung**: Im Schadensfall ist eine Begrenzung des Schadens erforderlich (z.B. durch Datensicherung, Ermittlung des Verursachers, etc.).

- **Überwälzung**: Schließlich kann ein Risiko auch versichert werden.

Man wird versuchen, Risiken nach diesen Kategorien zu minimieren. Eine absolute Sicherheit ist allerdings nicht zu gewährleisten, d.h. es verbleibt immer ein sog. *Restrisiko*. Bei der Einführung von Internet-Diensten verbleiben im wesentlichen die Möglichkeiten der Vermeidung und Reduzierung (vgl. Abbildung 1).

Abbildung 8 listet zu den vorhandenen Bedrohungen, die in Abbildung 7 eingeführt wurden, mögliche Gegenmaßnahmen auf. Damit ist der untere Teil der Abbildung 1, die Ableitung von Maßnahmen aus den Gefahren, abgedeckt.

Auch hier gilt wieder, daß die Tabelle nicht vollständig sein kann. Ihre Aufgabe ist, möglichst viele Risiken und Gefahren zu verknüpfen; die hier vorgestellte Tabelle liefert aus Platzgründen nur einen kleinen Teil von Maßnahmen und Gefahren.

	Abschottung von Netzsegmenten durch Firewalls	Verbot der Ausführung von Diensten und Dienstteilen	Einsatz digitaler Signaturen	Einsatz von Verschlüsselung	Redundante Einrichtungen schaffen	Protokollierung (Logfiles)
Stören der Verfügbarkeit von Netzkomponenten	●				●	
Ausführen von Prog. mit unerwünschter Wirkung (Systemanomalien)	●	●	●			
Leugnung der Kommunikation			●			
Leugnung des Empfangs						●
Manipulation der eigenen Daten		●	●			
Spionage interner Informationen	●	●		●		
Verkehrsflussanalyse	●	●			●	
gefälschte Authetifizierung	●		●			
Authentizität des Kommunikationspartners			●			

Abbildung 8 Zuordnung von Bedrohungen und Gegenmaßnahmen

Aus technischer Sicht gibt es vier grundlegende Maßnahmen zur Sicherung bei der Verwendung von Internet-Diensten, die getroffen werden können (vgl. [Wei97]):

- **Firewalls**: Der Abgrenzung von Teilen des Netzes und eine damit verbundene Beschränkung der netzübergreifenden Dienstnutzung ist für die kontrollierte Dienstnutzung sinnvoll.

- **Kryptographische Methoden**: Sie helfen bei der Bestimmung der Authentität von Personen und Dokumenten (digitale Signaturen), der Integrität von Dokumenten und bei der Wahrung der Vertraulichkeit (Verschlüsselung).

- **Redundante Einrichtungen**: Die Schaffung von Redundanzen bzgl. Rechnern, Netzverbindungen, etc. kann einen entstandenen Schaden begrenzen. Insbesondere kann die Verfügbarkeit des Netzes dadurch gesichert werden.

- **Situative Rechtevergabe**: Durch feine Rechtegranulate und flexiblere Rechteverteilung kann die Sicherheit vor Manipulation von Daten erheblich gesteigert werden. Rechte müssen aufgabenspezifisch vergeben werden können und nicht durch einen Superuser (vgl. auch Need-To-Know Ansatz bei [Hol96]).

Bei der Nutzung von Diensten ist daher darauf zu achten, welche Dienste genutzt werden sollen, wie sie genutzt werden sollen, und durch wen sie genutzt werden sollen.

Dies sind nötige Vorgaben, um eine Firewall korrekt zu konfigurieren. Verschlüsselung kann zudem eingesetzt werden, um vertrauliche Nachrichten über öffentliche Netze auszutauschen.

Es zeigt sich auch, daß bei der Nutzung von Internet-Diensten nicht alle Risiken adäquat abgedeckt werden können (z.B. Leugnung des Empfangs), z.B. gibt es kein allgemeines Mailsystem, daß über eine Funktionalität eines Einschreibebriefs mit Rückantwort verfügt. Die Auswertung von Protokolldaten kann diesen Vorgang nicht sinnvoll abdecken.

6 Zusammenfassung und Ausblick

Die sichere Nutzung von Internet-Diensten muß einerseits sorgfältig geplant werden, andererseits ist das vollständige Abarbeiten von Anforderungsspezifikation und anschließendem Initialisierung eines Sicherheitsmanagementprozesses in der Praxis gerade für kleinere Unternehmen zu komplex. Insbesondere dann, wenn die Nutzungspotentiale und damit die Nutzungsszenarien noch nicht definiert sind, ergeben sich zu häufig Änderungen, so daß es nicht sinnvoll ist, jedesmal eine komplette Risikoanalyse duchzuführen. Durch die integrative Betrachtung von Anforderungen und Risiken können jedoch sowohl die Funktionalität als auch die Sicherheit angemessen berücksichtigt werden. Dies schließt eine nachträgliche Überarbeitung und Harmonierung mit einem unternehmensweiten Sicherheitsansatz nicht aus.

Das Modell, das zu diesem RSD-Ansatz führt, ist aus einer Betrachtung von Sicherheitsmanagement-Ansätzen im SINUS Projekt entstanden. Es erwies sich als ein praktikabler Weg, der in der Praxis häufig beschritten wird. Durch eine Strukturierung dieser Betrachtung entstand das Modell.

Einige Annahmen bedürfen noch einer genaueren Überprüfung; insbesondere die in der Einleitung festgestellten Schwachstellen vieler Sicherheitskonzeptionen bedürfen einer genaueren Untersuchung. In unserem Fall treffen sie zu, aber in wie weit sind diese Feststellungen verallgemeinerbar? Vielfach angedeutet war auch die Übertragung dieses Vorgehensmodells auf allgemeine Informationssysteme, nicht nur für Internet-Dienste; dies muß ebenfalls überprüft werden. Schließlich ergeben sich noch Lücken bei der Ausgestaltung des Modells selbst. Insbesondere die hier vorgestellten Tabellen geben lediglich einen Eindruck vom Vorgehen, sie decken bei weitem nicht alle auftretenden Fälle ab.

Da allerdings der RSD-Ansatz eng verwandt ist mit TQM- und RPD-Ansätzen, die sich in der Praxis aufgrund ihrer geringen Komplexität bewährt haben, ist anzunehmen, daß der Einsatz dieses Modells im Sicherheitsmanagement allgemein sinnvoll wäre.

7 Referenzen

[Art93]   Lowell J. Arthur. *Improving Software Quality - An Insider's Guide to TQM.* John Wiley & Sons, Inc., New York, 1993.

[BFI95]   Bundesamt für Informatik. *Weisung Informatiksicherheit Nr. S02 - Grundschutz von Informatiksystemen und -anwendungen.* Bundesamt für Informatik, Bern, 1995.

[Bla94]   Günther Blaschek. *Object-Oriented Programming with Prototypes.* Springer Verlag, Berlin, 1994.

[Bor96]   Marco Borer. *Potentielle Bedrohungen und Schwachstellen für Unternehmen am Internet.* diploma thesis, Institut für Informatik, Universität Zürich, 1996.

[Bro92]   Rainer Brockhaus. *Informationsmanagement als ganzheitliche, informationsorientierte Gestaltung von Unternehmen.* Unitext Verlag, Göttingen, 1992.

[BSI96]   Bundesamt für Sicherheit in der Informationstechnik. *IT Grundschutzhandbuch 1996, Maßnahmenempfehlungen für den mittleren Schutzbedarf.* In: Schriftenreihe zur IT-Sicherheit, Volume 3. Bundesanzeiger Verlag, Köln, 1996.

[ChZw96] D. Brent Chapman, Elisabeth D. Zwicky. *Einrichten von Internet Firewalls: Sicherheit im Internet gewährleisten.* dt. Übersetzung von Katja Karsunke, Thomas Merz, O'Reilly/International Thomson Verlag, Bonn, 1996.

[HeIn88] Sharam Hekmatpour, Darrel Ince. *Software Prototyping, Formal Methods and VDM.* Addison-Wesley Publishing, Workingham, England, 1988.

[Hol96]   Ralp Holbein. *Secure Information Excange in Organisations – An approach for solving the information misuse problem.* Dissertation, Universität Zürich, Shaker Verlag, Aachen, 1996.

[KaSi96] André Kaufmann, Pascal Sieber. *Schweizer Unternehmen im Internet II, eine empirische Untersuchung.* Arbeitsbericht Nr. 87, Institut für Wirtschaftsinformatik, Universität Bern, Bern, Oktober 1996.

[Ker91]   Heinrich Kersten. *Einführung in die Computersicherheit.* Oldenbourg Verlag, München, 1991.

[KTW97]   Philipp Kirsch, Stephanie Teufel, Harald Weidner. *SINUS – Security in Usage of Online Services.* In: Jan HP Eloff, Rossouw von Solms. Information Security – from Small Systems to Management of Secure Infrastructures. Proccedings of WG 11.1 and WG 11.2 of IFIP/SEC'97, Kopenhagen, Mai 1997.

[LaLo95] Stefan Lang, Peter Lockemann. *Datenbankeinsatz.* Springer Verlag, Berlin, 1995.

[OzAs90] Kazuo Ozeki, Tetsuichi Asaka. *Handbook of quality tools, the Japanese approach,* Productivity Press, Inc., Cambridge, MA, 1990.

[Sch-B92] Ingeborg Schaumüller-Bichel. *Sicherheits-Management - Risikobewältigung in informationstechnologischen Systemen.* BI Wissenschaftsverlag, Mannheim, 1992.

[Var93]   Mario Devargas. *Network Security.* NCC Blackwell, Oxford, 1993.

[Var95]   Mario Devargas. *The Total Quality Management Approach to IT Security.* NCC Blackwell, Oxford, 1995.

[Wei97]   Harald Weidner. *Sicherheit im Internet - Stand der Technik.* Institutsbericht, Institut für Informatik, Universität Zürich, Zürich, 1997.

[Woj91]   Marek Wojcicki. *Sichere Netze: Analysen, Massnahmen, Koordination.* Hanser Verlag, München, 1991.

Sicherheitsmanagement

in großen und komplexen Anwendungsgebieten

Ray H. Matthews[*], Michael Miller[**], Angelika Steinacker[***]

[*]AEA Technology plc
Thomson House
Risley, Warrington, Cheshire WA3 6AT, United Kingdom
Email: ray.matthews@aeat.co.uk

[**]Industrieanlagen-Betriebsgesellschaft mbH, Abt. IS 21
Metzlerstraße 21, D-60594 Frankfurt
Email: mmiller@iabg.de

[***]Industrieanlagen-Betriebsgesellschaft mbH, Abt. CC33
Einsteinstr. 20, D-85521 Ottobrunn
Email: steinacker@iabg.de

Zusammenfassung

Unter Sicherheitsmanagement verstehen wir die Ermittlung, die Organisation und Kontrolle der Sicherheitsmaßnahmen eines komplexen Systems, eines technischen Gerätes oder eines ganzen Unternehmens (z.B. Verkehrsleitsysteme, Flugsicherungssysteme, Kernkraftwerke u.ä.), im weiteren „Anwendungsgebiet" genannt.

Unsere Erfahrungen bei der Erstellung von Sicherheitskonzepten haben gezeigt daß es nicht ausreicht, einzelne IT-Systeme, technische Anlagen oder ähnliches zu betrachten. Sicherheit eines Anwendungsgebietes erreicht man nur, wenn man es als ganzes erfaßt und eine systemübergreifende Bedrohungsanalyse durchführt. Für das Erkennen von Gefahren und Bedrohungen, für die Einschätzung möglicher Schäden und die Ausarbeitung geeigneter Gegenmaßnahmen ist ein systematisches Vorgehen – Risikomanagement oder positiv ausgedrückt: Sicherheitsmanagement – erforderlich. Ferner ist es notwendig, daß für ein Sicherheitsmanagement, das den Betrieb eines komplexen Anwendungsgebietes kontinuierlich begleitet, eine erhebliche Menge von Informationen bereitgehalten werden muß.

Grundlage unserer Vorgehensweise ist ein Modell, das die relevanten Gegebenheiten des Anwendungsgebietes unter Sicherheitsaspekten beschreibt und die Identifikation von Gefahren und Bedrohungen unterstützt.

Da die meisten Anwendungsgebiete einem kontinuierlichen Wandel unterworfen sind, ist davon auszugehen, daß auch die sicherheitsrelevanten Informationen ständigen Änderungen ausgesetzt sind. Aus diesem Grund ist es wichtig, eine Repräsentationsform zu finden, die mit vertretbarem Aufwand zu pflegen ist. Die Struktur des hier vorgestellten Modells genügt dieser Forderung.

1   Einleitung

Die grundlegende Vorgehensweise des Sicherheitsmanagements kann in folgende Schritte unterteilt werden:

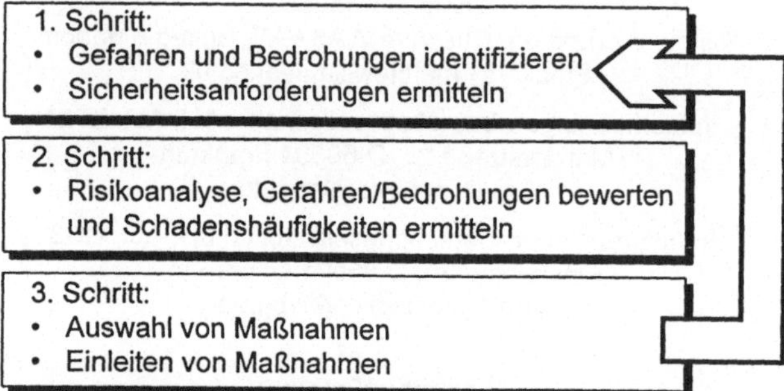

Abbildung 1: Grundlegende Vorgehensweise des Sicherheitsmanagements

Wir unterscheiden hierbei zwei Phasen:

• Erstellen eines Sicherheitskonzepts

Im günstigsten Fall wird das Sicherheitskonzept parallel zur Konzeption und zum Aufbau des Anwendungsgebietes erstellt. Es kommt aber auch vor, daß ein Sicherheitskonzept erst im laufenden Betrieb erstellt wird. In jedem Fall sollten die oben genannten Schritte durchgeführt werden.

• Pflege des Sicherheitskonzepts

Die Pflege des Sicherheitskonzeptes wird in Abbildung 1 durch den Pfeil von Schritt 3 nach Schritt 1 symbolisiert. Da die meisten Anwendungsgebiete einem kontinuierlichen Entwicklungsprozess unterworfen sind und auch durch die Einführung der Sicherheitsmaßnahmen eine Veränderung des Anwendungsgebietes herbeigeführt wird, ist eine regelmäßige, möglicherweise sogar kontinuierliche Überprüfung des erstellten Sicherheitskonzepts durch Überarbeiten der Bedrohungs- und Risikoanalyse notwendig.

Die Ausführung dieser Schritte ist mit zahlreichen Problemen und Schwierigkeiten verbunden, wie wir aus unseren Erfahrungen beim Erstellen und Pflegen von Sicherheitskonzepten in den verschiedensten Anwendungsgebieten wissen.

In den letzten Jahren sind eine Anzahl Methoden zur Lösung dieser Probleme entwickelt worden. Dabei sind insbesondere das „IT-Sicherheitshandbuch" des BSI [IT-SHB] und das „Safety Case"-Prinzip [SM-ATC] zu nennen, mit denen die Autoren in der Praxis umgingen.

Dabei hat es sich gezeigt, daß diese beiden „Standardmethoden" besonders dann zu Problemen führen, wenn die Anwendungsgebiete sehr komplex sind und sich kontinuierlich verändern. Einer der Gründe dafür liegt in der Repräsentation der Informationen für die Bedrohungs- und Risikoanalyse. Bei der Pflege des Sicherheitskonzepts müssen diese Informationen unter Umständen ständig überarbeitet werden.

Um die Bedrohungs- und Risikoanalyse durchführen zu können, muß das Anwendungsgebiet und seine Funktionen wenn möglich vollständig erfaßt und so strukturiert werden, daß die Bedrohungen und Gefahren daraus möglichst direkt erkennbar werden und mit der Struktur des Anwendungsgebiets verknüpft werden können. Für kleinere bis mittlere überschaubare Anwendungsgebiete ist dies noch relativ einfach zu bewältigen. Nimmt die Komplexität und Dynamik des Anwendungsgebietes zu, so ist eine Partitionierung und Strukturierung notwendig, die Änderungen des Informationsinhaltes mit vertretbarem Aufwand zulassen.

Um eine geeignete Strukturierung der notwendigen Informationen zu finden, haben wir die strukturierte Vorgehensweise des IT-Sicherheitshandbuches [IT-SHB] mit der Separierung eines Anwendungsgebietes in „Safety Cases" [SM-ATC] verknüpft und so ein Modell erhalten, das unsere Anforderungen an eine flexible Handhabung erfüllt. Das Modell haben wir zusätzlich um die Betrachtung der Beziehungen zwischen den Einheiten des Anwendungsgebietes erweitert.

Die Methode des IT-Sicherheitshandbuches und das Vorgehen bei der Erstellung von „Safety Cases" wurden ausgewählt, da beide Verfahren Ansätze zur Handhabung großer und komplexer Systeme zeigen, und nach beiden Verfahren von uns Sicherheitskonzepte für große und komplexe Anwendungsgebiete erstellt wurden, so daß vergleichbare Erfahrungen aus der Praxis existieren.

2 Informationsrepräsentation gemäß IT-Sicherheits-handbuch

Das Verfahren des IT-Sicherheitshandbuchs besteht aus vier Stufen:

- Ermittlung der Schutzbedürftigkeit:

  Hierbei wird aus der Sicht des Benutzers des Systems das ausgewählt und abgegrenzt, was das Anwendungsgebiet und somit der Gegenstand weiterer Untersuchungen ist. Dazu werden die Anwendungen, d.h. die Funktionen des Systems, wie sie sich dem Benutzer des Systems darstellen, und die Informationen, die verarbeitet werden, ermittelt und hinsichtlich ihrer Schutzbedürftigkeit bewertet.

- Bedrohungsanalyse:

  Ziel dieser Stufe ist es, alle vorstellbaren Bedrohungen zu ermitteln, die die in der ersten Stufe ermittelten Teile des Systems gefährden können. Im Mittelpunkt der Betrachtung stehen dabei die sogenannten „bedrohten Objekte", die in Kategorien (Infrastruktur, Hardware/Software, Personal, Organisation) eingeteilt werden.

- Risikoanalyse:

  In dieser Stufe wird bewertet, wie schädlich sich die ermittelten Bedrohungen auf den IT-Einsatz auswirken können und welche Risiken aktuell bestehen. Zu jeder Bedrohung werden drei Werte notiert:

  - Es wird festgehalten, welcher Schaden entstehen kann, wenn die entsprechende Bedrohung eintritt.

  - Ein Wert, der die Häufigkeit charakterisiert, mit der die Bedrohung eintritt.

  - Letztendlich wird noch festgehalten, ob das Risiko, das mit der entsprechenden Bedrohung verbunden ist, tragbar oder untragbar ist.

- Auswahl von Maßnahmen:

  In der letzten Stufe werden Maßnahmen gegen die Bedrohungen ausgewählt und ihre Wirkungen beurteilt. Dabei muß entschieden werden, welche Maßnahmen angemessen sind und welches Restrisiko tragbar ist.

Die Informationsrepräsentation gemäß IT-Sicherheitshandbuch kann man mit zwei Worten charakterisieren: „Tabellen" und „Verweise". Im Rahmen derartiger Risikoanalysen mußten bis zu 300 Seiten umfassende Kataloge mit den entsprechenden Tabellen angefertigt werden. In einigen Fällen wurden mehr als 500 Bedrohungen identifiziert und analysiert. Wegen der zahlreichen Querverweise in den Tabellen ist es fast unmöglich, diese manuell zu erstellen und zu pflegen.

Vorteile des Verfahrens des IT-Sicherheitshandbuchs:

- Die Methode führt auf einem strukturierten und definierten Weg zum Ziel.

- Die Informationsstruktur kann leicht auf ein relationales Datenbankmodell übertragen werden.

Nachteile:

- Ohne DV-Unterstützung ist ein kontinuierliches Sicherheitsmanagement, wie es weiter oben definiert wurde, nur für kleine und überschaubare Anwendungsgebiete möglich.

- Die Methode ist auf die Untersuchung von IT-Systemen ausgerichtet.

- Die Methode ist nicht geeignet, den Entwicklungsprozeß (Analyse, Anforderungen, Design, Realisierung, ...) eines Systems zu begleiten.

3 Informationsrepräsentation im „Safety-Case"

„Safety Cases" („Case" hier im Sinne von „Koffer, Kiste") wurden in den späten sechziger Jahren erstmals verwendet. Sie haben sich bei Sicherheitsanalysen für Kernkraftwerke und später auch bei Sicherheitsanalysen in der Öl-Industrie (Raffinerien, Bohrinseln, usw.) bewährt. Seit Anfang der neunziger Jahre werden „Safety Cases" für Sicherheitsuntersuchungen der Britischen Luftverkehrskontrolle eingesetzt.

Ein „Safety Case" enthält alle Informationen bezüglich einer Organisationseinheit oder eines Projekts, die sicherheitsrelevant sind. Welche Informationen als relevant angesehen werden und wie diese Informationen strukturiert werden, ist nicht anwendungsübergreifend standardisiert.

Ein „Safety Case" besteht in der Regel aus mehreren Teilen. Für die meisten Anwendungen bietet es sich an, die folgenden 4 Teile[1] zu unterscheiden.

- **„Safety Case", Teil 1** (Wie sicher muß es sein?):

  Im ersten Teil werden ausgehend von betrieblichen Anforderungen oder Gegebenheiten zunächst potentielle Gefahren und Bedrohungen identifiziert. Diese werden dann bezüglich ihrer Sicherheitsrelevanz kategorisiert. Anschließend wird für jede erkannte Gefahr oder Bedrohung eine Sicherheitsanforderung formuliert.

- **„Safety Case", Teil 2** (Berücksichtigt der Systementwurf die Sicherheitsanforderungen?):

  Es wird in diesem Teil geprüft, ob die grundlegende Struktur (Design) des Anwendungsgebietes den in Teil 1 definierten Anforderungen genügt. Für jede einzelne Anforderung wird mit einer geeigneten Methode nachgewiesen (plausibilisiert), daß diese erfüllt wird.

- **„Safety Case", Teil 3** (Ist das System sicher genug für den Betrieb?):

  Nachdem das betrachtete System realisiert wurde, wird systematisch getestet, ob die einzelnen Sicherheitsanforderungen auch tatsächlich erfüllt werden.

- **„Safety Case", Teil 4** (Wie wird die Sicherheit während des Betriebs gewährleistet?):

  Teil 4 des „Safety Case" enthält regelmäßige Berichte über Erfahrungen aus dem operationellen Betrieb.

Vorteile des „Safety Case"-Prinzips:

- Sicherheitsanforderungen werden in einer einheitlichen Form dokumentiert.

- Die Zusicherung, daß Sicherheitsanforderungen im laufenden Betrieb erfüllt werden, wird ebenfalls dokumentiert.

- Die Struktur der Dokumentation ist relativ einfach und läßt dem Anwender größtmögliche Freiheit beim Durchführen der Risikoanalyse.

---

[1] Es sei darauf hingewiesen, daß diese Zerlegung in 4 Teile nichts mit den 4 Stufen des IT-Sicherheitshandbuchs gemeinsam hat.

Nachteile:

- Die Methode ist nicht besonders geeignet für Anwendungsgebiete oder Systeme, die sich in ihrer Struktur kontinuierlich verändern, da es sehr aufwendig ist, die Sicherheitsuntersuchungen auf dem aktuellen Stand zu halten.

- Komplexe Systeme können in Teilsysteme zerlegt und für jedes Teilsystem ein „Safety Case" erzeugt werden. Die „Safety Case"-Methode bietet jedoch keine strukturierte Möglichkeit, Wechselwirkungen zwischen den einzelnen Teilsystemen zu erfassen.

- Es wird nicht detailliert festgelegt, wie und in welcher Form die sicherheitsrelevanten Informationen dokumentiert werden. Das führt bei großen Anwendungsgebieten zwangsläufig dazu, daß unterschiedliche Formen verwendet werden. Ein Vergleich zwischen einzelnen „Safety Cases" wird dadurch erschwert.

4 Unternehmensweites Sicherheitsmanagement

4.1 Einleitung

Um das Sicherheitsmanagement für ein komplexes Anwendungsgebiet erfolgreich durchführen zu können, ist es unbedingt notwendig, bestimmte Informationen über potentielle Gefahren und Bedrohungen verfügbar zu haben. Dieser Informationsbestand muß strukturiert und stets aktuell sein.

Im folgenden wird die Grundstruktur für ein Modell vorgestellt, das diesen Informationsbestand repräsentiert, das **Sicherheits-Analyse-Modell**. Es wird beschrieben,

- welche Informationen für das Sicherheitsmanagement wichtig sind,
- wie die Menge der Einzelinformationen zu strukturieren ist, und
- wie die Informationen zu aktualisieren sind.

Dabei gehen wir davon aus, daß ein Anwendungsgebiet erfaßt werden soll, das wegen seiner Komplexität und Größe notwendigerweise in Teilgebiete zerlegt werden muß. Ferner gehen wir davon aus, daß es sich um ein dynamisches Anwendungsgebiet handelt, d.h., daß Weiterentwicklungen permanent zu Veränderungen führen, und die Informationsbasis für das Sicherheitsmanagement entsprechend zu aktualisieren ist.

Die Menge der Informationen, die für das Sicherheitsmanagement benötigt werden, kann grob in fünf Bereiche unterteilt werden. Abbildung 2 zeigt, zwischen welchen Bereichen Querverweise existieren.

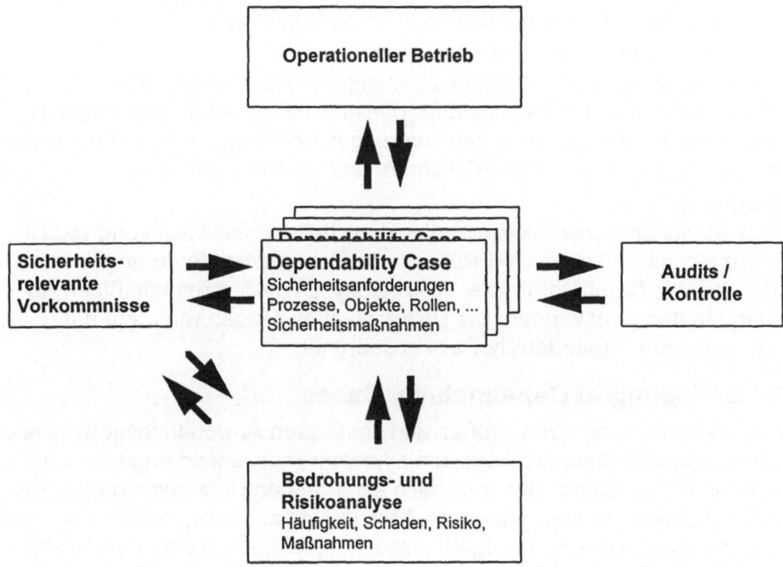

Abbildung 2: Grobes Modell der Informationsstruktur für das Sicherheitsmanagement

1. Operationeller Betrieb
Die Gegebenheiten im operationellen Betrieb bilden die Grundlage für die Zusammenstellung gefährlicher und bedrohlicher Situationen.

2. „Dependability Cases"
Dieser Bereich enthält alle Sicherheitsanforderungen und die betrieblichen Grundlagen für die Bedrohungsanalyse, d.h. eine strukturierte Zusammenstellung möglicher Situationen, in denen Gefahren und Bedrohungen auftreten können. Von diesen Situationen ausgehend, existieren bidirektionale Verweise auf einen Katalog, in dem alle konkret identifizierten Gefahren und Bedrohungen aufgeführt werden. Im nächsten Abschnitt wird dieser Bereich detailliert beschrieben.

3. Bedrohungs- und Risikoanalyse
Es handelt sich um die Zusammenstellung aller identifizierten Gefahren und
Bedrohungen. Diese sind mit den entsprechenden Häufigkeitswerten,
Schadenswerten und Risiken verknüpft. Ferner existiert ein Maßnah-
menkatalog. Zur Dokumentation dieser Informationen eignet sich die
Struktur, die im IT-Sicherheitshandbuch vorgegeben ist.

4. Sicherheitsrelevante Vorkommnisse
Im Rahmen des Sicherheitsmanagements wird ein Katalog aller sicherheitsrelevanten Vorkommnisse geführt. Dieser Katalog enthält bidirektionale
Verweise auf die Sicherheitsanforderungen und Situationsbeschreibungen
sowie auf die Bedrohungs-, Gefahren- und Maßnahmenkataloge.

5. Audits / Kontrolle
Das Einhalten sicherheitsrelevanter Vorschriften, die Umsetzung von Sicherheitsmaßnahmen, die Erfüllung von Sicherheitsanforderungen sowie
die korrekte Durchführung der Risikoanalysen muß in regelmäßigen Ab-
ständen überprüft werden. Die Dokumentation dessen wird dem entsprechenden Dependability Case zugeordnet.

4.2 Zerlegung in Dependability Cases

Für die Beschreibung eines großen und komplexen Anwendungsgebiets reicht
ein Dependability Case nicht aus, um eine dem Problemfeld angemessene
Beschreibung zu liefern. Daher ist nach einer groben Untersuchung des Anwendungsgebiets dieses in sinnvolle Teilbgebiete zu zerlegen. Alle sicherheitsrelevanten Informationen bezüglich eines Teilgebietes werden dann in einem
sogenannten „Dependability Case" zusammengefaßt.

Wie in Abbildung 3 angedeutet wird, werden diese in einer hierarchischen
Struktur angeordnet. Die Spitze dieser Hierarchie ist Dependability Case 1
(DC1), in dem das gesamte Anwendungsgebiet grob untersucht wird.
Im Rahmen dieser Untersuchung wird eine Zerlegung in detailliertere Dependability Cases vorgenommen (DC 2.1, DC 2.2, DC 2.3, ...). Außerdem wird im
ersten Schritt eine grobe Bedrohungs- und Risikoanalyse durchgeführt, aus
der dann Sicherheitsanforderungen resultieren, welche an die untergeordneten
Dependability Cases weitergereicht werden und deren Erfüllung dort nachgewiesen werden muß.

Sollte sich ein Dependability Case als immer noch zu groß und komplex her-
ausstellen, so ist mit diesem ebenso zu verfahren. Zunächst ist eine geeignete
Zerlegung zu finden, es ist eine Bedrohungs- und Risikoanalyse durchzuführen, und es sind Sicherheitsanforderungen für die untergeordneten Teilgebiete
zu definieren.

Die schrittweise Zerlegung wird solange fortgesetzt, bis ein angemessener
Detaillierungsgrad erreicht wird. Die Wahl dieser Zerlegung ist entscheidend

für die Qualität des Sicherheitsmanagements.

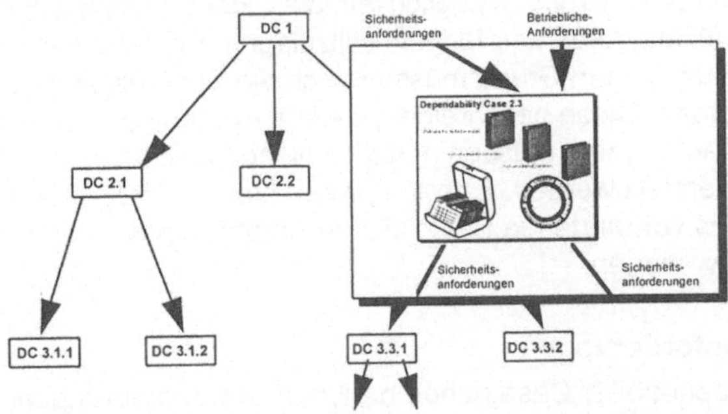

Abbildung 3: Zerlegung eines großen und komplexen Anwendungsgebiets

4.3 Informationsstruktur eines Dependability Case

Abbildung 4 veranschaulicht, welche Informationen jeweils für ein Teilgebiet vorzuhalten sind. Dieser Informationsbestand wird als „Dependability Case" für das entsprechende Teilgebiet bezeichnet.

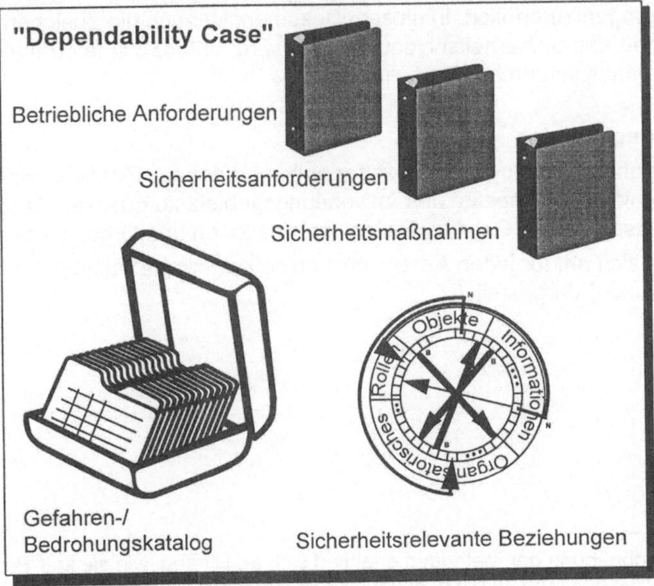

Abbildung 4: Informationen in einerm Dependability Case

Betriebliche Anforderungen

Häufig besteht die einfachste Möglichkeit, die Sicherheitsanforderungen für ein System zu erfüllen, darin, das System stillzulegen oder erst gar nicht in Betrieb zu nehmen. Aus diesem Grund müssen auch die Betrieblichen Anforderungen betrachtet werden. Diese beschreiben, welche Aufgabe das betrachtete System zu erfüllen hat, und müssen natürlich nicht speziell für das Sicherheitsmanagement erstellt werden. Es reicht, wenn hier eine Kopie oder ein Verweis auf die ohnehin vorhandenen Betrieblichen Anforderungen des entsprechenden Systems vorliegen.

Sicherheitsanforderungen

Zu jedem Dependability Case gehören Sicherheitsanforderungen, die detailliert beschreiben, welches Sicherheitsniveau von den entsprechenden Teilen des Anwendungsgebietes einzuhalten sind. Diese Sicherheitsanforderungen sind aus den Bedrohungs- und Risikoanalysen des übergeordneten Dependability Case hervorgegangen.

Sicherheitsmaßnahmen

Dieses Dokument beschreibt kurz das Resultat der Risikoanalyse und verweist auf die wesentlichen Maßnahmen, die eingeleitet wurden, um alle Sicherheitsanforderungen zu erfüllen. In diesem Dokument ist auch die Zusicherung enthalten, daß das Sicherheitsniveau erhalten wird, und daß eine ordnungsgemäße Risikoanalyse durchgeführt wurde.

Bedrohungskatalog

Der Bedrohungskatalog wird prozeßorientiert aufgebaut. Zunächst einmal sind die verschiedenen Aspekte des Anwendungsgebiets zu erfassen. Dazu gehören Arbeitsprozesse, Objekte, Rollen, Informationen und Organisatorisches.

Es bietet sich an, für jeden Arbeitsablauf eine Karteikarte[2] anzulegen, wie sie in Abbildung 5 dargestellt ist.

---

[2] Die Beschreibung der Methoden orientiert sich an Dingen, wie sie im ("Papier"-) Büro üblich sind, um so ein anschauliches Bild zu liefern. Natürlich ist es ab einer bestimmten Menge von Elementen im System günstiger, die Erfassung und weitere Verarbeitung computergestützt durchzuführen. Am Prinzip der Methode ändert sich dadurch allerdings nichts.

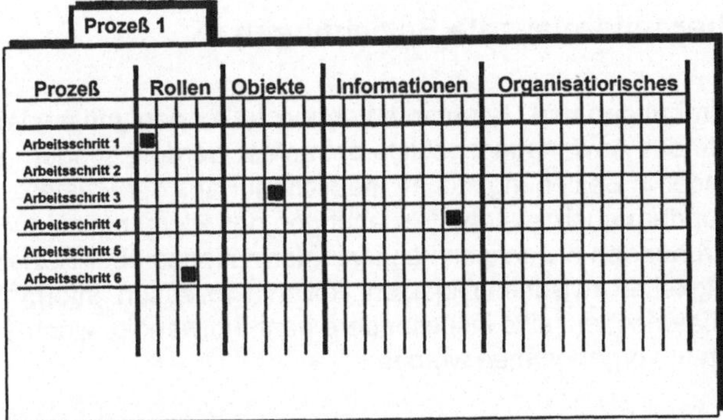

Abbildung 5: "Karteikarte" zur Beschreibung eines Prozesses

In die erste Spalte dieser Karte wird der Prozeß mit seinen einzelnen Ablaufschritten eingetragen. In den folgenden Spalten finden sich die Informationen, die Rollen, die Objekte und die organisatorischen Elemente des Anwendungsgebiets wieder. Sind alle wiedergegeben, so werden in die Felder der Karteikarte Verweise zu den Bedrohungen und Gefahren eingetragen, die sich in Verbindung des Prozeßschrittes mit dem jeweiligen Element ergeben können. Die Verweise auf potentielle Gefahren und Bedrohungen sind durch die Quadrate in der Karteikarte dargestellt. Welche weiteren Informationen zur Dokumentation von Gefahren und Bedrohungen festgehalten werden, wird weiter unten beschrieben.

4.3.1 Sicherheitsrelevante Beziehungen

Um alle Bedrohungen und Gefahren erkennen zu können, müssen nicht nur alle Elemente des Anwendungsgebiets betrachtet werden, sondern auch, welche miteinander in Beziehung stehen, da nicht nur aus der Existenz eines Elements, sondern auch aus den Beziehungen des Elements zu anderen Elementen Bedrohungen erwachsen können. Informationen über Beziehungen und Abhängigkeiten zwischen Objekten, Rollen, Prozessen, Informationen oder Organisatorischem sind insbesondere dann notwendig, wenn Änderungen am System vorgenommen werden.

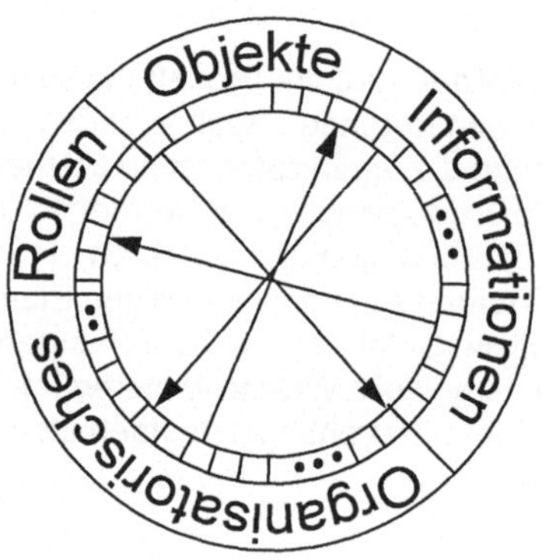

Abbildung 6: Beziehungen zwischen Elementen

Die Pfeile in Abbildung 6 geben die die Beziehungen zwischen den Elementen des entsprechenden Dependability Case wieder. Zeigt ein Pfeil von Objekt „A" nach Objekt „B", dann bedeutet dies, daß Änderungen an Objekt „A" sicherheitsrelevante Auswirkungen an Objekt „B" haben können.

Nun ist es nicht nur möglich, sondern bei einem komplexen System sehr wahrscheinlich, daß Beziehungen nicht erkannt oder deren Dokumentation vergessen werden, so daß die tatsächlich ermittelten Beziehungen, die durch die dikkeren Pfeile (mit „B" gekennzeichnet) in Abbildung 7 symbolisiert werden, nur eine Teilmenge aller existierenden Beziehungen darstellen.

Aus existierenden Beziehungen, die während der Bedrohungsanalyse nicht erkannt werden, entstehen Fehler bei der Abschätzung des Gesamtrisikos. Außerdem besteht die Möglichkeit, daß bei Änderungen des Systems die sicherheitsrelevanten Auswirkungen unterschätzt werden. Um dies zu verhindern, schlagen wir folgende Vorgehensweise vor:

Neben den erkannten und dokumentierten Beziehungen wird zusätzlich ermittelt und vermerkt, welche Elemente des Anwendungsgebiets definitiv <u>nichts</u> miteinander zu tun haben. Diese „Nicht-Beziehungen" sind in Abbildung 7 durch die äußeren Pfeile (mit „N" gekennzeichnet) symbolisiert.

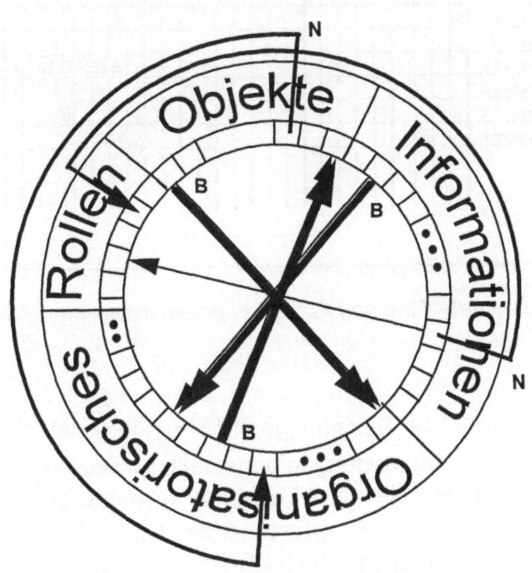

Abbildung 7: Beziehungen und „Nicht-Beziehungen" zwischen den Elementen eines Dependability Case

Auf den "Karteikarten" werden die Beziehungen zwischen den Arbeitsschritten eines Prozesses und den Elementen des Anwendungsgebiets mit „B" oder „N" repräsentiert. Dabei besagt ein **B** in der Zeile y und der Spalte z der Karteikarte, daß zwischen dem Prozeßteil y des beschriebenen Prozesses und dem Element z eine Beziehung besteht, daher bei Gefährdung oder Änderung eines Teils der andere möglicherweise mit betroffen ist. Ein **N** hingegen repräsentiert eine "Nicht"-Beziehung, d.h. daß das Element z in keiner Weise vom Prozeßteil y betroffen ist und daß der Prozeßteil y nicht auf das Element z angewiesen ist.

Eine "Karteikarte" dieser Art ist in Abbildung 8 beispielhaft gegeben.

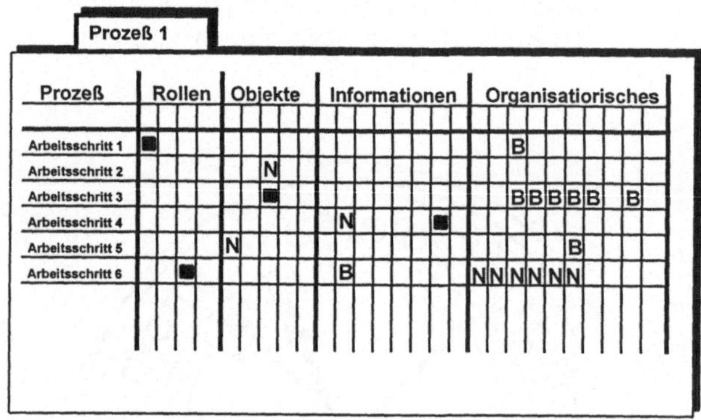

Abbildung 8: "Karteikarte" für einen Prozeß in einem komplexen Anwendungsgebiet

Die Bestimmung der "Nicht"-Beziehungen hat (mindestens) 2 Vorteile:

- Es werden genauere Überlegungen angestellt, ob Beziehungen zwischen Elementen bzw. Prozeßteilen und Elementen vorhanden sind.
- Für die Pflege der Bedrohungs- und Risikoanalyse bei Änderungen an einzelnen Elementen können die "Nicht"-Beziehungen genutzt werden, um den Arbeitsaufwand für die Überprüfung möglicher Auswirkungen auf andere Elemente einzuschränken.

4.4 Bedrohungs- und Risikoanalyse

Grundlage für die Bedrohungs- und Risikoanalyse ist eine "Tabelle" (Für viele Anwendungsgebiete ist eine Datenbank angemessen!) mit allen Gefahren und Bedrohungen, die im Rahmen der einzelnen Teilgebiete identifiziert wurden. Diese Tabelle ist durch die Verweise von den einzelnen Karteikarten aus (Quadrate) mit den Informationsbeständen der einzelnen Teilgebiete verknüpft.

Mit dieser "Bedrohungstabelle" wird verfahren, wie es im Schritt 3 und Schritt 4 des IT-Sicherheitshandbuchs vorgesehen ist (siehe Kapitel 3 Seite 12). Gemäß einer für das Anwendungsgebiet standardisierten Werteskala wird jeder Bedrohung ein „Schadenswert" zugeordnet. Dieser Wert dokumentiert die Höhe des Schadens, der entstehen kann wenn die Bedrohung eintritt. Ferner wird zu jeder Bedrohung ein „Häufigkeitswert" ermittelt, der grob beschreibt, wie groß die Wahrscheinlichkeit für die Enstehung des Schadensfalls ist. Die Struktur zur Ablage dieser Informationen kann im wesentlichen aus dem IT-Sicherheitshandbuch übernommen werden.

5 Ausblick

Standardisierte Modelle zur Unterstützung des Sicherheitsmanagements sind heute noch nicht "Stand der Technik", und es existieren nur wenige Erfahrungen beim Umgang mit Modellen dieser Art. Das hier vorgestellte Sicherheits-Analyse-Modell wurde bisher in kleineren Anwendungsgebieten erprobt und hat sich dort bewährt. In den nächsten Monaten wird das Modell in einem großen und komplexen Anwendungsgebiet in der Praxis getestet werden, dazu wird auch ein Datenbanksystem eingesetzt, um die anfallenden Informationsmengen angemessen zu verwalten. Die Erfahrungen, die hierbei gewonnen werden, sollen zu weiteren Verbesserungen des Modells führen.

Literaturverzeichnis

[IT-SHB]   IT-Sicherheitshandbuch: Handbuch für die sichere Anwendung der Informationstechnik, Bundesamt für Sicherheit in der Infor-mations-technik 1992.

[IT-GSH]   IT-Grundschutzhandbuch 1996, Bundesamt für Sicherheit in der Informationstechnik 1996.

[ISK]   Internationale Sicherheitskriterien, H. Pohl, Oldenbourg Verlag 1993.

[SSI]   Szenarien zur Sicherheit informationstechnischer Systeme, S. Herda, S. Mund, A. Steinacker, Oldenbourg Verlag 1993.

[IT-SEC]   Kriterien für die Bewertung der Sicherheit von Systemen der Informationstechnik, Bundesamt für Sicherheit in der Informati-onstechnik 1991.

[SRC]   A Guide to Security Risk Management for Information Techno-logy Systems, Communications Security Establishment Go-verment of Canada, August 1995.

[NIST-ST]   A Study on Hazard Analysis in High Integrity Software Stan-dards and Guidelines, L. M. Ippolito und D. R. Wallace, Natio-nal Institute of Standards and Technology, 1995.

[HAZOP]   HAZOP & HAZAN, Notes on the Identification and Assessment of Hazards, T. A. Kletz, The Institution of Chemical Engineers 1986.

[MoSo]   A methodology to include computer security, safety and resili-ence requirements as part of the user requirement. D.N.J Mo-stert und S.H. von Solms, Computers & Security, 13 (1994) 349-364

[SoMe]   The management of computer security profiles using a role-oriented approach. S.H. von Solms und I. van der Merwe, Computers & Security, 13 (1994) 673-680

[SM-ATC]   Systematic Safety Management in the Air Traffic Services, Ri-chard Profit, Euromoney Publications, 1995

BOOTSTRAP: FIVE YEARS OF ASSESSMENT EXPERIENCE

Hans Stienen, Hartmut Gierszal*
SYNSPACE AG, Rottmannsbodenstrasse 30, CH - 4102 Binningen
Phone: (41) 61 4220240, Fax: (41) 61 4220933, Email: stienen@synspace.ch
* SYNSPACE GmbH, Jechtingerstraße 11, D - 79111 Freiburg
Phone: (49) 761 4764565, Fax (49) 761 4764568, Email: hg@synspace.freinet.de

Abstract: The main characteristics of the BOOTSTRAP method are described, i.e. the reference framework, the assessment procedure, the structure of the questionnaires and the rating and scoring mechanisms used. Experiences from over 40 sites and 180 projects are reported. BOOTSTRAP proved to be a very efficient and effective means not only to assess the current status of software process quality but also to initiate appropriate improvement actions. The last part deals with future trends, i.e. the development of a common framework to perform software process assessments.

1 INTRODUCTION

A software product is always the result of a process. As it is still extremely difficult to determine the quality of the final product, a careful design and control of the software development process is essential. In order to prevent errors in the resulting work products and to produce quality instead of attempting to achieve it only by (late) testing and verification activities, there is an urgent need for sound software process improvement techniques (Stevens, 1991).

A *Software Process Assessment* is a comprehensive examination and evaluation of all engineering and management processes that appertain to *Application Development*, *Product Installation* and *Operational Support* including *Software Maintenance*. It results in a set of metrics (profiles), that encourage an organisation to recognise its major strengths and weaknesses. The resulting profiles enable the determination of feasible targets for software process improvements. An assessment provides a solid baseline for measuring progress.

BOOTSTRAP - has been developed in an ESPRIT project (Kuvaja et.al., 1993, 1994). The approach evolved on the basis of CMM (Paulk, 1993, 1995), the ISO 9000 Series and the European Space Agency's Software Engineering Standards. Since 1991, over 100 assessments have been performed, mostly by members of the *Bootstrap Institute* but also by *licensees* of the method. Based on these experiences, the assessment instruments have been revised and refined continuously. Recently, additional criteria that originated from the Total Quality Model issued by the European Foundation for Quality Management (Hakes, 1995) has been incorporated.

The resulting assessment and improvement method is characterised by the following points:

♦ It provides detailed maturity levels and capability profiles of both organisations and projects.

♦ It allows optionally to determine the degree of fulfillment of the 20 elements of ISO 9001.

♦ It allows benchmarking, i.e. the comparison with specific industrial sectors, countries and other assessment data stored in the BOOTSTRAP database.

♦ It includes through its standardised procedure immediate feedback to the organisation and support for initiating improvement actions.

From past experiences with software process assessment methods it appears that a major problem encountered is that deviations from a reference framework as e.g. advocated by CMM may be so large that few of the specified characteristics may be observed, although the organisation performs evidently well on the market. This is especially true for small software development organisations with short communication and reporting lines and an informal organisational structure. The basic assumptions that only one way to perform the development process exists are not accepted. A major challenge for an assessment method is therefore the incorporation of appropriate mechanisms for tailoring individual assessments to the actual needs and expectations of an organisation. BOOTSTRAP is flexible enough to account for the various application areas, the typical size and cultural differences across organisations and countries (Koch, 1995).

BOOTSTRAP has been initiated primarily as a project to encourage technology transfer. Assessments were identified as an excellent vehicle for that purpose. In order to identify good software engineering and management practices, it is necessary to have the instruments available for the classification, measurement and analysis of these practices. Therefore, an organisation and its process quality are assessed with respect to organisation, methodology and technology. BOOTSTRAP assumes that an organisation has to satisfy basic organisational requirements first, and then to take care of engineering methods before making investments in technology as third priority (Koch, 1993). The underlying model guides organisations to a mature, disciplined and stable way of producing software.

2   MAIN FEATURES OF BOOTSTRAP

Rather than to perform a plain survey of the status, the main objective of a BOOTSTRAP assessment remains to generate an improvement action plan. *Improving software process quality requires measuring it.* BOOTSTRAP recognises that transitions should take place step by step.

The fact that the method does not assume strict adherence to a distinct key practice model like CMM, but allows for alternative approaches, is a key factor to BOOTSTRAP's success. Mean features of BOOTSTRAP are:

- Uniform procedure, mandatory assessor qualification and training scheme
- Carefully designed questionnaires for both site and project evaluations
- Constructive, not normative approach
- Open questions, granular scoring and rating mechanism
- Immediate feedback and action planning

Even with sophisticated assessment instruments, rating and scoring answers given during the interviews always contain an element of judgement. For this reason strict adherence to the established procedures is expected in order to keep variance as small as possible. The *Bootstrap Institute* maintains a strict qualification scheme for BOOTSTRAP assessors. The curriculum encompasses at least one week of training, followed by ten days of apprenticeship in real assessments. The Bootstrap Institute founded by the original partners of the ESPRIT project also reviews all assessment data and maintains the BOOTSTRAP Database which is accessible for *licensees* of the method to provide benchmarks and comparative analyses to their clients.

2.1 The BOOTSTRAP Procedure

As initiation and continuous support for improvement actions is the prime goal of BOOTSTRAP, a co-operative approach is used in order to characterise and analyse a software organisation or project. In order to get full commitment and support on all levels, the perspectives of management and engineering staff are equally important. Therefore, a typical schedule of a full assessment includes several briefings and debriefing activities.

Main objective of the first steps of an assessment, the briefings, is to agree on the objectives and scope of the assessment and to obtain commitment from both senior and line management as well as engineering staff. During the initial phase, also a pre-assessment questionnaire is distributed to prepare the interviews and to collect that kind of information that is otherwise not readily available. Core activity of a BOOTSTRAP assessment is the on-site week. To emphasise the importance of team-oriented processes, group interviews and workshops are the central part of this week.

Two nearly symmetrical questionnaires are used for the group interviews, which take between three and six hours of time. Between 110 and 150 questions are discussed, which cover all key aspects of the software development process at that level of detail that is needed to determine candidate key areas for further improvement and to provide a baseline to be able to demonstrate progress.

Using computer-based tools, a preliminary evaluation is already made on-site, immediately after the interviews. First results are presented and discussed at the end of the on-site week. Confidentiality is an important aspect here. Results of individual projects are presented only to the project representatives during the debriefings. Beyond this, only aggregated figures are presented and discussed.

Improvement action planning starts at the end of the on-site week with the presentation of the preliminary results to all participants and is continued with a one-day consolidation and action-planning workshop with the improvement team.

Organisations appreciate that BOOTSTRAP reduces *initial costs and lead-time to a minimum* required establishing a solid baseline for necessary improvement actions and corresponding investments. They experience BOOTSTRAP as a very motivating exercise and as a real support in order *to make improvements really happen* without increasing the amount of paperwork.

2.2 Rating and Scoring

An important feature is that BOOTSTRAP uses open questions to assess the actual situation and to solicit opinions of the key players that will influence the success of the software improvement activities. By using well designed questionnaires it is guaranteed that no critical points are missed. A scheme with a fine granularity is applied to judge the relative importance of each aspect.

To achieve a common agreement upon improvement actions by all member of the organisation is rather difficult, as different opinions will occur inevitably. Hence, to generate confidence in the obtained results consistent scoring is vital. The purpose of a scoring mechanism is to reach consensus, so the algorithms for aggregating the large amount of data collected during the interviews must be robust enough to support intuition.

In order to obtain a clear and true picture of current practices responses must be evaluated in several dimensions. BOOTSTRAP uses mainly the measures:

♦ Overall process capability **and** Quality profile of the applied processes.

Mere existence of a process does not say too much about an organisation's capability to develop quality software within given time and budget constraints. BOOTSTRAP requires that existing processes are analysed in further detail during the interviews and responses are scored in one of the categories *basic*, *significant* or *excellent*. When, however, the process or activity addressed by a given question is not relevant, that question may be said to be *not-applicable* and, as a consequence, excluded from further evaluation. In a typical case about 10% of the generic processes that are identified by BOOTSTRAP may be not relevant, but also that about 35% of the relevant ones are not practised.

With each question BOOTSTRAP assigns a number indicating a capability level of the corresponding activity. Evaluation of the responses leads to a quantitative assessment of the fulfillment of specific requirements for each level. It is frequently observed that organisations pay attention to all processes, irrespective of their maturity.

Experience has shown that there may exist good reasons for an organisation to emphasise some processes that belong to the higher levels or to skip some lower level processes. Typical examples are on one hand processes that address quality, i.e. quality management and quality assurance and on the other hand, some of the product design processes. The first is often the case in organisations that concentrate on ISO 9000 certification. The latter is often observed in software development units that produce different kinds of systems. In this case, the responsibility for defining the individual engineering processes is delegated to each of the software engineering teams within such a unit. If during assessment, it was found that this approach works well, responses in the site questionnaire that address particular management concerns belonging to these aspects were set to 'not applicable'. By doing so, criteria associated with these questions were excluded from the computation of the overall BOOTSTRAP rating for the organisation.

Other interesting aspects of how software organisations work can be examined using the BOOTSTRAP database that contains the key data of a large variety of organisations. *Organisational stability, clear customer focus and permanent staff training seem to be the dominant factors for attaining excellent results.* A trend towards higher ratings for larger organisations can be observed. It may be true that in larger organisations project teams are larger, and therefore the supporting processes are more visible. However, this does not imply that productivity is lower. A detailed analysis of the available data reveals that in the larger organisations software engineers tend to be higher educated, have more years of practical experience and are working in the same area over a longer time. However, also excellent performing small engineering teams are present.

Not only the overall BOOTSTRAP rating varies from case to case, but it may be expected that also the existence and quality of the software processes will vary accordingly. Indeed most organisations concentrate on level 2 and 3 processes. A typical organisation with a BOOTSTRAP rating of 2 will perform not only nearly 80% of all level 2 processes, but also 60% of all level 3 processes and even 20% of the level 4 processes. However, the mere existence of a process is not enough to produce quality. It turns out that in a level 2 organisation about 80% of all level 2 processes are existent, but merely in a rudimentary form, as only 40% of all level 2 criteria are fulfilled.

It can also be observed that on the average, an organisation with a rating of 3 will perform in some way all basic processes on level 2 and about 80% processes on level 3. On that capability level, the importance of level 4 processes is growing fast. The observations made here are exactly in line with CMM, which assumes for the completion of each level the existence of 80% of the processes associated with that level.

It can be observed that within most organisations some individual projects perform far better than one would really expect from the organisational framework, but a general tendency to overestimate the effectiveness of the software development processes by senior management can be observed (Engelmann, 1995). The perception of executives and software engineers is often quite different, and this is exactly where capability profiles provide a solution.

A closer analysis of profiles revealed that most organisations had difficulties in areas, like process description, measurement and control as well as architectural design and testing. Also the lack of a proper quality system visibly supported by executive management was a major source of complaints. Weaknesses in these areas are frequently compensated (at high costs) by both extensive integration and acceptance testing as well as rigorous quality assurance cq. quality control procedures.

4 OUTLOOK

During the last five years, the BOOTSTRAP assessment instruments were refined several times. The establishment of a stable software development process within an organisation is equally important as it was when the first version of BOOTSTRAP was developed. However, in some areas questions could be dropped. Nowadays, questions addressing the working environment and the technical infrastructure are no longer relevant, as at least in Europe the conditions improved significantly. Also due to the proliferation of PC-based development the configuration management of both code and design documents is now state-of-practice in most organisations. Attention is shifting to other areas. Organisations frequently signalise problems in adopting client-server technology. Applications are no longer developed from scratch but assembled from standard components. Development cycles become shorter and shorter. Establishing good relations with customers and users and good product design are a major concern today. Assessment methods have to recognise that.

In 1993, a joint ISO/IEC working group established a project called SPICE, Software Process Improvement and Capability Evaluation (Dorling, 1993) aimed at the harmonisation of the various software process assessment methods. SPICE's Baseline Practices Guide currently contains a very detailed framework which is based on the recent ISO Standard 12 207. With such a detailed framework it is not a straightforward task to develop an assessment method (Paulk, 1995).

5   LITERATURE

Dorling, A.: Software Process Improvement and Capability Determination, Software Quality Journal, Vol. 2, Nr. 4, 1993, pp. 209-224.

Engelmann, F.; Schynoll, W.; Stienen, H.: BOOTSTRAP Software Process Assessment - Experiences and Further Developments, Proc. SafeComp '95, Belgirate, Italy.

Hakes, C.: The Corporate Self-Assessment Handbook. Chapman & Hall, 2nd ed., London 1995.

Koch, G.R.: Process Assessment: the BOOTSTRAP Approach, Information and Software Technology, Vol. 35, Nr. 6/7, 1993, pp. 387-403.

Koch, G.R.: Quality Improvement as a Multicultural Challenge - The European View, First World Congress of Software Quality, San Francisco, June, 1995.

Kuvaja, P., et. al.: BOOTSTRAP: Europe's assessment method, IEEE Software, Vol. 10, Nr. 3 (1993) pp. 93-95.

Kuvaja, P., et. al.: Software Process Assessment & Improvement - The BOOTSTRAP Approach, Blackwell, 1994, ISBN 0-631-19663-3.

Paulk, M.; Curtis, B.M; Crissis, M.; Weber, C.: Capability Maturity Model for Software, Version 1.1, Technical Report CMU/SEI-93-TR-24, Software Engineering Institute, Pittsburgh, Febr. 1993.

Paulk, M., et. al.: The Capability Maturity Model - Guidelines for Improving the Software Process, Addison Wesley, New York 1995

Paulk, M.; Conrad, M.; Garcia, S.:   CMM versus Spice architectures, IEEE Software Process Newsletter, Nr. 3, 1995, pp. 7 - 11.

Stevens, R.: Creating Software the Right Way. Byte, Aug. 1991, pp. 31 - 38.

Transparenter versus Sicherer Mobiler Informationszugriff – eine Anforderungsanalyse

Astrid Lubinski

Universität Rostock, FB Informatik

e-mail: lubinski@informatik.uni-rostock.de

Abstract

Die mobile Arbeit mit Computern bringt in unterschiedlichen Anwendungsbereichen große Vorteile gegenüber einer stationären. Die Mobilität verstärkt jedoch bekannte und produziert neue Datensicherheits- und Datenschutzprobleme. Das wohl bekannteste Risiko für die Persönlichkeitsrechte mobil arbeitender Nutzer stellt die Anfertigung von Bewegungsbildern dieser Personen dar.

Die negativen Effekte resultieren jedoch nicht allein aus der Nutzung von Mobilkommunikationsnetzen. Aus der Kenntnis darüber, wie ein Nutzer mobil auf welche Informationen und Dienste zugreift, können ebenso sein Aufenthaltsort und -dauer und andere Informationen über ihn abgeleitet werden. Als eine Ursache für die Probleme sind die automatisch erstellten und verwalteten Kontextdaten anzusehen, die für die Gewährleistung und die Anpassung der mobilen Arbeit und Kommunikation an die mobilen Gegebenheiten notwendig sind.

Dieser Artikel analysiert anhand der Schutzziele und der Kontexte in mobiler Umgebung die auftretenden Sicherheitsanforderungen. Es werden dann Vorschläge gemacht, wie den neuen Problemstellungen begegnet werden kann.

keywords: Sicherheit, Mobilität, Datenbanksysteme

1 Einleitung

1.1 Mobilität und Sicherheit

Die mobile Nutzung von Informationen ist ein Thema von wachsendem Interesse, zumal sich die Hardwaregrundlage hierfür rasch entwickelt. Mobiler Zugriff und mobile Kommunikation bergen neue Sicherheitsrisiken hauptsächlich dadurch in sich, daß neben den „Inhaltsdaten" eines Zugriffs bzw. einer Kommunikation „Kontextdaten" entstehen und verwaltet werden, die diesen Zugriff und die Kommunikation ermöglichen. Hier handelt es sich z.B. um Informationen zum Aufenthaltsort eines mobilen Gerätes oder um Festlegungen zu dessen Erreichbarkeit. Andererseits dienen Kontextdaten dazu, den Zugriff und die Kommunikation, die im mobilen Bereich durch Ortswechsel und dynamische und beschränkte Ressourcen gekennzeichnet sind, zu unterstützen und mit möglichst geringen Qualitätseinbußen zu realisieren. Diese Kontextdaten werden automatisch durch die entsprechenden Systemkomponenten verwaltet und verwendet. Der Nutzer muß sich um diese Aufgabe nicht kümmern, die Kontexte und ihre Anwendung sind für ihn **transparent**. Diese Transparenz als Verbergen der Informationen und ihrer Anwendung vor dem Nutzer stehen der Datenschutzbestrebung, Informationsverarbeitung transparent und im Sinne von Durchschaubarkeit sichtbar zu machen, entgegen. Die Lösung des Konfliktes zwischen den beiden Transparenzen und die vertrauliche, integre und verfügbare Verwaltung transparenter (verborgener) Kontextinformationen sind in mobiler Umgebung die Basissicherheitsanforderungen.

In diesem Artikel sollen die Kontextdaten für die Mobilkommunikation eine geringere Rolle spielen. Es handelt sich dabei um:

- statische Kontexte, das sind Bestandsdaten
- dynamische Kontexte, das sind
  - Verkehrsdaten,
  - Entgeltdaten,
  - Verbindungsdaten und
  - Verbindungsvorbereitungsdaten.

Den Schutz von Kommunikationskontexten, in denen Personen- und Ortsdaten in Zusammenhang gebracht werden können, haben verschiedene Artikel von Pfitzmann, Federrath u.a. ([Pfit93][FJKP95]) zum Gegenstand. Dort werden Schutzziele für die Übermittlung der Nutzdaten über Mobilkommunikationsnetze spezifiziert, für die dann Verfahren zu ihrer Erreichung gezeigt werden. Nutzdaten sollen durch kryptographische Verfahren vor Fälschung und Offenbarung geschützt werden. Die in den Vermittlungsdaten verborgenen Ortsinformationen, die das geschützte Bewegungsbild einer Person erzeugen lassen, werden geschützt, indem die Peilbarkeit eingeschränkt wird und die Lokalisierungsinformationen vertrauenswürdig gespeichert werden, um die Ortsinformation zu verschleiern und die Erreichbarkeit eines mobilen Teilnehmers zu regeln.

Als Grundlage für die Konkretisierung der Sicherheitsanforderungen, die durch die mobile Umgebung hinzukommen, sollen im nächsten Abschnitt zunächst das zugrunde liegende Mobilitätsverständnis und der Begriff des Kontextes näher beschrieben werden. Anhand der Kontextkomponenten und der Beziehungen zwischen ihnen werden dann im Abschnitt 3 Sicherheitsanforderungen gestellt und Hinweise gegeben, wie sie erfüllbar sind und ob die verfügbaren Mittel und Methoden ausreichend sind oder ob neue Ansätze gebraucht werden.

2 Mobilität und Kontexte und Transparenz

2.1 Der Mobilitätsbegriff

Mobilität bedeutet nach Imielinski und Badrinath [IB93], daß von überall und jederzeit Zugriffe möglich sind. Bei der konkreten Interpretation dieser Beschreibung unterscheiden sich jedoch die Arbeiten zum Mobile Computing. Für eine große Gruppe von Autoren, allen voran Imielinski und Badrinath, beinhaltet der obige Satz, daß jeder Nutzer sein eigenes mobiles Gerät nutzt und mit diesem eine "mobile unit" bildet. Das mobile Gerät kommuniziert über eine Funkverbindung. Ein Nutzer ist in diesem Szenario eindeutig über ein Gerät identifizierbar. Besondere Forschungsanstrengungen werden in die Probleme begrenzter Übertragungsbandbreiten bzw. unterbrochener Funkverbindungen investiert.

Dahingegen werden bei Wachowicz und Hild [WH96] Experimente durchgeführt, die den Standort des mobilen Nutzers mittels „active badges" über „location sensors" ermitteln. Der Nutzer ist nicht auf ein persönlich zugeordnetes Gerät fixiert. Die „totale Mobilität" als Mobilität von Nutzern oder Terminals erfordert natürlich spezielle, „eingeweihte" Computer, um sich auf die jeweils anwesende und nutzende Person einstellen zu können. Die Nutzermobilität ist durch die Abhängigkeit von der Reichweite dieser location sensors und active badges und von verfügbaren "eingeweihten" Computern auf kleinere, lokal abgegrenzte Bereiche beschränkt.

Im Projekt "Mobile Visualisierung" (MoVi) [*] werden generelle Konzepte für eine mobile Visualisierung, bestehend aus einer Informationsanforderung bzw. -erzeugung in einem mobilen Nutzerinterface, der Bereitstellung, Übertragung und Visualisierung multimedialer Informationen und Dienste, erarbeitet. Mobilität bedeutet in diesem Projekt Mobilität des Menschen, dessen Standort sich in der Zeit verändert und der für seine Aufgabenerledigung oder Interessendurchsetzung mittels irgendeinem Gerät und Netz beliebige zur Verfügung stehende Informationen und Services nutzen kann. Geräte und Netze können mobil sein, müssen es jedoch nicht.

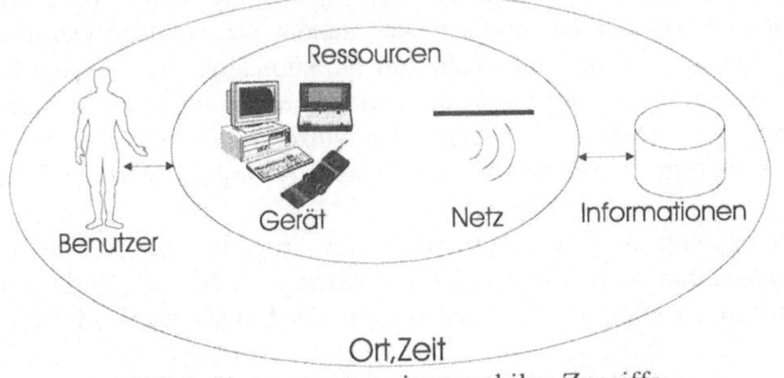

Abb.1: Komponenten eines mobilen Zugriffs

[*] Die hier vorgestellten Arbeiten entstanden im Projekt „Mobile Visualisierung", das durch die Deutsche Forschungsgemeinschaft (DFG) unterstützt wird.

2.2 Transparente mobile Verarbeitung

Eine wichtige Eigenschaft mobiler Verarbeitung besteht darin, daß diese vier mobilen Komponenten von der sich ändernden Umgebung, dem **mobilen Kontext**, abhängen. Dieser mobile Kontext besteht aus eben den vier genannten Komponenten und dem Ort und der Zeit, denn Mobilität entsteht durch Ortsveränderung in der Zeit. Ort und Zeit bilden im mobilen Umfeld den zentralen mobilen Kontext, mit dem die anderen Kontexte korrelieren. Die Ortsangabe, möglichst automatisch ermittelt, fließt in die Auswahl der Anwendung und der Informationen ein. Dies kann z.B. durch das Erweitern der Selektionsbedingung in einer Datenbankanfrage erfolgen.
Beispiel:
Eine Anfrage nach Kundendaten in einer Vertriebsfirma wird als Nachfrage nach den Kundendaten interpretiert, in deren Nähe der Vertreter sich gerade aufhält bzw. auf die er zusteuert.

Ein bestimmter Ort kann darüber hinaus bestimmend für die Auswahl von Anwendungen und Tools sein. Da mobile Geräte meist ressourcenbeschränkt z.B. bei der Bildschirmgröße, dem verfügbaren Speicher, dem Ein- und Ausgabemedium und der Energie sind, müssen sich Software und Informationen an diese Ressourcenkontexte durch Auswahl oder eine alternative Darstellung anpassen. Eine aktuelle Anfrage wird beispielsweise aufgrund des bei vorhergehenden Datenbankanfragen geltenden Kontextes optimiert und kann so besser an den aktuellen Kontext angepaßt werden. Das Optimierungsziel könnte z.B. die Schonung der verfügbaren Akkukapazität als Ressourcenkontext eines mobilen Gerätes sein.

Eine Anpassung an Netzpartitionierungen, die aus Kosten- und Zeitgründen unvermeidbar sind, sind z.B. durch ein rechtzeitiges Laden der wahrscheinlich benötigten Informationen und Funktionalität vorbereitbar, bergen jedoch auch die Gefahr, erforderliche Informationen und Anwendungen, auch die Sicherheit und den Datenschutz betreffend, bei Bedarf nicht verfügbar zu haben. Andererseits sind umfangreiche multimediale Informationen nicht oder kaum über mobile Netze übertragbar und auf kleinen mobilen oder ständig wechselnden Geräten darstellbar. Die mobil verfügbaren Informationen und Funktionalität, so auch die Funktionalität eines Datenbanksystems und damit die Anfragebearbeitung, sind dynamisch verteilt. Diese Verteilung kann gleichfalls durch den Nutzerkontext hervorgerufen werden, der Fähigkeiten, Präferenzen, Rollen und das personelle Umfeld des Nutzers beinhaltet.
Beispiel:
Ein Nutzer benötigt als Firmenvertreter genau die Unterlagen, die es ihm gestatten, abhängig davon, mit wem er gerade kommuniziert, entweder ein Verkaufsgespräch mit einem Kunden zu führen oder aber seinem Chef über die Verkaufsgespräche zu berichten.

Nicht jeder Kontext ist immer von Bedeutung. Es hängt von vielen Faktoren ab, welche Kontexte jeweils Beachtung finden. Ziel des Mobile Computing und speziell der Mobilen Visualisierung ist es, die Systemkomponenten an diesen dynamischen Kontext zu adaptieren.

Über die Transparenzforderungen verteilter Systeme hinaus sind bei der mobilen Verarbeitung die konkreten Parameter des mobilen Kontextes, wie z.B. der Standort und die lokal verfügbaren Ressourcen und die Adaptionen an den Kontext für den Nutzer weitestgehend transparent gehalten, wobei sie, wie gezeigt, großen Einfluß auf die Verarbeitung haben können.

Abbildung 2 zeigt die Komponenten des mobilen Kontextes und Beziehungen zwischen den Komponenten.

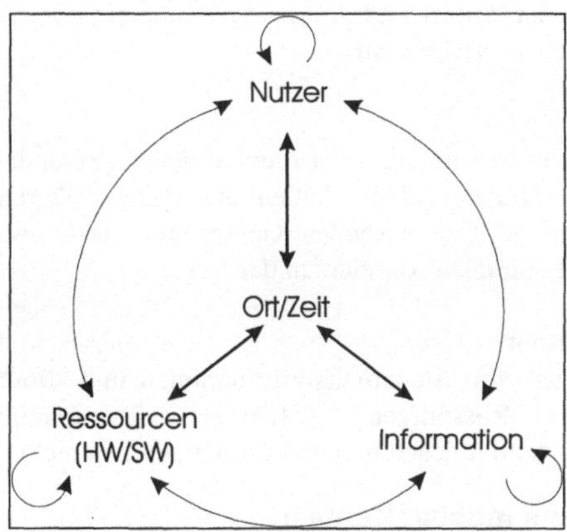

Abb.2: Kontexte und ihre Beziehungen

Ein **Kontextmanager**, der auf allen Geräten verfügbar sein muß, verwaltet in MoVi die Kontexte dynamisch und nutzertransparent, führt eine Kontextstatistik und ermittelt oder berechnet die aktuellen Kontexte und initiiert die Auswahl alternativer Funktionalität, Verfahren und Strategien zur Adaption an die dynamischen Kontexte.

3 Sicherer Mobiler Zugriff

3.1 Schutzziele von Sicherheit in mobiler Umgebung

Bestimmte Zustände des mobilen Kontextes können zu Gefährdungen der Schutzziele führen. Die Besonderheiten der Gefährdung der Schutzziele für öffentliche Mobilkommunikationnetze sind in [Pfit93] beschrieben. Im vorliegenden Artikel soll versucht werden, die drei Schutzziele für die mobile Arbeit zu untersetzen.

Schutzziel **Vertraulichkeit**:
Die Vertraulichkeit kann aus Sicht des Datenschutzes mit den Prinzipien der Datensparsamkeit, der Zweckbindung und Funktionentrennung durchgesetzt werden. Dieses Schutzziel besteht bei mobiler Arbeit aus folgenden Einzelzielen:

- Schutz der Informationen: Informationen und Dienste müssen unabhängig vom geltenden mobilen Kontext und Sicherheitsmodell zugriffsgeschützt sein,
- Schutz des Zugriffs: Der Zugriff muß so anonym wie möglich erfolgen. Die Information, wer auf welche Daten von wo zugegriffen hat, ist vertraulich zu behandeln, um die Persönlichkeitssphäre des Nutzers zu schützen,
- Schutz der Kontexte: Besonders sensibel ist der Ortskontext. Informationen über den Aufenthaltsort der Nutzer und dessen Änderung sind vertraulich zu verwalten, um Bewegungsbilder zu verhindern.

Schutzziel **Integrität**:
Der Schutz der Informationen vor unbefugter Veränderung wird durch Rücknahmefestigkeit, Innen- und Außentransparenz (Durchschaubarkeit) und Revisionsfähigkeit erreicht. Eine besondere Gefährdung für dieses Schutzziel stellen in mobiler Umgebung beschränkte Ressourcen dar.

Schutzziel **Verfügbarkeit**:
Der Schutz vor Verlust von Sicherheitsinformationen und -funktionen ist ebenfalls durch beschränkte Ressourcen gefährdet. Transparenz als einziges Durchsetzungsprinzip genügt deshalb in mobiler Umgebung nicht.

3.2 Sicherheit versus mobiler Kontext

Bei der Untersuchung von Sicherheitsanforderungen in mobiler Verarbeitungsumgebung sollen die Komponenten des mobilen Kontextes: Nutzer, Geräte und Netze, Informationen einzeln und in ihren Beziehungen betrachtet werden, um die Gefährdungen und resultierenden Sicherheitsanforderungen zu systematisieren. Ort und Zeit können implizit aus den zugegriffenen Informationen und den benutzten Geräten als auch explizit aus den Kommunikationskontexten gewonnen werden. Da das Problem des Schutzes des Bewegungbildes so vielschichtig ist, da der Orts-, Zeitkontext eng mit den anderen Komponenten zusammenhängt, werden die Einzelaspekte bei den jeweiligen Kontextkomponenten betrachtet.

3.2.1 Gefährdungen durch und für Nutzer in mobilen Umgebungen

Ist ein Vertreter mit einem PDA unterwegs, muß er mit geringen Speicherkapazitäten zurechtkommen. Diese erhöhen im Vergleich zu leistungsfähigeren Geräten den Kommunikationsbedarf, und damit die Wahrscheinlichkeit begrenzter Bandbreiten oder Verbindungsunterbrechungen. Um den begrenzten Ressourcen gerecht zu werden, wird nur das Wesentliche gespeichert oder übertragen. Die Wichtung legt der Nutzer oder die Anforderungen der Rolle, in der er agiert, fest. Da Sicherheit ein Systemaspekt ist, dessen Bedeutung mangels Problembewußtsein immer noch oft gering geschätzt wird, besteht die Gefahr, daß Sicherheitsaspekte bei Ressourcenbeschränkungen zuerst "unter den Tisch fallen". Damit könnten Zugriffskontrollsysteme, die Anwendung kryptographischer Verfahren etc. wegfallen, um Zeit und Kosten zu sparen. Alle Schutzziele von Sicherheit sind gefährdet, mißachtet zu werden.

Einzig mögliche Gegenmaßnahme ist die **Motivation** als personell - organisatorische Maßnahme. Es muß ein Problembewußtsein geschaffen werden, das Sicherheitsmaßnahmen eine angemessene Wichtung verleiht. Zielpublikum sind die Entwickler und Nutzer mobiler Hard- und Software, um die Sicherheitsanforderungen auch technisch umzusetzen (siehe weiter unten).

Eine weitere nutzerbezogene Forderung ist die nach **Transparenz** (Durchschaubarkeit) transparenter (verborgener) Kontexte und Kontextadaptionen. Dies wird im nächsten Abschnitt gesondert behandelt.

Ein stark gefährdetes Schutzziel in mobiler Umgebung stellt die Vertraulichkeit bezüglich der Ortsveränderung eines Nutzers dar, da sich das schutzwürdige Bewegungsbild einer Person aus dieser Information zeichnen läßt. Mit diesem Problem hängt ein weiteres, das der maximal möglichen Anonymität, zusammen. Das Prinzip der **Datensparsamkeit** fordert, daß die mobile Arbeit so anonym wie möglich erfolgt. Eine Identifikation und Authentifikation des Nutzers erfolgt so selten wie nötig. Ist Anonymität als rücknahmefeste Prozedur nicht möglich, gilt deren abgeschwächte Form der möglichst transaktionsorientierten Pseudonymisierung oder als dritte Stufe die Datenvermeidung. Insbesondere die Zuordenbarkeit von personenbezogenen Daten zu den anderen Kontexten ist sehr sensibel. In 3.3 wird ein Modell zur Bewertung der datensparsamen Verwendung von Kontexten vorgestellt.

Die wesentliche Voraussetzung für Zugriffsschutzmaßnahmen stellt die Identifikation und Authentifikation des Nutzers gegenüber dem Gerät, mit dem der Nutzer arbeitet, und des Gerätes gegenüber dem Netz dar. Dieser Prozeß wird erleichtert, wenn der Nutzer auf "trusted hardware" zurückgreifen kann, was bei dem mobilen Standardszenario der Fall ist. In [MST95] wird ein Authentifikationsverfahren für mobile Nutzer vorgeschlagen, in dem auch die Eingabe des Nutzernamens, Passwortes und der home-domain zur Identifikation und Authentifikation ausreicht. Eine eindeutige Authentifikation des Nutzers macht das Kontaktieren seiner home-domain erforderlich. Greift ein Vertreter jedoch zum Beispiel auf das örtliche Telefonverzeichnis zu, ist mit diesem Verfahren ein allzu hoher Aufwand verbunden und es werden unnötig personenbezogene Informationen über unterschiedliche Netze bewegt. Es ist abhängig von den Zieldaten und dem, was der Nutzer machen möchte, abzuwägen, ob eine Identifikation vonnöten ist oder eine der oben beschriebenen Formen der Datensparsamkeit eingesetzt wird.

3.2.2 Gefährdungen durch Geräte und Netze in mobiler Umgebung:

Ein Problem der Geräte, deren sich mobile Nutzer bedienen, sind öffentliche Kommunikationsgeräte, wie sie zur Zeit noch nicht verfügbar sind (ähnlich heutigen Telefonzellen). Die Nutzer an diesen Geräten wechseln häufig. Es resultiert ein besonderes Schutzerfordernis für die in einer Sitzung zugegriffenen Informationen. Da bei der Nutzung öffentlicher Geräte sowieso davon ausgegangen werden kann, daß ein Nutzer selten dasselbe öffentliche Gerät mehrfach nutzt, sollten bei Sitzungsende alle Informationen von oder über einen Nutzer gelöscht bzw. nur eine temporäre Informationsspeicherung zugelassen werden. Diese Forderung betrifft das Schutzziel Vertraulichkeit. Es gilt das Prinzip der **Datenvermeidung** für die personenbezogenen bzw. -beziehbaren Informationen.

In [HS95] sehen die Autoren bei der Nutzung von Funknetzen eine Gefahr in der Möglichkeit von Maskeraden bei dem Übergang zwischen verschiedenen „Levels of connection", wo sich ein Gerät als mobiles Gerät gegenüber einer Basisstation bzw. als Basisstation gegenüber einem Gerät ausgeben kann. Die zu ergreifenden Maßnahmen zur Integritätssicherung liegen auf der Ebene der Kommunikation und werden, wie bereits angedeutet, hier nicht weiter betrachtet.

Ein (semi-)mobiles Gerät stellt der Laptop dar. Bei einigen Fabrikaten besteht die Gefahr, daß sie Informationen und Systemzustände (einschließlich des zu ladenden Betriebssystems) bei knapp werdender Energie "vergessen". Das kann die Integrität und die Verfügbarkeit von Informationen und Systemfunktionalität stark beeinträchtigen. Solche Ressourcenkontexte, die bei stationärer Arbeit als Ausnahmen auftreten, wie eine nicht verfügbare Netzanbindung oder der Mangel an Strom, können im mobilen Bereich die Normalität darstellen. Bei Beschränkungen sind nur noch wenige Funktionen zur Aufrechterhaltung des Betriebs eines mobilen Gerätes abarbeitbar. Dieser Zustand darf nicht zum Verstoß gegen Integritätsziele führen oder für unerlaubte Aktionen ausnutzbar sein.

Es gilt jedoch für alle Ressourcen (Software- und Hardwareressourcen) und insbesondere für ressourcenbeschränkte mobile Geräte, daß eine Basissicherheit gewährleistet werden muß. Dieses Integritätsziel wird in „normalen" sicheren Systemen mit der Fehlerüberbrückungs-Komponente gewährleistet. In mobiler Umgebung sind solche ressourcenbedingten Fehler jedoch anders geartet und stellen darüber hinaus nicht die Ausnahme dar wie in rein stationären Systemen. Es gilt nicht allein Überbrückungs-funktionen festzulegen und technisch umzusetzen, sondern die grundlegenden Sicherheitsmaßnahmen, die auch durch knappe Ressourcen nicht beschränkbar sein dürfen, zu gewährleisten. Ohne solche **Basissicherheit**sfunktionalität darf die zu schützende Funktionaliät (z.B. Zugriffe auf Informationen, Kommunikation) nicht ausführbar sein.

Beispiel:

Eine Festlegung für eine Basissicherheitsbedingung könnte lauten: Genügen die Ressourcen nicht, Informationen vor einer Übetragung über ein Mobilkommunikationsnetz zu verschlüsseln, darf die Kommunikation nicht stattfinden, selbst wenn für die Übertragung selbst ausreichend Ressourcen verfügbar wären.

Basissicherheitsmaßnahmen müssen auf unterschiedlichen Ebenen, auf der Ebene der Anwendung, dem Datenbanksystem, dem Betriebssystem usw. festgelegt werden.

Beispiel:
Auf der Ebene eines Datenbanksystems lautet eine Basissicherheitsforderung: Ein Zugriff darf nur über ein Zugriffskontrollsystem erfolgen, um die Rechtmäßigkeit eines Zugriffs festzustellen oder den Zugriff abzuweisen. Stehen für dessen Speicherung oder Abarbeitung keine ausreichenden Ressourcen zur Verfügung, dürfen auch die Informationen und Ressourcen nicht zugreifbar sein. Ihre Integrität muß gewahrt bleiben.

Eine Problem, das für die Beziehung zwischen Geräten und Informationen gilt, ist das der Heterogenität der Sicherheitslevel unterschiedlicher Geräte bzw. Systeme. Die Mobilität eines Menschen, der verschiedene Geräte nutzt, verursacht eine Migration von Informationen und Funktionalität zwischen mehreren Geräten, denn ein Nutzer kann mobil auf eine Vielzahl von Informationen in unterschiedlichsten Systemen zugreifen und Informationen lesen, einfügen, modifizieren und löschen. Diese Systeme können vielfältige Schutzziele und -strategien verfolgen und sie mit recht unterschiedlichen Sicherheitsmodellen (MAC oder DAC) durchsetzen. Ein großes Problem stellt die mangelnde Kompatibilität der unterschiedlichen Mechanismen, Ziele und Modelle dar. Dieses Problem aller verteilten Systeme wird im mobilen Bereich dadurch verstärkt, daß die Daten auf sehr unterschiedlichen, nicht unbedingt vorhersagbaren Geräten und Systemen zugegriffen, verteilt oder repliziert verwaltet werden und daß diese Aussage neben Informationen gleichzeitig Systeme und Dienste betrifft, die in mobiler Umgebung migriert werden. Da der mobile Zugriff auf beliebige Informationen und Systeme unterstützt werden soll, ist nicht davon auszugehen, daß alle zugreifbaren Informationen für jedweden Nutzer öffentlich sind oder auf demselben Sicherheitsmodell basieren. Diese Gefährdung tangiert alle drei Schutzziele. Die Lösungen für diese Probleme müssen das Prinzip der Rücknahmefestigkeit erfüllen.

Da die Informationen und Softwareressourcen als föderiert anzusehen sind (es gibt kein globales Schema), gibt es keine zentrale Stelle zur Konfliktbereinigung zwischen den unterschiedlichen Sicherheitsmodellen. Eine Anpassung der Sicherheitsmodelle muß lokal vorgenommen werden. Das bedeutet, daß Sicherheits-informationen quasi huckepack mit den Informationen übermittelt und lokal angepaßt werden. Das dient der Sicherung für beide Seiten, der Informationsquelle und der Informationssenke, denn einerseits hat das den Zweck, daß

- die Informationssenke vor eingehenden, zugegriffenen Informationen, die in das lokale Sicherheitsmodell integriert werden, zu schützen und andererseits
- schützt dieses Verfahren die Informationsquelle vor Übertragungen in einen „ungeschützen Raum".

Nachteilig ist, daß gerade bei Verbindungsunterbrechungen Informationen und Sicherheitsinformationen schneller veralten können als der zugreifende Nutzer sie erhält.

Die Sicherheitsinformationen beinhalten Zugriffs- und Weitergaberechte. Das Zielsystem kann diese zugeordneten Rechte nur durchsetzen, wenn es sie auch versteht. Dazu muß das lokale Zugriffskontrollsystem außerdem mindestens das Sicherheitsmodell der Informationsquelle kennen. Es sind zwei Möglichkeiten denkbar, wie ein Zugriff auf derart angereicherte Informationen und Systeme erfolgen kann. Einerseits könnte das lokale Zugriffskontrollsystem so erweitert werden, daß es in der Lage ist, den Zugriff auf Informationen und Systeme unterschiedlicher Sicherheitsmodelle zu prüfen. Da der Aufwand zur Feststellung der Berechtigung eines Zugriffes dadurch erhöht wird, kann dies zu merklichen Leistungseinbußen vor allem bei häufigen Zugriffen auf wenige Informationen kommen. Das Zugriffskontrollsystem wird umfangreicher und dadurch weniger geeignet für ressourcenbeschränkte Geräte. Die andere Möglichkeit besteht in einer einmaligen Anpassung der Rechte an das lokale Sicherheitsmodell. Dieser Weg ist ineffizient, wenn viele Informationen nur gelesen, selten wiederholt zugegriffen oder gar nicht lokal gespeichert werden. Für eine Anpassung außerhalb des Zugriffskontrollsystems an das lokale Sicherheitsmodell bedarf es einer ausgezeichneten vertrauenswürdigen **Anpassungskomponente**, die auf dem mobil verwendeten Gerät oder, lassen das die lokalen Ressourcen nicht zu, auf der Basisstation ausgeführt werden kann. Eine solche Komponente stellt die bevorzugte Lösung dar, da sie in beiden Richtungen Vertraulichkeit unterstützen kann:

Der Informationsquelle wird zugesichert, daß die zugegriffenen Informationen weiterhin zugriffsgeschützt verwaltet werden. Verfügt das Zielsystem über keinen Zugriffsschutz, kann eine Übertragung verhindert werden. Für die Informationssenke werden die Informationen an das lokale Sicherheitsmodell angepaßt.

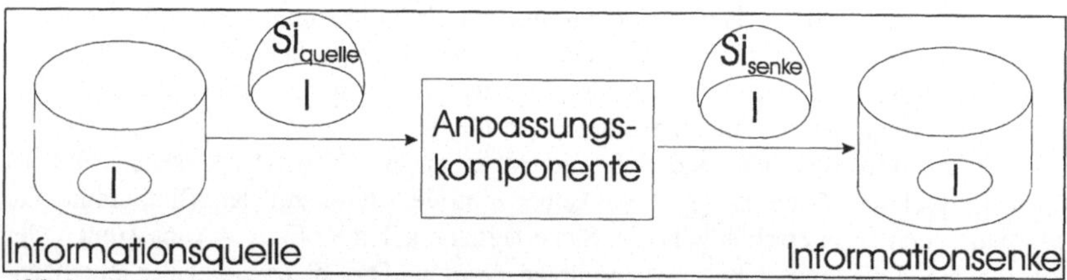

*Abb.3: Anpassung der Sicherheitsinformationen **Si** zur Information **I***

3.2.3 Gefährdungen von und durch Informationen:

Die Nutzung von Mobilnetzen und (auch geplante) Verbindungsunterbrechungen führen dazu, daß die angeforderten Informationen asynchron übermittelt werden. Dazu werden sie stationär zwischengespeichert und für den Nutzer je nach Standort zur nächsten Basisstation migriert, um für die asynchrone Übermittlung bereitzustehen. Dieses Verfahren der Informationsnachführung oder -vorausschickung läßt ebenfalls die

Konstruktion von Bewegungsbildern zu. Hier ist eine starke Zugriffskontrolle bezüglich der Migrationsinformationen aufgrund der **Zweckbindung** angeraten.

Auch im mobilen Bereich gilt, daß sensible Daten nicht über besonders unsichere Netze oder auf ungesicherte Geräte übertragen werden dürfen. Die Zugriffskontrollkomponente der informationsverwaltenden Instanz legt fest, welche Informationen außerhalb ihres Kontrollbereiches gelangen dürfen. Im Gegensatz zu üblichen Implementierungen darf sie nicht allein feststellen, ob ein bestimmter Nutzer auf das Datum zugreifen darf, sondern muß den Zugriff davon abhängig machen, von wo dieser Zugriffswunsch erfolgt.

Die Forderungen nach einer Anpassung des Zugriffsschutzes an das Schutzerfordernis des Datums und die Gewährleistung des Zugriffsschutzes auch bei ungünstigen Ressourcenzuständen wurden weiter oben erläutert.

Bei Informationseigenschaften ist noch anzumerken, daß gerade multimediale Informationen mit ihrer Größe und Komplexität die Ressourcen mobiler Geräte sprengen können und die Verfügbarkeit von Informationen und Funktionalität beeinträchtigen. Eine Abschätzung des Verhältnisses von gebrauchten und verfügbaren Ressourcen ist Voraussetzung für jede mobile Arbeit und beugt solchen Fehlerzuständen vor.

3.3 Sicherheit versus transparenter, adaptiver Zugriff

Die Verarbeitung einschließlich der verarbeiteten Informationen wird automatisch an die mobile Umgebung angepaßt, an den aktuellen Standort, an die verfügbaren Ressourcen und den Nutzer und sein personelles Umfeld. Diese transparenten automatischen Anpassungen können zu Sicherheitsproblemen führen. Während der Zugriff und die Zugriffskontexte vertraulich verwaltet werden sollen, muß es dem betroffenen Nutzer möglich sein, über ihn betreffende Kontexte Kenntnis zu erlangen. Die Transparenz, die dem Nutzer Aufwand abnimmt, bringt ihm mangelnde Durchschaubarkeit.

Beispiel:

Eine Anfrage nach Informationen über die Kunden wird abhängig vom Standort und eventuellen Informationen über anwesende Personen umgewandelt in eine Anfrage nach einem speziellen Kundenkreis. Der Vertreter erwartet ein Ergebnis auf die von ihm formulierte Anfrage, erhält jedoch ein davon abweichendes Resultat.

Fehlinterpretationen der Ergebnisse transparenter (nicht sichtbarer) Zugriffe durch uninformierter Nutzer sind die Konsequenz dieser Adaption.

Auf der anderen Seite werden die Kontexte für statistische Zwecke und ihre Wiederverwendung gespeichert. Die angefragten Informationen dienen der Wichtung zur Entscheidung über ihre Speicherung oder Löschung bei Ressourcenknappheit. Die Kontextparameter zur Übertragung dienen der Auswahl von Übertragungskanälen und der Übertragungsreihenfolge von Informationen bei nachfolgenden Kommunikationen. Die Energie- und Speicherinformationen gehen als Heurisitik in die Anfrageoptimierung ein. Es werden Kontexte, die vergangene Aktionen betreffen, aktuelle und für die Zukunft festgelegte bzw. geschätzte verwaltet. Die Speicherung erfolgt repliziert auf dem mobil benutzten und dem kontaktierten Gerät. Diese Kontextinformationen sind im Zusammenhang mit personenbezogenen Informationen sehr sensibel und bedürfen einer vertrauenswürdigen Verwaltung. Es gilt hier unbedingt das Prinzip der Datensparsamkeit.

Informationen zur Durchführung von Mobilkommunikation sind nicht die einzigen, die Hinweise auf den Standort von Nutzern geben. Die Anfragen, die der Nutzer selbst formuliert bzw. die automatisch an den (Orts-) Kontext angepaßt werden, geben Auskunft über den Aufenthaltsort einer Person in der Vergangenheit, Gegenwart und Zukunft.

Beispiel:
Die Anfragen des Vertreters nach Kunden geben ein gutes Bild, wo er sich wann aufgehalten hat.

So sind alle Zugriffe, egal ob lesend oder modifizierend, sensitiv im Sinne der Offenbarung von Lokalisierungsinformationen, in denen auf direkt oder indirekt ortsabhängige Informationen zugegriffen wird. Eine vertrauenswürdige Speicherung von Lokalisierungsdaten ist auf semantischer Ebene nicht unproblematisch zu realisieren, da zunächst ermittelt werden muß, welche Informationen solche Schlüsse zulassen und die Zugriffe auf verteilte Informationsbestände erfolgen können. Objektattribute wie Adresse oder Wohnort sind relativ leicht als Ortsangaben zu identifizieren, während aus dem Zugriff auf Kunden kaum automatisch auf Ortsinformationen geschlossen werden kann, diese Informationen jedoch stark ortsabhängig sind. Die Ortsinformationen sind jedoch bereits erforderlich zur Adaption der Anfragen und können zu Schutzzwecken wiederverwendet werden.

Die Hauptforderungen, die sich aus der obigen Problembeschreibung ergeben und technisch umgesetzt werden müssen, sind die Forderungen nach dem
1. **Schutz der Kontexte,**
2. **Transparenz der Kontexte,**
3. **Sparsame Verwendung von Kontexten**,
4. **Schutz der (aggregierten) Kontextsicht**, bestehend aus mehreren Teilforderungen:
   - **Schutz vor vertikaler Kontextaggregation,**
   - **Schutz vor horizontaler Kontextaggregation,**
   - **Schutz vor Aggregation der Kontextkomponenten.**

1. Schutz der Kontexte:
Kontexte sind sehr sensible Informationen und vertraulich zu verwalten. In der Regel werden sie automatisch erfaßt, verwertet und gelöscht, so daß kein expliziter Zugriff auf die Kontexte zugelassen werden muß und sie transparent verwaltet werden. Es werden bei der Informationsverarbeitung automatisch nur die jeweils aktuell relevanten Kontexte verwendet. Das bedeutet, daß bei ordnungsgemäßer Funktion des Kontextmanagers keine Kontexte als die zur angemeldeten Person gehörenden angewandt werden. Die Durchsetzung dieser Forderung unterstützt die Anonymität durch die Separierung der einzelnen Nutzerkontexte.

2. Transparenz der Kontexte:

Die verwalteten Kontexte müssen für den Nutzer durchschaubar verwaltet werden. Entgegen den rigorosen Schutzforderungen unter dem 1. Anstrich, muß ein expliziter lesender Zugriff aller Nutzer auf sie betreffende Kontexte ermöglicht werden, um die Transparenzforderung des Datenschutzes zu erfüllen. Außerdem obliegt die Entscheidung für den Grad der Einbeziehung der Kontexte in die Informationsverarbeitung letztendlich dem Nutzer. So muß ihm eine Eingriffsmöglichkeit in den Automatismus der Kontextadaptionen gegeben werden.

3. Sparsame Verwendung von Kontexten:

Da die Kombination von Kontexten außerordentlich sensibel ist, muß gefordert werden, daß nur so viele Kontexte wie nötig gespeichert, übertragen und benutzt werden. Darüber hinaus sollten sie nur dort und zu dem Zeitpunkt angefordert und verwaltet werden, wo und wenn sie unmittelbar zur Anwendung kommen. Eine Kontexthaltung auf Vorrat muß vermieden werden, indem bei der Verwaltung und Übertragung von vorn herein eine Löschung eingeplant wird. Es gelten somit die Methoden der Datensparsamkeit.

4. Schutz vor Kontextaggregationen:

Zur Unterstützung der letzten Forderung nach Kontextaggregationen dienen unterschiedliche Ebenentrennungen.

Schutz vor vertikaler Kontextaggregation

Die Kontexte für den Zugriff und die zugegriffenen Informationen auf mehreren Domains, Geräten, Systemen, etc. dürfen nicht kombiniert werden. Je mehr über die Zugriffe und die zugegriffenen (auch ortsabhängigen) Informationen zu einem Nutzer bekannt ist, desto besser lassen sich Persönlichkeits- und Bewegungsbilder erstellen. Daher müssen die Zugriffsinformationen separiert werden. Diese Forderung ist kompliziert zu erfüllen, da aus Abrechnungszwecken der home-domain Informationen über die Zugriffe übermittelt werden.

Schutz vor horizontaler Kontextaggregation

Informationen über die Kontexte müssen in den horizontalen Ebenen separiert werden. Das bedeutet, daß sie die horizontalen Ebenen nicht verlassen dürfen. Ortsinformationen der Datenbankebene (ortsabhängige Informationen) dürfen beispielsweise nicht in der Netzwerkebene zugreifbar sein (Schutz der Nutzdaten auf Netzwerkebene).

Schutz vor Aggregation der Kontextkomponenten

Ressourcen-, Nutzer-, Informations- und Ortskontext dürfen nur für den betroffenen Nutzer kombinierbar sein. Ansonsten sollten nur Zugriffe auf separate Kontextarten möglich sein. Erfolgt die Arbeit anonym, ist eine Kontextaggregation weniger sensibel.

Durch die Speicherung von Nutzerpräferenzen entstehen jedoch personenbeziehbare Informationen.

Systeme für mobile Arbeit lassen sich hinsichtlich ihrer Datensparsamkeit im Sinne der Kombination mit personenbezogenen Daten bewerten, indem für alle Kontextdaten einschließlich der Kommunikationskontexte der Grad der angewendeten Datensparsamkeit (Anonymität, transaktionsorientierte oder sonstige Pseudonymisierung, Datenvermeidung) geprüft wird. Die folgende Matrix verschafft einen guten Überblick über die Bewertungen.

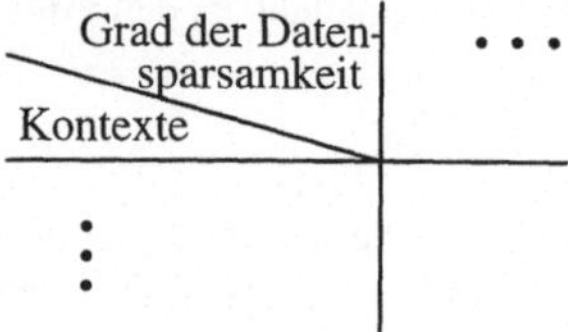

Abb.4: Matrix zur Einschätzung der Datensparsamkeit von Kontexten

Bei dem in 2.1 vorgestellten verallgemeinerten Mobilitätsmodell bilden Nutzer und Gerät <u>keine</u> Einheit, das Gerät läßt zunächst keinen Schluß auf die nutzende Person zu, da Nutzer und Gerät separate Modellkomponenten sind. Ein Vorgehen nach diesem Modell unterstützt einen anonymen Zugriff auf Informationen. Voraussetzung hierfür ist jedoch, daß die zugegriffenen Informationen keinen besonderen Schutzbedarf haben.

4 Zusammenfassung

In diesem Artikel wurde beschrieben, welche Sicherheitsprobleme beim mobilen Zugriff entstehen können. Diese Probleme beruhen einerseits auf den spezifisch mobilen Eigenschaften wie z.B. begrenzten Ressourcen, andererseits wird die Unterstützung, die einem mobilen Nutzer für einen transparenten Zugriff gegeben wird, zum Sicherheitsrisiko. Die automatischen Anpassungen entlasten den Nutzer, aber entmündigen ihn gleichermaßen, wenn ihm keine Möglichkeit gegeben ist, Kenntnis über die verwalteten Daten und Automatismen zu erlangen und in diese einzugreifen. Voraussetzung dafür ist jedoch eine ausreichende Motivation des Nutzers. Ansonsten beugen Basissicherheitsmechanismen Gefährdungen der Schutzziele, die neben dem Nutzereingriff durch Ressourcenknappheit bestehen, vor. Beim mobilen Zugriff als eine neue und sich schnell verbreitende Anwendung ist es wichtig, daß Sicherheitsmaßnahmen wie Durchschaubarkeit und Datensparsamkeit von vornherein eingeplant und durchgesetzt werden. Dazu gehört auch die Forderung nach dem Schutz der Kontexte und deren Aggregationen, resultierend aus ihrer Funktion zur Unterstützung des mobilen transparenten Zugriffes. Ein weiteres Problem mobiler Arbeit stellt die Heterogenität der Systeme dar, die eine Übertragung der relevanten

Sicherheitsinformationen und deren Umwandlung durch entsprechende Anpassungskomponenten verlangt.

5 Literatur

[IB93]   Imielinski,T., Badrinath, B.R.:

Data Management for Mobile Computing;

SIGMOD RECORD Vol. 22, No.1, 1993.

[FJKP95]   Federrath,H., Jerichow,A., Kesdogan,D., Pfitzmann,A.:

Security in Public Mobile Communication Networks;

Proc. of the IFIP TC6 Int. Workshop on Personal

Wireless Communications; Aachen, 1995, S. 105 -116.

[MST94]   Molva,R., Samfat,D., Tsudik,G.:

Authentication of Mobile Users

IEEE Network, Special Issue on Mobile Communication, March,April, 1994.

[HL96]   Heuer,A., Lubinski,A.:

Database Access in Mobile Environments;

Proc. of the Int. Conf. DEXA '96, Zürich, 1996.

[HS95]   Hardjono,T., Seberry,J.:

Information Security Issues in Mobile Computing

in: Eloff,J.H.P., von Solms,S.H.: Information Security - the Next Decade, Proc.
of the eleventh int. conf. on information security, IFIP/Sec '95, 1995, S.143-
151.

[WH96]   Wachowicz,M., Hild,S.G.:

Combining Location and Data Management in an Environment for Total

Mobility; Proc. Of the IMC'96, Rostock, 1996.

[Pfit93]   Pfitzmann,A.:

Technischer Datenschutz in öffentlichen Funknetzen;

Datenschutz und Datensicherheit DuD 17/8 (1993),

S.451-463.

Increasing Privacy in Mobile Communication Systems using Cryptographically Protected Objects*

Uwe G. Wilhelm

Laboratoire de Systèmes d'Exploitation

Ecole Polytechnique Fédérale de Lausanne, CH-1015 Lausanne

Abstract

Confidential information contained in the transactional data created in mobile communication systems is not very well protected. We present an approach to protect entire objects against manipulation and disclosure, which allows to retain some control over information handed to a different entity or can be used to prevent reverse engineering of an object to obtain its original source. The approach is subsequently used in the context of mobile communication systems to protect the current location of a mobile user against unauthorized access.

1 Introduction

Communication has always had a considerable impact on our society. This observation pertains especially to systems for mobile communication, such as GSM (Global System for Mobile Communications), due to their omnipresence and ever increasing success. The fact that everyone is reachable at any time gives this form of communication a new quality that did not exist in the past.

The need for protecting the *communication contents* (payload data) in telecommunication systems against manipulation and disclosure (integrity and confidentiality protection) is well established and constitutes essentially a solved problem [15]. The protection of *transactional data* (such as who is communicating when, how often, and from where with whom) that contains equally sensitive information, is much less understood. This data could be used to construct a complete trace of the movements of a user carrying a mobile phone in standby-mode or a list of the people that user called on a given day. There are many examples where this data might contain information that users might rather have protected.

---

*Research supported by a grant from the EPFL ("Privacy" project).

In existing telecommunication architectures, there are no means available to protect this transactional data, since it is essential for the operation of these systems (accounting, billing, or locating users to establish a communication channel). Thus, many of the employees of a telecommunication provider have potentially immediate access to the transactional data that is available in the telecommunication system. The only protection that is currently provided for this data are laws that prohibit its use for anything other than the intended purpose. Even though we generally assume that a telecommunication provider (such as a national phone company) is trustworthy and will adhere to these laws, we can not necessarily assume the same for each of its employees. This concern for the possibility of *malicious insiders* is founded on some recent publications that describe the vulnerability of systems towards their proper employees [12, 14].

The fact that laws provide threats to deter most people from abusing the data that is available to them, might not be enough to prevent data abuse in the case when the benefits for the offending person outweigh the threats. This problem is further magnified by the fact that data abuse is inherently hard to detect and that it is very difficult to prove that it happened in the first place. We assume that a malicious insider has legitimate access to all the data that is available within a given telecommunication system. Therefore, this attacker model subsumes an external attacker who gains access to the system or an operational error that makes the data of the system accidentally available to outsiders.

In this paper we will present a new approach for protecting transactional data in telecommunication systems from unauthorized access, which is based on the *cryptographically protected objects* (CryPO) protocol (see Section 4).

The remainder of the paper is organized as follows. First, in Section 2, we will give a more precise definition of the problem concerning location information in mobile communication systems. We will also identify a more general problem that we will resolve first. In Section 3 we introduce some basic definitions and notations that we need for the description of the protocol. Next, in Section 4, we introduce the CryPO protocol in detail, which solves the more general problem. Then Section 5 describes the application of the general solution to mobile communication systems and discusses some results. Finally, Section 6 concludes the paper with a summary of the main contributions.

2   The Problem

The basic problem that we are concerned with is that once some information is disclosed to another entity, the original owner of the information looses all control over this information. The receiver can use it without any constraints, duplicate it, or even disclose it to a third party. Sending the data in encrypted form or with an attached signature does not prevent anything of the above.

This loss of control is inevitable if the goal of the interaction is that the receiver uses the information for some specific purpose. However, there are possible interactions, where the information is disclosed to the receiver just in case it is needed at a later time. The receiver will store it in an easily accessible place, but most of the time the information is not actually needed and simply discarded or overwritten with a more recent information.

We will first rephrase this problem for the protection of location information[1] in a mobile communication system and then describe a more general form of the problem that is concerned with protecting the code and data of an object against manipulation and disclosure.

2.1   The Protection of Privacy

We want to protect the location information in mobile communication systems, such as GSM (Global System for Mobile Communications)[2].

This system divides the entire coverage area of the mobile communication system into small compartments called *cells*, in order to make better use of the scarce bandwidth available for the radio link between a mobile terminal and a base station (which serves as intermediary to the wired network). These cells could be as small as a single room within a building or as large as a few kilometers in an unobstructed area.

For a user, say Alice, to remain reachable (available to receive incoming calls), she has to register the cell in which she currently resides with a well known location register (LR); in GSM this is called Home Location Register (HLR). If another person, say Bob, tries to reach Alice, he

---

[1] We are particularly interested in the location information and will limit our discussion to this issue. However, the methods we propose for protecting it can easily be applied to other confidential data.

[2] GSM deals primarily with terminal mobility. Similar problems exists in the Intelligent Network service Universal Personal Telecommunications (UPT) that deals with personal mobility or the Universal Mobile Telecommunications System (UMTS) that deals with both forms of mobility. For reasons of brevity we will limit our discussion to terminal mobility.

(or rather the underlying mobile communication system) will first contact
Alice's LR, which is identified in her personal phone number, and obtain
the cell in which she is registered. Then Alice will be contacted (*paged*)
within this cell and informed about the communication request from Bob.
If she accepts the call, the communication with Bob is established.

In the described system, which uses simplified concepts from GSM, the
provider of the mobile communication system has complete access to the
data stored in the LR and can easily detect in which cell Alice is currently
located. However, for the operation of the system it is not necessary that
the telecommunication provider knows Alice's location at all times, but it
is sufficient to know her location when she has to be paged for an incoming
call. Since it is impossible to foresee when this will occur, Alice has to send
continuous updates on her current location, which will be stored in the LR
and subsequently overwritten with more recent information whenever she
moves to a new cell. This information could be used by the provider to
create a continuous trace of Alice's movements (a location profile).

The location information is confidential data that should not be available to anyone without a compelling reason (such as law enforcement
personnel with an order issued by a judge). Currently this data is solely
protected by legal means, which make it unlawful to use it for anything
other than the intended purpose (i.e., locating a user for an incoming
call). Since this is not sufficient in the presence of malicious insiders, we
want to add stronger protection based on technical means that allows only
authorized access to this confidential information.

An approach to solve this problem has been proposed in [9]. It is based
on a fixed terminal within the network (which we will call *HPC* for Home
PC) that is under Alice's exclusive control. She sends continuous location
updates to her HPC that are not accessible to any other entity. In the
case of an incoming call, the telecommunication provider can query her
HPC to obtain her current location. There are many possibilities how the
HPC can react to such a query. The simplest would be to always disclose
the current location of the user. This would not really protect the location information, but if the HPC keeps a log of all the queries, the user
can verify that his location was only disclosed for genuine communication
requests[3]. On the other extreme, the HPC could act as reachability manager [4] by requesting an authentication from each caller and accepting

---

[3]The HPC can detect systematic attacks, such as continuous probing that could be launched to obtain
a trace of the user's location.

only communication requests from a set of users defined by the owner of the HPC (other callers can be deviated to a voice mail system).

Since Alice trusts her HPC to never disclose her current location without proper authorization, the goal of protecting the access to confidential information is achieved. Nevertheless, the described approach has some disadvantages:

- unreliability: the HPC is a computer under the administration of the user, with all the advantages and problems this situation entails. For many users it could be a major task to keep the HPC running.

- inefficiency: the HPC is outside the domain of the telecommunication provider and, thus, probably connected to the core of the system via a relatively slow link. Since the HPC can not be migrated within the network, this could be a major problem if the user moves to a different network for an extended period of time.

- vulnerability to:
  - attacks: the HPC is not particularly protected against physical attacks within the home of the user. A determined attacker could probably gain access to it and obtain the confidential data stored in the HPC. Moreover, as the HPC is connected to a communication system, it might also be possible to attack it without physical access.
  - traffic flow analysis: since all the traffic passing through the HPC belongs with high probability to its owner, the approach is quite vulnerable to traffic flow analysis. In [9] this problem is countered with the introduction of MIXES (see [3]), which allows to hide communication connections.

- standardization: the approach described above is very different to those taken in most telecommunication systems. Telecommunication providers would probably be very reluctant to adopt it, primarily for reasons of reliability and efficiency.

An analysis of the approach reveals that the main idea is to encapsulate the confidential data in an object that provides access to the stored information only under the constraints of strict access control. The approach is valid if the object resides on a trusted environment that protects it from any access not supported by its interface. Is this not guaranteed, an attacker could obtain a copy of the object and analyse its contents[4]. We can therefore map the above problem to the following more general problem of protecting an entire object against manipulation and disclosure.

2.2 The Protection of Objects

We consider a *mobile object*[5] O, identified with its unique name O_{name}, that consists of code and data, which can be sent via a communication channel to a different host. There it can subsequently be executed on a virtual machine, which means that the object is platform independent.

A provider who sends an object to a receiver currently has no guarantee whatsoever concerning the confidentiality of the code or the data that comprise the object. If the communicated information constitutes a considerable economic value (e.g., proprietary algorithms) or contains confidential data, there is a potential problem. The receiver of the object could reverse engineer the code to obtain the original object source and reuse all or parts of the illicitly obtained code or he could analyse the object in order to obtain the confidential data.

This is not a new problem, yet it is perceived as a major threat in the context of *Java*, which seems to be particularly prone to reverse engineering. In the following Section 4, we will introduce the *cryptographically protected objects* (CryPO) protocol that can be used to overcome this problem. A rather similar approach for a different environment is described by Herzberg and Pinter in [6]. Their concern is for protecting software sold through regular distribution channels against piracy. The resulting protocols are quite similar, but the technical development has caused rather drastic changes to the computing environment so that a re-evaluation of these ideas and an adaption to a more current context seems necessary.

---

[4]The object might store the information in a form that is not easily accessible (e.g., encrypted or otherwise difficult to access). Since the object itself can access the information, all the necessary data has to be available in the object (e.g., the decryption key is coded in the object). A profound analysis (reverse engineering) of the object will hence, with high probability, allow to recover the information.

[5]Such objects have recently attracted a lot of attention in the context of Java applets (*Java* is a trademark of Sun Microsystems Ltd.) and ActiveX controls (*ActiveX* is a trademark of Microsoft Corp.). We will from now on simply refer to them as objects.

3 Preliminaries

In order to describe our system, we need to introduce some important concepts from cryptography and tamper resistance.

3.1 Cryptography and Notation

A detailed description of cryptography and the corresponding notations is not within the scope of this presentation. Several books [2, 11] deal extensively with these questions. We will only introduce some basic notations needed for the description of the CryPO protocol.

The protocol uses public key cryptography (such as RSA [10]), in which a *principal* P has a pair of keys (K_P, K_P^{-1}) where K_P is P's public key that is known to everyone and K_P^{-1} the private key that is exclusively known to P. Given these keys and the corresponding algorithm, it is possible to *encrypt* a message m, denoted $\{m\}_{K_P} = m'$, so that only P can *decrypt* m' using his private key: $\{m'\}_{K_P^{-1}} = m$.

Some public key cryptosystems (notably RSA) also allow to use the same keys for *signing* messages. A message m signed by P is denoted $\{m\}_{K_P^{-1}} = m_{sig}$. The signature on m can be verified by everyone using P's public key: $\{m_{sig}\}_{K_P} = m$. If a message, which confers certain rights, bears the signature of a trusted principal it is also referred to as *certificate*.

3.2 Tamper Proofedness

The concept of tamper proofedness usually applies to a well-defined module, sometimes called black-box, that executes a given task. The outside environment cannot interfere with the task of this module other than through a restricted interface, which is under the control of the tamper proof module.

The actual construction of such a tamper proof module in the real world is extremely difficult, nevertheless, there are many applications that rely on them (e.g., payphones, debit cards, or SIM cards for GSM). Given sufficient time and resources, it becomes very probable that an attacker can violate the protection of the module [1]. Therefore, a tamper proof module should not protect information that is more valuable than the estimated cost[6] for breaking its protection. For the remainder of this presentation,

---

[6]This cost can only be a rough estimation and is likely to decrease over time, when more efficient attacks become possible.

we make the explicit assumption that the investment necessary to break
our system is much greater than the possible gain.

4   The Overall Approach

In this Section, we will introduce the protocol that guarantees the integrity
of the execution environment and protects the code and the data of an
object executing in this environment.

We will first present the execution environment that we rely on and
then describe the CryPO (cryptographically protected objects) protocol
that uses it. Figure 1 gives an overview of the entities in the system and
their interactions.

4.1   The Execution Environment

In order to achieve our goal, we need a special execution environment
based on the notion of tamper proofedness, which we will call *tamper proof
environment* (TPE). The TPE provides a full execution environment for
objects, which can not be inspected or tampered with.

The TPE is a complete computer that consists of a CPU, RAM, ROM,
and non-volatile storage (e.g. hard-disk or flash RAM). It runs a virtual
machine (VM) that serves as execution environment for objects and an
operating system that provides the external interface to the TPE and
controls the VM (e.g., protection of objects from each other). Furthermore,
the TPE contains a private key K_{TPE}^{-1} that is known to no entity other
than the TPE – even the physical owner of the TPE has no information
concerning this private key (see 4.2).

The TPE is connected to a host computer that is under the control
of the TPE owner. This host computer can access the TPE exclusively
through a well defined interface that allows, for instance, the following
operations on the TPE:

- upload, migrate, or remove objects;

- facilitate interactions between host and object or between different
  objects on the TPE;

- verify certain properties of the TPE (such as which objects are currently executing).

Due to its implementation as tamper proof module and the restricted access via the operating system, it is impossible to directly access the information that is contained on the TPE.

This property is ensured and guaranteed by the TPE manufacturer[7] *TM*, which also provides the object user *OU* with a certificate (signed by TM). The certificate contains information about the TPE, such as its manufacturer, its type, the guarantees provided, and its public key. The object provider *OP* has to trust the TM that the TPE actually does provide the protection that is claimed in the certificate.

4.2 Key Generation

One of the most important properties of the TPE that has to be guaranteed with utmost certainty is that the private key of the TPE is never disclosed. To ensure this property, the TPE could create the key itself (based on some random input). Using this approach, the key is never available outside of the TPE and, thus, protected by the operating system and the tamper proofedness of the TPE. This initialization can be done by the TPE manufacturer and subsequently the TPE will be prevented from ever disclosing this key or accepting a new one.

Other, more sophisticated approaches to create the pair of keys could be envisaged, which could also incorporate key recovery mechanisms (e.g., escrowed key shares).

4.3 CryPO Protocol

The CryPO (cryptographically protected objects) protocol transfers objects exclusively in encrypted form over the network to a TPE. Therefore, it is impossible for anyone who does not know the proper key to obtain the code or data of such an encrypted object.

The protocol is divided into two distinct phases. The first phase consists of an initialization, which has to be executed once before the actual execution of the protocol. The second phase is concerned with the usage of the TPE. The protocol is based on the interactions given in figure 1.

---

[7]It would be possible to extend the model and introduce a certifier that takes some of the responsibilities (e.g. regular controls).

Initialization

In the initialization phase, the participants establish the trust relations
that are associated with the different keys:

- the TM publishes its certification key K_{TM};

- the TM sends the certificate $Cert_{TPE} = \{K_{TPE}\}_{K_{TM}^{-1}}$ to the OU.

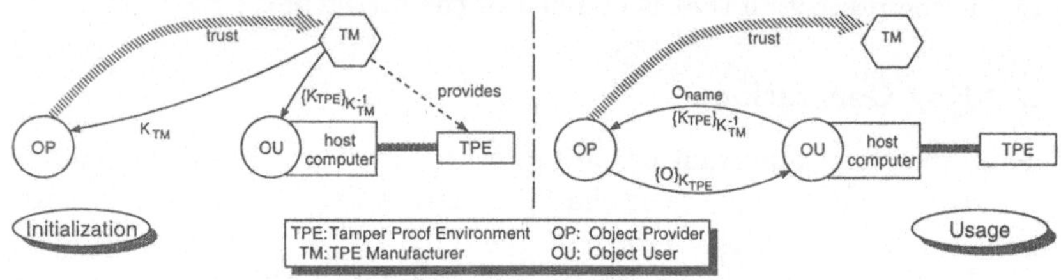

Figure 1: Overview of the CryPO protocol

TPE Usage

After the participants have finished the initialization, they can execute the
actual CryPO protocol:

- The OU sends a request to the OP that contains the name O_{name} of
  the object he wants to obtain together with the certificate $Cert_{TPE}$
  of his TPE.

- The OP verifies the certificate $Cert_{TPE}$ to check the manufacturer
  and the type of the TPE, in order to decide if it satisfies the security requirements of the OP. Then the OP verifies other information
  associated with the interaction (such as OU's identity, payment, or
  others). If the OP is satisfied with these checks, it will continue with
  the protocol.

- The OP sends the object encrypted under the public key of the TPE,
  $\{O\}_{K_{TPE}}$ to the OU.

- The OU cannot decrypt $\{O\}_{K_{TPE}}$ nor can he do anything other than
  upload it to his TPE.

- The TPE decrypts $\{O\}_{K_{TPE}}$ using its private key K_{TPE}^{-1} and obtains
  the executable object O. Depending on the directives of the OU,

O will eventually be started on the TPE and can interact with the local environment of the OU, other objects on the TPE, or even with remote applications.

4.4 Guarantees of the CryPO Protocol

The protocol described above guarantees the integrity of the execution environment. This guarantee is based on the trust relation between the OP and the TM. The OP trusts the TM to properly manufacture its TPEs so that the claimed guarantees hold, even without any physical control or access to the TPE. Thus, the objects are protected against manipulation and disclosure of code and data both in transit and during execution and are therefore safe from reverse engineering.

The OU is guaranteed that the object can only access the host computer through the specified interface. Any damage that could possibly be caused by an object is limited to the TPE and the restricted access it has to the user's host computer. By adding a digital signature on the object, the CryPO protocol could easily and transparently be extended to incorporate an object certification, such as those defined for Java or ActiveX [5, 7].

4.5 Additional Properties

Apart from the basic protection of the code and data of an object, the CryPO protocol can ensure other interesting properties that can be useful for certain applications.

Object Migration

The operating system of the TPE might allow an object to move to a different TPE. To achieve this, the TPE will obtain the certificate for another TPE, TPE′. The operating system will then check whether the other TPE satisfies the required properties. This is a complicated decision that depends on many non-technical issues, which are not within the scope of this presentation. The operating system will then stop the object (e.g., by calling a special method of the object that ensures that it is in a coherent state), marshal it into a transportable form (including all the code and the current data), and encrypt it with the public key $K_{TPE′}$ of TPE′. The object will then be sent to TPE′, where it will be loaded as usual and resume its execution.

The operating system of the original TPE can take further measures to ensure that the object will only exist as a single incarnation (if this condition is deemed necessary).

Message Confidentiality and Receiver Anonymity

The operating system of the TPE could allow objects on the TPE to receive messages. These messages can be encrypted with the public key of the TPE, so that no other entity can access the contents of such a message – not even the owner of the TPE. Thus, the confidentiality of the message content is under the exclusive control of the receiving object. It can make the content available to other entities or it can decide to not disclose it at all.

Furthermore, if the identity of the receiving object is also encrypted with the public key of the TPE, no entity outside the TPE can observe which of the objects currently residing on the TPE has actually received the message. Provided that the operating system of the TPE does not reveal any information about the result of the communication operation, then even if there is only a single object on the TPE, it might not be the receiver of the message. Consequently it is again the exclusive decision of the receiving object to identify itself as the receiver of the message. This approach can be used to implement receiver anonymity for the objects on the TPE. Possible attacks on the receiver anonymity by the owner of the TPE that consist of uploading an object, which pretends to be the legitimate receiver, but makes the message available to the user have not been studied yet.

Other interesting ideas are object expiration, where an object has a well defined time to life before it is eventually removed by the operating system of the TPE or the possibility to include a private key in an object so that it can sign the messages it sends.

4.6 Possible Problems

The proposed protocol has a weak point that we have to be very careful about. If the private key of a TPE becomes available to some user, that user can pretend to run a TPE (using the certificate for the TPE of which he has the private key), while he is actually capable to decipher everything that is sent to the TPE. This would violate all the security guarantees offered by the CryPO protocol with a slim chance of that user being discovered.

To alleviate this problem we could require regular inspections of the TPE hardware by some independent certifier. If we can ensure that an attacker cannot obtain the private key from the TPE without destroying it in the process, he would not be able to present the undamaged TPE to the certifier and thus, the duration during which damage could occur would be limited over time.

5 Application of CryPO to Mobile Communication Systems

Using the CryPO protocol introduced above, the solution to the problem of protecting the location information in mobile telecommunication systems as described in Section 2.1, is now straightforward.

We require the presence of a TPE in the underlying telecommunication system that satisfies all the properties introduced in Section 4.1 as well as a special object, which we will call *User Agent* (UA). A user, say Alice, can obtain such a UA from a trusted software vendor[8] and customize it to suit her special requirements. She will then encrypt the customized UA with the public key of the TPE and send it to her telecommunication provider. If Alice has previously established a contract with the telecommunication provider, the latter will upload the encrypted UA to its TPE, where the UA will subsequently start its execution.

The functionality of this UA is very similar to that of the HPC described in 2.1. Whenever Alice moves to a different cell, she will send a location update to her UA, encrypted with the public key of the TPE on which her UA resides (for optimization she can establish a shared session key with her UA and encrypt the data with a shared key algorithm). Hence, Alice's location is exclusively known to her UA, but not to the telecommunication provider. If another user, say Bob, tries to reach Alice, the telecommunication provider will query her UA in order to obtain her current location. The UA in turn can now apply the methods introduced in 2.1 to decide whether it should disclose the information.

Since we assume that Alice trusts the TPE to behave according to its specification (i.e., to not disclose any information about her UA) and that

---

[8]The telecommunication provider, which controls the TPE could require access to the source of the UA, in order to verify that the UA does not implement malicious behaviour (Trojan horse attacks).

she also trusts her UA to not disclose her current location without proper authorization, the goal of protecting the access to confidential information is again achieved.

5.1   Advantages of the Approach

Compared to the approach proposed in [9], the one introduced above has several advantages:

- the TPE is under the administration of the telecommunication provider, which will ensure its reliability. The telecommunication provider can also verify the code of a UA to ensure its reliable operation.

- the TPE is located in the core of the telecommunications system and can be interconnected with high speed links.

- the UA can be migrated to a different TPE (see 4.5) if the user moves to a different network for an extended period of time.

- the TPE can be physically protected at the site of the telecommunication provider, which can probably also put better software protection (e.g., firewalls) in place.

- the TPE will host many UAs (on the order of several thousand or more). If the TPE allows receiver anonymity (see 4.5), even the telecommunication provider would have to do heavy traffic flow analysis to identify the called user. A more detailed analysis of this issue is necessary to allow a quantification of the incurred protection. If the underlying telecommunication system provides MIXES [3], these can also be used to further protect the location information.

- a recent approach to define the next generation of telecommunication systems is TINA [8] (Telecommunications Information Networking Architecture), which is developed by a consortium of many important telecommunication providers and equipment manufacturers. In TINA there exists the notion of a User Agent [13], which is an object that has very similar tasks to those described above. However, the UA in TINA executes under the control of the telecommunications access provider (in TINA this entity is called *retailer*). We consider it possible to integrate our approach into TINA and will pursue this issue in our future work.

5.2 Feasibility

We assume that the TPE is a quite powerful computer with a special operating system. The TPE is physically protected with a special hardware that can effectively be sealed to detect tampering, is under continuous video surveillance, and subject to challenge inspections by an independent organization (such as the TÜV[9]). Since a telecommunication provider does only need a small number of these installations, we consider the cost for its realization quite realistic.

As explained in [1], such an installation is conceivable and can even resist massive attacks. Again, a thorough analysis of the remaining risks has to be undertaken, but this is not within the scope of this presentation.

Another problem exists with legislation that requires access to the information we try to protect. To resolve this dilemma, UAs could be equipped with a trapdoor that allows to access its internal data. Since the telecommunication provider has to approve of UA implementations, it could easily refuse those implementations that do not provide this trapdoor. Obviously, this functionality would have to be cryptographically protected with a highly secure key that would be used to sign such orders issued by a judge.

6 Conclusion

An approach to protect confidential information in mobile communication systems has been presented. The approach takes advantage of the fact that oftentimes, information is not actually needed and restricts the disclosure of the information to the required minimum. The underlying idea, which allows to retain some control over information handed to a different entity, is sufficiently general to be applicable to other problems with similar constraints.

The approach is illustrated with the example of protecting location information in a mobile communication system, which uses simplified concepts from GSM (Global System for Mobile Communications). It is compared to another approach from the literature [9].

We further define the problem of protecting mobile objects against manipulation and disclosure and present a solution, which consists of a tamper

---

[9]The *Technischer Überwachungs-Verein* is an independent, government-approved expert appraisal and inspection organization.

proof execution environment and a protocol that provides a comprehensive protection for the objects executed on this environment. This can be used to protect objects against reverse engineering, which is an interesting application in the context of Java and ActiveX.

Acknowledgments

I would like to thank Sebastian Staamann for the discussion that spawned the presented idea.

References

[1] R. Anderson and M. Kuhn. Tamper resistance — a cautionary note. In *The Second USENIX Workshop on Electronic Commerce Proceedings*, pages 1–11, Oakland, California, November 1996.

[2] G. Brassard. *Modern Cryptology – A Tutorial*, volume 325 of *Lecture Notes in Computer Science*. Springer Verlag, 1988.

[3] D. Chaum. Untraceable electronic mail, return addresses, and digital pseudonyms. *Communications of the ACM*, 24(2):84–88, February 1981.

[4] H. Damker, H. Federrath, M. Reichenbach, and A. Bertsch. Persönliches Erreichbarkeitsmanagement. In G. Müller and A. Pfitzmann, editors, *Sicherheit in der Kommunikationstechnik*. Addison-Wesley, 1997.

[5] J. S. Fritzinger and M. Mueller. Java security. White paper, Sun Microsystems, Inc., 1996.

[6] A. Herzberg and S. S. Pinter. Public protection of software. In *Advances in Cryptology: CRYPTO'85*, pages 158–179, Santa Barbara, California, August 1985.

[7] P. Johns. Signing and marking ActiveX controls. *Developer Network News*, November 1996.

[8] G. Nilsson, F. Dupuy, and M. Chapman. An overview of the telecommunicatinos information networking architecture. In *TINA'95*, pages 1–11, Melbourne, Australia, February 1995.

[9] A. Pfitzmann. Technischer Datenschutz in öffentlichen Funknetzen. *Datenschutz und Datensicherung (DuD)*, 17(8):451–463, 1993.

[10] RSA Data Security, Inc. *PKCS #1: RSA Encryption Standard*. RSA Data Security, Inc., November 1993.

[11] B. Schneier. *Applied cryptography*. Wiley, New York, 1994.

[12] New York Times. U.S. workers stole data on 11,000, agency says, April 6, 1996.

[13] TINA Consortium. *Service Architecture, Version 4.0*, October 1996.
     `http://www.tinac.com/96/sa96_public.ps`.

[14] I. S. Winkler. The non-technical threat to computing systems. *Computing Systems, USENIX Association*, 9(1):3–14, Winter 1996.

[15] P. R. Zimmermann. *PGPfone Owner's Manual*. Pretty Good Privacy, Inc., July 1996.
     `http://www.pgp.com/products/fone-docs/fone_01.cgi`.

Ein nachweisbares Authentikationsprotokoll am Beispiel von UMTS

Stefan Pütz

Institut für Nachrichtenübermittlung
Universität Siegen, Hölderlinstraße 3, D-57076 Siegen
Tel.: (0271) 740-2623, Fax: -2536, e-mail: puetz@nue.et-inf.uni-siegen.de

Zusammenfassung

Zunehmend vielfältigere Dienstleistungen, die digitale Mobilfunksysteme erbringen können, erfordern eine Überarbeitung bestehender Sicherheitsanforderungen und -dienste. Sollen beispielsweise zukünftig elektronische Zahlungsmittel über Mobiltelefone nachladbar sein, so sind starke Sicherheitsmechanismen, etwa eine gegenseitige Authentikation verknüpft mit Non-repudiation, erforderlich.

Ausgehend von bekannten Non-repudiation und Authentikationsmechanismen auf der Basis asymmetrischer Kryptoverfahren führt dieser Beitrag die *Non-repudiation authentication* ein. Als Anwendung des neuen *Non-repudiation of mutual agreement service* dokumentiert sie die Authentikation zweier Authentikationspartner unabstreitbar und für einen unabhängigen Dritten nachvollziehbar. Der Authentikationsnachweis, den die Kommunikationspartner während der eigentlichen Authentikation generieren, dient als Non-repudiation token. Somit wird beispielsweise die Authentikation zwischen einem Teilnehmer und einem beliebigen Mobilfunknetzbetreiber für den betreffenden Service Provider nachvollziehbar.

Das neue Authentikationsprotokoll wird in zwei Varianten vorgestellt, und dessen Einsatzumgebung, das Szenario digitaler Mobilfunksysteme, kurz skizziert. Eine Variante erfordert vertrauenswürdige Zeitstempel, da diese Nachvollziehbarkeit und Non-repudiation einer Authentikation und optionaler Datenfelder ermöglichen. Die andere Variante verzichtet auf vertrauenswürdige Zeitstempel. Jedoch erfordert sie lokale Realzeituhren bei den Authentikationspartnern. Allerdings ist die Nachvollziehbarkeit eingeschränkt und an bestimmte Voraussetzungen geknüpft. Eine ausführliche Darstellung der logischen Protokollstruktur erläutert sowohl die Verschachtelung als auch die Bedeutung einzelner Elemente innerhalb der Protokolle. Die Diskussion der zeitvarianten Parameter zeigt deren mehrfache Bedeutung während eines Protokollablaufs und bzgl. der verschiedenen Non-repudiation Eigenschaften.

Schlüsselwörter

Authentikation, nachweisbare, nachvollziehbare; Kryptoverfahren, asymmetrische; Mobilfunknetze; Mobilfunksysteme, digitale; Non-repudiation; Parameter, zeitvariante; Sicherheitsdienste; Signaturen, digitale, mehrfache; UMTS; Zeitstempel, vertrauenswürdige; Zeitstempeldienste.

1. Einleitung

Durch die Koexistenz mehrerer digitaler Mobilfunknetze, wie D_1, D_2, E-Plus und zukünftig E_2, desselben Standards nebeneinander entsteht alleine in Deutschland eine zunehmend heterogene Struktur in sich abgeschlossener und gleichberechtigter Einzelnetze, die miteinander über Gateways kommunizieren. Verschiedene Mobilfunknetze müssen nicht tatsächlich räumlich nebeneinander angeordnet sein. Ein und dasselbe Gebiet darf auch mehrfach durch die Netze verschiedener Betreiberorganisationen abgedeckt werden. Der Teilnehmer kann somit im gesamten Gebiet, das durch diese Netze aufgespannt wird, frei kommunizieren. Wird ein Gebiet von mehreren Anbietern versorgt, hat er zudem die Möglichkeit, an ein und derselben Stelle die Dienste verschiedener Mobilfunknetze zu nutzen oder über sie zu kommunizieren. So kann er beispielsweise verschiedene Leistungsmerkmale oder Gebührensätze ausnutzen, die angeboten werden. Um Maskeradeangriffe abzuwenden ist die Authentikation des Netzbetreibers erforderlich. Die Kontrolle der Netzzugangsrechte bedingt hingegen die Authentikation des Teilnehmers.

Heutige und zukünftige Mobilfunkstandards, wie GSM (Global System for Mobile communications), UMTS (Universal Mobile Telecommunications System) oder FPLMTS (Future Public Land Mobile Telecommunications System), erfordern für den Entwurf, die Entwicklung und die Erprobung neuer Authentikationsprotokolle eine flexible, zeitgemäße und an die Anforderungen heutiger Problemstellungen angepaßte Sicherheitsstruktur zugrunde zu legen.

Asymmetrische Kryptoverfahren bieten bzgl. dieser Forderungen eine Reihe von Vorteilen [Pütz_95]. Sie ermöglichen eine dezentrale Authentikation zwischen Teilnehmer und Mobilfunknetz ohne Online-Verfügbarkeit des eigenen Service Providers. Allerdings kann der Service Provider, wenn er nicht in den Authentikationsprozeß eingebunden ist, diesen – mit bisher bekannten Verfahren – auch nicht verifizieren. Der Wunsch nach Nachweisbarkeit wächst jedoch bei allen ordentlichen Mobilfunkparteien zunehmend, da somit momentan diskutierte Sicherheitsmechanismen wie *Fraud Detection* oder eine *Nachweisbare Gebührenabrechnung* bestens unterstützt würden. Neben bekannten Sicherheitsdiensten, wie gegenseitige Authentikation, Vertraulichkeit der Übertragungsdaten, der Teilnehmeridentität und des -aufenthaltortes, ermöglicht das neue Protokoll einer unabhängigen Partei eine beliebige dezentrale Authentikation anhand eines Nachweises zu verifizieren, der während der Authentikation durch die Authentikationspartner generiert worden ist. Dabei ist es für den Verifizierer bedeutungslos, welchem Mobilfunknetz die Authentikationspartner angehören. Ein Netzbetreiber kann somit nachvollziehen, wer sein Netz zu welcher Zeit benutzt hat. Ebenso kann ein Service Provider beispielsweise prüfen, welcher seiner Kunden in welchem Netz eingebucht ist bzw. war. Weiterhin tauschen die Authentikationspartner optionale Datenfelder miteinander aus, für die sie Sende- und Empfangsnachweise nach dem Non-repudiation Mechanismus erstellen. Werden in diesen Datenfeldern Gebühreninformationen übertragen, so würde eine nachweisbare und daher nachvollziehbare Abrechnungsgrundlage geschaffen. Die Nachweisbarkeit dieser Abrechnung schützt den Nutzer einer Dienstleistung vor zu Unrecht erhobenen Gebühren durch den Diensteanbieter. Gegenüber dem Diensteanbieter könnte der Dienstenutzer hingegen nicht leugnen, die nachgewiesenen Dienstleistungen auch tatsächlich genutzt zu haben [Pütz_96].

2. Mechanismen

2.1 Authentikation basierend auf asymmetrischen Kryptoverfahren

Authentikation ist die Bestätigung einer angegebenen Identität und erbringt mit vorhersagbarer Wahrscheinlichkeit den Nachweis, daß ein Kommunikationspartner auch derjenige ist, der er zu sein vorgibt [Rula_93, Schn_96].

Die Authentikation wird im Verlauf eines Authentikationsprotokolls durch einen Authentikator erbracht und dieser durch eine digitale Signatur über ein Parameterfeld, das speziellen Anforderungen genügen muß, generiert. Das Authentikationsprotokoll läuft zwischen den Kommunikationspartnern ab, die sich einseitig oder gegenseitig authentisieren wollen. Allgemein sind für eine starke Authentikation zwei Voraussetzungen zu erfüllen:

- Der Authentikator darf eindeutig und nachweislich nur durch jene Partei erstellt worden sein, die sich in diesem Prozeß authentisieren möchte.
- Der Authentikator muß geeignete Elemente enthalten, die seine Aktualität und Einmaligkeit gewährleisten und somit Wiedereinspielung und Unterdrückung eines gültigen Authentikators erkennen lassen.

Die erste Anforderung an eine Authentikation läßt sich aufbauend auf ein *asymmetrisches Kryptosystem* durch digitale Signaturen [9796, 14888-3] erfüllen. Solange der geheime Schlüssel eines asymmetrischen Schlüsselsystems ausschließlich dem rechtmäßigen Schlüsselinhaber bekannt ist, kann auch nur dieser eine gültige Signatur mit diesem Schlüssel anfertigen. Für den Verifizierer hingegen muß der zugehörige öffentliche Schlüssel authentisch und vertrauenswürdig sein. Die Authentizität des Schlüsselursprungs besagt, daß dieser öffentliche Schlüssel zu genau einem speziellen Schlüsselinhaber gehört. Die Vertrauenswürdigkeit versichert, daß der Schlüssel selbst korrekt ist.

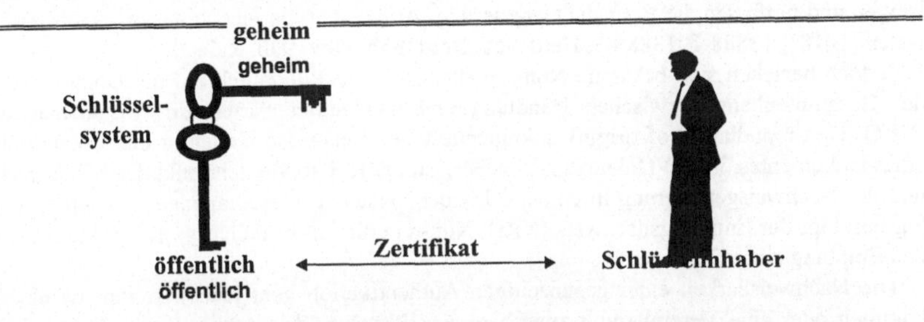

Bild 2.1: Zertifikat verbindet Schlüsselsystem und -inhaber

Verläuft die Verifikation einer digitalen Signatur erfolgreich, so bilden der geheime Schlüssel, mit dem die Signatur erstellt worden ist, und der öffentliche Schlüssel, der der Verifikation diente, ein Schlüsselpaar. Die Zusammengehörigkeit beider Schlüssel als ein Schlüsselpaar ist nachgewiesen. Die Authentizität des öffentlichen Schlüssels verbindet nun dieses asymmetrische Schlüsselpaar mit seinem rechtmäßigen Inhaber. Der Schlüsselinhaber authentisiert sich demnach indirekt durch den Besitz bzw. den Gebrauch seines geheimen Signaturschlüssels.

Zeitvariante Parameter (TVP, Time Variant Parameter), die durch (Pseudo-) Zufallszahlen, Reihenfolgenummern oder Zeitstempel gebildet werden, weisen die Einmaligkeit und Aktualität eines gültigen Authentikators nach. In Authentikationsprotokollen können sie den Protokollelementen, in denen sie eingesetzt werden, folgende Eigenschaften zusichern [BAN_90]:

- Einmaligkeit,
- Aktualität,
- Zusammengehörigkeit und Aneinanderbindung mehrerer Protokollelemente,
- Reihenfolge zusammengehörender (sequentieller) Protokollelemente und
- Verhinderung der Signatur vorgefertigter oder präparierter Nachrichten.

Hierbei gilt es zu beachten, daß zwar bestimmten TVPs differenzierte Eigenschaften zugeordnet werden können, die Einhaltung dieser Eigenschaften jedoch zusätzlich durch die Protokollstruktur beeinflußt wird. Um Sinn und Zweck eines TVPs in einem Protokoll zu bewerten, muß demnach sowohl der TVP-Typ als auch die Protokollstruktur berücksichtigt werden.

2.2 Non-repudiation

Aufgabe und Ziel der Non-repudiation Mechanismen liegen darin, abgelaufene Vorgänge bzw. Aktionen oder eingetretene Ereignisse eindeutig und nachvollziehbar auch für unabhängige und/ oder unbeteiligte Dritte in Form von rechtsgültigen Nachweisen unwiderrufbar und für die beteiligten Parteien unleugbar zu beschreiben, aufzuzeichnen und evtl. zu archivieren, so daß anhand dieser erstellten Beweismittel (Nachweise) Rechtsstreitigkeiten infolge von Unstimmigkeiten der in die betreffenden Aktionen oder Ereignisse involvierten Parteien eindeutig und zuverlässig sowie sachlich richtig und rechtsgültig geschlichtet bzw. entschieden werden können.

Die Beweismittel beziehen sich auf ein in der Vergangenheit eingetretenes Ereignis oder einen abgelaufenen Vorgang. Sie sollen in der Lage sein zu klären, ob und wenn ja in welcher Form eine Handlung tatsächlich stattgefunden hat. Dabei soll es den involvierten Parteien unmöglich sein, eine Beteiligung an dieser Handlung bzw. deren Existenz zu leugnen.

Non-repudiation Dienste finden ihre Anwendung in vielerlei Situationen und Zusammenhängen und umfassen die Bereiche Generierung, Aufbewahrung und Transport elektronischer Daten [10181, 13888-1, 13888-3, Herd_95a, Herd_95b, Pütz_97b, Rula_93].

Jedoch beziehen sich bekannte Non-repudiation Dienste lediglich auf die Dokumentation des Zusammenhanges zwischen Handlung und handelnder Partei. Ein Urhebernachweis (NRO, Non-repudiation of origin) dokumentiert beispielsweise eindeutig die Urheberschaft eines Dokumentes, indem Dokument, Urheber, Empfänger sowie Zeitpunkt der Urheberschaft und der Nachweisgenerierung in einem (Urheber-) Nachweis zusammengefaßt werden. Analog bestätigt der Empfangsnachweis (NRD, Non-repudiation of delivery) den ordnungsgemäßen Empfang eines Dokumentes.

Die Nachweisbarkeit einer gegenseitigen Authentikation geht jedoch weiter, da ein Abkommen oder eine Vereinbarung zwischen *zwei* Parteien, den Authentikationspartnern, für eine unabhängige dritte Partei nachweisbar und für die beteiligten Parteien unabstreitbar dokumentiert wird. Diese Eigenschaften werden durch den neuen *Non-repudiation of mutual agreement service* unterstützt [Pütz_97b]. Siehe Bild 2.2.

Eine Vorstufe hierzu bildet die Unabstreitbarkeit einer Partei, in einen Vorgang oder eine Aktion involviert gewesen zu sein. Diesen Non-repudiation Dienst nennt man *Non-repudiation of procedure involvement* [ETSI_1]. Das ETSI (European Telecommunications Standards Institute) spezifiziert diesen Dienst als Teil von UMTS, dem zukünftigen europaischen Mobilfunkstandard der dritten Generation.

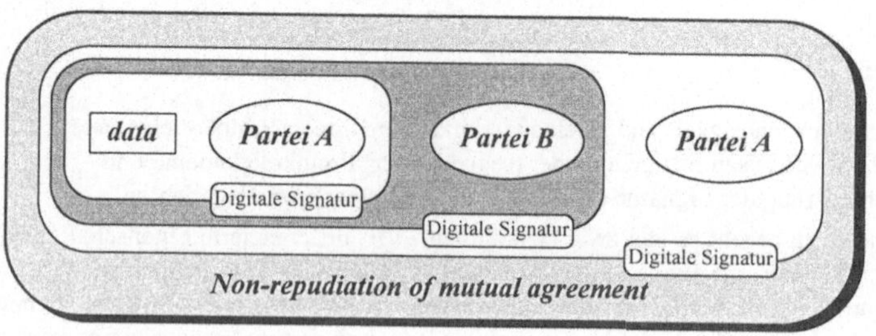

Bild 2.2: Non-repudiation of mutual agreement (NRMA)

2.3 Nachvollziehbarkeit und Non-repudiation einer Authentikation

Durch die Anwendung des Non-repudiation Mechanismus auf eine Authentikation entsteht begrifflich die *Non-repudiation authentication*. Sie dokumentiert nachweislich und unwiderruflich die Authentikation zwischen Kommunikationspartnern. Für die beteiligten Parteien wird die Teilnahme an der betreffenden Authentikation somit unabstreitbar und ihr Resultat unanfechtbar.

Der Nachweis selbst liegt nach Abschluß der eigentlichen Authentikation in Form eines Authentikationsnachweises vor. Als *gemeinsamen* Ausführungsnachweis der gegenseitigen Authentikation der Kommunikationspartner läßt sich dieser Nachweis, der Non-repudiation of mutual authentication token, in die Reihe der Non-repudiation token einfügen.

2.3.1 Definition des Authentikationsnachweises

Der Authentikationsnachweis wird wie folgt definiert:

Der Authentikationsnachweis beschreibt objektiv, eindeutig und für unabhängige Dritte nachvollziehbar, welche Kommunikationspartner sich wann gegenseitig erfolgreich authentisiert haben.

Nach dieser Definition wird die Verifizierbarkeit eines Authentikationsnachweises durch unabhängige Dritte garantiert, sobald

- Objektivität,
- Eindeutigkeit und
- Nachvollziehbarkeit

des Authentikationsnachweises erfüllt sind.

Als zusätzliches Leistungsmerkmal des Authentikationsnachweises wird die Vollständigkeit bzw. Reihenfolge einer Gruppe von Authentikationsnachweisen definiert. Dies ist im Hinblick auf *Fraud detection* oder neue Abrechnungs- und Zahlungssysteme elementar, da hier gewährleistet werden muß, daß keine Nachweise unterdrückt werden.

2.3.2 Sicherheitsanforderungen an den Authentikationsnachweis

Der Authentikationsnachweis muß folgende Sicherheitsanforderungen erfüllen:

Manipulation, Wiedereinspielung, Verzögerung und Unterdrückung des Authentikationsnachweises müssen erkannt und somit dessen Ursprünglichkeit, Einmaligkeit und Aktualität gewährleistet werden.

2.3.3 Anforderungen an das Authentikationsprotokoll

Das entworfene Authentikationsprotokoll soll die gegenseitige Authentikation der Authentikationspartner sicherstellen und darüber hinaus einen Authentikationsnachweis liefern.

Eine Alternative zur Generierung eines Authentikationsnachweises bietet die Aufzeichnung des Authentikationsprotokolls. Der entscheidende Unterschied zwischen einem Authentikationsnachweis und der Aufzeichnung eines Authentikationsprotokolls besteht in der Eigenschaft eines Authentikationsprotokolls, spezifizierte Sicherheitsanforderungen wie die Einmaligkeit seiner Protokollelemente nur gegenüber den jeweiligen Authentikationspartnern zu erbringen, diese Sicherheitsanforderungen bei einer Betrachtung durch einen unabhängigen Dritten jedoch nicht uneingeschränkt zu erfüllen.

In der Regel werden sicherheitsrelevante Eigenschaften wie die Einmaligkeit der Nachrichten eines Authentikationsprotokolls während der Entwicklung durch TVPs berücksichtigt und diese in geeigneter Weise in Protokollelemente und -struktur eingearbeitet.

Die Einmaligkeit eines vollständig aufgezeichneten Authentikationsprotokolls ist jedoch eine eigenständige Eigenschaft [Pütz_95], die während der Entwicklung und Standardisierung der Authentikationsprotokolle nach [9798-3] und [X.509] weder spezifiziert noch berücksichtigt worden ist.

Daher ist die Verifizierbarkeit eines aufgezeichneten Authentikationsprotokolls durch unbeteiligte Dritte bedingt durch die fehlende Eigenschaft *Einmaligkeit der Authentikation* bei den angeführten Protokollen nicht gegeben. Auch sind darüber hinaus keine Protokolle bekannt, die diese Eigenschaft erfüllen.

Wird die Integrität der Authentikation des jeweiligen Authentikationspartners oder des vollständig aufgezeichneten Authentikationsprotokolls zudem nicht von beiden Partnern z.B. durch eine digitale Signatur gewährleistet und bestätigt, läßt sich im nachhinein nicht feststellen, ob nicht einer der vermeintlichen Authentikationspartner Protokollelemente einer alten Authentikation wiederverwendet und das Authentikationsprotokoll durch Hinzufügen manipulierter oder neuer Protokollelemente zu einer gefälschten Aufzeichnung zusammengefügt hat.

Wird einem unabhängigen Dritten ein vollständig aufgezeichnetes Authentikationsprotokoll nach [9798-3] oder [X.509] vorgelegt *und* dessen Integrität nachgewiesen, kann dieser die Authentikation nachvollziehen. Er kann jedoch i. allg. nicht eindeutig feststellen, wann diese Authentikation stattgefunden hat. Weiterhin kann er nicht prüfen, ob er diese Aufzeichnung in der Vergangenheit bereits einmal verifiziert hat, ohne alle bisher verifizierten Aufzeichnungen aufzubewahren. Selbst wenn in diesem Authentikationsprotokoll aktuelle Zeitstempel als TVP verwendet werden, müssen diese für den unabhängigen Dritten vertrauenswürdig sein, um den Zeitpunkt der Authentikation für ihn nachvollziehbar zu dokumentieren [Pütz_95]. Die Aktualität der Authentikation und deren Protokollelemente wird i.d.R. nur für die beteiligten Authentikationspartner sichergestellt.

Aufzeichnungen vollständiger Authentikationsprotokolle enthalten hinsichtlich der Verifikation der Authentikation Redundanz, da Kennungen und TVPs i.d.R. mehrfach in verschiedenen Protokollelementen verwendet werden (müssen). Der Authentikationsnachweis hingegen enthält alle wesentlichen Elemente lediglich einmal.

2.4 Mehrfachsignaturen

Mehrfachsignaturen ermöglichen die Signatur derselben Nachricht durch mehrere Kommunikationspartner, während diese gleichzeitig, abhängig von der gewählten Struktur der Signatur, die gegenseitige Akzeptanz dieser Nachricht bestätigen. Zahlreiche Mehrfachsignaturverfahren auf der Basis von Fiat-Schamir, Diskreter Logorithmus und RSA sind in [HaZh_92, ItNa_83, OhOk_91, Okam_88] veröffentlicht worden. Ebenso sind Risiken und verschiedene Angriffe auf Mehrfachsignaturverfahren in [HoMi_96, HoMP_95] beschrieben.

Mehrfachsignaturen bezeichnen ineinander verschachtelte Signaturen, die aus einzelnen bzw. wiederum aus verschachtelten Signaturen bestehen können. Eine Doppelsignatur hat demnach folgenden allgemeinen Aufbau (zur allgemeinen Notation siehe Tabelle 4.1), wobei m_X bzw. m_Y die von den Parteien X bzw. Y signierten Nachrichten darstellen:

$$\mathrm{Sig}_Y(m_Y \parallel \mathrm{Sig}_X(m_X)).$$

Die *innere* Signatur $\mathrm{Sig}_X(m_X)$ besteht aus der Signatur von X über die Nachricht m_X; X signiert m_X. Die *äußere* Signatur $\mathrm{Sig}_Y(m_Y \parallel \mathrm{Sig}_X(m_X))$ besteht hingegen aus der Signatur von Y über die Nachricht $m_Y \parallel \mathrm{Sig}_X(m_X)$; Y signiert also m_Y und zusätzlich die Signatur von X über m_X.

Allgemein läßt sich daher aussagen, daß X die Existenz der Nachricht m_X, Y hingegen die Existenz der Nachricht m_Y und der Signatur von X über m_X bestätigt.

Daraus läßt sich jedoch weder schließen, daß Y die Nachricht m_X selbst kennt oder ihre Existenz bestätigt, noch daß Y durch seine eigene Signatur der Signatur von X über die Nachricht m_X auch implizit die Existenzbestätigung der Nachricht m_X durch X bestätigt. Y bestätigt lediglich, daß zum Zeitpunkt seiner Signatur zusätzlich zu der Nachricht m_Y auch die Bitkombination 'Sig$_X$(m_X)' vorgelegen hat, nicht jedoch deren Bedeutung oder Interpretation.

Soll Y durch seine eigene Signatur über Sig$_X$(m_X) bestätigen, daß X die Existenz der Nachricht m_X durch die Signatur Sig$_X$(m_X) bestätigt oder die Nachricht m_X selbst anerkannt hat, so müssen diese Forderungen in allgemeinen Konventionen festgehalten werden.

Für Y bedeuten diese Forderungen, daß Y selbst die Signatur von X über m_X verifizieren muß, bevor Y seine eigene Signatur über $m_Y \parallel$ Sig$_X$(m_X) berechnen darf.

Verifiziert Y die Signatur von X über die Nachricht m_X, kennt Y auch die Nachricht m_X und bestätigt deren Existenz mit seiner eigenen Signatur. Dies ist jedoch nicht gleichbedeutend mit der Anerkennung der Nachricht m_X durch Y, da hier lediglich die Kenntnis der Nachricht m_X bestätigt wird.

Soll Y durch seine Signatur über $m_Y \parallel$ Sig$_X$(m_X) auch die Nachricht m_X anerkennen, so muß folgende Beziehung zwischen m_X und m_Y gelten:

$$m_X \subseteq m_Y.$$

Gilt diese Beziehung, weist Y durch seine Signatur über $m_Y \parallel$ Sig$_X$(m_X) ebenfalls die Existenz der Nachricht m_X nach.

3. Mobilfunkszenario

Ein Mobilfunknetz gliedert sich grob in zwei Bereiche: das Festnetz und die Funkübertragungsstrecke. Das Festnetz besteht aus ortsfesten Einrichtungen und übernimmt die netzseitige Kommunikation und Signalisierung. Netzkomponenten kommunizieren miteinander über Festnetzkanäle. Die netzübergreifende Kommunikation zwischen den Festnetzen unterschiedlicher Betreiber ist über Gateways möglich. Den zweiten Bereich eines Mobilfunknetzes bildet die Funkübertragungsstrecke (Luftschnittstelle), welche eine flexible Anbindung eines Mobilfunkteilnehmers an das Festnetz bereitstellt. Eine direkte Teilnehmer-zu-Teilnehmer Kommunikation ist derzeit über Mobilfunknetze nicht möglich.

Aufgrund der Einteilung des Mobilfunksystems in Festnetz und Funkübertragungbereich ergeben sich für Festnetzkomponenten und Teilnehmer unterschiedliche Kommunikationspfade. Während Festnetzkomponenten untereinander kommunizieren können, hat der Teilnehmer immer eine Festnetzkomponente des Netzbetreibers zum Kommunikationspartner, bei dem er momentan eingebucht ist. Dem Teilnehmer ist somit keine direkte Kommunikation mit einer beliebigen Festnetzkomponente möglich.

Da der Netzbetreiber als Dienstleistungserbringer und der Teilnehmer als Dienstenutzer Geschäftspartner darstellen und diese i. allg. unterschiedliche Interessen verfolgen, ist weder der Netzbetreiber für den Teilnehmer noch der Teilnehmer für den Netzbetreibner vertrauenswürdig. Soll ein Teilnehmer beispielsweise, wie in dem nachfolgend vorgestellten Authentikationsprotokoll vorgesehen, mit einem vertrauenswürdigen Zeitstempel versorgt werden, so muß dieser Zeitstempel durch einen vertrauenswürdigen Zeitstempeldienst verbreitet und über das Netz des nicht vertrauenswürdigen Betreiber zum Teilnehmer transportiert werden. In diesem Fall ist ein gesichertes Übertragungsprotokoll vorzusehen, welches dem Zeitstempel Aktualität, Integrität und Authentizität garantiert [DeSa_81, HaSt_90, PiFr_96].

4. Authentikationsprotokoll

Aufbauend auf den Lösungsansatz aus [Pütz_95] wurde ein Authentikationsprotokoll entwik-
kelt, das sowohl den Sicherheitsanforderungen des bestehenden Standards GSM als auch der
bisherigen Spezifikation für Mobilfunksysteme der dritten Generation, wie UMTS [ETSI_1],
genügt und diese durch zusätzliche Merkmale noch übertrifft.

Im folgenden werden zwei Situationen unterschieden: Authentikation eines unbekannten
Teilnehmers sowie die Folgeauthentikation eines bereits bekannten Teilnehmers. Dabei bezeichnet UIM (User Identity Module) den Teilnehmer bzw. die Teilnehmerchipkarte, NO (Network
Operator) den Mobilfunknetzbetreiber, TTS (Trusted Timestamping Service) einen vertrauenswürdigen Zeitstempeldienst [HaSt_90] und TP (Third Party) eine beliebige dritte Partei.

Authentikation eines unbekannten Teilnehmers bedeutet die erstmalige Authentikation eines Teilnehmers in einer bestimmten Netzwerkdomäne, Folgeauthentikation entsprechend die
Authentikation eines bereits bekannten Teilnehmers. Die Folgeauthentikation wird durch die
Ausführung des Authentikationsmechanismus erreicht. Ist der Teilnehmer noch unbekannt,
wird zuvor ein Setup-Mechanismus eingeleitet. Der Setup-Mechanismus umfaßt Zertifikats-
und Schlüsselaustausch zwischen Teilnehmer und Netzbetreiber. Er bereitet beide Kommunikationspartner auf die anschließende Ausführung des Authentikationsmechanismus vor.

4.1 Authentikation eines unbekannten Teilnehmers

4.1.1 Eigenschaften und Ziele

Zu den wichtigsten Eigenschaften des neuen Authentikationsprotokolls zählen die gegenseitige Authentikation der Kommunikationspartner sowie die Nachweisbarkeit dieser Authentikation. Das Authentikationsprotokoll erbringt dazu die Non-repudiation authentication von UIM und NO nach Kapitel 2.3 und liefert einen Authentikationsnachweis, den jeder beliebige Dritte verifizieren kann. Zu diesem Zweck werden digitale Mehrfachsignaturen auf der Basis asymmetrischer Kryptoverfahren eingesetzt. Um digitale Signaturen zu verifizieren, werden zertifizierte öffentliche Schlüssel zwischen UIM und NO ausgetauscht. Eine Voraussetzung für die *Non-repudiation authentication* liegt in der Verfügbarkeit eines vertrauenswürdigen Zeitstempels [PiFr_96, Pütz_95].

Die Absprache eines symmetrischen Sitzungsschlüssels K zwischen UIM und NO ermöglicht Verschlüsselung auf der Luftschnittstelle, die wiederum Vertraulichkeit aller Übertra-

Tabelle 4.1: Legende aller verwendeten Abkürzungen

UIM	Teilnehmerchipkarte	(User Identity Module)
IMUI	Teilnehmerkennung	(International Mobile User Identity)
NO	Netzwerkbetreiber	(Network Operator)
TTS	Vertrauensw. Zeitstempeldienst	(Trusted Timestamping Service)
TP	Dritte Partei	(Third Party)
CA	Zertifizierungsinstanz	(Certification Authority)
RND_X	(Pseudo-) Zufallszahl von X	(RaNDom number)
SEQ_X	Reihenfolgenummer von X	(SEQuence number)
TS_X	Zeitstempel von X	(TimeStamp)
STS	Vertrauensw. Zeitstempel des TTS	(Secure TimeStamp)
dataN	Optionale Datenfelder, N = 1,2,3	(data field)
Sig_X	Signaturalgorithmus von X	
h(m)	Hashwert h(m), berechnet mittels einer kryptographischen kryptographischen Einweg-Hashfunktion h( ), hier über die Nachricht m	
Enc	Enc{K,data} bezeichnet die Verschlüsselung der Nachricht 'data' mit dem symmetrischen Verschlüsselungsalgorithmus Enc und dem Schlüssel K.	
K	Symmetrischer Sitzungsschlüssel	
L	Länge des symmetrischer Sitzungsschlüssels	
g	g ist Generator einer finiten Gruppe G, in der das Diskrete Logarithmus Problem (DLP) besteht und mit vertretbarem Aufwand in akzeptabler/verfügbarer Zeit nicht zu lösen ist.	
g^{tNX}	Öffentl. Schlüsselteil von X zur Absprache eines symmetr. Sitzungsschl. K	
tN_X	Geheimer Schlüsselteil von X zum öffentlichen Schlüsselteil g^{tNX}, N = 1,2	
$Cert_X$	Gültiges und verfügbares Zertifikat, ausgestellt von der Zertifizierungsinstanz CA über den öffentl. Schlüssel des asymmetrischen Schlüsselsystems von X	
X	X steht stellvertretend für UIM, NO oder TTS	
u || v	Aneinanderreihung beliebiger Datenelemente u und v	

Tabelle 4.2: Erforderliche Kenntnisse und Fähigkeiten der einzelnen Parteien

Partei	Erforderliche Kenntnisse, Fähigkeiten
UIM	Identität NO des Netzbetreibers
NO	Identität TTS des Zeitstempeldienstes
UIM, NO, TP	Authentischen und vertrauenswürdigen öffentlichen Schlüssel einer Zertifizierungsinstanz
UIM, NO, TTS	Zertifikat der CA über ihren eigenen öffentlichen Schlüssel
UIM, NO, TTS	Asymmetrischen Signaturalgorithmus
UIM, NO, TP	Asymmetrischen Verifikationsalgorithmus
UIM, NO	Symmetrischen Ver- und Entschlüsselungsmechanismus
UIM, NO, TTS, TP	Kollisionsfreie kryptographische Einweg-Hashfunktion
UIM, NO	(Pseudo-) Zufallszahlen
UIM	Reihenfolgenummern
TTS	Aktuelle, sichere und vertrauenswürdige Zeitstempel
UIM, NO, TP	Vertrauen in den Zeitstempeldienst TTS

gungsdaten, der Teilnehmeridentität und des -aufenthaltortes liefert. Das Key-Management garantiert Authentikation und Aktualität des ausgehandelten Sitzungsschlüssels sowie dessen Bestätigung. Durch eine Variante des Schlüsselaustauschprotokolls nach Diffie-Hellman [DiHe_76] wird Good forward secrecy [DiOW_92] des Sitzungsschlüssels erreicht.

UIM kennt die Identität des Netzbetreibers bereits zu Beginn des Setup-Mechanismus, NO die Teilnehmeridentität IMUI zu Beginn des Authentikationsmechanismus. Daher wird ein Sitzungsschlüssel nur zwischen bekannten Parteien ausgehandelt.

Weiterhin ermöglicht das Authentikationsprotokoll den Austausch optionaler Datenfelder. Diese Datenfelder unterliegen ebenfalls dem Non-repudiation Mechanismus und weisen spezielle Eigenschaften auf, die in Kapitel 4.1.5 behandelt werden.

Eine formale Analyse in [ETSI_2], die auf einer in [KeWe_96] eingeführten Variante der ursprünglichen BAN-Logik [BAN_90] basiert, weist die Sicherheit des neuen Protokolls nach. Diese Analyse zeigt, daß das Protokoll alle spezifizierten Sicherheitsziele erfüllt.

Eine ausführliche Evaluation befindet sich in [ETSI_3]. Dort sind Angaben über die Nachrichtenlängen der einzelnen Protokollelemente für verschiedene Signaturverfahren nach RSA [RSA_78], DSS [NIST_94] oder El Gamal [ElGa_85] auch auf elliptischen Kurven [FuHe_94, FoRö_96] zu finden. Somit läßt sich die Effizienz des Protokolls in einer bestimmten Umgebung abschätzen und der erforderliche Bandbreitebedarf bzw. die Ausführungszeit angeben.

Entsprechende Angaben über die verwendeten Verfahren und Mechanismen im GSM-Standard befinden sich in [BeDa_96, Pütz_95, Pütz_97a].

4.1.2 Voraussetzungen

Um das Authentikationsprotokoll ausführen zu können, müssen den einzelnen Parteien Kennungen, Zertifikate und Schlüssel entsprechend Tabelle 4.2 bekannt sowie die angegebenen Algorithmen implementiert sein. Tabelle 4.1 zeigt eine Legende aller Abkürzungen.

Beispielsweise muß UIM die Identität NO des Netzbetreibers kennen, um ihn selektieren und adressieren zu können. Da die Kennung von NO nicht vertraulich behandelt werden muß, wird sie üblicherweise als broad cast Nachricht über die Luftschnittstelle verbreitet.

4.1.3 Protokollbeschreibung

4.1.3.1 Setup-Mechanismus

Der Setup-Mechanismus erfüllt folgende Aufgaben:

- Bereitstellung des öffentlichen Schlüsselteils von NO beim UIM
  (als Vorbereitung zur Schlüsselvereinbarung),
- Vereinbarung eines symmetrischen Sitzungsschlüssels zwischen UIM und NO
  (zur Verschlüsselung während der Setup-Phase),
- Vertraulichkeit der Übertragungsdaten, der Teilnehmeridentität und des -aufenthaltortes
  auf der Luftschnittstelle und
- Austausch aller erforderlichen Zertifikate zwischen UIM und NO.

Bild 4.1 zeigt den Nachrichtenaustausch während der Setup-Phase. Die Protokollelemente S1 und S2 werden zwischen UIM und NO ausgetauscht.

Die Aushandlung des Sitzungsschlüssels erfolgt nach Diffie-Hellman [DiHe_76]. NO und UIM tauschen je einen öffentlichen Teilschlüssel miteinander aus.

Nachdem NO sein Geheimnis t_{NO} zufällig gewählt hat, berechnet er daraus den öffentlichen Teilschlüssel g^{tNO} sowie anschließend die Signatur $Sig_{NO}(g^{tNO})$ über diesen Teilschlüssel. Er generiert das Protokollelement S1 und sendet dieses an UIM. Nachricht S1 setzt voraus, daß NO das Zertifikat $Cert_{TTS}$ besitzt.

UIM wählt ebenfalls ein zufälliges Geheimnis $t1_{UIM}$ und berechnet den öffentlichen Teilschlüssel g^{t1UIM}. Nach dem Empfang von der Nachricht S1 verifiziert es die Zertifikate $Cert_{NO}$ und $Cert_{TTS}$ und die Signatur $Sig_{NO}(g^{tNO})$. Unter Verwendung des berechneten Sitzungsschlüssels $K_S = g^{(tNO\,*\,t1UIM)} \bmod L$ und des symmetrischen Verschlüsselungsalgorithmus Enc verschlüsselt es sein Zertifikat $Cert_{UIM}$. UIM sendet S2 an NO.

NO berechnet ebenfalls den Sitzungsschlüssel $K_S = g^{(t1UIM\,*\,tNO)} \bmod L$, entschlüsselt und verifiziert $Cert_{UIM}$. $Cert_{UIM}$ enthält IMUI, die Teilnehmerkennung des UIM.

Die Signatur $Sig_{NO}(g^{tNO})$ garantiert, daß das Geheimnis t_{NO}, aus dem der öffentliche Schlüsselteil g^{tNO} berechnet wurde und das NO zur Berechnung des Sitzungsschlüssels benötigt, tat-

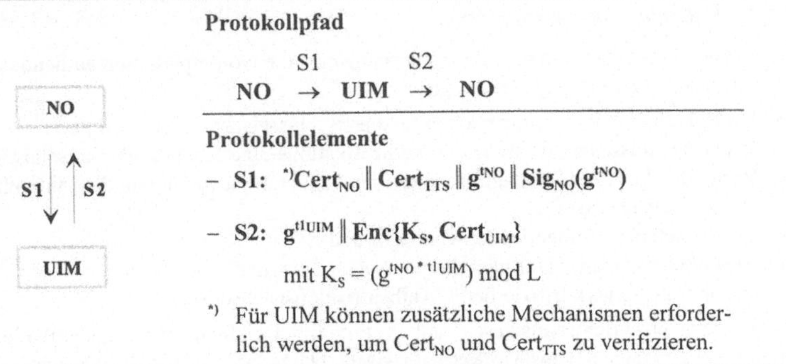

Protokollpfad

	S1		S2	
NO	→	UIM	→	NO

Protokollelemente

- S1: *)$Cert_{NO} \parallel Cert_{TTS} \parallel g^{tNO} \parallel Sig_{NO}(g^{tNO})$

- S2: $g^{t1UIM} \parallel Enc\{K_S, Cert_{UIM}\}$

  mit $K_S = (g^{tNO\,*\,t1UIM}) \bmod L$

*) Für UIM können zusätzliche Mechanismen erforderlich werden, um $Cert_{NO}$ und $Cert_{TTS}$ zu verifizieren.

Bild 4.1: Nachrichtenaustausch der Setup-Phase

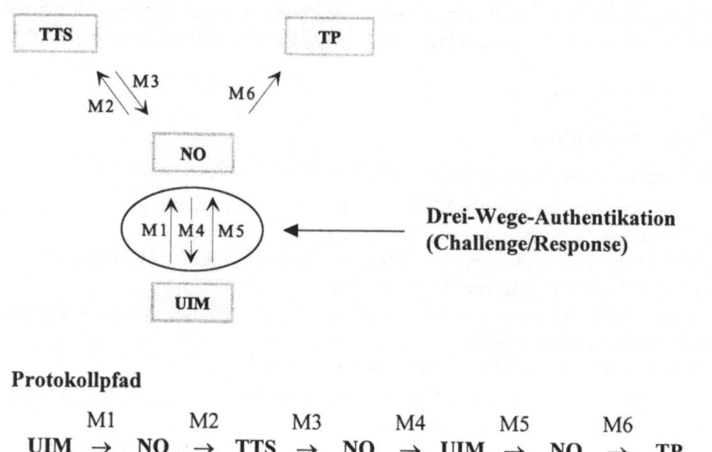

Bild 4.2: Nachrichtenaustausch der Authentikationsphase

sächlich von NO stammt. Diese Eigenschaft ist notwendig, um Vertraulichkeit der Teilnehmeridentität zu gewährleisten. Ist das Geheimnis t_{NO} nicht authentisch und würde beispielsweise g^{tX} von einer fremden Partei X verbreitet, die auch t_X kennt, so würde UIM in Nachricht S2 seine Teilnehmeridentität zwar verschlüsseln, jedoch mit einem Schlüssel, der auf dem Geheimnis t_X beruht. Partei X könnte den verwendeten Schlüssel ebenfalls berechnen ($g^{t|UIM}$ ist im Klartext übertragen worden und daher öffentlich bekannt!) und somit Nachricht S2 entschlüsseln. UIM würde die Vertraulichkeit seiner Teilnehmeridentität verlieren. Diesen Angriff verhindert die Signatur $Sig_{NO}(g^{tNO})$ über den Teilschlüssel g^{tNO}.

Existiert ein Authentication Server, von dem NO Teilnehmerzertifikate abrufen kann, würde das Zertifikat $Cert_{UIM}$ in Nachricht S2 durch die kürzere Teilnehmerkennung IMUI ersetzt. S2 erhält somit folgende Form: $g^{t|UIM} \parallel Enc\{K_S, IMUI\}$. Entschlüsselt NO die Teilnehmerkennung IMUI erfolgreich, fordert er mit ihrer Hilfe das zugehörige Zertifikat vom Authentication Server an. Diese Maßnahme reduziert den Umfang von Nachricht S2 deutlich.

4.1.3.2 Authentikationsmechanismus

Entsprechend liefert der Authentikationsmechanismus die Non-repudiation authentication von UIM und NO und erfüllt folgende Aufgaben:

- Vereinbarung eines symmetrischen Sitzungsschlüssels zwischen UIM und NO (zur Verschlüsselung während der Authentikations- und Kommunikationsphase),
- Vertraulichkeit der Übertragungsdaten, der Teilnehmeridentität und des -aufenthaltortes auf der Luftschnittstelle,
- Non-repudiation authentication von UIM und NO,
- Austausch optionaler Non-repudiation Nachrichten und
- Generierung eines verifizierbaren Authentikationsnachweises.

Bild 4.2 zeigt Nachrichtenaustausch und Protokollpfad während der Authentikationsphase. Die Protokollelemente M1 bis M6 werden zwischen UIM, NO, TTS und TP ausgetauscht.

Die eigentliche Authentikation zwischen UIM und NO erfolgt in den Nachrichten M1, M4 und M5 nach dem Prinzip der Drei-Wege-Authentikation. Die Protokollelemente M2 und M3 dienen der Kommunikation von NO und TTS. TTS liefert in M3 einen vertrauenswürdigen Zeitstempel. Nachricht M6 bildet den Authentikationsnachweis. Seine Übertragung ist im Gegensatz zu der Übertragung der übrigen Nachrichten nicht realzeitorientiert und kann daher im Batch-Betrieb erfolgen.

Tabelle 4.3 zeigt die Protokollelemente des Authentikationsmechanismus und deren logische Bedeutung innerhalb des Protokollablaufs.

UIM wählt ein unvorhersagbares Geheimnis $t2_{UIM}$ und berechnet daraus seinen öffentlichen Teilschlüssel zu g^{t2UIM} sowie den Sitzungsschlüssel $K_M = g^{(NO \, * \, t2UIM)} \bmod L$ für diese Authentikation. In den Sitzungsschlüssel fließt der öffentliche Teilschlüssel g^{tNO} von NO aus der vorhergehenden Setup-Phase ein.

Aus seiner Teilnehmerkennung IMUI, der inkrementierten Reihenfolgenummer SEQ_{UIM}, einem ersten optionalen Datenfeld data1 sowie den Schlüsseln g^{t2UIM} und K_M generiert UIM das Protokollelement M1 und sendet es als *Request for Authentication* an NO.

Die optionalen Datenfelder data1, data2 und data3 besitzen, da sie in die verschiedenen Signaturen des Authentikationsmechanismus eingebunden werden, eigene Non-repudiation Merkmale. Diese werden in Kapitel 4.1.5 gesondert betrachtet.

NO berechnet ebenfalls den Sitzungsschlüssel K_M und entschlüsselt den chiffrierten Teil der Nachricht M1. Mit diesem Input sowie seiner eigenen Kennung NO, einer Zufallszahl

Tabelle 4.3: Protokollelemente des Authentikationsmechanismus

---

- **M1:**   *Request for Authentication of NO*

  $g^{t2UIM} \parallel Enc\{K_M, IMUI \parallel SEQ_{UIM} \parallel data1\}$

- **M2:**   *Request for Timestamping Service*

  $NO \parallel h(IMUI \parallel SEQ_{UIM} \parallel NO \parallel RND_{NO} \parallel data1 \parallel data2)$

- **M3:**   *Response of Timestamping Service*

  $TTS \parallel STS \parallel Sig_{TTS}(*)$

- **M4:**   *Response of Authentication of NO / Request for Authentication of UIM*

  $Enc\{K_M, RND_{NO} \parallel TTS \parallel STS \parallel data2 \parallel Sig_{TTS}(*) \parallel Sig_{NO}(** \parallel Sig_{TTS}(*))\}$

- **M5:**   *Response of Authentication of UIM*

  $Enc\{K_M, data3 \parallel Sig_{UIM}(h(g^{t2UIM}) \parallel ** \parallel data3 \parallel Sig_{TTS}(*) \parallel Sig_{NO}(** \parallel Sig_{TTS}(*)))\}$

- **M6:**   *Mutual Authentication Token (Authentikationsnachweis)*

  $$h(g^{t2UIM}) \parallel ** \parallel data3 \parallel$$
  $$Sig_{TTS}(*) \parallel$$
  $$Sig_{NO}(** \parallel Sig_{TTS}(*)) \parallel$$
  $$Sig_{UIM}(h(g^{t2UIM}) \parallel ** \parallel data3 \parallel Sig_{TTS}(*) \parallel Sig_{NO}(** \parallel Sig_{TTS}(*))) \parallel$$
  $$Sig_{NO}(data3 \parallel Sig_{UIM}(h(g^{t2UIM}) \parallel ** \parallel data3 \parallel Sig_{TTS}(*) \parallel Sig_{NO}(** \parallel Sig_{TTS}(*))))$$

- verwendete Abkürzungen

  $* = TTS \parallel STS \parallel h(IMUI \parallel SEQ_{UIM} \parallel NO \parallel RND_{NO} \parallel data1 \parallel data2)$

  $** = TTS \parallel STS \parallel IMUI \parallel SEQ_{UIM} \parallel NO \parallel RND_{NO} \parallel data1 \parallel data2$

  $K_M = g^{(NO \, * \, t2UIM)} \bmod L$

RND_{NO} und einem zweiten optionalen Datenfeld data2 generiert er die Nachricht M2 und sendet diese als *Request for Timestamping Service* an TTS.

Der Hashwert in Nachricht M2 verhindert, daß Rückschlüsse aus den Zeitstempeldienstanforderungen, die NO an TTS sendet, gezogen werden können. Immerhin bildet TTS für einen Angreifer eine zentrale Stelle, um Informationen über Authentikationen (Authentikationspartner, Häufigkeit, Frequenz, Zeitpunkt, Verteilung, etc.) zu erhalten. Um TTS die korrekte Adressierung seiner Response zu ermöglichen, erweitert NO den berechneten Hashwert um seine eigene Kennung (im Klartext).

TTS berechnet eine Signatur über den Hashwert aus Nachricht M2, den er zuvor um seine eigene Kennung TTS sowie den aktuellen Zeitstempel STS erweitert hat. Er sendet Kennung, Zeitstempel und Signatur als *Response of Timestamping Service* in Nachricht M3 an NO.

NO rekonstruiert * und verifiziert die Signatur $Sig_{TTS}(*)$. Da * die Zufallszahl RND_{NO} enthält (Challenge/Response) und TTS für NO vertrauenswürdig ist, akzeptiert NO den Zeitstempel STS als aktuell, falls zwischen Absenden von M2 und Empfang von M3 nicht mehr als eine maximale Zeitspanne T_MAX_{NO} verstrichen ist. Dieses Zeitfenster beschreibt die aus der Übertragung resultierende zulässige Ungenauigkeit des Zeitstempels STS.

NO generiert die Nachricht M4 und sendet diese an UIM. Mit seiner eigenen Signatur über ** $\| Sig_{TTS}(*)$ bestätigt NO seine Akzeptanz des Zeitstempels sowie die Kenntnis aller weiteren in ** enthaltenen Elemente. Nachricht M4 authentisiert ihn gegenüber UIM, da ** die Reihenfolgenummer SEQ_{UIM} enthält. Um Vertraulichkeit zu gewährleisten, ist Nachricht M4 mit dem Sitzungsschlüssel K_M verschlüsselt.

UIM entschlüsselt Nachricht M4, rekonstruiert * und verifiziert die Signatur $Sig_{TTS}(*)$ des TTS. Da * die Reihenfolgenummer SEQ_{UIM} enthält (Challenge/Response) und der TTS für UIM vertrauenswürdig ist, akzeptiert UIM den Zeitstempel STS als aktuell, falls zwischen dem Absenden von M1 und dem Empfang von M4 nicht mehr als eine maximale Zeitspanne T_MAX_{UIM} verstrichen ist. Auch hier beschreibt dieses Zeitfenster die aus der Übertragung resultierende zulässige Ungenauigkeit des Zeitstempels STS.

UIM rekonstruiert ** und verifiziert $Sig_{NO}(** \| Sig_{TTS}(*))$. Da ** die Reihenfolgenummer SEQ_{UIM} enthält, authentisiert sich NO mit der Signatur $Sig_{NO}(** \| Sig_{TTS}(*))$ gegenüber UIM. Ebenfalls zeigt diese Signatur, daß NO den Zeitstempel STS kennt und akzeptiert hat.

UIM konstruiert Nachricht M5. Mit der Signatur über ** $\| Sig_{TTS}(*)$ bestätigt UIM seine Kenntnis und Akzeptanz des Zeitstempels STS (STS ist Element von **!), mit der Signatur über $Sig_{NO}(** \| Sig_{TTS}(*))$ die Authentikation von NO. Da ** die Zufallszahl RND_{NO} enthält (Challenge/Response), authentisiert sich UIM durch die Signatur über ** gegenüber NO. Gleiches gilt für den öffentlichen Schlüsselteil g^{t2UIM}, dessen Hashwert ebenfalls Bestandteil der Signatur ist. Optional kann ein drittes Datenfeld data3 eingefügt werden. Auch M5 wird mit K_M verschlüsselt und somit vertraulich über die Luftschnittstelle übertragen.

NO entschlüsselt Nachricht M5 und verifiziert die Signatur des UIM. Ist diese Signatur gültig, authentisiert sich UIM gegenüber NO, da ** die Zufallszahl RND_{NO} enthält (Challenge/Response). Neben data3 wird ebenfalls die Authentikation des öffentlichen Teilschlüssels g^{t2UIM} erreicht.

NO signiert die entschlüsselte Nachricht M5, um die Authentikation des UIM nachvollziehbar zu bestätigen. Werden alle bisher signierten Daten sowie die Signaturen selbst zur Nachricht M6 zusammengestellt, so entsteht der Authentikationsnachweis, der diese gegenseitige Authentikation von UIM und NO beschreibt. NO sendet M6 an TP.

4.1.3.3 Authentikationsnachweis

Anhand von Protokollelement M6, dem Authentikationsnachweis, kann die Authentikation zwischen UIM und NO durch eine beliebige dritte Partei TP nachvollzogen werden.

Im folgenden werden kurz die Bestandteile des Authentikationsnachweises sowie deren Bedeutung innerhalb des Authentikationsnachweises erläutert. Die Eigenschaften, welche die optionalen Datenfelder durch die Signaturen erhalten, werden in Kapitel 4.1.5 beschrieben.

IMUI und NO beschreiben die Kennungen der Kommunikationspartner, die diese Authentikation durchgeführt haben. TTS und STS sagen aus, welcher vertrauenswürdige Zeitstempeldienst welchen Zeitstempel generiert hat. SEQ_{UIM} beschreibt die absolute Anzahl aller bisher durchgeführten Authentikationen des UIM. Liegen dem Verifizierer mehrere Authentikationsnachweise vor, beschreibt SEQ_{UIM} zusätzlich deren Reihenfolge.

TP rekonstruiert * und **. Anschließend verifiziert er sequentiell die einzelnen Signaturen des Authentikationsnachweises.

Die Signatur des TTS weist nach, daß zum Zeitpunkt STS der Hashwert über Kennungen, TVPs und Daten der Authentikationspartner (aus Nachricht M2) beim TTS vorgelegen hat.

Die erste Signatur von NO über ** $\|$ $Sig_{TTS}(*)$ zeigt, daß NO UIM als Authentikationspartner ebenso akzeptiert hat wie den vertrauenswürdigen Zeitstempel STS des TTS. Zusätzlich zeigt diese Signatur die Bereitschaft von NO, sich gegenüber UIM zu authentisieren, und welche Challenge (RND_{NO}) dem UIM für dessen Authentikation übergeben wurde.

Die Signatur von UIM über $h(g^{t2UIM})$ $\|$ ** $\|$ data3 $\|$ $Sig_{TTS}(*)$ $\|$ $Sig_{NO}(** \| Sig_{TTS}(*))$ zeigt, daß UIM die Authentikation von NO ebenso akzeptiert hat wie den vertrauenswürdigen Zeitstempel STS des TTS. Zusätzlich zeigt diese Signatur die Bereitschaft des UIM, sich gegenüber dem NO zu authentisieren.

Die zweite Signatur von NO und gleichzeitig letzte Signatur des Authentikationsprotokolls über data3 $\|$ $Sig_{UIM}(h(g^{t2UIM}) \| ** \| data3 \| Sig_{TTS}(*) \| Sig_{NO}(** \| Sig_{TTS}(*)))$ läßt erkennen, daß NO die Authentikation vom UIM akzeptiert hat.

Damit ist die gegenseitige Akzeptanz der beiderseitigen Authentikation gezeigt.

4.1.4 Zeitvariante Parameter

Nach Kapitel 2.1 existieren drei verschiedene TVP-Typen: Zufallszahlen, Reihenfolgenummern und Zeitstempel. In diesem Protokoll werden alle Typen eingesetzt. SEQ_{UIM}, RND_{NO} und STS sichern den Nachrichten des Authentikationsprotokolls unterschiedliche Eigenschaften zu. Jedoch wechselt die Bedeutung einzelner TVPs innerhalb des Protokolls. Während die TVPs in den ersten fünf Protokollelementen nur für UIM und NO bedeutsam sind, so müssen sie in Nachricht M6 ihre Aufgabe gegenüber einem unabhängigen Dritten erfüllen.

4.1.4.1 Bedeutung im Authentikationsmechanismus

In den Protokollelementen M1 bis M5 fungieren SEQ_{UIM} und RND_{NO} als Challenges mehrerer separater Challenge/Response Vorgänge zwischen den Parteien UIM, NO und TTS.

Insgesamt läßt sich das Authentikationsprotokoll in vier Challenge/Response Vorgänge zerlegen. Die Authentikation von NO erfolgt mit dem Nachrichtenpaar M1/M4. Als Challenge wird in M1 die Reihenfolgenummer SEQ_{UIM} übergeben. Entsprechend authentisiert sich UIM mit dem Nachrichtenpaar M4/M5. Challenge ist hier die Zufallszahl RND_{NO}. Die übrigen zwei Challenge/Response Vorgänge werden von UIM und NO mit TTS abgewickelt. Mit ihrer Hilfe erhalten UIM und NO den aktuellen und vertrauenswürdigen Zeitstempel STS. Während zwischen NO und TTS die Nachrichten M2/M3 ausgetauscht werden, erhält UIM seinen Zeitstempel vom TTS über NO als Zwischenstation. Diese Kommunikation umfaßt die Nachrichten M1 bis M4. Als Challenges an den TTS dienen mit RND_{NO}, SEQ_{UIM} dieselben TVPs, die auch für die gegenseitige Authentikation verwendet werden.

Der vertrauenswürdige Zeitstempel STS zeigt den Authentikationspartnern die aktuelle Uhrzeit an. Ansonsten ist er für die gegenseitige Authentikation bedeutungslos. Ebenso wirkt

SEQ_{UIM} in ihrer bisherigen Funktion lediglich als Challenge. Ihre Bedeutung als Reihenfolgenummer ist ohne Belang; eine Zufallszahl wäre an dieser Stelle ausreichend.

4.1.4.2 Bedeutung im Authentikationsnachweis

Diese Bedeutungen ändern sich in Nachricht M6. Einem unabhängigen Dritten, der den Authentikationsnachweis (Nachricht M6) verifiziert, fehlt jeder zeitliche Bezug, wann die Authentikation stattgefunden hat. Daher beschreibt der Zeitstempel STS diesen Zeitpunkt vertrauenswürdig. Weiterhin garantiert er diesem Protokollelement Einmaligkeit, da ein und derselbe Zeitstempel in Verbindung mit dem Hashwert aus Nachricht M2 nur einmal existiert. Anhand des Zeitstempels läßt sich die korrekte Reihenfolge mehrerer Authentikationsnachweise prüfen. Optional könnte ein weiterer vertrauenswürdiger Zeitstempel $STS_{GÜLTIG}$ die Gültigkeitsdauer eines Authentikationsnachweises beschreiben, falls diese nicht einem festen Zeitintervall entspricht.

Unter der Voraussetzung, daß die Reihenfolgenummer SEQ_{UIM} für den unabhängigen Dritten vertrauenswürdig ist, unterstützt sie das Merkmal *Vollständigkeit einer Gruppe von Authentikationsnachweisen.*

* Wiederholt sich eine Reihenfolgenummer bei gleichem Zeitstempel, so ist Nachricht M6 erneut an TP gesendet worden. TP erkennt die Wiedereinspielung eines alten Authentikationsnachweises.
* Unterscheiden sich die Zeitstempel zweier Authentikationsnachweise bei gleicher Reihenfolgenummer, liegt ein Angriff vor. UIM versucht, eine weitere Authentikation durchzuführen, ohne seine Reihenfolgenummer zu inkrementieren.
* Tritt statt dessen ein Sprung der Reihenfolgenummern auf, fehlt ein Authentikationsnachweis von diesem UIM. Der Versuch, eine Reihenfolgenummer zu überspringen bzw. einen Authentikationsnachweis zu unterdrücken, wird somit durch TP erkannt.

4.1.4.3 Vertrauenswürdigkeit

Die Vertrauenswürdigkeit des Zeitstempels STS wird durch den TTS garantiert. Seine Signatur authentisiert den Zeitstempel, weist seine Integrität nach und schützt ihn somit vor unbemerkten Manipulationen. Daher ermöglicht sie eine nachweislich korrekte Übertragung. Die Aktualität wird durch das verwendete Challenge/Response Verfahren erreicht.

Die Vertrauenswürdigkeit der Reihenfolgenummer SEQ_{UIM} läßt sich wie folgt begründen: Da die Reihenfolgenummer SEQ_{UIM} auf der Teilnehmerchipkarte UIM gespeichert, inkrementiert und signiert wird und eine Manipulation der UIM durch deren eigenen hohen Sicherheitsstandard praktisch ausgeschlossen werden kann, wird die fortschreitende Inkrementierung der Reihenfolgenummer bei jeder Authentikation als zuverlässig angenommen. Eine vom UIM generierte und durch den TTS signierte Reihenfolgenummer ist daher (auch für einen unabhängigen Dritten) vertrauenswürdig.

4.1.5 Non-repudiation der optionalen Datenfelder

Die optionalen Datenfelder sind bereits in Kapitel 4.1.1 erwähnt worden. Aus der Protokollbeschreibung in Kapitel 4.1.3 ist ihre Plazierung innerhalb des Authentikationsmechanismus ersichtlich. Hier wird nun erläutert, welche Signaturen, Parameter und Protokollelemente diesen Datenfeldern ihre Non-repudiation Merkmalen zusichern. IMUI und NO bezeichnen die Kennungen des Urhebers bzw. Empfängers des jeweiligen Datenfeldes. Der Zeitstempel STS gibt den Zeitpunkt an, zu dem der Non-repudiation Nachweis generiert wurde. Die Einbindung der gültigen Signatur des vertrauenswürdigen Zeitstempeldienstes TTS bewirkt, daß der Zeitpunkt, zu dem der Nachweis generiert wurde, auch für einen unabhängigen Verifizierer eines Nachweises vertrauenswürdig ist. Diese Signatur erstreckt sich u.a. über die Datenfelder data1 und

Tabelle 4.4: Non-repudiation Eigenschaften der optionalen Datenfelder

data1:	Urhebernachweis (NRO) durch Signatur von UIM in Nachricht M5
	Empfangsnachweis (NRD) durch Signatur von NO in Nachricht M4
data2:	Urhebernachweis (NRO) durch Signatur von NO in Nachricht M4
	Empfangsnachweis (NRD) durch Signatur von UIM in Nachricht M5
data3:	Urhebernachweis (NRO) durch Signatur von UIM in Nachricht M5
	Empfangsnachweis (NRD) durch Signatur von NO in Nachricht M6
	(Der Empfangsnachweis selbst wird nicht an UIM zurückgesendet!)

data2. Da alle Nachweise aus dem Authentikationsprotokoll hervorgehen, dessen Ausführungszeit exakt begrenzt ist, wird hier nicht zwischen dem Zeitpunkt, zu dem das Datenfeld generiert wurde, und dem Zeitpunkt, zu dem der Nachweis erstellt wurde, unterschieden (vgl. [10181, 13888-1]).

Die Signatur des UIM in Nachricht M5 bildet den Urhebernachweis für data1. Hier signiert UIM **, worin u.a. IMUI (Kennung des Urhebers) und NO (Kennung des Empfängers), der Zeitstempel STS und das eigentliche Datenfeld data1 enthalten sind, sowie die Signatur des Zeitstempeldienstes über *. Den zugehörigen Empfangsnachweis für data1 stellt NO bereits in Nachricht M4 aus, da er dort dieselben relevanten Elemente wie UIM in Nachricht M5 signiert. Lediglich die Bedeutungen der Kennungen ist vertauscht, da IMUI nun den Empfänger des Nachweises darstellt und NO den Urheber.

Entsprechend wird der Urhebernachweis für das Datenfeld data2 durch die Signatur von NO in Nachricht M4 gebildet, der Empfangsnachweis analog durch die Signatur von UIM in Nachricht M5.

Das Datenfeld data3 wird, wie data1, durch UIM generiert. Es unterscheidet sich jedoch von data1, da es nicht durch den vertrauenswürdigen Zeitstempeldienst signiert worden ist. UIM signiert dieses Datenfeld in Nachricht M5 u.a. zusammen mit den Kennungen, dem Zeitstempel und dessen Signatur. Die neue Signatur schafft den notwendigen nicht-abstreitbaren Zusammenhang zwischen den genannten Elementen und bildet den Urhebernachweis für data3. Den zugehörigen Empfangsnachweis für data3 stellt NO in Nachricht M6 aus, da er dort dieselben relevanten Elemente wie zuvor UIM in Nachricht M5 signiert. Allerdings wird dieser Empfangsnachweis nicht an UIM übermittelt. Er steht daher nur einem unabhängigen Dritten, der den Authentikationsnachweis verifiziert, zur Verfügung. Da NO jedoch zu dem Zeitpunkt, zu dem er den Empfangsnachweis generiert, den zukünftigen Verifizierer i.d.R. nicht kennt, kann er auch dessen Kennung nicht in den Empfangsnachweis einbinden.

Zusammenfassend zeigt Tabelle 4.4, welche Signaturen den optionalen Datenfeldern data1 bis data3 in welchen Protokollelementen ihre Non-repudiation Eigenschaften zusichern.

4.2 Folgeauthentikation (eines bekannten Teilnehmers)

Ist der Teilnehmer dem Mobilfunknetz bereits bekannt, reicht innerhalb derselben Netzwerkdomäne eine Folgeauthentikation aus, um UIM und NO zu authentisieren. Hierzu wird lediglich der Authentikationsmechanismus, wie in Kapitel 4.1.3.2 beschrieben, ausgeführt. Die Voraussetzungen sind durch die einmalige Ausführung des Setup-Mechanismus gegeben.

Tabelle 5.1: Erforderliche Kenntnisse und Fähigkeiten der einzelnen Parteien

Partei	Erforderliche Kenntnisse, Fähigkeiten
UIM	Identität NO des Netzbetreibers
UIM, NO, TP	Authentischen und vertrauenswürdigen öffentlichen Schlüssel einer Zertifizierungsinstanz
UIM, NO	Zertifikat der CA über ihren eigenen öffentl. Schlüssel
UIM, NO	Asymmetrischen Signaturalgorithmus
UIM, NO, TP	Asymmetrischen Verifikationsalgorithmus
UIM, NO	Symmetrischen Ver- und Entschlüsselungsmechanismus
UIM, NO, TP	Kollisionsfreie kryptographische Hashfunktion
UIM, NO	Realzeituhr, Zeitstempel generieren bzw. verifizieren
UIM, NO	(Pseudo-) Zufallszahlen
UIM	Reihenfolgenummern

5. Authentikationsmechanismus ohne TTS

In diesem Kapitel wird eine Variante des Authentikationsmechanismus aus Kapitel 4.1.3.2 vorgestellt, der auf die Einbindung vertrauenswürdiger Zeitstempel und somit auf die Notwendigkeit eines vertrauenswürdigen Zeitstempeldienstes verzichtet. Allerdings benötigt jede Komponente (UIM, NO) eine lokale Realzeituhr, um Zeitstempel abzuleiten bzw. zu verifizieren. An dieser Stelle sei lediglich auf die Schwierigkeiten hingewiesen, die sich aus dem erforderlichen Gleichlauf der lokalen Realzeituhren aller beteiligten Komponenten ergeben. Entsprechende Mechanismen zur Synchronisation sind vorzusehen. Einsparungen liegen hingegen im Rechenaufwand (geringere Komplexität), in der erforderlichen Übertragungsbandbreite sowie in der Bereitstellung des Zeitstempeldienstes selbst. Im Gegenzug erbringt der Mechanismus jedoch *keine* der Non-repudiation Eigenschaften, die der Authentikationsmechanismus aus Kapitel 4.1.3.2 liefert, weder die der optionalen Datenfelder noch die der Authentikation selbst.

Die verwendeten Abkürzungen aus Tabelle 4.1 bleiben unverändert gültig. Da die Protokollvariante ohne vertrauenswürdigen Zeitstempel keine Non-repudiation Dienste unterstützt, ändern sich die Eigenschaften und Ziele aus Kapitel 4.1.1 entsprechend. Die notwendigen Voraussetzungen zeigt Tabelle 5.1.

Der Setup-Mechanismus aus Kapitel 4.1.3.1 bleibt unverändert. Der Authentikationsmechanismus hingegen erfüllt nun folgende Aufgaben:

- Vereinbarung eines symmetrischen Sitzungsschlüssels zwischen UIM und NO,
- Vertraulichkeit der Übertragungsdaten, der Teilnehmeridentität und des -aufenthaltortes auf der Luftschnittstelle,
- Authentikation von UIM und NO,
- Austausch optionaler Datenfelder und
- Generierung eines Authentikationsnachweises.

Tabelle 5.2: Protokollelemente des Authentikationsmechanismus (Variante ohne TTS)

- **M1':** *Request for Authentication of NO*

  $g^{t2UIM} \parallel Enc\{K_M, IMUI \parallel SEQ_{UIM} \parallel data1\}$

- **M4':** *Response of Authentication of NO / Request for Authentication of UIM*

  $Enc\{K_M, TS_{NO} \parallel data2 \parallel Sig_{NO}(***)\}$

- **M5':** *Response of Authentication of UIM*

  $Enc\{K_M, Sig_{UIM}(h(g^{t2UIM}) \parallel *** \parallel Sig_{NO}(***))\}$

- **M6':** *Mutual Authentication Token (Authentikationsnachweis)*

$$h(g^{t2UIM}) \parallel *** \parallel$$
$$Sig_{NO}(***) \parallel$$
$$Sig_{UIM}(h(g^{t2UIM}) \parallel *** \parallel Sig_{NO}(***)) \parallel$$
$$Sig_{NO}(Sig_{UIM}(h(g^{t2UIM}) \parallel *** \parallel Sig_{NO}(***)))$$

- verwendete Abkürzung

  $*** = IMUI \parallel SEQ_{UIM} \parallel NO \parallel TS_{NO} \parallel data1 \parallel data2$

  $K_M = g^{(tNO * t2UIM)} \bmod L$

Bild 5.1 zeigt den Nachrichtenaustausch und den Protokollpfad während der Authentikations-phase. Der Authentikationsmechanismus besteht aus insgesamt vier Protokollschritten. Die Protokollelemente werden zwischen UIM, NO und TP ausgetauscht und sind in Anlehnung an Kapitel 4.1.3.2 mit M1', M4', M5' und M6' gekennzeichnet. Die grundlegende Bedeutung der einzelnen Protokollelemente bleibt erhalten.

Die eigentliche Authentikation zwischen UIM und NO erfolgt durch die Nachrichten M1', M4' und M5' ebenfalls nach dem Prinzip der Drei-Wege-Authentikation. Nachricht M6' bil-det den Authentikationsnachweis. Seine Übertragung ist im Gegensatz zu der Übertragung der übrigen Nachrichten nicht realzeitorientiert.

Tabelle 5.2 zeigt die Protokollelemente dieses Authentikationsmechanismus sowie deren logische Bedeutung innerhalb des Protokollablaufs.

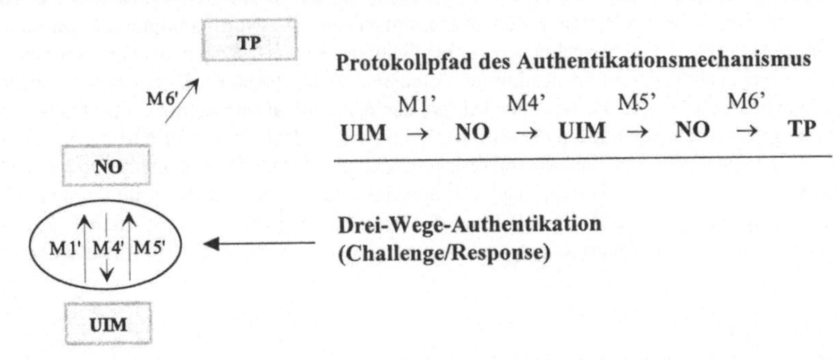

Bild 5.1: Nachrichtenaustausch der Authentikationsphase (Variante ohne TTS)

Die Beschreibung des Authentikationsmechanismus gleicht der aus Kapitel 4.1.3.2. Abweichend fehlen hier die Protokollelemente M2 und M3 sowie deren Einbindung in die folgenden Protokollelemente. Anstatt der Zufallszahl RND_{NO} fügt NO in Schritt M4' seinen aktuellen Zeitstempel TS_{NO} ein. Diesen Zeitstempel leitet NO von seiner lokalen Realzeituhr ab. UIM verifiziert diesen Zeitstempel durch Vergleich des Zeitstempels mit der Zeit seiner lokalen Realzeituhr. Ist die Differenz der beiden Zeitstempel kleiner als eine maximal zulässige (übertragungsbedingte) Ungenauigkeit, erkennt UIM den Zeitstempel durch seine Signatur in Nachricht M5' als den tatsächlichen Zeitpunkt der Authentikation an. Da ohne TTS kein Unterschied zwischen den Datenfeldern data1 und data3 besteht, verzichtet diese Variante auf data3. Die Bedeutung von data1 und data2 bleibt unverändert. Jedoch weisen sie keine Non-repudiation Merkmale auf.

6. Bewertung

In der Vergangenheit sind bereits mehrfach Sicherheitsprotokolle entwickelt worden [AzDi_94, BeCJ_93, HoMü_96, Zhen_96], welche die Vorteile asymmetrischer Kryptographie nutzen und dennoch für den bandbreitebegrenzten und realzeitorientierten Einsatz in mobilen Kommunikationseinrichtungen geeignet sind. Jedoch bietet keines der bisher bekannten Protokolle die speziellen Merkmale des neuen Authentikationsprotokolls aus Kapitel 4.

Diese liegen einerseits in der Nachweisbarkeit einer Authentikation und andererseits in den umfangreichen Non-repudiation Eigenschaften der optionalen Datenfelder.

Die Non-repudiation authentication bietet neue Möglichkeiten in den Bereichen Betrugserkennung und -nachweisbarkeit (*Stichwort: Fraud Detection*) sowohl für Netzbetreiber und Service Provider als auch für den Teilnehmer selbst. Durch die gegenseitigen Urheber- und Empfangsnachweise erhalten die optionalen Datenfelder Eigenschaften, die denen eines herkömmlichen Vertragsabschlusses gleichen. Der Besitz der Daten, die in diesen Feldern enthalten sind, kann von keinem der beiden Authentikationspartner geleugnet oder abgestritten werden (*Stichwort: Grundlage für eine nachweisbare Gebührenabrechnung oder nachladbare elektronische Zahlungsmittel*). Somit wird der Betrug einer Einzelpartei unterbunden und gemeinsamer Betrug der Kommunikationspartner erkannt.

Um diese Vorteile nutzen zu können, ist jedoch die Einbindung eines vertrauenswürdigen Zeitstempels erforderlich, der den Authentikationspartnern durch einen vertrauenswürdigen und ggf. unabhängigen Zeitstempeldienst bereitgestellt wird.

Das vereinfachte Protokoll aus Kapitel 5 verzichtet hingegen auf den vertrauenswürdigen Zeitstempeldienst. Es ist kürzer und weniger komplex, erfordert dafür jedoch lokale Realzeituhren bei den einzelnen Parteien. Zudem erbringt es keine der Non-repudiation Eigenschaften. Zwar bietet diese Variante auch einen Authentikationsnachweis. Jedoch ist seine Nachvollziehbarkeit eingeschränkt und an bestimmte Voraussetzungen geknüpft. Durch den Verzicht auf den vertrauenswürdigen Zeitstempel verliert der Authentikationsnachweis alle Non-repudiation Eigenschaften, sowohl die der Authentikation selbst als auch die der optionalen Datenfelder. Der Verifizierer muß sich darauf verlassen, daß die Authentikationspartner den eingebundenen Zeitstempel durch Zeitvergleich mit ihrer lokalen Realzeituhr kontrollieren und bei Unregelmäßigkeiten, also evtl. Einzelbetrug, das Protokoll abbrechen. Gemeinsamer Betrug der Authentikationspartner wird jedoch weder unterbunden noch durch den Verifizierer erkannt.

Literatur

9796   ISO/IEC 9796: *Information technology – Security techniques – Digital signature scheme giving message recovery.* 1991.

9798-3   ISO DIS 9798: *Entity authentication mechanisms – Part 3: Entity authentication using a public-key algorithm.* 1992.

10181   ISO/IEC DIS 10181: *Information technology – Open Systems Interconnecion – Security frameworks in Open Systems – Part 4: Non-repudiation.* 1995.

13888-1   ISO/IEC CD 13888: *Information technology – Security techniques – Non-repudiation – Part 1: General Model.* 1995.

13888-3   ISO/IEC CD 13888: *Information technology – Security techniques – Non-repudiation – Part 3: Using asymmetric techniques.* 1996.

14888-3   ISO/IEC CD 14888: *Information technology – Security techniques – Digital signatures with appendix – Part 3: Certificate-based mechanisms.* 1995.

AzDi_94   Aziz, A.; Diffie, W.: *Privacy and Authentication for Wireless Local Area Networks.* IEEE Personal Communication, First Quarter 1994, Vol. 1, No. 1.

BAN_90   Burrows, Michael; Abadi, Martin; Needham, Roger: *A Logic of Authentication.* ACM Transactions on Computer Systems, Vol. 8, No. 1, 2/90, S. 18-36.

BeCJ_93   Beller, M. J.; Chang, L.; Jacobi, Y.: *Privacy and Authentication on a Portable Communications System.* IEEE Journal on Selected Areas in Communications, Vol. 11, No. 6, 1993.

BeDa_96   Benkner, Thorsten; David, Klaus: *Digitale Mobilfunksysteme.* Teubner, Stuttgart, 1996.

DeSa_81   Denning, Dorothy E.; Sacco, Giovanni Maria: *Timestamps in key distribution protocols.* Communications of the ACM, Vol. 24, No. 8, 1981, S. 533-536.

DiHe_76   Diffie, Whitfield; Hellman, Martin E.: *New Directions in Cryptography.* IEEE Transactions on Information Theory, Bd. IT-22, Nr. 6, 1976, S. 644-654.

DiOW_92   Diffie, Whitfield; Oorschot, Paul C. van; Wiener, Michael J.: *Authentication and Authenticated Key Exchange.* Designs, Codes & Cryptography, Nr. 2, 1992, S. 107-125.

ElGa_85   El Gamal, Taher: *A Public Key Cryptosystem and Signature Scheme Based on Discrete logarithms.* IEEE Transactions on Information Theory, Bd. IT-31, Nr.4, 7/1985, S. 469-472.

ETSI_1   ETSI Technical Report: *Security Principles for the UMTS.* Draft ETR (09.01), Version 2.5.0, June 1996.

ETSI_2   ETSI SMG SG Doc 117/96: *Formal Analysis of the Public Key Protocol for UMTS.* Source: Gürgens, S., GMD Darmstadt, Juni 1996.

ETSI_3   ETSI SMG SG Doc 118/96: *Evaluation of the new authentication protocol introduced by DeTeMobil/UNI Siegen in TD 80/96.* Source: DeTeMobil/UNI Siegen, June 1996

FoRö_96   Fox, Dirk; Röhm, Alexander W.: *Effiziente Digitale Signatursysteme auf der Basis Elliptischer Kurven.* In: Horster, P. (Hrsg.): Digitale Signaturen. Proc. der Arbeitskonferenz Digitale Signaturen '96, vieweg-Verlag, Braunschweig 1996, S. 201-220.

FuHe_94   Fumy, Walter; Hess, Erwin: *What are Today's Alternatives for a Digital Signature Scheme?* Proc. of Securicom '94, Paris, pp. 23-32.

HaSt_90   Haber, Stuart; Stornetta, W. Scott: *How to time-stamp a digital document.* In: Menezes, A.J.; Vanstone, S.A. (Hrsg.): Proc. of Crypto '90, Springer Verlag, Berlin, 1991, S. 437-455.

HaZh_92   Hardjono, Thomas; Zheng, Yuliang: *A practical digital multisignature scheme based on discrete logariths.* Proc. of Asiacrypt '92, Springer Verlag, Berlin, 1992, S. 122-132.

Herd_95a   Herda, Siegfried: *Non-repudiation: Constituting evidence and proof in digital cooperation.* Computer Standards & Interfaces, Vol. 17, Elsevier, Amsterdam, 1995, S. 69-79.

Herd_95b   Herda, Siegfried: *Nichtabstreitbarkeit (Non-repudiation): Stand der Standardisierung.* In: Horster, P. (Hrsg.): Tagungsband Trust Center, DuD-Fachbeiträge, vieweg-Verlag, Braunschweig, 1995, S.271-282.

HoMi_96   Horster, Patrick; Michels, Markus: *On the risk of disruption in several multiparty signature schemes.* Proc. of Asiacrypt '96, Springer, Berlin, 1996, pp. 334-345.

HoMP_95   Horster, Patrick; Michels, Markus; Petersen, Holger: *Meta-multisignature schemes based on the discrete logarithm problem.* Proc. of IFIP/SEC '95, Chapman & Hall, London, 1995, S. 128-142.

HoMü_96   Horn, G.; Müller, K.: *Ein effizientes Sicherheitsprotokoll zur Authentikation und Schlüsselvereinbarung am Beispiel des Mobilfunksystems UMTS.* Proc. der SIS '96, vdf Verlag, 1996.

ItNa_83   Itakura, K; Nakamura, K.: *A public key cryptosystem suitable for digital multisignatures.* NEC Research and Development, Vol. 71, 1983.

ITSEC   *Kriterien für die Bewertung der Sicherheit von Systemen in der Informationstechnik.* EGKS-EWG-EAG, Brüssel, Luxemburg 1991, ISBN 92-826-3003-X.

KeWe_96   Kessler, Volker; Wedel, Gabriele: *Formal Semantics for Authentication Logics.* Proc. of Computer Security – Esorics '96, LNCS 1146, Springer Verlag, Berlin, 1996, S. 219-241.

NIST_94   National Institute of Standards and Technology (NIST): *Digital Signature Standard (DSS)*. Federal Information Processing Standards Publication 186 (FIPS-PUB), 19. Mai 1994.

OhOk_91   Ohta, Kazuo; Okamoto, Tatsuaki: *A digital multisignature scheme based on the fiat-shamir scheme*. Proc. of Asiacrypt '91, Springer Verlag, Berlin, 1991, pp. 139-146.

Okam_88   Okamoto, Tatsuaki: *A digital multisignature scheme using bijective public-key cryptosystems*. ACM Transactions on Computer Systems, Vol. 6, No. 8, 1988, pp. 432-441.

PiFr_96   Pinto, Fernando; Freitas, Vasco: *Digital time-stamping to support non-repudiation in Electronic Communications*. Proc. of Securicom '96 – 14th Worldwide Congress on Computer and Communications Security and Protection, MCI (Manifestations & Communications Internationales), CNIT, Paris, France, 1996, pp. 397-406.

RSA_78   Rivest, Ronald L.; Shamir, Adi; Adleman, Leonard: *A Method for obtaining Digital Signatures and Public Key Cryptosystems*. Communications of the ACM, Bd. 21, Nr. 2, 1978, S. 120-126.

Pütz_95   Pütz, Stefan: *Neue Lösungsansätze für Authentikation in künftigen Mobilfunksystemen*. 2. ITG-Fachtagung „Mobile Kommunikation '95". ITG-Fachbericht 135, vde-Verlag, Berlin, 1995, S. 411-422, Hrsg.: Walke, B.

Pütz_96   Pütz, Stefan: *Secure Billing – Incontestable Charging*. „Communications and Multimedia Security II". Proc. of IFIP TC 6 / TC 11 International Conference on Communications and Multimedia Security '96, Chapman & Hall, London, 1996, S. 208-221, Hrsg.: Horster, P.

Pütz_97a Pütz, Stefan: *Zur Sicherheit digitaler Mobilfunksysteme*. Datenschutz und Datensicherheit (DuD), 6/97, S. 321-327.

Pütz_97b Pütz, Stefan: A new security *service: Non-repudiation of mutual agreement*. „Communications and Multimedia Security III". Proc. of IFIP TC 6 / TC 11 International Conference on Communications and Multimedia Security '97, Chapman & Hall, London, 1997.

Rula_93   Ruland, Christoph: *Informationssicherheit in Datennetzen*. DataCom-Verlag, Bergheim, 1993.

Schn_96   Schneier, Bruce: *Applied Cryptography*. John Wiley & Sons, New York 1996.

X.509   ITU (ehemals CCITT) Recommendation X.509: *The Directory: Authentication Framework*. Genf 1989.

Zhen_96   Zheng, Y.: *An Authentication and Security Protocol for Mobile Computing*. IFIP World Conference on Mobile Communications, Chapman & Hall, 1996.

IT-Sicherheit

Grundlagen und Umsetzung in der Praxis

von Rolf Oppliger
1997. XXIV, 541 S. Kart.
(DuD-Fachbeiträge; hrsg. von Pfitzmann, Andreas/
Reimer, Helmut/ Rihaczek, Karl/ Roßnagel, Alexander)
ISBN 3-528-05566-9

Aus dem Inhalt: Kryptologische Grundlagen, Kryptosysteme - Anwendungen, Schlüsselverwaltung - Allgemeine Schutzmaßnahmen - Zugangs- und Zugriffskontrollen - Evaluation und Zertifikation - Softwareanomalien und -manipulationen - Offene Systeme, lokale Netze und Weitverkehrsnetze - Internet, elektronische Nachrichtenvermittlungssysteme - Authentifikations- und Schlüsselverteilsysteme

Das Buch bietet eine umfassende und aktuelle Einführung in das Gebiet der IT-Sicherheit. In drei getrennten Teilen werden Fragen der Kryptologie, bzw. der Computer- und Kommunikationssicherheit thematisiert. Der Leser wird dabei schrittweise in die jeweiligen Sicherheitsprobleme eingeführt und mit den zur Verfügung stehenden Lösungsansätzen vertraut gemacht.
Das Buch kann sowohl zum Eigenstudium als auch als Begleitmaterial für entsprechende Vorlesungen, Kurse und Seminare verwendet werden.